Chemical Biology of Sterols, Triterpenoids and Other Natural Products

Chemical Biology of Sterols, Triterpenoids and Other Natural Products

A Themed Issue in Honor of Professor W. David Nes on the Occasion of His 65th Birthday

Special Issue Editors

Wenxu Zhou

De-an Guo

MDPI • Basel • Beijing • Wuhan • Barcelona • Belgrade

Special Issue Editors
Wenxu Zhou
Texas Tech University
USA

De-an Guo
Chinese Academy of Sciences
China

Editorial Office
MDPI
St. Alban-Anlage 66
4052 Basel, Switzerland

This is a reprint of articles from the Special Issue published online in the open access journal *Molecules* (ISSN 1420-3049) from 2018 to 2019 (available at: https://www.mdpi.com/journal/micromachines/special_issues/MEMS_Accelerometers)

For citation purposes, cite each article independently as indicated on the article page online and as indicated below:

LastName, A.A.; LastName, B.B.; LastName, C.C. Article Title. *Journal Name* **Year**, *Article Number*, Page Range.

ISBN 978-3-03897-416-1 (Pbk)
ISBN 978-3-03897-417-8 (PDF)

Contents

About the Special Issue Editors

Wenxu Zhou studies sterol metabolism and regulations of a variety of organisms including higher plants, marine algae, pathogenic fungi, parasitic protozoa, and animals. His research focuses on identifying the unique niches of sterol metabolism among different organisms and utilizing them for new chemotherapeutic targets and new strategies for plant pest control. He has extensive experience in natural product chemistry, chromatography technologies, GC/MS base metabolomics, and enzymology. Dr. Zhou received his PhD in chemistry from Texas Tech University. He completed postdoctoral studies in metabolomics and sterol metabolism at Iowa state University and the University of Western Australia.

De-an Guo s a research professor at the Shanghai Institute of Materia Medica, Chinese Academy of Sciences (Shanghai, China) and the director of the Shanghai Research Center for TCM Modernization (2005–present). His scientific focus is on the development of methods for phytochemical analysis and quality control of Chinese herbal medicines. He has published over 650 original papers with an H index of 46 and supervised 70 PhD candidates. In 2012, he was awarded the National Natural Science Prize in China and ABC Norman Farnsworth Excellence in Botanical Research Award for his outstanding scientific research. He was the editor-in-chief of the 2010 and 2015 English editions of Chinese Pharmacopoeia. He is an associate editor, editor, or editorial board member of 16 international scientific journals, including Phytochemistry, Fitoterapia, and Phytochemistry Letters.

Preface to "Chemical Biology of Sterols, Triterpenoids and Other Natural Products"

Sterols and other isoprenoids are an extremely large and diverse group of natural products that, while structurally complex, have a common biosynthetic origin based on C5-isoprenoid units derived from pyruvate–deoxyxyulose-phosphate or acetate-mevalonate. In this Festschrift Special Issue of Molecules in honor of the 65th birthday of Prof. W. David Nes, who made significant contributions to the sterol–isoprenoid field, 16 papers from an international cast of chemists and biochemists are presented. The first set of papers covers aspects of the biosynthesis, developmental significance, and evolution of isoprenoids and sterol biosynthesis in bacteria and eukaryotes. The second set of papers begins with an overview aimed at filling in the gaps in microbial sterolomics as a tool in chemical biology. Subsequent papers in this section put this area into perspective by describing work on the role of oxysterols in retinal degeneration and cholesterol effects on Chinese hamster ovary cells. Additionally, the biological effects of pentacyclic triterpenoids and related natural products on the biology and biochemistry of cholesterol in native and cancer cells is examined. In the next two papers, natural triterpenoids and medical azoles are evaluated against protozoan and fungal steroidogenesis. Finally, the concluding set of papers details the renewed interest in methods of development for C27 bile alcohols, along with a review of the synthesis of sterol biosynthesis inhibitors that target sterol C24-methyltransferase and sterol C14-demethylase (CYP51) in pathogenic organisms and a paper reporting an improved synthesis of N-methylcadaverine. We are grateful for the breadth and depth of each of the contributions and recognize that the various research topics addressed will no doubt be an area for considerable research in the future.

Wenxu Zhou, De-an Guo

Special Issue Editors

Editorial

More than 40 Years Active in Steroid and Isoprenoid Research—A Personal Note on W. David Nes' Career and His Multiple Achievements in this Field

Thomas J. Bach

IBMP, CNRS UPR 2357, Université de Strasbourg, Institut de Botanique, F-67000 Strasbourg, France; bach@unistra.fr

Received: 25 February 2019; Accepted: 1 March 2019; Published: 5 March 2019

W. David Nes has worked on steroids and related fields (lipids!) for more than 40 years, with an enormous scientific output, which is ever increasing and has never been interrupted regardless of where he worked. He is a perfect candidate for the dedication of articles, not only for his "pure science" and related achievements, but also his activity in organizing meetings within societies like the American Chemical Society and especially the American Oil Chemists' Society (for instance, several "Steroid meetings!"), as co-organizer of plant lipid meetings, and as an active speaker in the "TERPNET" community. He is also author or editor of standard works like *Lipids in Evolution* (with William R. Nes) and *Biochemistry and Function of Sterols* (with E.J. Parish; CRC Press 1997), and a leading member of evaluation panels (e.g., as deputy at the National Science Foundation in Washington, DC for about two years).

I came into contact with David Nes in 1982, when we detected a common interest in a pathway first described by George Popják (FRS) for mammalian tissue; specifically, the retro-conversion of isopentenyl diphosphate (IPP) into acetyl-CoA via methylcrotonyl-CoA, *trans*-methylglutaconyl-CoA and hydroxymethylglutaryl-CoA, a pathway the amino acid leucine also enters during its degradation. Within essentially six weeks of intense work at the USDA Western Regional Research Center (WRRC) in Albany in early 1983 we could provide clear evidence for such a "mevalonate shunt" pathway in wheat seedlings. A funny remembrance worth mentioning is our sitting side-by-side in darkness at the bench, with green lamps on our heads, to work up etiolated wheat seedlings for incorporation and other experiments, as plants don't "see" green light.

As a DAAD postdoc fellow I was certainly the first foreign collaborator at a senior postdoc level working with David Nes already in early 1983, being considerably older than he was at that time. It was clear that as a rising star within the USDA Western Regional Research Center in Albany, he had the full support of its director, especially when it came to the use of analytical platforms, from which I greatly profited too.

From Albany, David Nes then moved to the USDA Richard B. Russell Research Center in Athens, where I could visit him shortly and again discuss scientific matters of interest. A former Humboldt fellow with me (Dr. M. Venkatramesh), at that time still at the Technical University of Karlsruhe (Germany) before my transfer to Strasbourg, joined his lab in Athens and stayed for a while with David Nes when he left the USDA and joined Texas Tech University (TTU) in Lubbock, where he continued his career and successful work until today.

What is always impressive to see from the outside is David Nes' combination of organic chemistry, the refined analysis of steroids and triterpenoids plus their derivatives, with a deep understanding of biological processes in plants, microorganisms and parasites, including the use of modern methods in molecular biology to elucidate enzymatic reaction mechanisms, and the rational design and synthesis of mechanism-based inhibitors of sterol biosynthesis. These studies led to important clues for sterol evolution across kingdoms, and to chemotherapeutic leads to prevent disease by opportunistic parasites dependent on an intact ergosterol pathway. Furthermore, such studies afforded success in engineering soybean plants with modified sterol seed compositions to benefit human health.

In some more detail, the main focus of David Nes' research has been to unravel relationships between sterol biosynthesis and function in a range of organisms by unearthing the molecular libraries (genome–metabolome congruence) associated with phyla-specific reaction sequences that regulate the sterol patterning in organisms. Particular emphasis is given on the characterization of intracellular metabolites and enzyme specificities involved in sterol production and processing, and factors regulating the carbon flux and sterol homeostasis. Of course, such studies require collaboration with colleagues and their laboratories (Michael R. Waterman (Nashville), Steven L. Kelly (Swansea), Jonathan Gershenzon (Jena), and Henry T. Nguyen (Columbia). Those collaborations are attested by numerous co-authored publications, the fruits of sabbaticals abroad.

David Nes is not only known as a very productive scientist and teacher at the university level, but is instrumental in all that concerns "organization and administration", which is already visible through his functions as Director of the TTU Center for Chemical Biology and Chair of Biochemistry Division, and as Professor of Immunology and Molecular Microbiology at the TTU Health Sciences Center School of Medicine in Lubbock, Texas. This in addition to his activities as associate editor of a number of journals in the field. I would also stress his success in attracting research money for equipment and (international) personnel, not an easy task nowadays.

I wish David Nes all the best for his 65th birthday and for the successful continuation of his research.

Conflicts of Interest: The author declares no conflict of interest.

Review

Metabolism and Biological Activities of 4-Methyl-Sterols

Sylvain Darnet [1,*] and Hubert Schaller [2,*]

1 CVACBA, Instituto de Ciências Biológicas, Universidade Federal do Pará, Belém, PA 66075-750, Brazil
2 Plant Isoprenoid Biology (PIB) team, Institut de Biologie Moléculaire des Plantes du CNRS, Université de Strasbourg, Strasbourg 67084, France
* Correspondence: sylvain@ufpa.br (S.D.); hubert.schaller@ibmp-cnrs.unistra.fr (H.S.); Tel.: +55-91-3201-7456 (S.D.); +33-3-6715-5265 (H.S.)

Academic Editor: Wenxu Zhou
Received: 24 December 2018; Accepted: 23 January 2019; Published: 27 January 2019

Abstract: 4,4-Dimethylsterols and 4-methylsterols are sterol biosynthetic intermediates (C4-SBIs) acting as precursors of cholesterol, ergosterol, and phytosterols. Their accumulation caused by genetic lesions or biochemical inhibition causes severe cellular and developmental phenotypes in all organisms. Functional evidence supports their role as meiosis activators or as signaling molecules in mammals or plants. Oxygenated C4-SBIs like 4-carboxysterols act in major biological processes like auxin signaling in plants and immune system development in mammals. It is the purpose of this article to point out important milestones and significant advances in the understanding of the biogenesis and biological activities of C4-SBIs.

Keywords: sterol; C4-demethylation complex (C4DMC); 4-methylsterol; hormone; steroid; development; genetic disease

1. An Introduction to 4-Methylsterols

Post-squalene sterol biosynthesis consists in the enzymatic conversion of $C_{30}H_{50}O$ steroidal triterpene precursors such as lanosterol or cycloartenol into pathway end-products among which the most popular are cholesterol, ergosterol, poriferasterol, sitosterol, and many others distributed among eukaryotes. Several dozens of sterol structures may be detected and identified in given organisms or tissues [1–5]. Biosynthetic relationships between all these sterol structures have been extensively documented [6–8]. Sterol structural differences between eukaryotic kingdoms involve the number of exocyclic carbon atoms at position C24 and unsaturations in the B cycle of the cholestane backbone (Figure 1A,B). Cholesterol is a Δ^5-sterol bearing the eight carbon side chain at position C17, which is a structural feature resulting from the cyclization of 2,3-oxidosqualene ($C_{30}H_{50}O$) into a protosteryl cationic reaction intermediate and then into lanosterol or cycloartenol [9]. In plants, campesterol and sitosterol are Δ^5-sterols with one and two methyl groups at position C24, respectively. In yeast, ergosterol is a $\Delta^{5,7}$-sterol with one extra methyl group at C24. Sterol pathways are markedly different between eukaryotes depending on the cyclization of 2,3-oxidosqualene into lanosterol in fungi and mammals or cycloartenol in some protists and plants (Figure 2). In fact, this dichotomy generates the particular series of 9β,19-cyclopropylsterols derived from cycloartenol, the biosynthetic and functional features of which have been discussed [6–8].

The enzymatic conversion of lanosterol or cycloartenol into pathway end-products (cholesterol, ergosterol, and phytosterols) implies crucial demethylation steps at C14 and C4 positions. Here again, substrates of these reactions in the eukaryotic kingdom differ. Mammals and fungi perform two consecutive C4-demethylations of 30-nor-lanosterol occurring right after the mandatory C14-demethylation of lanosterol, whereas plants carry out two distinct and nonconsecutive

C4-demethylations, the first one applying to a 4,4,14-trimethylsterol and the second one to a 4,14-dimethylsterol or a 4-methylsterol (Figure 2).

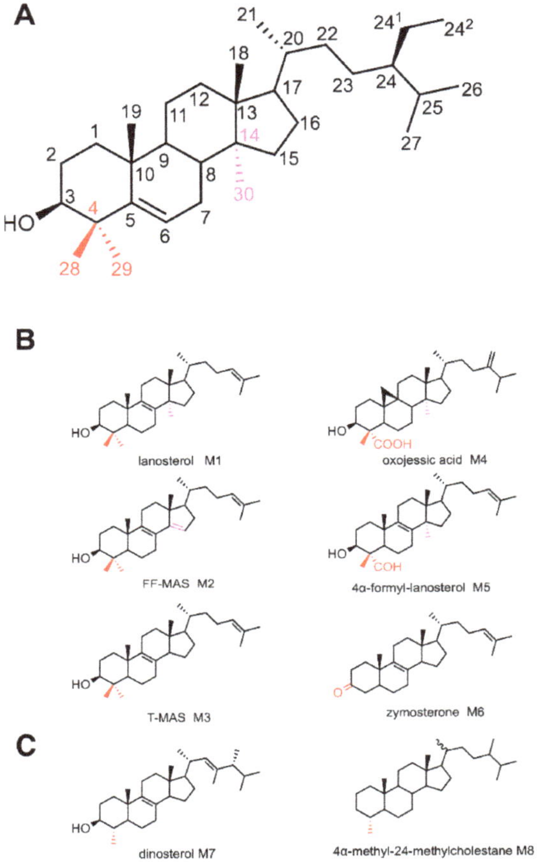

Figure 1. Sterol and 4-methylsterol structures. (**A**) carbon numbering. (**B**) some compounds described in this article. (**C**) dinosterol and a sterane, a biogeological marker.

Both sterol demethylations at C4 and C14 require molecular oxygen for the oxidative cleavage of carbon-carbon bonds, but enzymes at play are different. Demethylation at C14 is catalyzed by a 14α-methylsterol-14α-methyl-demethylase, which is a cytochrome P450—dependent mono-oxygenase also known as CYP51 in mammals [10–12], in yeast [13] and in plants [14–16]. A Δ^{14}-sterol-14-reductase catalyzes the reduction of the resulting $\Delta^{8,14}$-diene (Figure 2 and Figure S1A). This enzyme is encoded by a single gene in plants and yeast [17,18], while in human two distinct genes were characterized [19,20]. In the same organisms, the demethylation at C4 leads to the production of 4α-carboxysterols by an oxygen-dependent process followed by an oxygen-independent C-C cleavage that generates 3-ketosterols (Figure S1B) [21]. It is now established that sterol-C4-demethylation implies the consecutive action of three enzymes: a sterol-4α-methyl oxidase (SMO), a 3β-hydroxysteroid dehydrogenase/C4-decarboxylase (C4D), and a sterone ketoreductase (SKR) [22] (Figure 2 and Figure S1B). A protein called ERG28 was shown to tether all three enzymes as a complex in the endoplasmic reticulum [23].

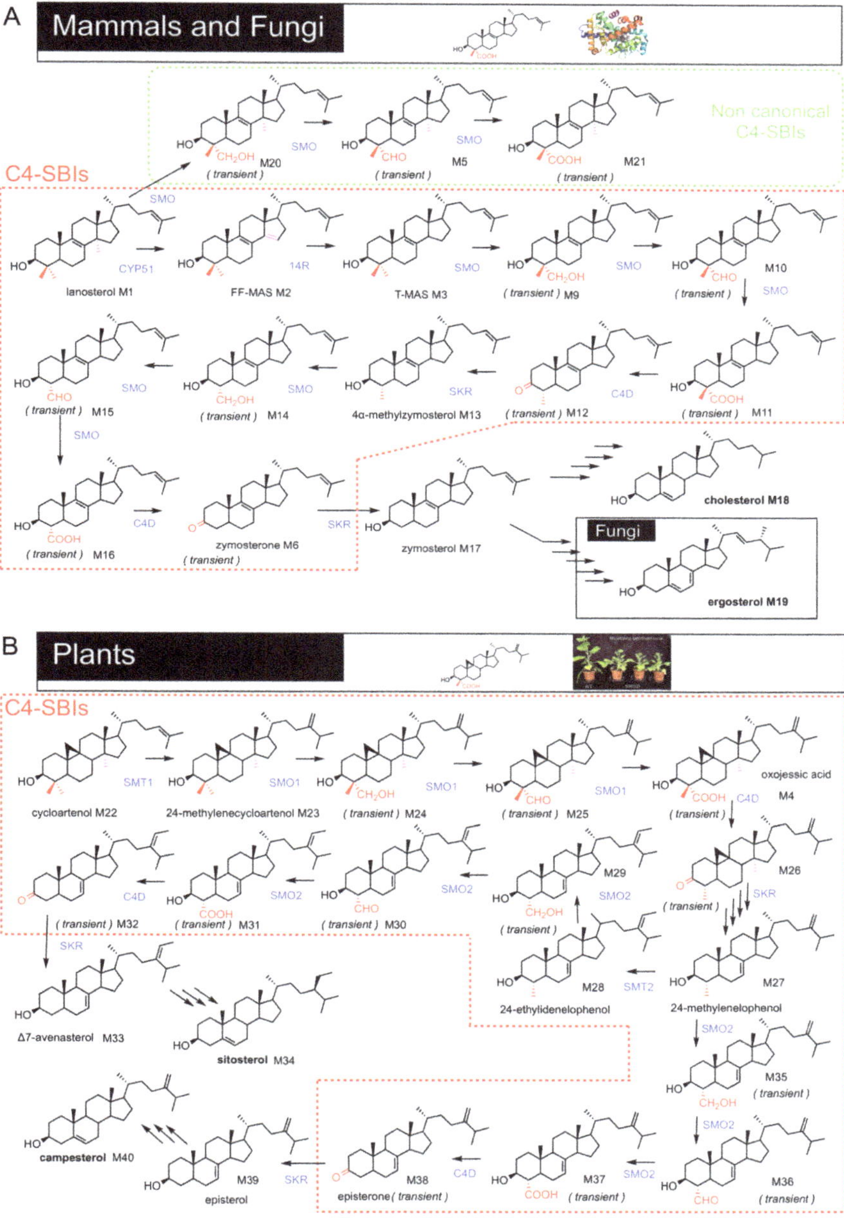

Figure 2. C4-demethylation pathways in mammals and fungi, and plants. Sterol nomenclature is given in Table 1. (**A**) pathways in mammals and fungi; (**B**) pathway in plants; C4-demethylation in eukaryotes: SMO, sterol-4α-methyl-oxidase; C4D, 3β-hydroxysteroid dehydrogenases/C-4 decarboxylase; SKR, sterone ketoreductase, C14-demethylation: CYP51, lanosterol-C14 demethylase. SMT, sterol methyltransferase; 14R, sterol-14-reductase. Each arrow represents an enzymatic step. Graphical insets are from references ([24] in top and [25] bottom panels).

A prominent category of sterol biosynthetic intermediates is 4-methylsterols (including 4,4-dimethylsterols) hereafter collectively named C4-Sterol Biosynthetic Intermediates (C4-SBIs). These molecules with one or two methyl groups at position C4 are precursors of 4-desmethylsterols, like cholesterol in animals, ergosterol in yeast, poriferasterol in some algae, and phytosterols in plants, as stated above.

The C4-SBIs in mammals and fungi are compounds which follow each other in a biosynthetic segment joining lanosterol to zymosterol. In plants, the pathway is different. The C4-SBIs include compounds which succeed each other in biosynthetic segments joining cycloartenol to episterol or Δ^7-avenasterol, two 24-alkyl-4-desmethylsterols (Figure 2). C4-SBIs are amphiphilic molecules with a rigid structure just like 4-desmethylsterols. Four rings (A, B, C, D) form a quasi-planar tetracyclic nucleus, with a hydroxy or keto group at the C3 position, one or two methyl groups at the C4 position, methyl groups at the C10 and C13 positions, and an aliphatic side chain of 8 to 10 carbon atoms at C17 (Figure 1A). C4-SBIs display a variety of structural motifs: unsaturations at different positions of the B ring ($\Delta^{7(8)}$ or $\Delta^{8(9)}$ or $\Delta^{9(11)}$ or $\Delta^{8,14}$), a cyclopropanic cycle on the B ring, a methyl group at C14 and a methyl (or methylene) or ethyl (or ethylidene) at C24 on the side chain (Figure 1B). C4-SBIs are generally in low abundance contrasting with cellular amounts of pathway end-products. There are however organisms that contain substantial amounts of 4-methylsterols such as dinosterol implied in cold adaptation (Figure 1C) in dinoflagellates [26], or 4,4-dimethylsterols and 4-methyl-Δ^7-sterols in the prokaryote *Methyloccocus capsulatus* [27]. Because dinosterol is restricted to dinoflagellates its sterane derivatives are used as biogeological markers of Phanerozoic sediments [28]. Alternatively, 4α-methyl-24-ethylcholestane may derive from C4-methylsterols from other yet unrecognized Proterozoic eukaryotic organisms that used C4-SBIs as membrane components (Figure 1C) [29]. The capacity of C4-SBIs like for instance cycloartenol to act as an efficient membrane structural component in primitive organisms has been discussed [30]. In this respect, a yeast mutant *erg7* deficient in lanosterol synthase (ergosterol-auxotrophic) could, however, live on C4-SBIs such as cycloartenol upon expression of a cycloartenol synthase [31,32].

C4-SBIs may be classified from an operational point of view according to amounts detected in an organism or tissue, as major C4-SBIs and transient C4-SBIs. Major C4-SBIs are present in few percents of total sterols, about several $\mu g \cdot g^{-1}$ dry weight, like for instance lanosterol in yeast, cycloartenol in *Artocarpus integrifolia* [33], or cycloeucalenol and obtusifoliol in plant tissues [34], and 24-ethylidenelophenol in *Hordeum vulgare* [35]. Transient C4-SBIs are intermediates of the sterol-C4-demethylation process catalyzed by a complex of enzymes (C4-DeMethylation Complex, C4DMC) and are generally not detected in sterol profiles under normal physiological conditions. These compounds are 4-hydroxymethylsterols, 4-formylsterols, 4-carboxysterols, canonical and non-canonical C4-SBIs and 3-ketosterols (Figure 2A).

The effectiveness of 4,4-dimethylsterols such as lanosterol (compared to cholesterol) in regulating membrane fluidity and supporting cellular functions in *Mycobacterium capricolum* was assessed by measuring microviscosity of membranes and establishing their capacity to promote prototrophic growth. Membranes of *M. capricolum* grown on medium containing 4,4-dimethylsterols or 4-methylsterols have microviscosity values found in between those of lanosterol (low value) and cholesterol (high value). These experiments demonstrated that the successive carbon removals at C14 of lanosterol then at C4 of 4,4-dimethylzymosterol and 4-methylzymosterol *en route* to cholesterol biosynthesis (Figure 2) progressively shaped a sterol molecule in order to sustain optimal cell growth [36]. This is in agreement with the identification of 4-methysterols in ancestral organisms [29,37,38].

Physiological roles of C4-SBIs have been described. Lanosterol in the brain is associated with a neuroprotective effect in Parkinson's disease [39]. An increase of oligodendrocyte formation and remyelination was observed in the presence of C4-SBIs [40]. In mammal reproductive biology, Meiosis Activating Sterols (MAS) are major C4-SBIs found in follicular fluid (FF-MAS) and testicular tissue (T-MAS) (Figure 1B) [41–43]. FF-MAS are crucial for proper meiosis and for oocyte maturation in vitro [43,44]. Sterol biosynthetic flux analyzed in mice revealed a high rate of FF-MAS and T-MAS synthesis that defines cell-type specific pathways and also raised new hypothesis about the fate of T-MAS in testes (forming zymosterol, another sterol, a steroid hormone, or an excreted product) [45]. Synthetic FF-MAS and T-MAS were developed for further biological studies [46,47]. Human genetic diseases known as sterolosis are characterized by a dramatic accumulation of sterol intermediates including the immediate cholesterol precursors lathosterol and desmosterol (their accumulation causing lathosterolosis and desmosterolosis, respectively) but also of C4-SBIs causing severe alterations in development at early (embryo malformation) or later stages (skin anatomical changes) [48–50]. In *Caenorhabditis elegans*, 4-methylsterols are generated from cholesterol by an unusual C4-methylation enzyme that is only found in worms (Figure 3) [51]. In plants and mammals, transient C4-SBIs bearing a 4-formyl or 4-carboxy group were functionally linked to critical biological processes: the accumulation of 4α-carboxy-4β-methyl-24-methylenecycloartanol (oxojessic acid, Figure 1B) was shown to hamper proper auxin signaling in the model plant *Arabidopsis thaliana* [52], and 4α-formyl-lanosterol (Figure 1B) was described as a physiological ligand of RORγ, a protein that regulates lymphoid cell development [25].

2. Some Crucial Milestones in Deciphering the Sterol-Demethylation Process and Functions of C4-SBIs in Mammals

In mammals, the first demethylation step occurs at C14 position (Figure 2). This is achieved by a lanosterol-14α-methyl-demethylase (CYP51) [10–12] which removes the 14α-methyl group as formic acid resulting in a $\Delta^{8,14}$-diene product (Figure S1A) [53]. This reaction requires NADPH and generates a 14α-formyloxysterol reaction intermediate on which CYP51 acts as a lyase in cleaving the C-C bond (Figure 2 and Figure S1A). The CYP51A1 gene was identified in human and characterized by heterologous expression in bacteria [54]. The subsequent $\Delta^{8,14}$-sterol-Δ^{14}-reduction (Figure 2 and Figure S1A) has been the focus of considerable research effort over the last decade. In human, two different genes encode products that bear sterol-14-reductase activity, namely, LBR and TM7SF2 genes [19,20]. The LBR protein is bifunctional; it has a lamin B receptor (LBR) and sterol-14-reductase domains and is mainly acting on the cholesterol biosynthetic flux. The TM7SF2 protein, although exhibiting sterol-14-reductase activity, has not a well-defined function in cholesterol biosynthesis. The intracellular localization of these two proteins is different: LBR is addressed to the nuclear envelope, it bears a chromatin-binding N-terminus; TM7SF2 resides in the endoplasmic reticulum membranes [55–58].

The removal of C4 methyl groups as carbon dioxide during the conversion of lanosterol into cholesterol was shown years ago, suggesting that the demethylation reaction implied an β keto acid intermediate [59]. Another experimental evidence was provided by Bloch and co-workers who showed that the aerobic incubation of labeled 4-hydroxy-methylene-cholest-7-en-3-one in a rat liver homogenate resulted in a marked release of carbon dioxide from the reaction medium [60,61]. The subsequent isolation of 3-keto and 4α-acid reaction products supported the proposed mechanistic hypothesis for the C4 demethylation reaction: in rat liver microsomes, the incubation of ^{14}C-labeled 4,4-dimethyl-5α-cholest-7-en-3-ol in the absence of NADPH led to the production of labeled carbon dioxide and mono-methylated products 3-keto-5α-cholest-7-en-3-one and 4-methyl-5α-cholest-7-en-3-one [62]. Similarly, the incubation of a 4-methylsterol produced carbon dioxide and a demethylated ketone at the C4 position [63]. In the presence of NADPH, these ketones are reduced to the corresponding 3α-alcohol by a 3-ketosteroid reductase [64]. The early stages of cholesterol biosynthesis studies and especially the identification of associated enzyme activities raised the question of the formation of the C4-carboxyl group preceding the carbon-carbon cleavage and loss of carbon dioxide, this based on a partial purification of an NAD+ decarboxylase [65,66]. Gaylor and co-workers showed that the oxidation of the methyl group at C4 to the corresponding acid required molecular oxygen and NADH and was sensitive to cyanide [67,68]. Also, the inhibition of C4-demethylation by snake venom phospholipases suggested the involvement of an NADH-dependent cytochrome b5 reducing system [63]. Finally, in recent decades, the complete set of genes coding the enzymes implied in the sterol-C4-demethylation step of mammalian cholesterol biosynthesis was identified particularly in deciphering some human genetic diseases; enzymes were thereafter biochemically characterized in heterologous systems [50,69,70]. The non-enzymatic protein ERG28 necessary for the activity of the C4DMC was lastly identified in human based on its yeast orthologs [71].

Functional studies of C4-SBIs have underlined critical biological properties of lanosterol. Lanosterol and oxysterols affect human cataracts [72]. A functional screening of molecules that bind alpha-crystallins (cryAA and cryAB) in vitro and reversed their aggregation identified 5-cholesten-3β,25-diol as an active compound, based on improved lens transparency in cataract models [73]. In another study, the direct relationship between congenital cataracts and lanosterol was shown by the elucidation of two causal mutations in the gene encoding lanosterol synthase [74]. The role of lanosterol in arresting cataract development was furthermore ascertained by its positive effect on protein disaggregation and the increase of lens transparency, both in vitro and in vivo, in rabbit and dog [74]. Further studies provided additional evidence to establish lanosterol firmly as an anti-cataract drug [74–77]. Although the molecular mechanism is not described, Quinlan [72] et al. have suggested that C4-SBIs, like lanosterol, could interact with small heat shock proteins, which function as sterol sensors regulating cellular and developmental processes. Lanosterol also has a tremendous impact on innate immunity [78]. The activation of Toll-Like Receptor 4 (TLR4) in macrophages is responsible for the transcriptional repression of CYP51, resulting in the accumulation of lanosterol. Such an accumulation of lanosterol, by genetic or by chemical inhibition, has a regulatory action on the immune response, membrane fluidity, ROS production and potentialize phagocytosis [78]. Considering cellular sterol homeostasis, lanosterol and 24,25-dihydrolanosterol are known to interact with the Insig signaling pathway that promotes the degradation of HMGR, a key enzyme of the mevalonate pathway [79]. Lanosterol and 24,25-dihydrolanosterol may also act as an oxygen sensor: in hypoxic conditions, the C14 and C4 demethylations rate is reduced, and consequently promote HMGR degradation, lowering thus the cholesterol biosynthetic flux [80].

The critical importance of C4-SBIs that are the reaction products of LBR and TM7SF2, two proteins bearing sterol-14-reductase domains, has emerged recently [55,56,58]. LBR and TM7SF2 act as regulators of TNFα expression in human, and skin papilloma development in mice [58,81–85]. The Greenberg skeletal dysplasia, the Renolds syndrome and Pelger–Huët anomaly are severe genetic diseases due to mutations in the LBR gene, causing a reduction in sterol-14-reductase activity and therefore promoting the accumulation FF-MAS, the substrate of the enzyme [86–89]. The molecular mechanism that is most probably at play in these diseases may be very close to an enhanced lipogenesis and the inhibition of cell proliferation mediated by the liver X receptor alpha (LXRα), to which binds the C4-SBI molecule FF-MAS [57]. Interestingly, a BODIPY-FF-MAS molecular probe was localized in nuclear lipid droplets of HepG2 cells. Such localization of FF-MAS is in line with the proposed regulatory role [57].

Functional genomics targeting components of the C4DMC led to highlights in human cholesterol biology. In a cancer cell line, the increased sensitivity to antagonists of an oncogenic epidermal growth factor receptor was revealed upon siRNA-based inactivation of SC4MOL and NSDHL leading to 4,4-dimethylzymosterol, 4-methylzymosterol, or 4-carboxysterol accumulation [90]. The inhibition of CYP51A1 suppressed the accumulation of these C4-SBIs and reversed the EGFR inhibitor sensibilization, rescuing cancer cell viability and EGFR degradation [90]. In human development, a hypomorphic temperature-sensitive allele of NSDHL causing the overaccumulation of 4-methylsterols in the cerebrospinal fluid was the cause of brain malformations typical of the CK syndrome (CKS) [49]. The SC4MOL-deficiency is an autosomal recessive lesion causing psoriasiform dermatitis, arthralgias, congenital cataracts, microcephaly, and developmental delay. Plasma sterol analysis showed a different cholesterol content in healthy individuals (140–176 $mg \cdot dL^{-1}$) versus patients (85–93 $mg \cdot dL^{-1}$). Most importantly, a ten-fold increase was obtained when measuring 4-methylsterols: 41–42 $mg \cdot mL^{-1}$ in patients plasma compared to 2.8–3.2 $mg \cdot mL^{-1}$ in healthy individuals [50]. In total, C4-SBIs presented a huge 500-fold increase in diseased individuals compared to healthy ones. No 4-carboxylmethylsterols neither 4-methylsterones were however detected. In such patients, fibroblasts had a 3-fold reduced rate of cell division, and immunocytes were abnormal, this was mimicked by applying aminotriazole, an inhibitor of SC4MOL/SMO. A causal relationship between the accumulation of C4-SBIs and skin barrier function, cell proliferation and immune regulation was then established [50]. Furthermore, the same authors demonstrated that C4-SBIs negatively regulate the epidermal growth factor receptor (EGFR), signaling and vesicular trafficking [91].

The Congenital Hemidysplasia with Ichthyosiform nevus and Limb Defects (CHILD) syndrome is a rare X-linked dominant disease with lethality for male embryos, sensorineural hearing loss, normal intelligence in females and one-sided cerebral hypoplasia [48]. More than 20 different alleles of the NSDHL gene were described [48,92,93].

Sterol analysis were performed in *nsdhl* mice: skin fibroblasts of bare patches of such mice contained about 20% of C4-SBIs in total sterols (71.4% of cholesterol, 18.2% of 4-methylsterols and 1.1 of 4,4-methylsterols), while control male mice had less than 0.1% of C4-SBIs and 99.9% of cholesterol [70]. The CKS consists of mild to severe intellectual disability in males, microcephaly, CNS malformation, seizures, hypotonia, dysphasia/speech delay, behavioral problems and possible psychopathological issues in female carriers. The CKS is lethal in females (whereas CHILD is lethal to males). Cerebrospinal fluid from CKS patients is enriched in 4-methylsterols and is low in cholesterol. It is also reported that CKS patients display a deficient hedgehog signaling [49]. No mutation (and associated human genetic disease) was reported in the case of C4D and ERG28. In mouse, the Rudolph mutant carries an allele of the C4D/HSD17B17 gene causing defective growth and patterning of the CNS, skeleton malformation, and an altered hedgehog signaling associated to an accumulation of zymosterone and 4-methylzymosterone [94]. The study of a conditional *nsdhl* mouse allele enabled a refined understanding of the link between cholesterol homeostasis and CNS at various developmental stages of pups. NSDHL deficiency and its associated accumulation of 4-methylsterols was responsible for defects in the cerebellum, hippocampus, cerebral cortex and led to early postnatal lethal phenotype [95]. At the cellular level, these defects were a thinner layer of granule cell precursors, which play a critical role in cerebral, cortical and hippocampal neuronal proliferation, differentiation and migration before birth. Using this *nsdhl* mouse line, an in vitro cell system was established from granule cell precursors to test the effect of 4-methylsterols on sonic hedgehog signaling (SHH). The obtained cell lines were cultivated with LDL supplementation and also ketoconazole treatment, in order to restore a cholesterol content, and to block the accumulation of 4-methylsterols, respectively. A hampered SHH signaling was correlated with the accumulation of 4-methylsterols. T-MAS (a functional 4-methylsterol), when added to wild-type cells obtained from granule precursors, mimicked perfectly the biogenetic accumulation otherwise noticed in conditional nsdhl cells, however no effect on SHH signaling was observed, most probably due to a mislocalization of T-MAS, or to the lack of bioconversion of T-MAS into an active yet unknown sterol-derived inhibitor of the SHH pathway.

C4-SBIs were described as essential players in the immune system. The binding capacity of C4-SBIs to the nuclear hormone receptor RORγt, an active component of lymphoid cells in thymus, was tested in vitro and in vivo [25]. 4-methylsterol biosynthetic intermediates in between the lanosterol to 4α-methylcholesta-8,24-dien-3-one (the substrate of C4D/HSD17B7) segment (Figure 2) exhibited the properties of ligands of RORγt albeit with significant affinity variations. 4-Methylsterols displayed the weaker affinity while oxygenated C4-SBIs like 4α-carboxy-4β-methylzymosterol (Figure 1) had a higher affinity. This study highlighted the regulatory role of bona-fide cholesterol biosynthetic intermediates upon immune system development and lymphoid functions. C4-SBIs have also a positive influence on mice oligodendrocyte formation and remyelination, as shown using sterol biosynthesis inhibitors. Inhibitors of C4-demethylation and of C14-reduction and Δ^8-Δ^7 isomerization (that promote the accumulation of C4-SBIs indirectly) led to the inactivation of a transcriptional program via the SREBP nuclear hormone receptors [40,96]. Further studies are required to identify firmly which C4-SBIs activate the SREBP machinery.

Table 1. IUPAC sterol nomenclature [97].

ID	Common Name	IUPAC
M1	lanosterol	lanosta-8,24-dien-3β-ol
M2	FF-MAS	4,4-dimethyl-5α-cholesta-8,14,24-trien-3β-ol
M3	T-MAS, 4,4-dimethylzymosterol	4,4-dimethyl-5α-cholesta-8,24-dien-3β-ol
M4	oxojessic acid, CMMC	4α-carboxy-4β,14α-dimethyl-9β,19-cyclo-5α-ergosta-24(24^1)-en-3β-ol
M5	4α-formyl-lanosterol	4α-formyl-4β,14α-methyl-cholesta-8,24-dien-3β-ol
M6	zymosterone	5α-cholesta-8,24-dien-3-one
M7	dinosterol	4α,23,24-trimethyl-5α-cholesta-22-en-3β-ol
M8	4α-methyl-24-ethylcholestane	4α,24-methyl-cholestan-3β-ol
M9	4α-hydroxymethyl-4β-methyl-zymosterol	4α-hydroxymethyl-4β-methyl-cholesta-8,24-dien-3β-ol
M10	4α-formyl-4β-methylzymosterol	4α-formyl-4β-methyl-cholesta-8,24-dien-3β-ol
M11	4α-carboxy-4β-methylzymosterol	4α-carboxy-4β-methyl-cholesta-8,24-dien-3β-ol
M12	3-keto-4α-methylzymosterol	4α-methyl-5α-cholesta-8,24-dien-3-one
M13	4α-methylzymosterol	4α-methyl-5α-cholesta-8,24-dien-3β-ol
M14	4α-hydroxymethylzymosterol	4α-hydroxymethyl-5α-cholesta-8,24-dien-3β-ol
M15	4α-formylzymosterol	4α-formyl-5α-cholesta-8,24-dien-3β-ol
M16	4α-carboxyzymosterol	4α-carboxy-5α-cholesta-8,24-dien-3β-ol
M17	zymosterol	5α-cholesta-8,24-dien-3β-ol
M18	cholesterol	cholest-5-en-3β-ol
M19	ergosterol	ergosta-5,7,22E-trien-3β-ol
M20	-	4α-hydroxymethyl-4β,14α-methyl-cholesta-8,24-dien-3β-ol
M21	-	4α-carboxy-4β,14α-methyl-cholesta-8,24-dien-3β-ol
M22	cycloartenol	9β,19-cyclo-lanost-24-en-3β-ol
M23	24-methylenecycloartanol	24-methylene-9β,19-cyclo-lanost-3β-ol
M24	4-hydroxymethyl-24-methylenecycloartanol	4α-hydroxymethyl-24-methylene-9β,19-cyclo-lanost-3β-ol
M25	4-formyl-24-methylenecycloartanol	4α-formyl-24-methylene-9β,19-cyclo-lanost-3β-ol
M26	cycloeucalenone	24-methylene-9β,19-cyclo-lanost-3-one
M27	24-methylenelophenol	4α-methyl-24-methylene-cholest-7-en-3β-ol
M28	24-ethylidenelophenol	4α-methyl-24Z-ethylidene-cholest-7-en-3β-ol
M29	4-hydroxymethyl-24-ethylidenelophenol	4α-hydroxymethyl-24Z-ethylidene-cholest-7-en-3β-ol
M30	4-formyl-24-ethylidenelophenol	4α-formyl-24Z-ethylidene-cholest-7-en-3β-ol
M31	4-carboxy-24-ethylidenelophenol	4α-carboxy-24Z-ethylidene-cholest-7-en-3β-ol
M32	avenasterone	24Z-ethylidene-cholest-7-en-3-one
M33	Δ7-avenasterol	24Z-ethylidene-cholest-7-en-3β-ol
M34	sitosterol	stigmast-5-en-3β-ol
M35	4-hydroxymethyl-24-methylenelophenol	4α-hydroxy-24Z-methylene-cholest-7-en-3β-ol
M36	4-formyl-24-methylenelophenol	4α-formyl-24Z-methylene-cholest-7-en-3β-ol
M37	4-carboxy-24-methylenelophenol	4α-carboxy-24Z-methylene-cholest-7-en-3β-ol
M38	episterone	24-methylene-cholest-7-en-3β-one
M39	episterol	24Z-methylene-cholest-7-en-3β-ol
M40	campesterol	campest-5-en-3β-ol
M41	lathosterone	cholest-7-en-3-one
M42	Δ7-dafachronic acid	(25s)-3-oxocholest-7-en-26-oic acid
M43	-	cholest-4-en-3-one
M44	Δ4-dafachronic acid	(25s)-3-oxocholest-7-en-26-oic acid
M45	lophenol	4α-methyl-cholest-7-en-3β-ol
M46	4α-methylcholest-8(14)-enol	4α-methyl-5α-cholest-8(14)-en-3β-ol
M47	4β -hydroxymethyl-4α -methyl-zymosterol	4β -hydroxymethyl-4α -methyl-cholesta-8,24-dien-3β-ol
M48	4β -formyl-4α -methyl-zymosterol	4β -formyl-4α -methyl-cholesta-8,24-dien-3β-ol
M49	4β -carboxy-4α -methyl-zymosterol	4β -carboxy-4α -methyl-cholesta-8,24-dien-3β-ol
M50	-	4α-methyl-5α-cholesta-8-en-3β-ol

3. *Saccharomyces cerevisiae*, a Versatile Model for Sterol Genetics and Auxotrophy Studies

The yeast *S. cerevisiae* has established itself as a privileged model for the identification of sterol biosynthesis genes [98,99]. The advantages of yeast are plentiful: a sterol biosynthesis pathway similar to that of animals or plants enabling metabolic interferences, the possibility of homologous recombination to create loss-of-function mutants, its ability to have an uptake of exogenous sterols, to mention a few. The identification of the yeast SMO gene was published independently in 1996 by two teams. The Kaplan team screened a yeast mutant deficient in SMO activity based on its limited heme biosynthetic capacities [100]. The Bard team isolated the *erg25/smo* mutant by screening for SMO activity deficiency and identified the ERG25 gene (Figure 2) [101]. The yeast C4D was identified based on its functional homology with an NAD(P)-dependent cholesterol dehydrogenase gene of Nocardia sp. [102]; it complemented a corresponding deficient yeast (*erg26*) and *Candida albicans* mutants [103,104]. The yeast SKR gene encoding ERG27/SKR complemented a null mutant *erg27* deficient in 3-ketosteroid reductase (Figure 2) [105]. Gene expression analysis pointed out ERG28 and ergosterol biosynthetic genes within the same levels of expression [106]. The disruption of ERG28 induced a loss of C4-demethylation activity [107]. Protein interaction studies showed that ERG25, ERG26, ERG27, and ERG28 proteins are assembled in a complex tethered by ERG28 [108]. Although ergosterol biosynthesis was tremendously studied, some components of the machinery like ERG29 (an ER-associated protein) were unveiled just very recently [109].

The yeast *erg25* mutant contains high amounts of 4,4-dimethylsterols that are more effective than 4,4,14-trimethylsterols (like lanosterol) to disrupt growth. The lethality of *erg25* was overcome by mutations in ERG11 (lanosterol-14-demethylase) and SLU (suppressor of lanosterol utilization) to prevent the accumulation of 4,4-dimethylsterols and consequently ergosterol auxotrophy [110]. In the fission yeast *Schizosaccharomyces pombe*, the overexpression of ERG25 affected proper cytokinesis: the accumulation of 4,4-dimethylzymosterol-downstream products and further compositional changes in sterol/lipid-rich membrane domains led to defects in actomyosin ring positioning and maintenance [111]. The isolation of a yeast thermosensitive mutant *erg26-1* defective in the decarboxylation of 4-carboxy-4-methylsterols revealed the inefficiency of these C4-SBIs to support growth as bulk components. Protein-protein interaction studies pointed out a function for ERG26 in ERG7 regulation [112] and also in lipid homeostasis [107,112,113]. ERG29 was identified as an interactant or modulator of SMO/ERG25. The loss of ERG29 resulted in the accumulation of C4-SBIs and affected cell viability. In these yeast cells, an increase of mitochondrial oxidants and the degradation of the mammalian frataxin ortholog involved in mitochondrial iron-sulfur (Fe-S) cluster synthesis showed a link between sterol composition and iron metabolism in the mitochondrial compartment [109]. The expression of a gene cluster for helvolic acid production into *Aspergillus oryzae* NSAR1 has revealed the identification of C4-SBIs bearing anti-*Staphylococcus aureus* properties and unsual C4-demethylation enzymes [114]. In *S. pombe*, C4-SBIs have been identified as signaling molecules acting as oxygen sensor by interacting with the SRE1/SCP1 complex, which is equivalent to the mammalian SREBP regulatory pathway responsible for cholesterol homeostasis. Under conditions of low oxygen and cell stress, C4-SBIs accumulate and activate the transcription factor SRE1 [115].

4. The Plant-Specific Sterol-C4-Demethylation Process and Its Influence upon Development

Sterol-4α-methyl oxidase (SMO) is the enzyme of the C4DMC that acts first in the sequence of reactions. SMO enzymatic activities were initially studied with microsomal fractions of *Zea mays* coleoptiles incubated with radioactively labeled sterol substrates. This led to the clear-cut identification and characterization of two distinct SMO activities. These two SMO activities were shown to occur in a non-consecutive manner in the sterol pathway: a first SMO oxidizes 4,4,14-trimethylcyclopropylsterols such as 24-methylenecycloartanol, and a second SMO oxidizes 4α-methyl-Δ^7-sterols such as 24-ethylidene lophenol (Figure 2 and Figure S1) [21]. It was also demonstrated that electrons are supplied via NADH to the oxygenase by the cytochrome b5/cytochrome b5 reductase system [116]. These distinct subcellular SMO activities corresponded in planta to the expression of distinct plant orthologs of the yeast SMO belonging to the SMO1 and SMO2 gene families [117]. When expressed in a yeast *erg25* mutant, SMO1 and SMO2 conferred different sterol biosynthetic capacity to their host [117,118]. Recently, the characterization of a cholesterol-specific biosynthetic segment in the Solanaceae (containing solanine or tomatine, which are steroidal glycoalkaloid derived from cholesterol) unveiled the function of additional SMO1 and SMO2 orthologs (named SMO3 and SMO4) that act specifically on C4-SBIs bearing cholesterol-type side chains [119]. Consequently, each SMO1 or SMO2 define distinct C4DMC comprising C4D, SKR, and ERG28. Two C4D (3β-hydroxysteroid dehydrogenase/C4-decarboxylase) were found to act redundantly in both types of C4DMC [120]. The plant SKR and ERG28 genes were functionally identified in protein-protein interaction assays and planta with the implementation of RNA silencing (RNAi) or knock-out T-DNA insertion lines [52]. Interestingly, the Arabidopsis SMO1-1 and SMO1-2 isoforms were identified as interactants of an Acyl-CoA-Binding Protein 1 (ACBP1) in yeast double hybrid assays, strongly suggesting a role for a SMO/ACBP1 complex in the regulation of lipid metabolism and particularly the activity of acyltransferases governing the production of triacylglycerols and sterol esters [121,122]. Also, it is proposed that the SMO1/ACBP1 complex controls plant development via an unknown lipid ligand that activates transcription factors like GLABRA2, HDG5, HDG10 [121,122]. The regulatory action of SMO1, possibly as a limiting step in phytosterol biosynthesis or by an unknown signaling activity of C4-SBIs, was illustrated in *A. thaliana* expressing jointly 3-hydroxy-3-methylglutaryl-coenzyme A reductase (HMGR) and SMO1 increasing by 54% in biomass [123].

The physiological functions of C4-SBIs were investigated using the elegant virus-induced gene silencing strategy (VIGS) in *Nicotiana benthamiana*, as already established in the case of the C14-demethylation step (CYP51; [124]). VIGS of SMO1 and SMO2 indicated deficiencies in distinct entities based on distinct sterol profiles: SMO1-silenced plants exhibited 4,4-dimethyl-9β,19-cyclopropylsterols as major sterols whereas SMO2-silenced plants had 4α-methyl-Δ^7-sterols [117]. The same approach was successfully implemented to characterize a C4D gene in *N. benthamiana*: the dramatic reduction in gene expression resulted in the accumulation of the 4-carboxymethyl-4-methylsterol substrate of C4D in silenced leaves [120]. In SMO2 silenced plants, changes in the activity of the C4DMC caused a subsequent increase (compared to wild-type) of the ratio of C24-methylsterols to C24-ethylsterols in the sterol profiles. 24-Methylenelophenol is the substrate of SMO2 and also of the sterol-C24-methyltransferase SMT2 [125–127], and consequently defines a branching point in plant sterol biosynthesis (Figure 2). Therefore, the down-regulation or overexpression of SMO2 indirectly modulate the ratio of 24-methylsterols to 24-ethylsterols (mainly, campesterol to sitosterol), causing deleterious effects on growth [24]. Biotic interactions at the sterol metabolism interface were also studied in the context of silenced SMO genes in *N. benthamiana* to investigate the replication of tombusviruses (TBSV, tomato bushy stunt virus), a group of viruses depending on cellular membranes for replication. The authors also implemented a chemical treatment of plants with 6-amino-2-*n*-pentylthiobenzothiazole (APB), an inhibitor of the fungal SMO [128]. Silencing of SMOs and APB treatment reduced virus replication. Notably, APB was effective in slowing down virus replication in *N. benthamiana* protoplasts. Exogenous addition of campesterol and sitosterol in the medium rescued replication of the virus. The authors have also tested the effect of sterol biosynthesis inhibition by APB on tobacco mosaic virus (TMV) replication and showed that TMV accumulation was sterol-independent. The authors proposed two explanations accounting for the difference in replication between the TBSV and TMV in their host plant: i) tombusviruses proteins are integrated into membranes and interact with sterols; ii) each type of virus replicate in distinct subcellular compartments having specific sterol composition [128].

Functional aspects of C4-SBIs were investigated in *A. thaliana* by overexpressing or knocking-out genes of interest and scrutinizing their associated phenotype. *A. thaliana* overexpressing C4D displayed a short internode phenotype that was not rescued by brassinosteroids. The authors suggested that the accumulation of 3-ketosterols, the products of C4D like 22-hydroxy-5β-ergostan-3-one would alter membrane properties, auxin transporter activity and consequently growth and development (Figure 2) [129]. This conclusion is also in line with possible modification of the sterol composition of membrane microdomains, which are tremendously important in cellular homeostasis and signaling [130]. The characterization of loss-of-function *smo2* alleles in *A. thaliana* required double null mutants of both SMO2-1 and SMO2-2, to deal with genetic redundancy [117]. The complete loss of SMO2 was lethal or at least resulted in an early arrest in embryogenesis [131]. However, heterozygote (*smo2-1/smo2-1, smo2-2/+*) were dwarfs and late-flowering plants, with phenotypic features like small round dark green leaves reminiscent of some other sterol biosynthetic mutants bearing genetic defects in the conversion of Δ^7-sterol intermediates to Δ^5-sterols (campesterol and sitosterol) [132–136]. The sterol profiles of heterozygote *smo2-1/smo2-2* lines showed a marked accumulation of C4-SBIs such as 24-ethylidenelophenol up to 20% of the total and a decrease in campesterol and stigmasterol [131]. A careful examination of the phenotypic traits of *smo2* plants pointed out very clearly their impaired response to auxin [131]. The exogenous application of auxin or the introgression of *smo2* mutations in auxin overproducer lines such as those overproducing free IAA upon enhancement of the YUCCA gene expression resulted in the rescue a wild-type developmental phenotype in *smo2* loss-of-function mutants [131]. It is conceivable that the accumulation of C4-SBIs alters plasma membrane properties, particularly the proper localization of auxin efflux PIN proteins, as shown earlier [137,138]. Alternatively, C4-SBIs may act as components of auxin signaling. This was proposed by independent studies consisting in altering the expression of an enzyme of C4DMC (C4D; [129]) or of ERG28, the non-enzymatic protein that tethers the C4DMC [52]. In the latter study, several *erg28* knocked-down Arabidopsis lines displayed an abnormal accumulation of the transient C4-SBI oxojessic acid (in $\mu g \cdot g^{-1}$ fresh weight amount compared to undetectable signals in wild-type plants). Phenotypes of such plants were reminiscent of an auxin disrupted homeostasis: in fact, experimental evidence supports the function of oxojessic acid as an inhibitor of polar auxin transport [52]. Taken together these results point out a novel critical role for C4-SBIs on growth and development that is distinct from the status of sterol end-products or brassinosteroids.

5. Caenorhabditis elegans: A Sterol Auxotroph with an Extraordinary C4-Methylation Capacity

Nematodes are sterol auxotrophs, just like insects and some other invertebrates. These organisms live on exogenous sterols provided by their diet. They also convert a proportion of cholesterol into steroid hormones known as the dafachronic acids, which bind the nuclear hormone receptor DAF12 responsible for reproductive development (Figure 3) [139]. In fact, the biogenesis of these compounds requires C4-desmethylsterols as substrates (i.e., cholesterol, with a free C4 position) and the action of a 3-hydroxysteroid dehydrogenase/Δ^5/Δ^4 isomerase (HSD-1) for the conversion of cholesterol to cholest-4-en-3-one en route to dafachronic acid (Figure 3) [51,139]. The arrest of the reproductive cycle upon environmental stress requires the inactivation of dafachronic acid biogenesis that enables unbound DAF12-mediated larval entry into the dauer stage, a particular diapause. Quite uncommon in the eukaryotic tree of life, a sterol-C4-methyltransferase named STRM-1 catalyzes the addition of a single methyl group provided by *S*-adenosyl-methionine onto the sterol tetracyclic moiety. The products of the methylation reaction like lophenol or its isomeric 4α-methyl-5α-cholest-8(14)-en-3β-ol are sterol biosynthesis end-products rather than C4-SBIs in this particular context (Figure 3) [51]. The enzymatic reaction catalyzed by STRM-1 has not been investigated into much detail. This sterol methylation restricted to nematodes regulate the biologically active amounts of dafachronic acids, pointing out the tremendous importance of 4-methylsterols in development since it is the C4-methylated product that triggers the entry of the worm into the dauer stage [139,140].

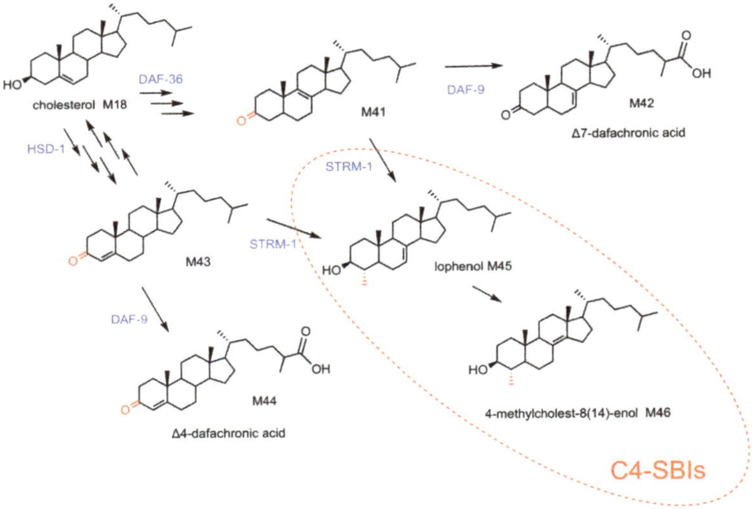

Figure 3. Dafachronic acid synthesis in *Caenorhabditis elegans*. Sterol nomenclature is given in Table 1. C4-demethylation in eukaryotes: HSD-1, 3-hydroxysteroid dehydrogenase/Δ5/Δ4 isomerase (HSD-1); STRM-1, Sterol 4-C-methyltransferase; DAF-9, steroid cytochrome P450 hydroxylase; DAF-36, cholesterol 7-desaturase. Each arrow represents an enzymatic step.

6. Bacteria Evolved Their Specific C4-Demethylation Enzymes

The capacity to synthesize sterols is usually not a prokaryotic feature. However, genes encoding the steroidal triterpene forming enzyme 2,3-oxidosqualene cyclase (OSC) were found in 34 bacterial genomes from several phyla (myxobacteria, methylococcales, rhizobiales, planctomycetes, and some others), thus predicting putative or minimal sterol pathway comprising 2,3-oxidosqualene cyclization products and subsequent C14 and C4 demethylations of those (Figure 4) [141]. Interestingly, γ-proteobacterial aerobic methanotrophs like *Methylococcus capsulatus* are characterized by a C4 demethylation process removing one single methyl group at C4-position of 4,4-dimethylsterols, whereas δ-proteobacterial myxobacteria can remove both methyl groups at C4 like it is the case in eukaryotes [141]. The single C4 demethylation that is typical of *M. capsulatus* is catalyzed by the consecutive action of two enzymes (Figure 4 and Figure S1C). These sterol demethylation (Sdm) enzymes are strikingly different from the eukaryotic C4-demethylation enzymes described above. SdmA is a Rieske-type oxygenase that catalyzes three successive oxidations of the C4β methyl group of 4,4-dimethylsterols, whereas the non-heme oxygenase SMO performs the successive oxidation reactions of the C4α methyl group of 4,4-dimethylsterols in eukaryotes. Rieske-type oxygenases have been described in sterol pathways of dafachronic acids in *C. elegans* [142] and of the protist *Tetrahymena thermophila* [143,144] where that type of enzymes acts as a cholesterol-7-desaturase. SdmB is the second enzyme responsible for both decarboxylation and ketoreduction steps. The reversibility of the last step has been discussed [141]. These findings demonstrate that a sterol-C4-demethylation process has evolved twice independently and that the bacterial Sdm enzymes are functionally restricted to demethylate at C4β without any further oxidation at C4α, explaining thus the production of C4-SBIs as pathway end-products in methanotrophs otherwise used as geological biomarkers. The function of 4-methylsterols in bacteria is not clearly understood. A role in adaptation to environmental constraints like water salinity or limitation in oxygen has been suggested [141].

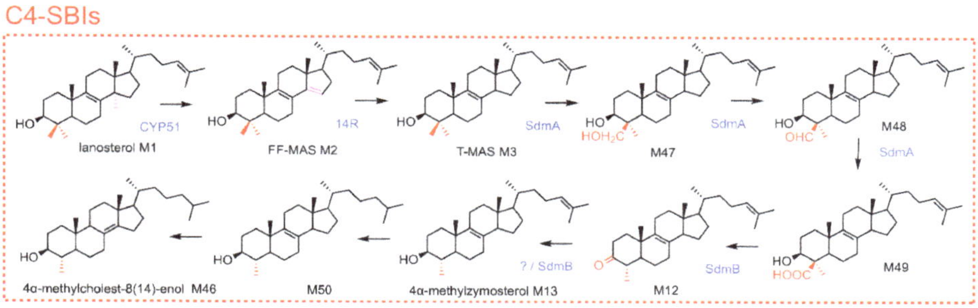

Figure 4. Sterol pathways in *Methylococcus capsulatus*. Sterol nomenclature is given in Table 1. CYP51, lanosterol-C14 demethylase; 14R, sterol-14-reductase; Sdm, sterol demethylase. Each arrow represents an enzymatic step.

7. Inhibitors of C4-SBIs Accumulation In Vivo, Canonical and Non-Canonical C4-SBIs, and Conjugated forms

The overall chemical or genetic inhibition studies of C4-demethylation steps of cholesterol (plants and mammals), ergosterol (fungi, algae) or phytosterol biosynthesis demonstrate that the removal of both methyl groups at C4-position are necessary for proper growth or development. The accumulation of lanosterol results from the inhibition of the sterol-C14-demethylase, that is a P450-dependent mono-oxygenase. The class of 'azoles', that includes imidazoles and triazoles, is widely used as therapeutic and agricultural antifungal drugs [145,146]. For instance, clotrimazole is used to monitor C4-SBI accumulation in yeast and human (Figure 5) [40,147]. It is worth noting that the accumulation of C4-SBIs may be caused by inhibitors acting in fact on the sterol-C14-reduction and sterol-C8-isomerization steps like the morpholine derivative amorolfine or the compound AY9944 [40,57,148].

It is therefore relevant to envision the SMO, C4D, and SKR enzymes as interesting target sites for new fungicides or herbicides. In yeast, APB (Figure 5) inhibited sterol-C4-demethylation [149]. APB was assayed in vitro on maize coleoptile microsomal SMO1 and SMO2 enzymatic activities: surprisingly, APB did not affect SMO1 whereas it displayed a limited inhibition of SMO2 (compared to the strong effect observed on the yeast SMO) [150–152]. Other compounds like PF1163A and PF1163B were isolated from *Penicillium* sp. PF1163A (Figure 5) caused a steady accumulation of 4,4-dimethylzymosterol in yeast indicating SMO as the target of these new antifungal antibiotics [153–155]. Garlic extract and 17-hydroxyprogesterone inhibited human SMO [57,156,157], 3-amino-1,2,3-triazole (ATZ) was described as a potent SMO inhibitor in mice [91,158,159], the cholesterol-lowering oxysteroid FR171456 (Figure 5) was recently characterized for its inhibitory property on C4-decarboxylation enzymes (NSDHL in human, ERG26 in yeast) [160]. Fenhexamid (a hydroxyanilide) and fenpyrazamine (an aminopyrazolinone) are antifungal agents presently on the market. The fungal sterol profiles established in the presence of fenhexamid displayed an accumulation of 3-ketosterols (zymosterone), showing that the inhibition of SKR was most probably the reason of fungitoxicity [161].

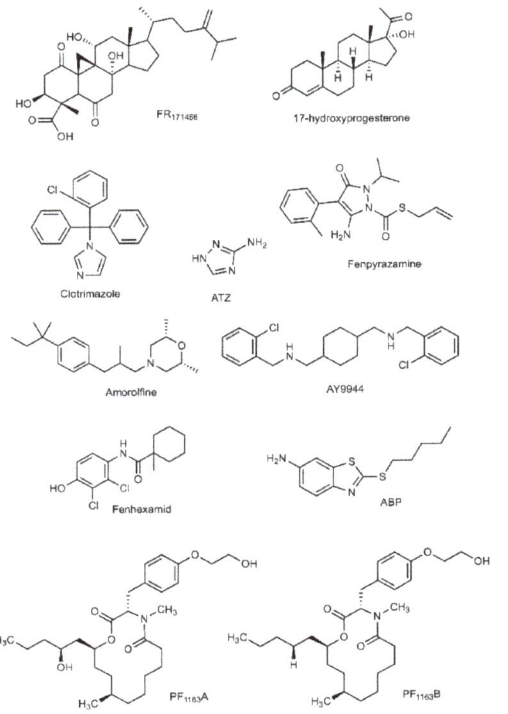

Figure 5. Chemical inhibitors for C4-SBI accumulation.

The classification of canonical and non-canonical C4-SBIs was proposed by the WD Nes (Texas Tech University, Lubbock, TX, USA) and the Littman (Howard Hugues Medical Institute, Chevy Chase, MD, USA) groups [25]. This definition is based on a thorough chemical analysis of oxysterols binding the nuclear receptor RORγ, this in several genetic backgrounds of mice carrying loss-of-function mutations of the enzymes CYP51 (sterol-C14-demethylation) or in SC4MOL/SMO (sterol-C4-demethylation). Canonical oxygenated metabolites derived from the major C4-SBIs T-MAS and 4α-methylzymosterol (Figure 2A) are generated by three successive SMO-catalyzed oxidations of the methyl group at C4 yielding a 4-hydroxymethylsterol, a 4-formylsterol and a 4-carboxysterol transient sterol biosynthesis intermediates (Figure 2A). These C4-SBIs are usually not detected in routine sterol profiling of given organs or tissues and for this reason, could be even considered as cryptic. However, an inhibitor of SMO fed to a yeast microsomal fraction enabled the identification of 4-hydroxymethylsterols, namely, 4β-methyl-4α-hydroxymethyl-5α-cholesta-8,24-dien-3β-ol and 4α-hydroxymethyl-5α-cholesta-8,24-dien-3β-ol [152]. Likewise, carboxysterols and ketosterols were identified in yeast, plants, and mammals following chemical or genetic inhibition. In the case of plants, C4D gene silencing in *N. benthamiana* led to a remarkable accumulation of 3β-hydroxy-4β,14-dimethyl-5α-ergosta-9β,19-cyclo-24(28)-en-4α-carboxylic acid [120].

In mammals, non-canonical oxygenated C4-SBIs are produced by the action of SMO on lanosterol before its demethylation at C14 (Figure 2A). Lanosterol was shown to be a substrate of the *S. cerevisiae* SMO [152]. Non-canonical oxygenated compounds are therefore 4-hydroxymethyl-14-methylsterols, 4-formyl-14-methylsterols, and 4-carboxymethyl-14-methylsterols [25]. Mice thymus contained concentrations of about 60 nM 4-hydroxymethyl-4,14-dimethylcholesta-8,24-dien-3β-ol [25]. Non-Canonical C4-SBIs bearing a 14-hydroxymethyl or 14-carboxymethyl group were identified in previous studies on the C14-demethylation reaction.

14-hydroxymethyl-4,4-dimethylcholesta-8,24-dien-3β-ol accounted for about 1% of total cellular sterol in hepatocytes [162]. The range of non-canonical C4-SBIs is therefore due to the versatility of SMOs that can react as 4α-methylsterol-oxidases on a variety of 4,4-dimethyl- and 4-methylsterol substrates [152].

Conjugated forms of C4-SBIs have been over-looked in biology. Many of these compounds belong to the so-called specialized metabolites (of plants, of protists, of bacteria). Lanosterol glycosides were reported in *Muscari paradoxum* [163]. Cycloartenol esters of fatty acids were found in *Ixora coccinea* [164]. In the marine diatom *Skeletonema marinoi,* sterol sulfates were associated with programmed cell death that occurs as a mechanism regulating phytoplankton blooms [165]. In mammals, the inhibition of SMO by an aminotriazole drug fed to rats resulted in a peroxisomal accumulation of 4α-methylcholest-7-en-3β-ol and its corresponding ester of fatty acids (18% of esters and 82 % of free 3β-OH form). In the same tissues, 4,4-dimethylcholest-8-en-3β-ol was found in its free form only, whereas the total cholesterol included 12% of cholesterol esters [158,159]. C4-SBIs were found as sulfates in patients suffering familial hypercholesterolemia and treated with partial ileal bypass surgery [166]. The function of these sulfates was not well perceived until recent studies in mice proposed for sterol sulfates the role of agonists of the endogenous retinoic acid receptor-related orphan receptor γ (RORγ). This receptor plays a crucial role in the differentiation of lymphocytes and autoimmune diseases [167]. The limited current understanding of the physiological role of conjugated C4-SBIs (glycosides, lipid esters, sulfolipids) as signaling molecules will require further research initiatives.

8. Concluding Remarks

Genes and their products responsible for sterol-C4-demethylation in mammals, yeast, plants, and bacteria have been quite well described by several groups over the last years, as shortly reviewed above. There are striking differences between bacteria and other organisms (protists, metazoans) regarding C4-demethylation mechanisms recruited during evolution. Plants use distinct C4DMC defined by substrate specificity: SMO1-based demethylation complex of 4,4-dimethylsterols and SMO2-based demethylation complex of 4-methylsterols, whereas other organisms demethylate 4,4-dimethylsterols and 4-methylsterols consecutively with a single SMO-based complex (of three enzymes and a tethering protein ERG28). In eukaryotes, genetic or chemical inhibition of the sterol-C4-demethylation may lead to the accumulation of significant amounts of C4-SBIs or transient C4-SBIs, but also of their oxygenated derivatives classified as canonical and non-canonical. The review of biological activities of 4-methylsterols characterized so far in different kingdoms shows clear common features. In yeast, in plants or mammals, the accumulation of C4-SBIs (including oxygenated derivatives) has deleterious effects on growth and development. In yeast and mammals, the role of C4-SBIs in cell division was shown. In plants and mammals, 4-methylsterols and 4-carboxysterols act as signaling molecules interfering with major pathways like auxin in plants and immune system in mammals. The major challenge remains the identification of physical interactions of sterol ligands with their targets. Another critical issue is the analytical scale of those biosynthetic intermediates: just like some oxysterols or brassinosteroids, 4-carboxysterols may be present at very low concentration, e.g., at "hormone-dose" and are therefore not detected in sterol profiles. For example in *A. thaliana*, bulk sterols were quantified 100–200 $\mu g \cdot g^{-1}$ fresh weight, canonical C4-SBIs 0.1–0.5 $\mu g \cdot g^{-1}$ fresh weight, but brassinolide 4×10^{-5} $\mu g \cdot g^{-1}$ fresh weight. Finally, the fate of C4-SBIs as metabolic products requires further investigations, regarding the enzymes implied in this process, and the type of formed products, like for instance hydroxysteroids or sulfates as shown in a study of RORγt receptors [167].

Supplementary Materials: Available online, Figure S1: C14 and C4 demethylation enzymatic reactions in eukaryotes and C4 demethylation in *Methylococcus capsulatus*. A, C14-demethylation: C14DM, lanosterol-C14 demethylase, 14R, sterol-14-reductase. B, C4-demethylation in eukaryotes: SMO, sterol-4α-methyl-oxidase; C4D, 3β-hydroxysteroid dehydrogenases/C-4 decarboxylase; SKR, sterone ketoreductase. C, bacterial C4-demethylation: sdm, sterol demethylase.

Funding: SD and HS are grateful to Centre National de la Recherche Scientifique for supporting an LIA initiative (International Associated Laboratory n°1170, 2017-2020) between UFPA, Belém, and IBMP, Strasbourg.

Conflicts of Interest: The authors declare no conflict of interest. The funders had no role in the design of the study; in the collection, analyses, or interpretation of data; in the writing of the manuscript, or in the decision to publish the results.

References

1. Itoh, T.; Tamura, T.; Matsumoto, T. Methylsterol compositions of 19 vegetable oils. *J. Am. Oil Chem. Soc.* **1973**, *50*, 300–303. [CrossRef]
2. Phillips, K.M.; Ruggio, D.M.; Toivo, J.I.; Swank, M.A.; Simpkins, A.H. Free and Esterified Sterol Composition of Edible Oils and Fats. *J. Food Compost. Anal.* **2002**, *15*, 123–142. [CrossRef]
3. Villette, C.; Berna, A.; Compagnon, V.; Schaller, H. Plant Sterol Diversity in Pollen from Angiosperms. *Lipids* **2015**, *50*, 749–760. [CrossRef] [PubMed]
4. Moreau, R.A.; Nystrom, L.; Whitaker, B.D.; Winkler-Moser, J.K.; Baer, D.J.; Gebauer, S.K.; Hicks, K.B. Phytosterols and their derivatives: Structural diversity, distribution, metabolism, analysis, and health-promoting uses. *Prog. Lipid Res.* **2018**, *70*, 35–61. [CrossRef] [PubMed]
5. Guo, D.A.; Venkatramesh, M.; Nes, W.D. Developmental regulation of sterol biosynthesis in *Zea mays*. *Lipids* **1995**, *30*, 203–219. [CrossRef] [PubMed]
6. Benveniste, P. Biosynthesis and accumulation of sterols. *Annu. Rev. Plant Biol.* **2004**, *55*, 429–457. [CrossRef]
7. Nes, W.D. Biosynthesis of cholesterol and other sterols. *Chem. Rev.* **2011**, *111*, 6423–6451. [CrossRef]
8. Schaller, H. Sterol and steroid biosynthesis and metabolism in plants and microorganisms. In *Comprehensive Natural Products II: Chemistry and Biology*; Mander, L.N., Hung-Wen, L., Eds.; Elsevier: Amsterdam, The Netherlands, 2010; Volume 1, pp. 755–787.
9. Abe, I.; Rohmer, M.; Prestwich, G.D. Enzymatic cyclization of squalene and oxidosqualene to sterols and triterpenes. *Chem. Rev.* **1993**, *93*, 2189–2206. [CrossRef]
10. Gaylor, J.L.; Moir, N.J.; Seifried, H.E.; Jefcoate, C.R. Assay and isolation of a cyanide-binding protein of rat liver microsomes. *J. Biol. Chem.* **1970**, *245*, 5511–5513.
11. Trzaskos, J.M.; Bowen, W.D.; Fisher, G.J.; Billheimer, J.T.; Gaylor, J.L. Microsomal enzymes of cholesterol biosynthesis from lanosterol: A progress report. *Lipids* **1982**, *17*, 250–256. [CrossRef]
12. Trzaskos, J.M.; Bowen, W.D.; Shafiee, A.; Fischer, R.T.; Gaylor, J.L. Cytochrome P450-dependent oxidation of lanosterol in cholesterol biosynthesis. Microsomal transport and C32-demethylation. *J. Biol. Chem.* **1984**, *259*, 13402–13412. [PubMed]
13. Aoyama, Y.; Okikawa, T.; Yoshida, Y. Evidence for the presence of cytochrome P-450 functional in lanosterol 14α-demethylation in microsomes of aerobically grown respiring yeast. *Biochim. Biophys. Acta* **1981**, *665*, 596–601. [CrossRef]
14. Rahier, A.; Taton, M. The 14α-demethylation of obtusifoliol by a cytochrome P-450 monooxygenase from higher plants microsomes. *Biochem. Biophys. Res. Commun.* **1986**, *140*, 1064–1072. [CrossRef]
15. Rahier, A.; Taton, M. Plant Sterol Biosynthesis Inhibitors—The 14-Demethylation Steps, Their Enzymology and Inhibition. *Biochem. Soc. Trans.* **1990**, *18*, 52–56. [CrossRef] [PubMed]
16. Kim, H.B.; Schaller, H.; Goh, C.H.; Kwon, M.; Choe, S.; An, C.S.; Durst, F.; Feldmann, K.A.; Feyereisen, R. *Arabidopsis* cyp51 mutant shows postembryonic seedling lethality associated with lack of membrane integrity. *Plant Physiol.* **2005**, *138*, 2033–2047. [CrossRef]
17. Schrick, K.; Mayer, U.; Horrichs, A.; Kuhnt, C.; Bellini, C.; Dangl, J.; Schmidt, J.; Jurgens, G. FACKEL is a sterol C-14 reductase required for organized cell division and expansion in *Arabidopsis* embryogenesis. *Genes Dev.* **2000**, *14*, 1471–1484.
18. Lorenz, R.T.; Parks, L.W. Cloning, sequencing, and disruption of the gene encoding sterol C-14 reductase in *Saccharomyces cerevisiae*. *DNA Cell Biol.* **1992**, *11*, 685–692. [CrossRef]
19. Roberti, R.; Bennati, A.M.; Galli, G.; Caruso, D.; Maras, B.; Aisa, C.; Beccari, T.; Della Fazia, M.A.; Servillo, G. Cloning and expression of sterol Δ14-reductase from bovine liver. *Eur. J. Biochem.* **2002**, *269*, 283–290. [CrossRef]

20. Waterham, H.R.; Koster, J.; Romeijn, G.J.; Hennekam, R.C.; Vreken, P.; Andersson, H.C.; FitzPatrick, D.R.; Kelley, R.I.; Wanders, R.J. Mutations in the 3β-hydroxysterol Δ24-reductase gene cause desmosterolosis, an autosomal recessive disorder of cholesterol biosynthesis. *Am. J. Hum. Genet.* **2001**, *69*, 685–694. [CrossRef]
21. Pascal, S.; Taton, M.; Rahier, A. Plant sterol biosynthesis. Identification and characterization of two distinct microsomal oxidative enzymatic systems involved in sterol C4-demethylation. *J. Biol. Chem.* **1993**, *268*, 11639–11654.
22. Bouvier, F.; Rahier, A.; Camara, B. Biogenesis, molecular regulation and function of plant isoprenoids. *Prog. Lipid Res.* **2005**, *44*, 357–429. [CrossRef] [PubMed]
23. Mo, C.; Bard, M. Erg28p is a key protein in the yeast sterol biosynthetic enzyme complex. *J. Lipid Res.* **2005**, *46*, 1991–1998. [CrossRef] [PubMed]
24. Darnet, S. *Biosynthèse des stérols: Biochimie et analyse moléculaire des stérol-4α-méthyl-oxydases de plantes*; Université de Strasbourg: Strasbourg, France, 2004.
25. Santori, F.R.; Huang, P.; van de Pavert, S.A.; Douglass, E.F., Jr.; Leaver, D.J.; Haubrich, B.A.; Keber, R.; Lorbek, G.; Konijn, T.; Rosales, B.N.; et al. Identification of natural RORγ ligands that regulate the development of lymphoid cells. *Cell Metab.* **2015**, *21*, 286–298. [CrossRef] [PubMed]
26. Volkman, J.K.; Rijpstra, W.I.C.; de Leeuw, J.W.; Mansour, M.P.; Jackson, A.E.; Blackburn, S.I. Sterols of four dinoflagellates from the genus *Prorocentrum*. *Phytochemistry* **1999**, *52*, 659–668. [CrossRef]
27. Bouvier, P.; Rohmer, M.; Benveniste, P.; Ourisson, G. Δ8(14)-steroids in the bacterium *Methylococcus capsulatus*. *Biochem. J.* **1976**, *159*, 267–271. [CrossRef] [PubMed]
28. Volkman, J.K. Sterols and other triterpenoids: Source specificity and evolution of biosynthetic pathways. *Org. Geochem.* **2005**, *36*, 139–159. [CrossRef]
29. Fowler, M.G.; Douglas, A.G. Saturated hydrocarbon biomarkers in oils of Late Precambrian age from Eastern Siberia. *Org. Geochem.* **1987**, *11*, 201–213. [CrossRef]
30. Ourisson, G. Pecularities of Sterol Biosynthesis in Plants. *J. Plant Physiol.* **1994**, *143*, 434–439. [CrossRef]
31. Bloch, K.E. Sterol structure and membrane function. *CRC Crit. Rev. Biochem.* **1983**, *14*, 47–92. [CrossRef]
32. Gas-Pascual, E.; Berna, A.; Bach, T.J.; Schaller, H. Plant oxidosqualene metabolism: Cycloartenol synthase-dependent sterol biosynthesis in *Nicotiana benthamiana*. *PLoS ONE* **2014**, *9*, e109156. [CrossRef]
33. Barton, D.H.R. Triterpenoids. Part III. cycloartenone, a triterpenoid ketone. *J. Chem. Soc.* **1951**, 1444. [CrossRef]
34. Goad, L.J. Aspects of Phytosterol Biosynthesis. In *Terpenoids in Plants*; Pridham, J.B., Ed.; Academic Press: London, UK, 1967; p. 159.
35. Lenton, J.R.; John Goad, L.; Goodwin, T.W. Sitosterol biosynthesis in *Hordeum vulgare*. *Phytochemistry* **1975**, *14*, 1523–1528. [CrossRef]
36. Dahl, C.E.; Dahl, J.S.; Bloch, K. Effect of alkyl-substituted precursors of cholesterol on artificial and natural membranes and on the viability of *Mycoplasma capricolum*. *Biochemistry* **1980**, *19*, 1462–1467. [CrossRef] [PubMed]
37. Volkman, J.K. Sterols in microorganisms. *Appl. Microbiol. Biotechnol.* **2003**, *60*, 495–506. [CrossRef] [PubMed]
38. Kawashima, H.; Ohnishi, M.; Ogawa, S. Distribution of Unusual Cholesterol Precursors, 4-Methyl- and 4, 4-Dimethylsterols with Δ8 Unsaturation, in Gonads of Marine Archaeogastropods. *J. Oleo Sci.* **2013**, *62*, 465–470. [CrossRef] [PubMed]
39. Lim, L.; Jackson-Lewis, V.; Wong, L.C.; Shui, G.H.; Goh, A.X.; Kesavapany, S.; Jenner, A.M.; Fivaz, M.; Przedborski, S.; Wenk, M.R. Lanosterol induces mitochondrial uncoupling and protects dopaminergic neurons from cell death in a model for Parkinson's disease. *Cell Death Differ.* **2012**, *19*, 416–427. [CrossRef] [PubMed]
40. Hubler, Z.; Allimuthu, D.; Bederman, I.; Elitt, M.S.; Madhavan, M.; Allan, K.C.; Shick, H.E.; Garrison, E.; Karl, M.T.; Factor, D.C.; et al. Accumulation of 8,9-unsaturated sterols drives oligodendrocyte formation and remyelination. *Nature* **2018**, *560*, 372–376. [CrossRef] [PubMed]
41. Rozman, D.; Cotman, M.; Frangež, R. Lanosterol 14α-demethylase and MAS sterols in mammalian gametogenesis. *Mol. Cell. Endocrinol.* **2002**, *187*, 179–187. [CrossRef]
42. Byskov, A.G.; Andersen, C.Y.; Leonardsen, L. Role of meiosis activating sterols, MAS, in induced oocyte maturation. *Mol. Cell. Endocrinol.* **2002**, *187*, 189–196. [CrossRef]
43. Byskov, A.G.; Andersen, C.Y.; Nordholm, L.; Thogersen, H.; Xia, G.; Wassmann, O.; Andersen, J.V.; Guddal, E.; Roed, T. Chemical structure of sterols that activate oocyte meiosis. *Nature* **1995**, *374*, 559–562. [CrossRef]

44. Grondahl, C.; Hansen, T.H.; Marky-Nielsen, K.; Ottesen, J.L.; Hyttel, P. Human oocyte maturation in vitro is stimulated by meiosis-activating sterol. *Hum. Reprod.* **2000**, *15* (Suppl. 5), 3–10. [CrossRef] [PubMed]
45. Mitsche, M.A.; McDonald, J.G.; Hobbs, H.H.; Cohen, J.C. Flux analysis of cholesterol biosynthesis in vivo reveals multiple tissue and cell-type specific pathways. *Elife* **2015**, *4*, e07999. [CrossRef] [PubMed]
46. Alonso, F.; Cirigliano, A.M.; Davola, M.E.; Cabrera, G.M.; Garcia Linares, G.E.; Labriola, C.; Barquero, A.A.; Ramirez, J.A. Multicomponent synthesis of 4,4-dimethyl sterol analogues and their effect on eukaryotic cells. *Steroids* **2014**, *84*, 1–6. [CrossRef] [PubMed]
47. Blume, T.; Guttzeit, M.; Kuhnke, J.; Zorn, L. Two syntheses of FF-MAS. *Org. Lett.* **2003**, *5*, 1837–1839. [CrossRef] [PubMed]
48. Konig, A.; Happle, R.; Bornholdt, D.; Engel, H.; Grzeschik, K.H. Mutations in the NSDHL gene, encoding a 3β-hydroxysteroid dehydrogenase, cause CHILD syndrome. *Am. J. Med. Genet.* **2000**, *90*, 339–346. [CrossRef]
49. McLarren, K.W.; Severson, T.M.; du Souich, C.; Stockton, D.W.; Kratz, L.E.; Cunningham, D.; Hendson, G.; Morin, R.D.; Wu, D.; Paul, J.E.; et al. Hypomorphic temperature-sensitive alleles of NSDHL cause CK syndrome. *Am. J. Hum. Genet.* **2010**, *87*, 905–914. [CrossRef] [PubMed]
50. He, M.; Kratz, L.E.; Michel, J.J.; Vallejo, A.N.; Ferris, L.; Kelley, R.I.; Hoover, J.J.; Jukic, D.; Gibson, K.M.; Wolfe, L.A.; et al. Mutations in the human SC4MOL gene encoding a methyl sterol oxidase cause psoriasiform dermatitis, microcephaly, and developmental delay. *J. Clin. Investig.* **2011**, *121*, 976–984. [CrossRef]
51. Hannich, J.T.; Entchev, E.V.; Mende, F.; Boytchev, H.; Martin, R.; Zagoriy, V.; Theumer, G.; Riezman, I.; Riezman, H.; Knolker, H.J.; et al. Methylation of the sterol nucleus by STRM-1 regulates dauer larva formation in *Caenorhabditis elegans*. *Dev. Cell* **2009**, *16*, 833–843. [CrossRef]
52. Mialoundama, A.S.; Jadid, N.; Brunel, J.; Di Pascoli, T.; Heintz, D.; Erhardt, M.; Mutterer, J.; Bergdoll, M.; Ayoub, D.; Van Dorsselaer, A.; et al. *Arabidopsis* ERG28 tethers the sterol C4-demethylation complex to prevent accumulation of a biosynthetic intermediate that interferes with polar auxin transport. *Plant Cell* **2013**, *25*, 4879–4893. [CrossRef]
53. Fisher, R.T.; Trzaskos, J.M.; Magolda, R.L.; Ko, S.S.; Brosz, C.S.; Larsen, B. Lanosterol 14α-Methyl Demethylase. *J. Biol. Chem.* **1991**, *10*, 6124–6132.
54. Stromstedt, M.; Rozman, D.; Waterman, M.R. The ubiquitously expressed human CYP51 encodes lanosterol 14α-demethylase, a cytochrome P450 whose expression is regulated by oxysterols. *Arch. Biochem. Biophys.* **1996**, *329*, 73–81. [CrossRef] [PubMed]
55. Bennati, A.M.; Castelli, M.; Della Fazia, M.A.; Beccari, T.; Caruso, D.; Servillo, G.; Roberti, R. Sterol dependent regulation of human TM7SF2 gene expression: Role of the encoded 3β-hydroxysterol Δ14-reductase in human cholesterol biosynthesis. *Biochim. Biophys. Acta* **2006**, *1761*, 677–685. [CrossRef] [PubMed]
56. Bennati, A.M.; Schiavoni, G.; Franken, S.; Piobbico, D.; Della Fazia, M.A.; Caruso, D.; De Fabiani, E.; Benedetti, L.; Cusella De Angelis, M.G.; Gieselmann, V.; et al. Disruption of the gene encoding 3β-hydroxysterol Δ14-reductase (Tm7sf2) in mice does not impair cholesterol biosynthesis. *FEBS J.* **2008**, *275*, 5034–5047. [CrossRef]
57. Gatticchi, L.; Cerra, B.; Scarpelli, P.; Macchioni, L.; Sebastiani, B.; Gioiello, A.; Roberti, R. Selected cholesterol biosynthesis inhibitors produce accumulation of the intermediate FF-MAS that targets nucleus and activates LXRalpha in HepG2 cells. *Biochim. Biophys. Acta Mol. Cell. Biol. Lipids* **2017**, *1862*, 842–852. [CrossRef] [PubMed]
58. Tsai, P.L.; Zhao, C.; Turner, E.; Schlieker, C. The Lamin B receptor is essential for cholesterol synthesis and perturbed by disease-causing mutations. *Elife* **2016**, *5*, e16011. [CrossRef] [PubMed]
59. Olson, J.A., Jr.; Lindberg, M.; Bloch, K. On the demethylation of lanosterol to cholesterol. *J. Biol. Chem.* **1957**, *226*, 941–956. [PubMed]
60. Lindberg, M.; Gautschi, F.; Bloch, K.E. Ketonic Intermediates in the demethylation of lanosterol. *J. Biol. Chem.* **1957**, *238*, 1661–1664.
61. Pudles, J.; Bloch, K. Conversion of 4-hydroxymethylene-Δ7-cholesten-3-one to cholesterol. *J. Biol. Chem.* **1960**, *235*, 3417–3420.
62. Swindell, A.C.; Gaylor, J.L. Investigation of the component reactions of oxidative sterol demethylation. Formation and metabolism of 3-ketosteroid intermediates. *J. Biol. Chem.* **1968**, *243*, 5546–5555.
63. Miller, W.L.; Gaylor, J.L. Investigation of the component reactions of oxidative sterol demethylation. Oxidation of a 4,4-dimethyl sterol to a 4 beta-methyl-4 alpha-carboxylic acid during cholesterol biosynthesis. *J. Biol. Chem.* **1970**, *245*, 5375–5381.

64. Miller, W.L.; Brady, D.R.; Gaylor, J.L. Investigation of the component reactions of oxidative demethylation of sterols: Metabolism of 4alpha-hydroxymethyl steroids. *J. Biol. Chem.* **1971**, *246*, 5147–5153. [PubMed]
65. Rahimtula, A.D.; Gaylor, J.L. Investigation of the component reactions of oxidative sterol demethylation. Partial purification of a microsomal sterol 4α-carboxylic acid decarboxylase. *J. Biol. Chem.* **1972**, *247*, 9–15. [PubMed]
66. Billheimer, J.T.; Alcorn, M.; Gaylor, J.L. Solubilization and partial purification of a microsomal 3-ketosteroid reductase of cholesterol biosynthesis. *Arch. Biochem. Biophys.* **1981**, *211*, 430–438. [CrossRef]
67. Nelson, J.A.; Kahn, S.; Spencer, T.A.; Sharpless, K.B.; Clayton, R.B. Some Aspects of Substrate-Specificity in Biological Demethylation at C4 of Steroids. *Bioorg. Chem.* **1975**, *4*, 363–376. [CrossRef]
68. Gaylor, J.L.; Miyake, Y.; Yamano, T. Stoichiometry of 4-methyl sterol oxidase of rat liver microsomes. *J. Biol. Chem.* **1975**, *250*, 7159–7167. [PubMed]
69. Marijanovic, Z.; Laubner, D.; Moller, G.; Gege, C.; Husen, B.; Adamski, J.; Breitling, R. Closing the gap: Identification of human 3-ketosteroid reductase, the last unknown enzyme of mammalian cholesterol biosynthesis. *Mol. Endocrinol.* **2003**, *17*, 1715–1725. [CrossRef] [PubMed]
70. Liu, X.Y.; Dangel, A.W.; Kelley, R.I.; Zhao, W.; Denny, P.; Botcherby, M.; Cattanach, B.; Peters, J.; Hunsicker, P.R.; Mallon, A.M.; et al. The gene mutated in bare patches and striated mice encodes a novel 3β-hydroxysteroid dehydrogenase. *Nat. Genet.* **1999**, *22*, 182–187. [CrossRef] [PubMed]
71. Ottolenghi, C.; Daizadeh, I.; Ju, A.; Kossida, S.; Renault, G.; Jacquet, M.; Fellous, A.; Gilbert, W.; Veitia, R. The genomic structure of c14orf1 is conserved across eukarya. *Mamm. Genome* **2000**, *11*, 786–788. [CrossRef] [PubMed]
72. Quinlan, R.A. DRUG DISCOVERY. A new dawn for cataracts. *Science* **2015**, *350*, 636–637. [CrossRef] [PubMed]
73. Makley, L.N.; McMenimen, K.A.; DeVree, B.T.; Goldman, J.W.; McGlasson, B.N.; Rajagopal, P.; Dunyak, B.M.; McQuade, T.J.; Thompson, A.D.; Sunahara, R.; et al. Pharmacological chaperone for alpha-crystallin partially restores transparency in cataract models. *Science* **2015**, *350*, 674–677. [CrossRef] [PubMed]
74. Zhao, L.; Chen, X.J.; Zhu, J.; Xi, Y.B.; Yang, X.; Hu, L.D.; Ouyang, H.; Patel, S.H.; Jin, X.; Lin, D.; et al. Lanosterol reverses protein aggregation in cataracts. *Nature* **2015**, *523*, 607–611. [CrossRef] [PubMed]
75. Shanmugam, P.; Barigali, A.; Kadaskar, J.; Borgohain, S.; Mishra, D.C.; Ramanjulu, R.; Minija, C.K. Effect of lanosterol on human cataract nucleus. *Indian J. Ophthalmol.* **2015**, *63*, 888–890. [CrossRef] [PubMed]
76. Chen, X.J.; Hu, L.D.; Yao, K.; Yan, Y.B. Lanosterol and 25-hydroxycholesterol dissociate crystallin aggregates isolated from cataractous human lens via different mechanisms. *Biochem. Biophys. Res. Commun.* **2018**, *506*, 868–873. [CrossRef] [PubMed]
77. Shen, X.; Zhu, M.; Kang, L.; Tu, Y.; Li, L.; Zhang, R.; Qin, B.; Yang, M.; Guan, H. Lanosterol Synthase Pathway Alleviates Lens Opacity in Age-Related Cortical Cataract. *J. Ophthalmol.* **2018**, *2018*, 4125893. [CrossRef] [PubMed]
78. Araldi, E.; Fernandez-Fuertes, M.; Canfran-Duque, A.; Tang, W.; Cline, G.W.; Madrigal-Matute, J.; Pober, J.S.; Lasuncion, M.A.; Wu, D.; Fernandez-Hernando, C.; et al. Lanosterol Modulates TLR4-Mediated Innate Immune Responses in Macrophages. *Cell Rep.* **2017**, *19*, 2743–2755. [CrossRef] [PubMed]
79. Song, B.L.; Javitt, N.B.; DeBose-Boyd, R.A. Insig-mediated degradation of HMG CoA reductase stimulated by lanosterol, an intermediate in the synthesis of cholesterol. *Cell Metab.* **2005**, *1*, 179–189. [CrossRef]
80. Nguyen, A.D.; McDonald, J.G.; Bruick, R.K.; DeBose-Boyd, R.A. Hypoxia stimulates degradation of 3-hydroxy-3-methylglutaryl-coenzyme A reductase through accumulation of lanosterol and hypoxia-inducible factor-mediated induction of insigs. *J. Biol. Chem.* **2007**, *282*, 27436–27446. [CrossRef]
81. Schiavoni, G.; Bennati, A.M.; Castelli, M.; Della Fazia, M.A.; Beccari, T.; Servillo, G.; Roberti, R. Activation of TM7SF2 promoter by SREBP-2 depends on a new sterol regulatory element, a GC-box, and an inverted CCAAT-box. *Biochim. Biophys. Acta* **2010**, *1801*, 587–592. [CrossRef]
82. Subramanian, G.; Chaudhury, P.; Malu, K.; Fowler, S.; Manmode, R.; Gotur, D.; Zwerger, M.; Ryan, D.; Roberti, R.; Gaines, P. Lamin B receptor regulates the growth and maturation of myeloid progenitors via its sterol reductase domain: Implications for cholesterol biosynthesis in regulating myelopoiesis. *J. Immunol.* **2012**, *188*, 85–102. [CrossRef]
83. Bellezza, I.; Roberti, R.; Gatticchi, L.; Del Sordo, R.; Rambotti, M.G.; Marchetti, M.C.; Sidoni, A.; Minelli, A. A novel role for Tm7sf2 gene in regulating TNFalpha expression. *PLoS ONE* **2013**, *8*, e68017. [CrossRef]

84. Bellezza, I.; Gatticchi, L.; del Sordo, R.; Peirce, M.J.; Sidoni, A.; Roberti, R.; Minelli, A. The loss of Tm7sf gene accelerates skin papilloma formation in mice. *Sci. Rep.* **2015**, *5*, 9471. [CrossRef] [PubMed]
85. Bartoli, D.; Piobbico, D.; Bellet, M.M.; Bennati, A.M.; Roberti, R.; Della Fazia, M.A.; Servillo, G. Impaired cell proliferation in regenerating liver of 3β-hydroxysterol Δ14-reductase (TM7SF2) knock-out mice. *Cell Cycle* **2016**, *15*, 2164–2173. [CrossRef] [PubMed]
86. Offiah, A.C.; Mansour, S.; Jeffrey, I.; Nash, R.; Whittock, N.; Pyper, R.; Bewley, S.; Clayton, P.T.; Hall, C.M. Greenberg dysplasia (HEM) and lethal X linked dominant Conradi-Hünermann chondrodysplasia punctata (CDPX2): Presentation of two cases with overlapping phenotype. *J. Med. Genet.* **2003**, *40*, e129. [CrossRef] [PubMed]
87. Greenberg, C.R.; Rimoin, D.L.; Gruber, H.E.; DeSa, D.J.; Reed, M.; Lachman, R.S. A new autosomal recessive lethal chondrodystrophy with congenital hydrops. *Am. J. Med. Genet.* **1988**, *29*, 623–632. [CrossRef] [PubMed]
88. Giorgio, E.; Sirchia, F.; Bosco, M.; Sobreira, N.L.M.; Baylor-Hopkins Center for Mendelian Genomics; Grosso, E.; Brussino, A.; Brusco, A. A novel case of Greenberg dysplasia and genotype-phenotype correlation analysis for LBR pathogenic variants: An instructive example of one gene-multiple phenotypes. *Am. J. Med. Genet. A* **2018**. [CrossRef] [PubMed]
89. Gaudy-Marqueste, C.; Roll, P.; Esteves-Vieira, V.; Weiller, P.J.; Grob, J.J.; Cau, P.; Levy, N.; De Sandre-Giovannoli, A. LBR mutation and nuclear envelope defects in a patient affected with Reynolds syndrome. *J. Med. Genet.* **2010**, *47*, 361–370. [CrossRef]
90. Sukhanova, A.; Gorin, A.; Serebriiskii, I.G.; Gabitova, L.; Zheng, H.; Restifo, D.; Egleston, B.L.; Cunningham, D.; Bagnyukova, T.; Liu, H.; et al. Targeting C4-demethylating genes in the cholesterol pathway sensitizes cancer cells to EGF receptor inhibitors via increased EGF receptor degradation. *Cancer Discov.* **2013**, *3*, 96–111. [CrossRef]
91. He, M.; Smith, L.D.; Chang, R.; Li, X.; Vockley, J. The role of sterol-C4-methyl oxidase in epidermal biology. *Biochim. Biophys. Acta* **2014**, *1841*, 331–335. [CrossRef]
92. Mi, X.B.; Luo, M.X.; Guo, L.L.; Zhang, T.D.; Qiu, X.W. CHILD Syndrome: Case Report of a Chinese Patient and Literature Review of the NAD[P]H Steroid Dehydrogenase-Like Protein Gene Mutation. *Pediatr. Dermatol.* **2015**, *32*, e277–e282. [CrossRef] [PubMed]
93. Seeger, M.A.; Paller, A.S. The role of abnormalities in the distal pathway of cholesterol synthesis in the Congenital Hemidysplasia with Ichthyosiform erythroderma and Limb Defects (CHILD) syndrome. *Biochim. Biophys. Acta* **2014**, *1841*, 345–352. [CrossRef]
94. Stottmann, R.W.; Turbe-Doan, A.; Tran, P.; Kratz, L.E.; Moran, J.L.; Kelley, R.I.; Beier, D.R. Cholesterol metabolism is required for intracellular hedgehog signal transduction in vivo. *PLoS Genet.* **2011**, *7*, e1002224. [CrossRef] [PubMed]
95. Cunningham, D.; DeBarber, A.E.; Bir, N.; Binkley, L.; Merkens, L.S.; Steiner, R.D.; Herman, G.E. Analysis of hedgehog signaling in cerebellar granule cell precursors in a conditional Nsdhl allele demonstrates an essential role for cholesterol in postnatal CNS development. *Hum. Mol. Genet.* **2015**, *24*, 2808–2825. [CrossRef] [PubMed]
96. Sato, R. Sterol metabolism and SREBP activation. *Arch. Biochem. Biophys.* **2010**, *501*, 177–181. [CrossRef] [PubMed]
97. Moss, G.P. Nomenclature of steroids (Recommendations 1989). *Pure Appl. Chem.* **1989**, *61*, 1783–1822. [CrossRef]
98. Daum, G.; Lees, N.D.; Bard, M.; Dickson, R. Biochemistry, cell biology and molecular biology of lipids of *Saccharomyces cerevisiae*. *Yeast* **1998**, *14*, 1471–1510. [CrossRef]
99. Lees, N.D.; Bard, M.; Kirsch, D.R. Biochemistry and molecular biology of sterol synthesis in *Saccharomyces cerevisiae*. *Crit. Rev. Biochem. Mol. Biol.* **1999**, *34*, 33–47.
100. Li, L.; Kaplan, J. Characterization of yeast methyl sterol oxidase (ERG25) and identification of a human homologue. *J. Biol. Chem.* **1996**, *271*, 16927–16933. [CrossRef] [PubMed]
101. Bard, M.; Bruner, D.A.; Pierson, C.A.; Lees, N.D.; Biermann, B.; Frye, L.; Koegel, C.; Barbuch, R. Cloning and characterization of *ERG25*, the *Saccharomyces cerevisiae* gene encoding C-4 sterol methyl oxidase. *Proc. Natl. Acad. Sci. USA* **1996**, *93*, 186–190. [CrossRef] [PubMed]
102. Horinouchi, S.; Ishizuka, H.; Beppu, T. Cloning, nucleotide sequence, and transcriptional analysis of the NAD(P)-dependent cholesterol dehydrogenase gene from a *Nocardia* sp. and its hyperexpression in *Streptomyces* spp. *Appl. Environ. Microbiol.* **1991**, *57*, 1386–1393. [PubMed]

103. Gachotte, D.; Barbuch, R.; Gaylor, J.; Nickel, E.; Bard, M. Characterization of the *Saccharomyces cerevisiae* ERG26 gene encoding the C-3 sterol dehydrogenase (C-4 decarboxylase) involved in sterol biosynthesis. *Proc. Natl. Acad. Sci. USA* **1998**, *95*, 13794–13799. [CrossRef]
104. Aaron, K.E.; Pierson, C.A.; Lees, N.D.; Bard, M. The *Candida albicans* ERG26 gene encoding the C-3 sterol dehydrogenase (C-4 decarboxylase) is essential for growth. *FEMS Yeast Res.* **2001**, *1*, 93–101. [CrossRef] [PubMed]
105. Gachotte, D.; Sen, S.E.; Eckstein, J.; Barbuch, R.; Krieger, M.; Ray, B.D.; Bard, M. Characterization of the *Saccharomyces cerevisiae* ERG27 gene encoding the 3-keto reductase involved in C-4 sterol demethylation. *Proc. Natl. Acad. Sci. USA* **1999**, *96*, 12655–12660. [CrossRef] [PubMed]
106. Gachotte, D.; Eckstein, J.; Barbuch, R.; Hughes, T.; Roberts, C.; Bard, M. A novel gene conserved from yeast to humans is involved in sterol biosynthesis. *J. Lipid Res.* **2001**, *42*, 150–154. [PubMed]
107. Baudry, K.; Swain, E.; Rahier, A.; Germann, M.; Batta, A.; Rondet, S.; Mandala, S.; Henry, K.; Tint, G.S.; Edlind, T.; et al. The effect of the *erg26-1* mutation on the regulation of lipid metabolism in *Saccharomyces cerevisiae*. *J. Biol. Chem.* **2001**, *276*, 12702–12711. [CrossRef] [PubMed]
108. Mo, C.; Valachovic, M.; Randall, S.K.; Nickels, J.T.; Bard, M. Protein-protein interactions among C-4 demethylation enzymes involved in yeast sterol biosynthesis. *Proc. Natl. Acad. Sci. USA* **2002**, *99*, 9739–9744. [CrossRef] [PubMed]
109. Ward, D.M.; Chen, O.S.; Li, L.; Kaplan, J.; Bhuiyan, S.A.; Natarajan, S.K.; Bard, M.; Cox, J.E. Altered sterol metabolism in budding yeast affects mitochondrial iron-sulfur (Fe-S) cluster synthesis. *J. Biol. Chem.* **2018**, *293*, 10782–10795. [CrossRef] [PubMed]
110. Gachotte, D.; Pierson, C.A.; Lees, N.D.; Barbuch, R.; Koegel, C.; Bard, M. A yeast sterol auxotroph (erg25) is rescued by addition of azole antifungals and reduced levels of heme. *Proc. Natl. Acad. Sci. USA* **1997**, *94*, 11173–11178. [CrossRef]
111. Wachtler, V.; Rajagopalan, S.; Balasubramanian, M.K. Sterol-rich plasma membrane domains in the fission yeast *Schizosaccharomyces pombe*. *J. Cell Sci.* **2003**, *116*, 867–874. [CrossRef]
112. Germann, M.; Gallo, C.; Donahue, T.; Shirzadi, R.; Stukey, J.; Lang, S.; Ruckenstuhl, C.; Oliaro-Bosso, S.; McDonough, V.; Turnowsky, F.; et al. Characterizing sterol defect suppressors uncovers a novel transcriptional signaling pathway regulating zymosterol biosynthesis. *J. Biol. Chem.* **2005**, *280*, 35904–35913. [CrossRef]
113. Swain, E.; Baudry, K.; Stukey, J.; McDonough, V.; Germann, M.; Nickels, J.T., Jr. Sterol-dependent regulation of sphingolipid metabolism in *Saccharomyces cerevisiae*. *J. Biol. Chem.* **2002**, *277*, 26177–26184. [CrossRef]
114. Lv, J.M.; Hu, D.; Gao, H.; Kushiro, T.; Awakawa, T.; Chen, G.D.; Wang, C.X.; Abe, I.; Yao, X.S. Biosynthesis of helvolic acid and identification of an unusual C-4-demethylation process distinct from sterol biosynthesis. *Nat. Commun.* **2017**, *8*, 1644. [CrossRef] [PubMed]
115. Hughes, A.L.; Lee, C.Y.; Bien, C.M.; Espenshade, P.J. 4-Methyl sterols regulate fission yeast SREBP-Scap under low oxygen and cell stress. *J. Biol. Chem.* **2007**, *282*, 24388–24396. [CrossRef] [PubMed]
116. Rahier, A.; Smith, M.; Taton, M. The role of cytochrome b5 in 4α-methyl-oxidation and C5(6) desaturation of plant sterol precursors. *Biochem. Biophys. Res. Commun.* **1997**, *236*, 434–437. [CrossRef] [PubMed]
117. Darnet, S.; Rahier, A. Plant sterol biosynthesis: Identification of two distinct families of sterol 4α-methyl oxidases. *Biochem. J.* **2004**, *378*, 889–898. [CrossRef] [PubMed]
118. Darnet, S.; Bard, M.; Rahier, A. Functional identification of sterol-4α-methyl oxidase cDNAs from *Arabidopsis thaliana* by complementation of a yeast *erg25* mutant lacking sterol-4α-methyl oxidation. *FEBS Lett.* **2001**, *508*, 39–43. [CrossRef]
119. Sonawane, P.D.; Pollier, J.; Panda, S.; Szymanski, J.; Massalha, H.; Yona, M.; Unger, T.; Malitsky, S.; Arendt, P.; Pauwels, L.; et al. Plant cholesterol biosynthetic pathway overlaps with phytosterol metabolism. *Nat. Plants* **2016**, *3*, 16205. [CrossRef]
120. Rahier, A.; Darnet, S.; Bouvier, F.; Camara, B.; Bard, M. Molecular and enzymatic characterizations of novel bifunctional 3β-hydroxysteroid dehydrogenases/C-4 decarboxylases from *Arabidopsis thaliana*. *J. Biol. Chem.* **2006**, *281*, 27264–27277. [CrossRef]
121. Lung, S.C.; Liao, P.; Yeung, E.C.; Hsiao, A.S.; Xue, Y.; Chye, M.L. Arabidopsis ACYL-COA-BINDING PROTEIN1 interacts with STEROL C4-METHYL OXIDASE1-2 to modulate gene expression of homeodomain-leucine zipper IV transcription factors. *New Phytol.* **2018**, *218*, 183–200. [CrossRef]

122. Lung, S.C.; Liao, P.; Yeung, E.C.; Hsiao, A.S.; Xue, Y.; Chye, M.L. Acyl-CoA-Binding Protein ACBP1 Modulates Sterol Synthesis during Embryogenesis. *Plant Physiol.* **2017**, *174*, 1420–1435. [CrossRef]
123. Lange, I.; Poirier, B.C.; Herron, B.K.; Lange, B.M. Comprehensive Assessment of Transcriptional Regulation Facilitates Metabolic Engineering of Isoprenoid Accumulation in *Arabidopsis*. *Plant Physiol.* **2015**, *169*, 1595–1606. [CrossRef]
124. Burger, C.; Rondet, S.; Benveniste, P.; Schaller, H. Virus-induced silencing of sterol biosynthetic genes: Identification of a *Nicotiana tabacum* L. obtusifoliol-14α-demethylase (CYP51) by genetic manipulation of the sterol biosynthetic pathway in *Nicotiana benthamiana* L. *J. Exp. Bot.* **2003**, *54*, 1675–1683. [CrossRef] [PubMed]
125. Nakamoto, M.; Schmit, A.C.; Heintz, D.; Schaller, H.; Ohta, D. Diversification of sterol methyltransferase enzymes in plants and a role for beta-sitosterol in oriented cell plate formation and polarized growth. *Plant J.* **2015**, *84*, 860–874. [CrossRef] [PubMed]
126. Schaller, H.; Bouvier-Navé, P.; Benveniste, P. Overexpression of an *Arabidopsis* cDNA Encoding a Sterol-C241-Methyltransferase in Tobacco Modifies the Ratio of 24-Methyl Cholesterol to Sitosterol and Is Associated with Growth Reduction. *Plant Physiol.* **1998**, *118*, 461–469. [CrossRef] [PubMed]
127. Schaeffer, A.; Bronner, R.; Benveniste, P.; Schaller, H. The ratio of campesterol to sitosterol with modulates growth in *Arabidopsis* is controlled by STEROL METHYLTRANSFERASE 2-1. *Plant J.* **2001**, *25*, 605–615. [CrossRef] [PubMed]
128. Sharma, M.; Sasvari, Z.; Nagy, P.D. Inhibition of sterol biosynthesis reduces tombusvirus replication in yeast and plants. *J. Virol.* **2010**, *84*, 2270–2281. [CrossRef]
129. Kim, B.; Kim, G.; Fujioka, S.; Takatsuto, S.; Choe, S. Overexpression of 3beta-hydroxysteroid dehydrogenases/C-4 decarboxylases causes growth defects possibly due to abnormal auxin transport in *Arabidopsis*. *Mol. Cells* **2012**, *34*, 77–84. [CrossRef]
130. Simon-Plas, F.; Perraki, A.; Bayer, E.; Gerbeau-Pissot, P.; Mongrand, S. An update on plant membrane rafts. *Curr. Opin. Plant Biol.* **2011**, *14*, 642–649. [CrossRef]
131. Zhang, X.; Sun, S.; Nie, X.; Boutte, Y.; Grison, M.; Li, P.; Kuang, S.; Men, S. Sterol Methyl Oxidases Affect Embryo Development via Auxin-Associated Mechanisms. *Plant Physiol.* **2016**, *171*, 468–482. [CrossRef]
132. Klahre, U.; Noguchi, T.; Fujioka, S.; Takatsuto, S.; Yokota, T.; Nomura, T.; Yoshida, S.; Chua, N.H. The *Arabidopsis* DIMINUTO/DWARF1 gene encodes a protein involved in steroid synthesis. *Plant Cell* **1998**, *10*, 1677–1690. [CrossRef]
133. Choe, S.; Dilkes, B.P.; Fujioka, S.; Takatsuto, S.; Sakurai, A.; Feldmann, K.A. The DWF4 gene of *Arabidopsis* encodes a cytochrome P450 that mediates multiple 22α-hydroxylation steps in brassinosteroid biosynthesis. *Plant Cell* **1998**, *10*, 231.
134. Choe, S.W.; Noguchi, T.; Fujioka, S.; Takatsuto, S.; Tissier, C.P.; Gregory, B.D.; Ross, A.S.; Tanaka, A.; Yoshida, S.; Tax, F.E.; et al. The Arabidopsis dwf7/ste1 mutant is defective in the Δ7 sterol C-5 desaturation step leading to brassinosteroid biosynthesis. *Plant Cell* **1999**, *11*, 207–221. [PubMed]
135. Choe, S.; Dilkes, B.P.; Gregory, B.D.; Ross, A.S.; Yuan, H.; Noguchi, T.; Fujioka, S.; Takatsuto, S.; Tanaka, A.; Yoshida, S.; et al. The Arabidopsis *dwarf1* mutant is defective in the conversion of 24-methylenecholesterol to campesterol in brassinosteroid biosynthesis. *Plant Physiol.* **1999**, *119*, 897–907. [CrossRef]
136. Silvestro, D.; Andersen, T.G.; Schaller, H.; Jensen, P.E. Plant Sterol Metabolism. Δ7-Sterol-C5-Desaturase (STE1/DWARF7), Δ5,7-Sterol-Δ7-Reductase (DWARF5) and Δ24-Sterol-Δ24-Reductase (DIMINUTO/DWARF1) Show Multiple Subcellular Localizations in *Arabidopsis thaliana* (Heynh) L. *PLoS ONE* **2013**, *8*, e56429. [CrossRef] [PubMed]
137. Willemsen, V.; Friml, J.; Grebe, M.; van den Toorn, A.; Palme, K.; Scheres, B. Cell polarity and PIN protein positioning in *Arabidopsis* require STEROL METHYLTRANSFERASE1 function. *Plant Cell* **2003**, *15*, 612–625. [CrossRef] [PubMed]
138. Men, S.; Boutte, Y.; Ikeda, Y.; Li, X.; Palme, K.; Stierhof, Y.D.; Hartmann, M.A.; Moritz, T.; Grebe, M. Sterol-dependent endocytosis mediates post-cytokinetic acquisition of PIN2 auxin efflux carrier polarity. *Nat. Cell Biol.* **2008**, *10*, 237–244. [CrossRef]
139. Butcher, R.A. Small-molecule pheromones and hormones controlling nematode development. *Nat. Chem. Biol.* **2017**, *13*, 577–586. [CrossRef] [PubMed]
140. Matyash, V.; Entchev, E.V.; Mende, F.; Wilsch-Brauninger, M.; Thiele, C.; Schmidt, A.W.; Knolker, H.J.; Ward, S.; Kurzchalia, T.V. Sterol-derived hormone(s) controls entry into diapause in *Caenorhabditis elegans* by consecutive activation of DAF-12 and DAF-16. *PLoS Biol.* **2004**, *2*, e280. [CrossRef]

141. Lee, A.K.; Banta, A.B.; Wei, J.H.; Kiemle, D.J.; Feng, J.; Giner, J.L.; Welander, P.V. C-4 sterol demethylation enzymes distinguish bacterial and eukaryotic sterol synthesis. *Proc. Natl. Acad. Sci. USA* **2018**, *115*, 5884–5889. [CrossRef]
142. Wollam, J.; Magomedova, L.; Magner, D.B.; Shen, Y.; Rottiers, V.; Motola, D.L.; Mangelsdorf, D.J.; Cummins, C.L.; Antebi, A. The Rieske oxygenase DAF-36 functions as a cholesterol 7-desaturase in steroidogenic pathways governing longevity. *Aging Cell* **2011**, *10*, 879–884. [CrossRef]
143. Poklepovich, T.J.; Urtasun, N.; Miranda, M.V.; Nusblat, A.D.; Nudel, C.B. Expression and functional characterization of a C-7 cholesterol desaturase from Tetrahymena thermophila in an insect cell line. *Steroids* **2015**, *96*, 132–139. [CrossRef]
144. Najle, S.R.; Nusblat, A.D.; Nudel, C.B.; Uttaro, A.D. The Sterol-C7 desaturase from the ciliate Tetrahymena thermophila is a Rieske Oxygenase, which is highly conserved in animals. *Mol. Biol. Evol.* **2013**, *30*, 1630–1643. [CrossRef] [PubMed]
145. Odds, F.C.; Brown, A.J.P.; Gow, N.A.R. Antifungal agents: Mechanisms of action. *Trends Microbiol.* **2003**, *11*, 272–279. [CrossRef]
146. Warrilow, A.G.; Parker, J.E.; Kelly, D.E.; Kelly, S.L. Azole affinity of sterol 14α-demethylase (CYP51) enzymes from *Candida albicans* and *Homo sapiens*. *Antimicrob. Agents Chemother.* **2013**, *57*, 1352–1360. [CrossRef] [PubMed]
147. Muller, C.; Binder, U.; Bracher, F.; Giera, M. Antifungal drug testing by combining minimal inhibitory concentration testing with target identification by gas chromatography-mass spectrometry. *Nat. Protoc.* **2017**, *12*, 947–963. [CrossRef] [PubMed]
148. Jachak, G.R.; Ramesh, R.; Sant, D.G.; Jorwekar, S.U.; Jadhav, M.R.; Tupe, S.G.; Deshpande, M.V.; Reddy, D.S. Silicon Incorporated Morpholine Antifungals: Design, Synthesis, and Biological Evaluation. *ACS Med. Chem. Lett.* **2015**, *6*, 1111–1116. [CrossRef] [PubMed]
149. Kuchta, T.; Bartkova, K.; Kubinec, R. Ergosterol depletion and 4-methyl sterols accumulation in the yeast *Saccharomyces cerevisiae* treated with an antifungal, 6-amino-2-n-pentylthiobenzothiazole. *Biochem. Biophys. Res. Commun.* **1992**, *189*, 85–91. [CrossRef]
150. Kuchta, T.; Leka, C.; Farkas, P.; Bujdakova, H.; Belajova, E.; Russell, N.J. Inhibition of sterol 4-demethylation in *Candida albicans* by 6-amino-2-n-pentylthiobenzothiazole, a novel mechanism of action for an antifungal agent. *Antimicrob. Agents Chemother.* **1995**, *39*, 1538–1541. [CrossRef] [PubMed]
151. Fabry, S.; Gaborova, S.; Bujdakova, H.; Klobusicky, M.; Vollekova, A.; Kuchta, T. Inhibition of germ tube formation, filamentation and ergosterol biosynthesis in *Candida albicans* treated with 6-amino-2-n-pentylthiobenzothiazole. *Folia Microbiol. (Praha)* **1999**, *44*, 523–526. [CrossRef]
152. Darnet, S.; Rahier, A. Enzymological properties of sterol-C4-methyl-oxidase of yeast sterol biosynthesis. *Biochim. Biophys. Acta* **2003**, *1633*, 106–117. [CrossRef]
153. Nose, H.; Fushimi, H.; Seki, A.; Sasaki, T.; Watabe, H.; Hoshiko, S. PF1163A, a Novel Antifungal Agent, Inhibit Ergosterol Biosynthesis at C-4 Sterol Methyl Oxidase. *J. Antibiot.* **2002**, *55*, 969–974. [CrossRef]
154. Ekhato, I.V.; Robinson, C.H. Synthesis of New Nitro and Amino Sterols—Potential Inhibitors of 4-Methyl Sterol Oxidase. *J. Chem. Soc.-Perkin Trans. 1* **1988**, 3239–3242. [CrossRef]
155. Ekhato, I.V.; Robinson, C.H. Synthesis of Novel 4α-Substituted Sterols. *J. Org. Chem.* **1989**, *54*, 1327–1331. [CrossRef]
156. Singh, D.K.; Porter, T.D. Inhibition of sterol 4α-methyl oxidase is the principal mechanism by which garlic decreases cholesterol synthesis. *J. Nutr.* **2006**, *136*, 759s–764s. [CrossRef]
157. Lindenthal, B.; Holleran, A.L.; Aldaghlas, T.A.; Ruan, B.; Schroepfer, G.J., Jr.; Wilson, W.K.; Kelleher, J.K. Progestins block cholesterol synthesis to produce meiosis-activating sterols. *FASEB J.* **2001**, *15*, 775–784. [CrossRef] [PubMed]
158. Hashimoto, F.; Hayashi, H. Identification of intermediates after inhibition of cholesterol synthesis by aminotriazole treatment in vivo. *Biochim. Biophys. Acta* **1991**, *1086*, 115–124. [CrossRef]
159. Hashimoto, F.; Hayashi, H. Peroxisomal cholesterol synthesis in vivo: Accumulation of 4-methyl intermediate sterols after aminotriazole inhibition of cholesterol synthesis. *Biochim. Biophys. Acta* **1994**, *1214*, 11–19. [CrossRef]
160. Helliwell, S.B.; Karkare, S.; Bergdoll, M.; Rahier, A.; Leighton-Davis, J.R.; Fioretto, C.; Aust, T.; Filipuzzi, I.; Frederiksen, M.; Gounarides, J.; et al. FR171456 is a specific inhibitor of mammalian NSDHL and yeast Erg26p. *Nat. Commun.* **2015**, *6*, 8613. [CrossRef] [PubMed]

161. Debieu, D.; Bach, J.; Hugon, M.; Malosse, C.; Leroux, P. The hydroxyanilide fenhexamid, a new sterol biosynthesis inhibitor fungicide efficient against the plant pathogenic fungus *Botryotinia fuckeliana* (*Botrytis cinerea*). *Pest Manag. Sci.* **2001**, *57*, 1060–1067. [CrossRef]
162. Trzaskos, J.M.; Fischer, R.T.; Favata, M.F. Mechanistic studies of lanosterol C-32 demethylation. Conditions which promote oxysterol intermediate accumulation during the demethylation process. *J. Biol. Chem.* **1986**, *261*, 16937–16942. [PubMed]
163. Ori, K.; Koroda, M.; Mimaki, Y.; Sakagami, H.; Sashida, Y. Lanosterol and tetranorlanosterol glycosides from the bulbs of *Muscari paradoxum*. *Phytochemistry* **2003**, *64*, 1351–1359. [CrossRef]
164. Ragasa, C.Y.; Tiu, F.; Rideout, J.A. New cycloartenol esters from *Ixora coccinea*. *Nat. Prod. Res.* **2004**, *18*, 319–323. [CrossRef] [PubMed]
165. Gallo, C.; d'Ippolito, G.; Nuzzo, G.; Sardo, A.; Fontana, A. Autoinhibitory sterol sulfates mediate programmed cell death in a bloom-forming marine diatom. *Nat. Commun.* **2017**, *8*, 1292. [CrossRef] [PubMed]
166. Massé, R.; Huang, Y.S.; Eid, K.; Laliberté, C.; Davignon, J. Plasma methyl sterol sulfates in familial hypercholesterolemia after partial ileal bypass. *Can. J. Biochem.* **1982**, *60*, 556–563. [CrossRef] [PubMed]
167. Hu, X.; Wang, Y.; Hao, L.Y.; Liu, X.; Lesch, C.A.; Sanchez, B.M.; Wendling, J.M.; Morgan, R.W.; Aicher, T.D.; Carter, L.L.; et al. Sterol metabolism controls T(H)17 differentiation by generating endogenous RORgamma agonists. *Nat. Chem. Biol.* **2015**, *11*, 141–147. [CrossRef] [PubMed]

Article

Squalene Cyclases and Cycloartenol Synthases from *Polystichum polyblepharum* and Six Allied Ferns

Junichi Shinozaki *, Takahisa Nakene and Akihito Takano

Faculty of Pharmaceutical Sciences, Showa Pharmaceutical University, Machida, Tokyo 194-8543, Japan; nakane@ac.shoyaku.ac.jp (T.N.); takano@ac.shoyaku.ac.jp (A.T.)
* Correspondence: sinozaki@ac.shoyaku.ac.jp; Tel.: +81-42-721-1575

Received: 29 June 2018; Accepted: 23 July 2018; Published: 24 July 2018

Abstract: Ferns are the most primitive of all vascular plants. One of the characteristics distinguishing them from flowering plants is its triterpene metabolism. Most cyclic triterpenes in ferns are hydrocarbons derived from the direct cyclization of squalene by squalene cyclases (SCs). Both ferns and more complex plants share sterols and biosynthetic enzymes, such as cycloartenol synthases (CASs). *Polystichum* belongs to Dryopteridaceae, and is one of the most species-rich of all fern genera. Several *Polystichum* ferns in Japan are classified as one of three possible chemotypes, based on their triterpene profiles. In this study, we describe the molecular cloning and functional characterization of cDNAs encoding a SC (PPH) and a CAS (PPX) from the type species *Polystichum polyblepharum*. Heterologous expression in *Pichia pastoris* revealed that PPH and PPX are hydroxyhopane synthase and CAS, respectively. By using the PPH and PPX sequences, we successfully isolated SC- and CAS-encoding cDNAs from six *Polystichum* ferns. Phylogenetic analysis, based on SCs and oxidosqualene cyclase sequences, suggested that the *Polystichum* subclade in the fern SC and CAS clades reflects the chemotype—but not the molecular phylogeny constructed using plastid molecular markers. These results show a possible relation between triterpenes and their biosynthetic enzymes in *Polystichum*.

Keywords: squalene cyclase; cycloartenol synthase; triterpene; fern; *Polystichum*

1. Introduction

Ferns are the most primitive of all vascular plants. They produce specialized characteristic metabolites that are not found in flowering plants [1]. These include triterpenoids and phloroglucinols with various biological activities [2–4]. Fern triterpenoids are biosynthesized in a manner that is significantly different from those in flowering plants. Most ferns produce triterpene hydrocarbons by the cyclization of squalene, a C30 acyclic substrate (**1**). Their physiological function remains largely elusive. The enzymes responsible for these reactions are squalene cyclases (SCs). Fern SCs are homologous to those in bacteria. While bacterial SCs generate hop-22(29)-ene (**2**) [5–10], fern SCs (discovered thus far) yield various cyclic triterpenes, such as migrated hopane and germanicane (pentacycles), dammarane and tirucallane (tetracycles), bicyclic polypodane skeletons, and hopane skeletons (**2**) [11–14]. However, several fern phytosterols are identical to those found in flowering plants. At the gene/enzyme level, the sterol pathway might be conserved between ferns and flowering plants. Cycloartenol synthases (CASs), which are oxidosqualene cyclases (OSCs) involved in sterol metabolism, have been successfully cloned from ferns [11,13]. Because ferns lie between bacteria and flowering plants on the evolutionary scale, they produce triterpenes that are similar to those produced by bacteria and flowering plants. Therefore, the study of ferns might be useful in investigating the molecular evolution of triterpene biosynthesis from bacterial to flowering plants.

Polystichum Roth (Dryopteridaceae) is one of the most species-rich of all fern genera. It is distributed in both temperate and subtropical regions and is found in lowlands, montane, and alpine areas. About >200 *Polystichum* species are estimated to occur worldwide [15]. There are 32 known species in Japan [16], which is one of the most species-rich of all areas. *Polystichum* species have diverse morphological traits and produce several different triterpenoids. These include the bicyclic γ- and α-polypodatetraenes (**3**, **4**) which were the first bicyclic triterpenes identified [17]. The major constituents are pentacyclic hopane and migrated hopane triterpenoids such as compound **2**, 22-hydroxyhopane (**5**), fern-9(11)-ene (**6**), dryocrassol (**7**), and dryocrassyl acetate (**8**) (Figure 1) [18]. The fern triterpenoids have been classified into three chemotaxonomic groups based on the profiles of 16 Japanese *Polystichum* species [19]: Group 1 includes *P. polyblepharum*, *P. fibrilloso-paleaceum*, and *P. pseudo-makinoi*. These produce compound **3** (Figure 1). Group 2 includes *P. rigens*, which produces compound **4**. Group 3 includes *P. longifrons*, *P. makinoi*, and *P. lepidocaulon*. None of these produces either **3** or **4**.

To elucidate the molecular basis of triterpenoid biosynthesis, we cloned SC and OSC from *P. polyblepharum*, the type species, and functionally analyzed their cDNAs by heterologous expression in yeast. By using these results, we attempted to isolate triterpene synthases from six other *Polystichum* species to determine whether molecular phylogeny based on triterpene synthase sequences was correlated with their chemotaxonomic classification.

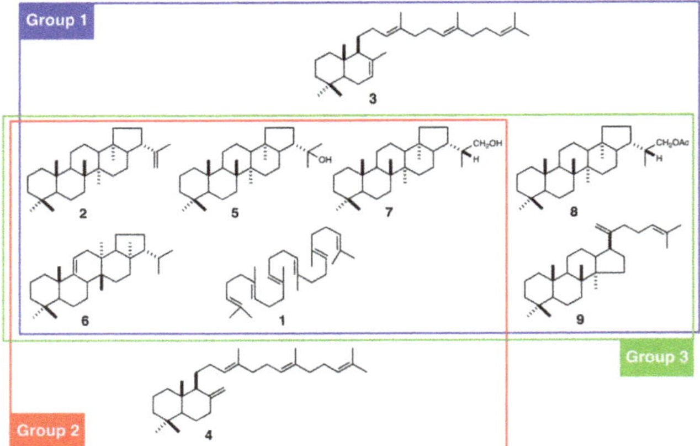

Figure 1. Triterpenoids isolated from *Polystichum* ferns and chemotaxonomic classification based on the triterpenoid profile. Group 1 includes *P. polyblepharum*, *P. fibrilloso-paleaceum*, and *P. pseudo-makinoi*; Group 2 includes *P. rigens*; and Group 3 includes *P. longifrons*, *P. makinoi*, and *P. lepidocaulon*. **1**, squalene; **2**, hop-22(29)-ene; **3**, γ-polypodatetraene; **4**, α-polypodatetraene; **5**, 22-hydroxyhopane; **6**, fern-9(11)-ene; **7**, dryocrassol; **8**, dryocrassyl acetate; and **9**, dammara-18(28),21-diene.

2. Results

2.1. Isolation of cDNAs Encoding SC and OSC

We cloned full-length cDNAs from *P. polyblepharum* encoding SC and OSC by using a combination of homology-based PCR and rapid amplification of cDNA ends (RACE). A DNA fragment encoding the putative SC was obtained by reverse-transcription PCR (RT-PCR) by using the degenerate primers SC-306S-1, SC-306S-2, SC-358S, SC-494A, and SC-537A [11,12] targeting highly conserved consensus sequences of bacterial SCs. The PCR product with the expected size of 450 bp was purified using gel electrophoresis, cloned, and amplified in *Escherichia coli*. Sixteen plasmids with inserts of the expected length were extracted from the *E. coli* transformants and sequenced. They were all identical and showed a high sequence identity to known fern SCs. The remaining sequences of the putative SC cDNA were acquired using 5′- and 3′-RACE PCR by using the gene-specific primers designed from the sequence obtained. The full-length cDNA was designated PPH. It consisted of 2530 bp with a 2058 bp coding region, 32 bp 5′ untranslated region (UTR), and 365 bp 3′ UTR region. The open reading frame (ORF) of PPH was a polypeptide 685 residues long with an estimated molecular weight of 77.1 kDa. The amino acid sequence of PPH showed 65–90% and 35–41% identity with functionally characterized SCs from ferns and bacteria, respectively. It most closely matched a hydroxyhopane synthase from Adiantum capillus-veneris (ACH).

A nested PCR was run with two sets of degenerate primers (OSC-162S, OSC-463S, OSC-467S, OSC-623A, and OSC-711A) [20,21] and cDNA prepared from template RNA. It yielded a 400 bp DNA fragment. A full-length cDNA sequence, including clonal 5′ and 3′ UTR, was designated as PPX. It was obtained using the RACE method by using gene-specific primers derived from the 400 bp DNA fragment. The full-length cDNA of PPX was 2596 bp in length and contained a 76 bp 5′ UTR, 234 bp 3′ UTR, and 2286 bp ORF corresponding to a deduced protein sequence of 761 amino acid residues (86.7 kDa). The amino acid sequence of PPX had a 66–89% identity with the CASs of other plants, including ferns. In contrast, it had relatively low identity (42–62%) with other OSCs. The highest amino acid identity score was 89% for a CAS derived from *Polypodiodes niponica* (PnCAS).

2.2. Heterologous Expression and Biochemical Characterization of PPH and PPX

The biochemical functions of PPH and PPX were elucidated by heterologously expressing them in the methylotrophic yeast *Pichia pastoris*. Yeast cells overexpressing PPH generated one product not present in control yeast cells. It carried an empty vector pPIC6 (Figure 2A). The addition of the squalene epoxidase inhibitor terbinafine to a post-induction culture revealed that PPH could use the accumulated endogenous squalene as an in vivo substrate. The PPH product was purified using silica-gel column chromatography and eluted as a hexane-ethyl acetate (100:3, v/v %) fraction. It did not elute as a hexane fraction. Therefore, it was a triterpene mono-alcohol. It was identified as 22-hydroxyhopane (**5**) by comparing its ^{1}H NMR spectrum with that of an authentic standard [22]. The results were highly suggestive that PPH encodes hydroxyhopane synthase in *P. polyblepharum*.

The ORF of PPX was subcloned into the expression vector pPIC3.5. The construct was transformed into *P. pastoris* strain GS115. A cell-free extract prepared from the induced transformant culture was used to test the in vitro reaction. The recombinant PPX cyclized racemic 2,3-oxidosqualene (**10**) to a product not detected in the negative control yeast cell homogenate carrying the empty vector pPIC3.5. Gas chromatography (GC) analysis showed that the hexane extract of the in vitro reaction of PPX was superimposed onto that of PnCAS (Figure 2B) [13]. Therefore, the cDNA sequence of PPX encodes *P. polyblepharum* CAS.

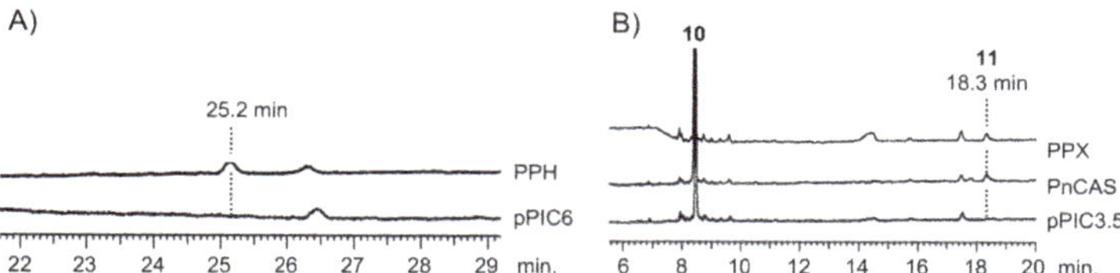

Figure 2. GC analysis of PPH (**A**) and PPX (**B**) products. (**A**) Gas chromatograms of triterpene mono-alcohol fraction from yeast transformants with PPH and pPIC6 (empty vector). Retention time of the PPH product was 25.2 min. (**B**) Gas chromatograms of triterpene mono-alcohol fraction from in vitro reaction of racemic 2,3-oxidosqualene (**10**) with recombinant PPX, CAS derived from *Polypodiodes niponica* (PnCAS), and empty vector (pPIC3.5). Retention time of the PPX and PnCAS product was 18.3 min.

2.3. Comparison of Active Site Residues between Hopene and Hydroxyhopane Synthases

We compared the active sites of PPH and ACH [11], CEH1 and CEH2 from *Colysis elliptica* (hopene synthases) [14], and AaSHC from *Alicyclobacillus acidocaldarius* (a bacterial hopene synthase; Figure 3). Mutagenesis experiments with AaSHC identified 10 amino acid residues (Thr41, Glu45, Glu93, Arg127, Trp133, Gln262, Pro263, Tyr267, Phe434, and Phe437) responsible for the final cation elimination reaction [23]. Four of them (Thr41, Arg127, Tyr267, and Phe434 in AaSHC) were strictly conserved among all five enzymes (Supplementary Figure S2). Glu93 in AaSHC was also conserved among all enzymes except ACH, which had a Gly at that location (Figure 3). PPH, ACH, and AaSHC had Pro263, but the corresponding residue in the fern hopene synthases (CEH1 and CEH2) was Ala (Figure 3). The following residues were conserved among the fern SCs: Glu45 in AaSHC was replaced by His (Gln in ACH); Gln262 in AaSHC was replaced by Tyr; and Phe437 in AaSHC was replaced by Leu in PPH and ACH and by Ile in CEH1 and CEH2 (Figure 3). Trp133 in AaSHC was not conserved among the fern SCs. The corresponding residues were Cys in PPH and ACH and Tyr in CEH1 and CEH2 (Figure 3).

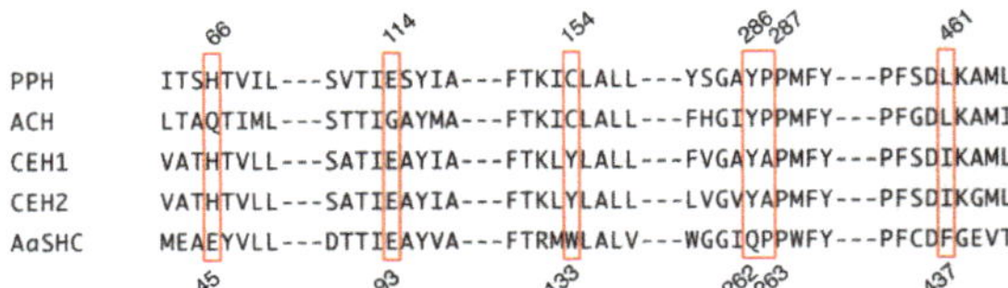

Figure 3. Partial amino acid sequence alignment of hydroxyhopane/hopene synthases. Numbers above the sequence show positions in PPH, and numbers below the sequence show those in AaSHC. Red boxes indicate active site residues, which might influence the final cation elimination reaction.

2.4. Phylogenetic Analysis of SCs and OSCs

A phylogenetic tree was constructed based on the neighbor-joining method (Figure 4) to elucidate the phylogenetic relationships of the deduced amino acid sequences of PPH and PPX with other known SC and OSC members. The sequences were separated into three divergent clades (such as SC and OSC). In the SC clade, the fern SCs were clearly divided into those for bacteria. The PPH of a hydroxyhopane synthase clustered with DCD (a dammaradiene synthase from *Dryopteris crassirhizoma*), but not with ACH; Figure 4B). The OSC clade consisted of a plant and the remaining eukaryotic OSC subclades. The plant OSC subclade was resolved as a monophyletic family separating into CAS and the other OSC clusters. The fern CASs, including PPX, were grouped together and clearly separated from those of higher plant origin. PPX was found to be closely related to PNX (*Polypodiodes niponica*), but not to ACX (*A. capillus-veneris*).

2.5. Phylogenetic Relationships between Polystichum Triterpene Synthases

The SC clade in the phylogenetic tree reflects a taxonomic relationship rather than the molecular evolution of enzyme functions. In this case, the tree topology reconstituted from the SC and/or CAS sequences is expected to reveal a chemotaxonomic relationship based on the triterpenoid profiles of various *Polystichum* ferns. We performed an additional phylogenetic analysis focusing on this relationship. We isolated SC- and CAS-encoding cDNAs from six other *Polystichum* ferns to determine whether the tree topology, based on the SC and/or CAS sequences, reflects a taxonomic relationship. The fern species used were *P. fibrillosopaleaceum*, *P. pseudo-makinoi* (chemotaxonomic group 1), *P. rigens* (chemotaxonomic group 2), *P. longifrons*, *P. makinoi*, and *P. lepidocaulon* (chemotaxonomic group 3).

The ORFs of the SC and OSC from these six ferns were successfully amplified. Primer sets were designed for the ORF amplification of PPH and PPX. The ORFs were ligated into the pT7 Blue T-Vector of a cloning vector, sequenced, and used to construct a neighbor-joining tree. In the fern CAS clade, the tree topology reflected chemotaxonomic classification except for PleCAS (*P. lepidocaulon* CAS, group 3), which was placed at the basal branch in the *Polystichum* subclade (Figure 5A). In the fern SC clade, the *Polystichum* SCs formed a polyphyletic group, including *Dryopteris* SC (DCD; Figure 5B). Those belonging to group 3 were grouped together. PriSC1 belonged to group 2 and was distantly related to the other two groups. However, the remaining three SCs, belonging to group 1, did not form a branch.

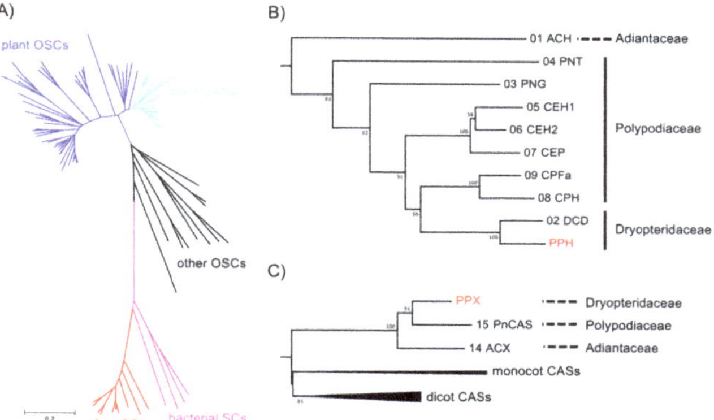

Figure 4. Phylogenetic analysis of PPH and PPX. (**A**) Unrooted neighbor-joining phylogenetic tree of OSC and SC homologs. Plant OSC and CAS clades are in blue and pale blue, respectively. Fern and bacterial SC clades are in red and magenta, respectively. For clarity, the branches have no taxon labels. The tree showing all taxon names is presented in Supplementary Figure S1. The scale represents 0.1 amino acid substitution per site. (**B**) Expanded fern SC clade from the phylogenetic tree in (**A**). Bootstrap values with 1000 replicates are shown at the nodal branches (cutoff value = 50%). The enzyme function and species of the sequences are as follows: ACH, hydroxyhopane synthase from *Adiantum capillus-veneris;* CEH1/2, hopene synthase from *Colysis elliptica;* CEP, α-polypodatetraene synthase; CPFa, fern-9(11)-ene synthase from *C. pothifolia;* CPH, hop-17(21)-ene synthase from *C. pothifolia;* DCD, dammaradiene synthase from *Dryopteris crassirhizoma;* PNT, tiriucalladiene synthase from *Polypodiodes niponica;* PNG, germanicene synthase from *P. niponica.* (**C**) Expanded plant CAS clade from the phylogenetic tree in (**A**). Sequences in monocot and dicot branches are collapsed for clarity. Bootstrap values with 1000 replicates are shown at the nodal branches (cutoff value = 50%). The enzyme function and species of the sequences are as follows: ACX, cycloartenol synthase from *A. capillus-veneris;* PnCAS, cycloartenol synthase from *P. niponica.* The accession numbers, abbreviations, enzyme functions, and species are listed in Supplementary Table S1.

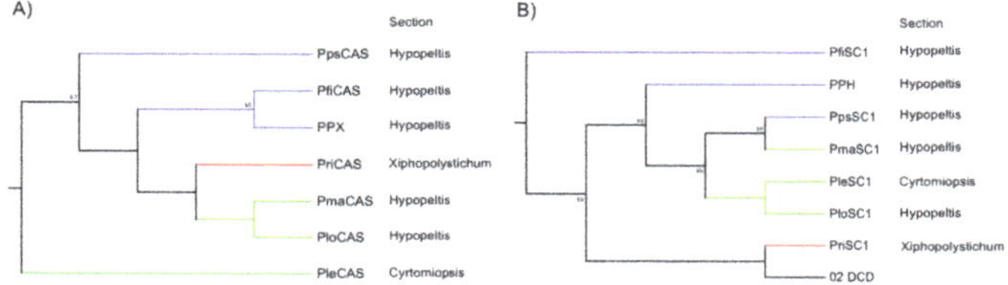

Figure 5. Phylogenetic analysis of *Polystichum* triterpene synthases. (**A**) Expanded *Polystichum* CAS clade. (**B**) Expanded Dryopteridaceae SC clade. Bootstrap values with 1000 replicates are shown at the nodal branches (cutoff value = 50%). Blue, red, and green branches indicate groups 1, 2, and 3, respectively. These are categorized by their triterpene-based chemotaxonomy. DCD is a dammaradiene synthase from *Dryopteris crassirhizoma.* PPX and PPH are a CAS and hydroxyhopane synthase from *Polystichum polyblepharum,* respectively. Species of other clones are as follows: Pfi, *P. fibrilloso-paleaceum;* Ple, *P. lepidocaulon;* Plo, *P. longifrons;* Pma, *P. makinoi;* Pps, *P. pseudo-makinoi;* Pri, *P. rigens.*

3. Discussion

3.1. Hydroxyhopane Synthases and Hopene Synthases

In the present study, a homological approach successfully isolated the SC cDNA sequence of PPH. Its functional expression in yeast revealed that PPH produced only 22-hydroxyhopane (5). This enzyme is the second reported example of a monofunctional hydroxyhopane synthase. PPH had the highest amino acid identity (90%) with ACH [11]. However, phylogenetic analysis showed that PPH was only distantly related to ACH despite the similarity of their product profiles. It was, in fact, closely related to *D. crassirhizoma* dammaradiene synthase (DCD; Figure 4) [12]. The fern SC clade in the phylogenetic tree reflects their taxonomic relationship rather than the similarity of their enzymatic functions.

Synthesis of 22-hydroxyhopane (**5**) consists of squalene (**1**) activation by cationic attack, followed by a cascade of cation-olefin cyclizations and quenching of the pentacyclic carbocation by the addition of water. The hop-22(29)-ene (**2**) reaction pathway is mostly shared by that of (**5**). They diverge at the final step in which their hopanyl cation quenching modes differ (Figure 6). Deprotonation of C29 from hopanyl cation produces (**2**). Bacterial hopene synthases (squalene-hopene cyclases; SHCs) also produce (**2**) and (**5**) [24,25]. The (**2**):(**5**) ratio is 84:16 for *Alicyclobacillus acidocaldarius* SHC [26]. In contrast, all fern hydroxyhopane- and hopene-synthases, including PPH, are monofunctional enzymes [11,14]. These findings suggest that fern hydroxyhopane- and hopene-synthases have more finely tuned catalytic capabilities than bacterial ones, with respect to the final cation elimination step.

The functions of 10 active site residues in AaSHC might not be strictly conserved in all fern SCs (Figure 3). Cys154 in PPH (and the corresponding Cys in ACH) might participate in the addition of water to the hopanyl cation. A previous study proposed that Trp133 in AaSHC is associated with the formation of a hydrogen-bonded water molecule network at the active site serving as a catalytic base [23]. The corresponding Tyr residue in CEH1 and CEH2 might have a function analogous to Trp with respect to aromaticity. In contrast, the Cys in PPH and ACH might not participate in the hydrogen-bonded network and could thus affect the water molecule as the catalytic base. Therefore, the final cation elimination reaction in PPH and ACH might favor the addition of water to the cation rather than its deprotonation.

Hoshino and colleagues proposed that Gln262 and Pro263 in AaSHC situate the catalytic base of the water molecule at its appropriate position, thereby contributing to the production of (**2**) and minimizing the formation of (**5**). However, these functions did not apply to fern SCs. Gln262 in AaSHC was replaced by Tyr in fern SCs (Figure 3). Therefore, Tyr was not responsible for the reaction selectivity of the catalytic base (deprotonation or water addition). Nevertheless, it might have, to some extent, contributed to the hydrogen bonding of water to the catalytic base. One of the active site residues, Pro263 in AaSHC, was identical to the corresponding residues in PPH and ACH. However, this identity does not explain the differences in the product profiles of AaSHC and PPH/ACH (Figure 3). In PPH and ACH, Pro263 might contribute to hydrogen bonding to the catalytic base, but does not influence product selectivity.

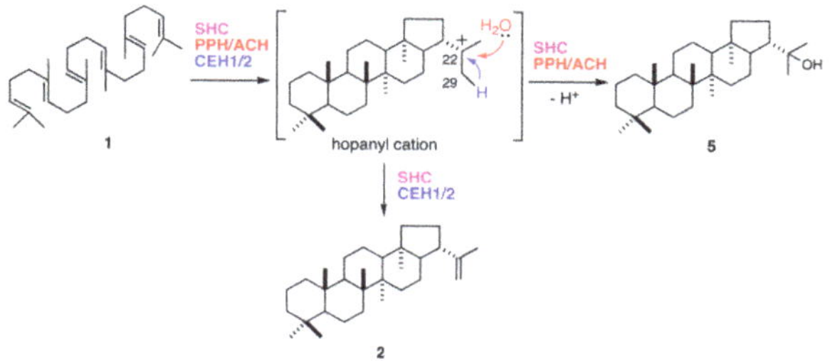

Figure 6. Cyclization of squalene (**1**) into hop-22(29)-ene (**2**) and 22-hydroxyhopane (**5**) by hopene and/or hydroxyhopane synthases.

3.2. Relationship between Phylogeny and Taxonomy

Hopene synthases biosynthesize bacterial hopanoids. These compounds maintain membrane integrity by serving as sterol surrogates in several bacteria. Therefore, hopene synthases and/or hydroxyhopane synthases in ferns were speculated to be evolutionarily basal SCs. PPH from *P. polyblepharum* is a second hydroxyhopane synthase, assuming that PPH is clustered with the ACH of another hydroxyhopane synthase in the phylogenetic tree. However, a phylogenetic analysis based on the SC and OSC sequences revealed that PPH was only distantly related to ACH. The latter is the most basal in the fern SC clade (Figure 4). Therefore, we hypothesized that the fern SC clade reflects taxonomic relationships among ferns, but not the molecular evolutionary history of their enzyme functions. The phylogenetic tree in Figure 4 shows that PPH and DCD are clustered in the same branch despite the differences in their enzyme functions. This observation corroborates the hypothesis since *Polystichum* and *Dryopteris* both belong to Dryopteridaceae.

To test the hypothesis, we isolated cDNAs encoding SC and CAS from six *Polystichum* ferns. Since SC and CAS synthesize triterpenes, a phylogenetic tree reconstituted from these sequences could reflect the triterpene profile-based chemotaxonomic relationship. Therefore, the six chosen *Polystichum* plants had different triterpene profiles and were classified into three different groups (Figure 1). The phylogenetic analysis showed that *Polystichum* CASs were entirely separate from the other fern CASs (ACH and PnCAS). The phylogenetic tree topology of this subclade reflected chemotaxonomic classifications except for PleCAS (Figure 5A). A recent classification performed by Zhang and Barrington [27] arranged *Polystichum* into two subgenera whose identities were verified by molecular phylogenetic analysis with multiple plastid loci [28]. Of the seven species surveyed in the present study, *P. lepidocaulon* (PleCAS) belongs to the subgenus *Haplopolystichum*, whereas the others belong to the subgenus *Polystichum*. The relationship between PleCAS and the other six CASs implies that these two subgenera are distantly related. This finding might also be consistent with those reported for certain phylogenetic studies based on plastid markers. The clade consisting of the subgenus *Polystichum* appears to represent the chemotaxonomic relationship alone, rather than the previous classification supported by morphological features and conventional molecular data (plastid loci). Five of the six species belong to the *Hypopeltis* section, whereas *P. rigens* is a member of the *Xiphopolystichum* section. The fact that these two sections are grouped together suggests that the phylogenetic tree based on CASs and SCs might resolve relationships other than those revealed by plastid marker-based analyses.

In the fern SC clade, the phylogenetic tree topology seemed to correlate molecular phylogeny with chemotaxonomy despite the inclusion of DCD from *D. crassirhizoma* in the *Polystichum* subclade (Figure 5B). The member of group 3 and PpsSC1 in group 1 fell into the same subclade. PriSC1 in group 2 formed a branch with DCD and was clearly separated from the other groups. In contrast, no clear relationship among group 1 species was established. Unlike the CAS clade, each enzyme in the fern SC clade might have a different function. Enzyme function might influence the phylogenetic tree topology of the SC clade which, in turn, could reflect the combined chemotaxonomic classification and molecular evolution of SC functions. In that case, an SC subclade formed by a congener (*Polystichum*) could provide insight into the molecular evolution of the SC enzymes because the clade would no longer be influenced by taxonomical factors. The parameters responsible for the topology can be identified by characterizing the SCs isolated in the present work. Further experiments are also required to validate the hypothesis that SC- and/or OSC-based phylogenetic trees reflect triterpene profile-based chemotaxonomic classification.

The present study showed that the primer set designed for PPH and PPX enabled the isolation of SC and OSC genes from the other six congeners. This approach might contribute to the rapid isolation of SC-encoding genes from other taxa in the future. To our knowledge, this is also the first study to focus on the comparison between the triterpene profile-based chemotaxonomy and conventional molecular phylogeny: Both of which were connected to each other, by using the sequences encoding SC and OSC. The present phylogenetic analysis showed that the triterpene profile might be related to the corresponding gene sequences. However, this interpretation might be altered because the physiological roles of triterpenes were not considered in the present study. The function of metabolites apparently affects the evolutionary history of organisms. Nonetheless, drawing any conclusive answer to the intriguing question of the relationship between chemistry and phylogeny is still difficult. Nevertheless, the application of the present approach to other fern genera is expected to provide both a deeper and more general understanding of triterpene biosynthesis and chemotaxonomic classification in ferns.

4. Materials and Methods

4.1. RNA Extraction and cDNA Synthesis

Polystichum polyblepharum, *P. fibrillosopaleaceum*, *P. pseudo-makinoi*, *P. rigens*, *P. longifrons*, *P. makinoi*, and *P. lepidocaulon* were propagated in the Medicinal Plant Garden of Showa Pharmaceutical University. Their fronds were harvested for total RNA extraction, which was performed using the RNeasy Plant Mini Kit (Qiagen, Hilden, Germany) according to manufacturer's instruction. First-strand cDNA was synthesized from the RNA by using SuperScript III reverse transcriptase (Thermo Fisher Scientific, Waltham, MA, USA) and an oligo (dT) primer (RACE 32 [20], 5′-GACTCGAGTCGACATCGATTTTTTTTTTTTTTTT-3′) according to manufacturer's instruction. For 3′- and 5′-RACE, an adaptor-ligated double-strand cDNA was synthesized using a SMARTer RACE 5′/3′ Kit (TaKaRa Bio, Shiga, Japan) according to manufacturer's instruction.

4.2. Cloning of Squalene Cyclase cDNAs

Homology-based nested PCR was performed using the single-strand cDNA and five degenerate primers [11,12] (SC-306S-1, SC-306S-2, SC-358S, SC-494A, and SC-537A). PCR was performed using Ex *Taq* DNA polymerase Hot Start Version (TaKaRa Bio, Shiga, Japan) in a final volume of 50 μL PCR conditions were identical to those reported in our previous study [11]. The primer-specific amplicon was purified from an agarose gel, cloned into pT7 Blue T-Vector (Merck KGaA, Darmstadt, Germany), and transformed into *E. coli* strain DH5α. Plasmid DNA was purified from transformed cells by using an Illustra PlasmidPrep Mini Spin Kit (GE Healthcare, Chicago, IL, USA) and sequenced using a BigDye terminator cycle sequence kit v.3.1 (Thermo Fisher Scientific, Waltham, MA, USA) and an ABI PRISM 3130 genetic analyzer (Thermo Fisher Scientific, Waltham, MA, USA). The sequence was used to design two specific primers (PPH-481A for first PCR, 5′-TTGCTCTCTCCTGAGATATGTCAAG-3′; PPH-468A for second PCR, 5′-ATTCCGGGGGCAGCACATTGGCCTC-3′). Nested PCR was run using the specific primers, SC306S-1, and the same PCR conditions as described above. The PCR product was purified, subcloned, and sequenced.

Based on the sequenced fragment, we designed gene-specific primers for 5′- and 3′-RACE: For 5′-RACE, PPH-451A for first PCR, 5′-AAGCTAGTACCTTGCAGTTTCAACCTG-3′;

PPH-345A for second PCR, 5′-TGAATGACCAATCTCCATGCTTTGTGA-3′; for 3′-RACE, PPH-414S for first PCR, 5′-ATGGATATCGTCTATGCAGGCCAGAGG-3′; PPH-461S for second PCR, 5′-CTGCAAGGTACTAGCTTTGATGAGG-3′. RACE PCR was performed using the adaptor-ligated double-strand cDNA as a template and gene-specific primers by using SMARTer RACE 5′/3′ Kit (TaKaRa Bio, Shiga, Japan) according to manufacturer's instruction. The full-length nucleotide sequence was named PPH.

To isolate the ORF of the SC from the six *Polystichum* ferns, we performed PCR by using PPH-specific primers designed for ORF amplification (N-PPH and C-PPH described below) and single-strand cDNA as a template. The first PCR was performed using KOD-Plus v. 2 (Toyobo, Osaka, Japan) in a final volume of 50 μL. PCR conditions were identical to those reported in our previous study [13]. The second PCR was the same as the first except the first PCR product was a template for putative SC-encoding cDNAs designated PfiSC1 from *P. fibrillosopaleaceum*, PpsSC1 from *P. pseudo-makinoi*, PriSC1 from *P. rigens*, PloSC1 from *P. longifrons*, PmaSC1 from *P. makinoi*, and PleSC1 from *P. lepidocaulon*. The primer-specific amplicon was purified on agarose gel, cloned into pT7 Blue T-Vector (Merck KGaA, Darmstadt, Germany), and transformed into *E. coli* strain DH5α. Plasmid DNA was purified from transformed cells and sequenced.

The seven nucleotide sequences were deposited in GenBank/EMBL/DDBJ under accession numbers LC389069 for PPH, LC389071 for PloSC1, LC389072 for PfiSC1, LC389073 for PmaSC1, LC389074 for PriSC1, LC389075 for PleSC1, and LC389076 for PpsSC1.

4.3. PPH Expression in Pichia pastoris and Product Analysis

The coding region of PPH was amplified using nested PCR by using primers (5′-PPH, 5′-CGCCCGGGCAGGTATTGATGTTAGG-3′ and 3′-PPH, 5′-AGGCTGCTTGCTATGAAGCTTGCAG-3′ for the first PCR; N-PPH, 5′-TTAGGCCTCGAGATGCTGCCATACAACCAAGAT-3′, and C-PPH, 5′-CAAGGCGCGGCCGCTTATGGAATTGGAGGCTTGAT-3′. The method used was the same as that described in our previous study [11]. After treatment with XhoI and NotI, the PCR product was inserted into the corresponding pPIC6B restriction sites (Thermo Fisher Scientific, Waltham, MA, USA). *Pichia pastoris* strain X-33 was transformed using the resultant plasmid by using the method reported in our previous study [13].

The transformed *P. pastoris* was grown in yeast extract peptone dextrose (YPD) medium and induced using methanol in glucose-free YPD medium. The cells were resuspended in potassium phosphate (0.1 M, pH 7.0) supplemented with glucose and terbinafine [29]. They were saponified with ethanolic KOH and extracted with *n*-hexane. The culture conditions and process were the same as those reported in our previous study [14].

The hexane extract was passed through a BondElut-Si column (500 mg, 3 mL; Varian, Santa Clara, CA, USA) and eluted with *n*-hexane/ethyl acetate (9:1, *v*/*v*) to yield the triterpene hydrocarbon and triterpene mono-alcohol fractions (Fraction-A). Fraction A was analyzed using gas chromatography on a Hitachi G-6000 instrument (Hitachi Hi-Technologies, Tokyo, Japan) equipped with a DB-5HT column (30 m × 0.25 mm; Agilent Technologies, Santa, Clara, CA, USA) under the same conditions as those reported in our previous study [13]. A PPH product was purified from Fraction A by using a BondElut-Si column (500 mg, 3 mL; Varian, Santa Clara, CA, USA) and eluted with a 0–3% ethyl acetate gradient in *n*-hexane. The ^{1}H NMR spectrum of the purified product was measured in $CDCl_3$ (Bruker AV600; Billerica, MA, USA), by using tetramethylsilane as an internal standard.

22-Hydroxyhopane (**5**): ^{1}H NMR (600 MHz, $CDCl_3$): δ 0.76 (s, 3H), 0.79 (s, 3H), 0.81 (s, 3H), 0.85 (s, 3H), 0.96 (s, 6H), 1.18 (s, 3H), 1.21 (s, 3H).

4.4. Cloning of Oxidosqualene Cyclase cDNA

A homology-based nested PCR was performed using the single-strand cDNA and five degenerate primers [20,21] (OSC-162S, OSC-463S, OSC-467S, OSC-623A, and OSC-711A). The PCR, subcloning, and sequencing were the same as those described above. The sequence was used to design two specific primers (PPX-504A for the first PCR, 5′-TTGTATTCACTGCATCGTAGAAGCG-3′; PPX-482A for the second PCR, 5′-GACTCAACGCTAGTGCAGCCTTAAA-3′). Nested PCR was performed using the specific primers and OSC-162S. PCR conditions were the same as those described above. The PCR product was purified, subcloned, and sequenced.

Based on the sequenced fragment, we designed gene-specific primers for 5′- and 3′-RACE: For 5′-RACE, PPX-206A for the first PCR, 5′-TTTTCCCCATGAGGGAATGGCTGTAGC-3′; PPX-189A for the second PCR, 5′-TCTTCCTCTCTCCATGGCTTGATCCAC-3′; for 3′-RACE, PPX-566S for the first PCR, 5′-AGTCATCCAAGGCTTAGCAGCCTTC-3′; PPX-588S for the second PCR, 5′-ATGCATTGAGCGTGCTGCTTGCTAC-3′. RACE PCR was performed using a SMARTer RACE 5′/3′ Kit (TaKaRa Bio, Shiga, Japan) according to manufacturer's instruction by using the adaptor-ligated double-strand cDNA as a template and gene-specific primers.

To isolate the ORF of SC from the six *Polystichum* ferns, we performed PCR by using PPX-specific primers designed for ORF amplification (N-PPX and C-PPX described below) and single-strand cDNA as a template. The PCR conditions, plasmid construction, and sequencing were the same as those described in Section 4.2. Putative CAS-encoding cDNAs were designated PfiCAS from *P. fibrillosopaleaceum*, PpsCAS from *P. pseudo-makinoi*, PriCAS from *P. rigens*, PloCAS from *P. longifrons*, PmaCAS from *P. makinoi*, and PleCAS from *P. lepidocaulon*.

The seven nucleotide sequences were deposited in GenBank/EMBL/DDBJ under accession number LC389070 for PPX, LC389077 for PloCAS, LC389078 for PfiCAS, LC389079 for PmaCAS, LC389080 for PriCAS, LC389081 for PleCAS, and LC389082 for PpsCAS.

4.5. Expression of PPX in Pichia pastoris and Analysis of Product

The coding region of PPX was amplified using nested PCR by using primers (5′-PPX, 5′-ACCAAAAGCGTGTAGAGAGAGAGAG-3′ and 3′-PPX, 5′-AGCTTGCCAACAATGATGCTGGATG-3′ for the first PCR; N-PPX, 5′-GAGAGAGAATTCGAAATGTGGAGCTTGAAGACAGCA-3′, and C-PPX, 5′-CACCTAGCGGCCGCTCAATGATGATGATGATGATGTTTATAGCTCAAAACGCTTCG-3′ for the second PCR) by using the same method as that described above. After treatment with EcoRI and NotI, the PCR product was inserted into the corresponding pPIC3.5 restriction sites (Thermo Fisher Scientific, Waltham, MA, USA). *Pichia pastoris* strain GS115 was transformed with the resultant plasmid by using the same method as that described above.

The transformed *P. pastoris* was grown in 25 mL minimal glycerol medium (1.34% w/v yeast nitrogen base with ammonium sulfate and no amino acids (YNB), 1% w/v glycerol, 4×10^{-5} biotin) at 30 °C for 24 h with shaking. The cells were then induced in 100 mL minimal methanol medium (1.34% YNB, 0.5% v/v methanol, 4×10^{-5} biotin) at 30 °C for 24 h with shaking. They were suspended in 10 mL extraction buffer [50 mM sodium phosphate (pH 7.4); 1 mM dithiothreitol, 5% w/v glycerol, and 0.8% w/v Protease Inhibitor Cocktail (Merck KGaA, Darmstadt, Germany)]. Glass beads (ø 0.35–0.5 mm, 5 mL) were added to the suspension, followed by 10 cycles of vortexing for 30 s cycle^{-1} at 30-s intervals and 4 °C. After cell disruption, Triton X-100 (final concentration, 0.2% w/v) was added to the homogenate. It was then centrifuged to separate the glass beads and cellular debris. The resulting supernatant was used as crude enzyme in the following procedure.

The crude enzyme solution (2.5 mL) was incubated at 30 °C for 20 h, with 250 μg (3*RS*)-oxidosqualene (**10**) in a total volume of 5 mL containing: 50 mM sodium phosphate buffer (pH 7.4), 0.2% w/v Triton X-100, and 1 mM dithiothreitol. The reaction was stopped by refluxing it with 5 mL of 20% w/v KOH in 50% v/v EtOH. After two extractions with 10 mL n-hexane, the organic layer was concentrated. The hexane extract was subjected to BondElut-Si (500 mg, 3 mL; Varian, Santa Clara, CA, USA) and eluted with n-hexane/ethyl acetate (4:1, v/v) to yield a triterpene mono-alcohol fraction (Fraction B). Fraction B was analyzed using gas chromatography performed on a Hitachi G-6000 instrument (Hitachi Hi-Technologies, Tokyo, Japan) equipped with a DB-5HT column (30 m × 0.25 mm; Agilent Technologies, Santa Clara, CA, USA) under the same conditions as those described above. Substrate **10** was chemically synthesized according to the literature [30].

4.6. Phylogenetic Analysis

The deduced amino acid sequences of PPH and PPX were phylogenetically analyzed against the SC and OSC sequences from plants, bacteria, fungi, and mammals as well as 90 amino acid sequences retrieved from the GenBank/EMBL/DDBJ database. The SC and OSC sequences are shown in Supplementary Table S1. The sequences were aligned using CLUSTAL W [31] by using default parameters. The evolutionary history was inferred using the neighbor-joining method [32]. The phylogenetic tree is drawn to scale. Its branch lengths are in the same units as those of the evolutionary distances used to infer the phylogenetic tree itself. The evolutionary distances were computed using the Poisson correction method [33] and are expressed as the number of amino acid substitutions per site. The estimated reliability of the phylogenetic tree was tested using the bootstrap method (500 replications) [34]. Evolutionary analyses were conducted in MEGA7 [35].

Phylogenetic analysis was also conducted for the dataset, including the additional six *Polystichum* SCs (PfiSC1, PpsSC1, PriSC1, PloSC1, PmaSC1, and PleSC1), the six OSCs (PfiCAS, PpsCAS, PriCAS, PloCAS, PmaCAS, and PleCAS), and the above 92 sequences. The methods and conditions used for sequence alignment and phylogenetic inference were the same as those described above.

Supplementary Materials: The supplementary materials are available online; Figure S1: Phylogenetic tree of triterpene synthases, Figure S2: Amino acid sequence alignment of PPH and hydroxyhopane and hopene synthases from ferns and a bacterium, Figure S3: Phylogenetic tree of CAS clade (A) and fern SC clade (B), Table S1: Sequence information used in the phylogenetic analysis.

Author Contributions: Conceptualization, J.S.; Investigation, J.S. and A.T.; Formal Analysis, J.S. and T.N.; Writing-Original Draft Preparation, J.S.

Funding: This research received no external funding.

Acknowledgments: The authors wish to thank Kazuo Masuda who contributed to the phytochemical investigations of fern triterpenoids. We also thank Kazuhiro Hatori, Yuka Kato, Hitomi Isobe, and Yuuto Koyama (Showa Pharmaceutical University) for their assistance in the experiments. We would also like to thank Editage (http://www.editage.jp) for English language editing.

Conflicts of Interest: The authors declare no conflict of interest.

References

1. Murakami, T.; Tanaka, N. *Progress in the Chemistry of Organic Natural Products*; Springer-Verlag: New York, NY, USA, 1988; Volume 54.
2. Konoshima, T.; Takasaki, M.; Tokuda, H.; Masuda, K.; Arai, Y.; Shiojima, K.; Ageta, H. Antitumor-promoting activities of triterpenoids from ferns. I. *Biol. Pharm. Bull.* **1996**, *19*, 962–965. [CrossRef] [PubMed]
3. Na, M.K.; Jang, J.P.; Min, B.S.; Lee, S.J.; Lee, M.S.; Kim, B.Y.; Oh, W.K.; Ahn, J.-S. Fatty acid synthase inhibitory activity of acylphloroglucinols isolated from *Dryopteris crassirhizoma*. *Bioorg. Med. Chem. Lett.* **2006**, *16*, 4738–4742. [CrossRef] [PubMed]
4. Lee, S.M.; Na, M.K.; An, R.B.; Min, B.S.; Lee, H.K. Antioxidant activity of two phloroglucinol derivatives from *Dryopteris crassirhizoma*. *Biol. Pharm. Bull.* **1996**, *26*, 1354–1356. [CrossRef]
5. Ochs, D.; Kaletta, C.; Entian, K.-D.; Beck-Sickinger, A.; Poralla, K. Cloning, expression, and sequencing of squalene-hopene cyclase, a key enzyme in triterpenoid metabolism. *J. Bacteriol.* **1992**, *174*, 298–302. [CrossRef] [PubMed]
6. Reipen, I.; Poralla, K.; Sahm, H.; Sprenger, G.A. *Zymomonas mobilis* squalene-hopene cyclase gene (*shc*): Cloning, DNA sequence analysis, and expression in *Escherichia coli*. *Microbiology* **1995**, *141*, 155–161. [CrossRef] [PubMed]
7. Perzl, M.; Müller, P.; Poralla, K.; Kannenberg, E.L. Squalene-hopene cyclase from *Bradyrhizobium japonicum*: Cloning, expression, sequence analysis and comparison to other triterpenoid cyclases. *Microbiology* **1997**, *143*, 1235–1242. [CrossRef] [PubMed]
8. Tippelt, A.; Jahnke, L.; Poralla, K. Squalene-hopene cyclase from *Methylococcus capsulatus* (Bath): A bacterium producing hopanoids and steroids. *Biochem. Biophys. Acta* **1998**, *1391*, 223–232. [CrossRef]

9. Ghimire, G.P.; Oh, T.J.; Lee, H.C.; Sohng, J.K. Squalene-hopene cyclase (Spterp25) from *Streptomyces peucetius*: Sequence analysis, expression and functional characterization. *Biotechnol. Lett.* **2009**, *31*, 565–569. [CrossRef] [PubMed]
10. Seitz, M.; Klebensberger, J.; Siebenhaller, S.; Breuer, M.; Siedenburg, G.; Jendrossek, D.; Hauer, B. Substrate specificity of a novel squalene-hopene cyclase from *Zymomonas mobilis*. *J. Mol. Cat. B Enzym.* **2012**, *84*, 72–77. [CrossRef]
11. Shinozaki, J.; Shibuya, M.; Masuda, K.; Ebizuka, Y. Squalene and oxidosqualene cyclase from a fern. *FEBS Lett.* **2008**, *582*, 310–318. [CrossRef] [PubMed]
12. Shinozaki, J.; Shibuya, M.; Masuda, K.; Ebizuka, Y. Dammaradiene synthase. A squalene cyclase from *Dryopteris crassirhizoma* Nakai. *Phytochemistry* **2008**, *69*, 2559–2564. [CrossRef] [PubMed]
13. Shinozaki, J.; Shibuya, M.; Takahata, Y.; Masuda, K.; Ebizuka, Y. Molecular evolution of fern squalene cyclases. *ChemBioChem* **2010**, *11*, 426–433. [CrossRef] [PubMed]
14. Shinozaki, J.; Hiruta, M.; Okada, T.; Masuda, K. Migrated hopene synthase from *Colysis pothifolia* and identification of a migration switch controlling the number of 1,2-hydride and methyl shifts. *ChemBioChem* **2016**, *17*, 65–70. [CrossRef] [PubMed]
15. Little, D.P.; Barrington, D.S. Major evolutionary events in the origin and diversification of the fern genus *Polystichum* (Dryopteridaceae). *Am. J. Bot.* **2003**, *90*, 508–514. [CrossRef] [PubMed]
16. Iwatsuki, K. *Dryopteridaceae. Flora of Japan*; Iwatsuki, K., Yamazaki, T., Boufford, D.E., Ohba, H., Eds.; Kodansha: Tokyo, Japan, 1995; Volume 1, pp. 120–173. ISBN 9784061546035.
17. Shiojima, K.; Arai, Y.; Masuda, K.; Kamada, T.; Ageta, H. Fern constituents: Polypodatetraenes, novel bicyclic triterpenoids, isolated from Polypodiaceous and Aspidiaceous plants. *Tetrahedron Lett.* **1983**, *24*, 5733–5736. [CrossRef]
18. Ageta, H.; Shiojima, K.; Arai, Y.; Kasama, T.; Kajii, K. Fern constituents: Dryocrassol and dryocrassyl acetate isolated from the leaves of aspidiaceous fern. *Tetrahedron Lett.* **1975**, *16*, 3297–3298. [CrossRef]
19. Shiojima, K.; Arai, Y.; Masuda, K.; Kamada, T.; Suzuki, M.; Ageta, H. Chemotaxonomy of fern plants (V). Japanese *Polystichum* species. *Nat. Med.* **1997**, *51*, 523–527.
20. Kushiro, T.; Shibuya, M.; Ebizuka, Y. Amyrin synthase. Cloning of oxidosqualene cyclase that catalyzes the formation of the most popular triterpene among higher plants. *Eur. J. Biochem.* **1998**, *256*, 238–244. [CrossRef] [PubMed]
21. Morita, M.; Shibuya, M.; Lee, M.-S.; Sankawa, U.; Ebizuka, Y. Molecular cloning of pea cDNA encoding cycloartenol synthase and its functional expression in yeast. *Biol. Pharm. Bull.* **1997**, *20*, 770–775. [CrossRef] [PubMed]
22. Ageta, H.; Shiojima, K.; Suzuki, H.; Nakamura, S. NMR spectra of triterpenoids. I. Conformation of the side chain of hopane and isohopane, and their derivatives. *Chem. Pharm. Bull.* **1993**, *41*, 1939–1943. [CrossRef]
23. Seckler, B.; Poralla, K. Characterization and partial purification of squalene-hopene cyclase from *Bacillus acidocaldarius*. *Biochem. Biophys. Acta* **1986**, *881*, 356–363. [CrossRef]
24. Kleemann, G.; Kellner, R.; Poralla, K. Purification and properties of the squalene-hopene cyclase from *Rhodopseudomonas palustris*, a purple non-sulfur bacterium producing hopanoids and tetrahymanol. *Biochem. Biophys. Acta* **1990**, *1210*, 317–320. [CrossRef]
25. Sato, T.; Hoshino, T. Catalytic function of the residues of phenylalanine and tyrosine conserved in squalene-hopene cyclases. *Biosci. Biotechnol. Biochem.* **2001**, *65*, 2233–2242. [CrossRef] [PubMed]
26. Sato, T.; Kouda, M.; Hoshino, T. Site-directed mutagenesis experiments on the putative deprotonation site of squalene-hopene cyclase from *Alicyclobacillus acidocaldarius*. *Biosci. Biotechnol. Biochem.* **2004**, *68*, 728–738. [CrossRef] [PubMed]
27. Zhang, L.-B.; Barrington, D.S. *Flora of China*; Wu, Z.-Y., Raven, P.H., Hong, D.-Y., Eds.; Science Press: St. Louis, MO, USA, 2013; Volume 2–3, pp. 629–713.
28. Le Péchon, T.; He, H.; Zhang, L.; Zhou, X.-M.; Gao, X.-F.; Zhang, L.-B. Using a multilocus phylogeny to test morphology-based classifications of *Polystichum* (Dryopteridaceae), one of the largest fern genera. *BMC Evol. Biol.* **2016**, *16*, 55. [CrossRef] [PubMed]
29. Sakakibara, J.; Watanabe, R.; Kanai, Y.; Ono, T. Molecular cloning and expression of rat squalene epoxidase. *J. Biol. Chem.* **1995**, *270*, 17–20. [CrossRef] [PubMed]
30. Nadeau, R.G.; Hanzlik, R.P. Synthesis of labeled squalene and squalene 2,3-oxide. In *Methods in Enzymology*; Clayton, R.B., Ed.; Academic Press: New York, NY, USA, 1969; Volume 15, pp. 346–349.

31. Thompson, J.D.; Higgins, D.G.; Gibson, T.J. CLUSTAL W: Improving the sensitivity of progressive multiple sequence alignment through sequence weighting, position-specific gap penalties and weight matrix choice. *Nucleic Acid Res.* **1994**, *22*, 4673–4680. [CrossRef] [PubMed]
32. Saitou, N.; Nei, M. The neighbor-joining method: A new method for reconstructing phylogenetic trees. *Mol. Biol. Evol.* **1987**, *4*, 406–425. [CrossRef] [PubMed]
33. Zuckerkandl, E.; Pauling, L. Evolutionary divergence and convergence in proteins. In *Evolving Genes and Proteins*; Bryson, V., Vogel, H.J., Eds.; Academic Press: New York, NY, USA, 1965; pp. 97–166.
34. Felsenstein, J. Confidence limits on phylogenies: An approach using the bootstrap. *Evolution* **1985**, *39*, 783–791. [CrossRef] [PubMed]
35. Kumar, S.; Stecher, G.; Tamura, K. MEGA7: Molecular Evolutionary Genetics Analysis version 7.0 for bigger datasets. *Mol. Biol. Evol.* **2016**, *33*, 1870–1874. [CrossRef] [PubMed]

Sample Availability: Samples of the compounds 22-hydroxyhopane are available from the authors.

Article

Phytosterol Composition of *Arachis hypogaea* Seeds from Different Maturity Classes

Wenxu Zhou [1], William D. Branch [2], Lissa Gilliam [3] and Julie A. Marshall [4,*]

1 Department of Chemistry and Biochemistry, Texas Tech University, Lubbock, TX 79409, USA; wenxu.zhou@ttu.edu
2 Crop & Soil Sciences, University of Georgia, Tifton, GA 30602, USA; wdbranch@uga.edu
3 Biochemical Research Lab, Lubbock Christian University, Lubbock, TX 79407, USA; lissa.gilliam@lcu.edu
4 Department of Chemistry and Biochemistry, Lubbock Christian University, Lubbock, TX 79407, USA
* Correspondence: julie.marshall@lcu.edu; Tel.: +1-806-720-7629

Received: 5 November 2018; Accepted: 20 December 2018; Published: 29 December 2018

Abstract: The seeds of cultivated peanut, *Arachis hypogaea*, are an agronomically important crop produced for human nutrition, oilseed and feed stock. Peanut seed is the single most expensive variable input cost and thus producers require seed with excellent performance in terms of germination efficiency. During the maturation process, triglycerides are stored in oil bodies as an energy resource during germination and seedling development. The stability of oil body membranes is essential for nutrient mobilization during germination. This study focused on evaluating the phytosterol composition in seed components including the kernel, embryo (heart), and seed coat or skin. Samples of different maturity classes were analyzed for macronutrient and phytosterol content. The three biosynthetic end products in the phytosterol pathway, β-sitosterol, campesterol and stigmasterol, comprised 82.29%, 86.39% and 94.25% of seed hearts, kernels and seed coats, respectively. Stigmasterol concentration was highest in the seed kernel, providing an excellent source of this sterol known to have beneficial effects on human health. Peanut hearts contained the highest concentration of sterols by mass, potentially providing protection and resources for the developing seedling. The amount of α-tocopherol increases in peanut hearts during the maturation process, providing protection from temperature stress, as well as stability required for seedling vigor. These results suggest that phytosterols may play a significant role in the performance of seeds, and provide a possible explanation for the poor germination efficiency of immature seeds.

Keywords: phytosterols; mesocarp; oilseed; maturity; pod-blast; α-tocopherol; oil bodies; campesterol; stigmasterol; β-sitosterol

1. Introduction

The seeds of cultivated peanut, *Arachis hypogaea*, store proteins, lipids and starch required for energy and growth upon germination. The seed can be harvested to serve as human nutrition, stock feed and biofuels [1–3]. Peanuts and other legumes have the ability to fix nitrogen and thus increase the sustainability of agricultural systems [4]. The pressure for peanut seed as plant-based protein and oilseed source is increasing with a growing world population. Due to its agronomic importance, improvement of peanut seed performance is necessary to meet the demand.

Lipid (oil) is the predominant macro component, and generally increases as the peanut seed matures [5]. For mature peanuts, total oil was reported to average about 50% on a fresh weight basis [6]. Oilseeds, such as peanuts, store most of their lipids in small, intracellular organelles commonly called oil bodies [7,8]. Triglycerides form the majority of these oil bodies, and the interior triglycerides are encapsulated by a phospholipid bilayer and embedded oleosin protein. The oil bodies provide a stable energy reserve that can be accessed upon germination [9].

Phytosterols are a special class of structural lipids that provide stability and fluidity in cell membranes including specialized encapsulation of lipids, such as oil bodies [10]. The primary function of phytosterols are as membrane reinforcers and precursors for brassinosteroids, an important phytohormone in plants. Unlike animal and fungi counterparts with cholesterol and ergosterol as the dominant sterol, the plant usually synthesizes an array of sterols with different alkyl group substitutions at the sterol side chain. In addition to the primary function, the correct sterol composition is necessary for many aspects of plant biology such as embryonic pattern formation, cell division, cell elongation, cell polarity, cellulose accumulation, and interactions with other signaling pathways [11].

Physiological maturity impacts seed quality through a variety of mechanisms including desiccation tolerance, preparation of storage reserves and establishment of dormancy [12]. Thus, the impact of seed maturity on germination efficiency is of primary importance to peanut producers. To best assess physiological maturation, a method of classification based on color and morphological differences of the mesocarp was described for determining the developmental stages of fresh peanut [13]. Maturity determination by this method requires removal of a portion of the exocarp to expose the pod mesocarp. The characteristic darkening of the pericarp is part of a progressive change in colors resulting in mesocarp colors from white (immature) to black (mature) [14]. The outer layer of the pod can be removed, and seeds sorted by maturity class (white, yellow, orange, brown or black) in a nondestructive manner allowing for physiological and chemical studies pertaining to the stage of development [12].

Seed maturity proceeds with the thickening of cell walls and addition of oil bodies [15]. For mature peanuts of the highest classification (black mesocarp), the cytoplasm of the parenchyma cells is essentially full of oil bodies. The specific aim of this project seeks to clarify the role of phytosterols in the formation of oil bodies as the seed matures. To accomplish this aim, seeds were organized into different maturity classes and the phytosterols extracted and identified. It is hypothesized that establishment of a critical mass of phytosterols provides the membrane stability for proliferation of oil bodies during maturation.

2. Results

2.1. Stage of Peanut Maturity and Macronutrient Composition in Different Maturity Classes

Peanut seed development can be classified into seven classes with four incremental stages in each class [13]. Based on the color of the mesocarp, the last three classes are described and named as "orange", "brown" or "black". Pooled samples from each maturity classes were analyzed to determine macronutrient percentages, and results are summarized in Table 1. The data reported is a check for comparison against the Biochemistry Research Laboratory database of crop samples from 2011–2018. The reported results in Table 1 are consistent with observed trends in peanut crop analysis conducted by the Biochemistry Research Laboratory at Lubbock Christian University (LCU) during the 2011–2018 CYs. The USDA National Nutrient Database for Standard Reference (Release 27, Basic Report 16087) reports macronutrient composition as 49.2% fat, 25.8% protein and 4.7% sugars. The reported results are within expected ranges as compared to the standard values. Any slight variation could be attributed to a specific variety versus average trends across all market types of cultivated peanut for a specific crop year.

Table 1. Macronutrient Composition of *Arachis hypogaea* Seeds of Different Maturation Classes.

Pod Color	Fat (%)	Protein (%)	Sugar (%)
Orange	50.95	20.90	3.43
Brown	51.49	19.90	3.63
Black	51.96	18.00	4.42

Method error ± 0.33% fat, 0.648–0.798% protein, and −0.33–0.52% sugar.

2.2. Isoprenoids and Phytosterol Composition in Different Tissues

Using GC/MS, we analyzed three classes of peanut samples for their isoprenoid and sterol composition. We have positively identified 11 compounds (Figure 1) based on the mass spectra and retention time relative to the authentic standards.

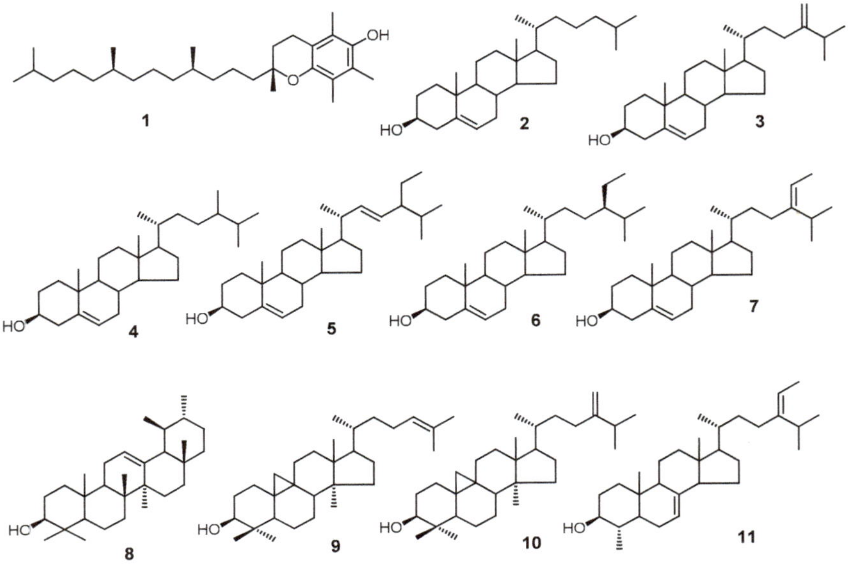

Figure 1. Structures of compounds identified by GC/MS. The structures are **1**. α-tocopherol; **2**. cholesterol; **3**. 24-methylenecholesterol; **4**. campesterol; **5**. stigmasterol; **6**. sitosterol; **7**. isofucosterol; **8**. α-amyrin; **9**. cycloartenol; **10**. 24-methylenecycloartanol; and **11**. citrostadienol.

We found that of the three seed component classes analyzed, the kernel and seed coat generally contained the same percentage of α-tocopherol (0.63% and 0.64%, respectively), decreasing slightly with maturity in brown versus black mesocarp classes, 0.69% to 0.63% kernel composition and 0.71% to 0.64% seed coat composition (Figure 2 and Tables 2–4). In contrast, the hearts contained significantly less tocopherol by percentage (Tables 2–5) as compared to the kernel and seed coat. Also, the amount of α-tocopherol increased from 0.09% in the orange mesocarp class, to 0.19% in the brown mesocarp class, and 0.24% in the black mesocarp class. Vitamin E comprises eight structurally related molecules including four forms of tocopherols. Peanuts, like many other oilseeds, contain tocopherols [16]. Within plants, these molecules are found in cell membranes and possess antioxidant activity which protects organelles from reactive oxygen species (ROS) [17]. Like phytosterols, tocopherols contribute to maintaining membrane fluidity [17].

Many phytosterols are present in nature [18]. In plants, the three biosynthetic end products in the phytosterol pathway are β-sitosterol, campesterol and stigmasterol. In 2004, it was reported that these three sterols account for approximately 95% of total peanut sterols [18]. Our analysis confirmed that these sterols are major components of total peanut sterols in each component, 82.29% of hearts, 86.39% of kernels and 94.25% seed coats (Tables 2–4), but at lower percentages as compared to the 2004 reported values. Other minor component phytosterols were detected (Tables 2–4) including those not previously identified most likely due to advances in technology. Squalene and γ-tocopherol as reported by Maguire et al. [18] were not detected. This discrepancy of results can be explained in addition to sample variations by two possible explanations. Firstly, squalene is the substrate for phytosterol, whereas γ-tocopherol can be converted to α-tocopherol via a methylation reaction. Depending on the harvest time, these two compounds may be converted to their final products and therefore are undetectable. Secondly, in the references, these sterols and tocopherols were determined by an HPLC method. Because of the differences in UV absorption between the two classes of compounds, tocopherol can be easily overestimated. Also, there were no internal standards used in the literature to correct the extraction and HPLC injection errors, as well as a reference for quantitation. Some variation may also be due to varietal differences, as these samples were isolated from Runner-type peanuts as compared to Spanish, Valencia and Virginia market types.

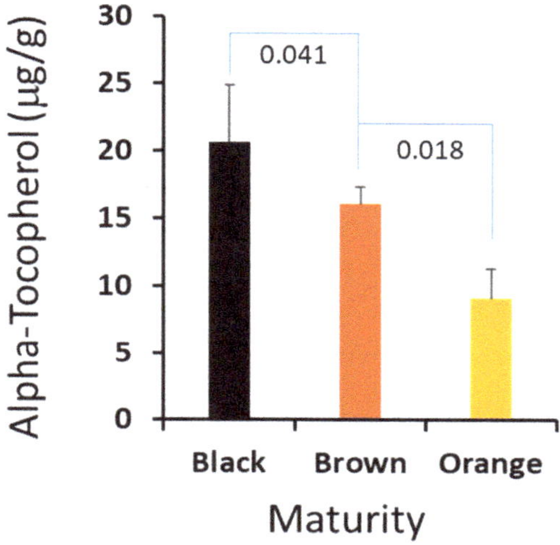

Figure 2. The α-tocopherol contents in the peanuts' hearts are correlated to the maturity. The error bars are standard deviation of 5 independent measurements and the p-values of Student's t-test between the maturity stages.

Table 2. Phytosterol and Isoprenoid Composition of *Arachis hypogaea* Seed Hearts of Different Classes.

Sterols and Isoprenoids	Heart		
	Black	Brown [1]	Orange
α-Tocopherol	0.24 ± 0.05	0.19 ± 0.02	0.09 ± 0.08
Cholesterol	0.19 ± 0.01	0.17 ± 0.01	0.19 ± 0.03
24-Methylenecholesterol	0.85 ± 0.06	0.86 ± 0.03	0.92 ± 0.13
Campesterol	17.42 ± 0.35	16.65 ± 0.53	16.78 ± 0.61
Stigmasterol	2.91 ± 0.25	2.87 ± 0.14	2.63 ± 0.1
β-Sitosterol	61.99 ± 0.48	62.03 ± 0.62	62.33 ± 2.15
Isofucosterol	9.75 ± 0.25	10.2 ± 0.15	11.01 ± 1.03
alpha-Amyrin	0.38 ± 0.05	0.38 ± 0.06	0.36 ± 0.23
Cycloartenol	2.35 ± 0.13	2.45 ± 0.14	1.95 ± 0.79
24-Methylenecycloartanol	1.95 ± 0.14	2.1 ± 0.27	1.65 ± 0.68
Citrostadienol	1.97 ± 0.11	2.12 ± 0.03	2.1 ± 0.19
Total (μg/mg)	8.68 ± 0.82	8.47 ± 0.61	10.25 ± 1.7

Sterol and isoprenoid content is AVE %; STD was calculated on each sterol for the sample; $n = 5$, [1] $n = 4$. Units are μg (sterol and isoprenoid)/mg (dry tissue).

Table 3. Phytosterol and Isoprenoid Composition of *Arachis hypogaea* Seed Kernels of Different Classes.

Sterols and Isoprenoids	Kernel		
	Black	Brown	Orange
Alpha-Tocopherol	0.63 ± 0.12	0.69 ± 0.38	0.24 ± 0.05
Cholesterol	0.21 ± 0.13	0.22 ± 0.11	0.19 ± 0.01
24-Methylenecholesterol	0.27 ± 0.2	0.86 ± 0.86	0.85 ± 0.06
Campesterol	12.14 ± 0.21	11.12 ± 1.42	17.42 ± 0.35
Stigmasterol	11.63 ± 0.49	11.61 ± 0.69	2.91 ± 0.25
β-Sitosterol	62.61 ± 1.9	58.7 ± 2.73	61.99 ± 0.48
Isofucosterol	10.7 ± 1.35	14.32 ± 1.42	9.75 ± 0.25
alpha-Amyrin	0.61 ± 0.43	0.65 ± 0.49	0.38 ± 0.05
Cycloartenol	0.7 ± 0.43	0.44 ± 0.29	2.35 ± 0.13
24-Methylenecycloartanol	0.33 ± 0.25	1.18 ± 0.84	1.95 ± 0.14
Citrostadienol	0.17 ± 0.12	0.21 ± 0.27	1.97 ± 0.11
Total (μg/mg)	1.04 ± 0.06	0.93 ± 0.08	8.68 ± 0.82

Sterol and isoprenoid content is AVE %; STD was calculated on each sterol for the sample; $n = 5$. Units are μg (sterol and isoprenoid)/mg (dry tissue).

Table 4. Phytosterol and Isoprenoid Composition of *Arachis hypogaea* Seed Coats of Different Classes.

Sterols and Isoprenoids	Seed Coat		
	Black	Brown	Orange
Alpha-Tocopherol	0.64 ± 0.12	0.71 ± 0.14	0.66 ± 0.14
Cholesterol	0.54 ± 0.07	0.57 ± 0.18	0.46 ± 0.25
24-Methylenecholesterol	0.13 ± 0.12	0.12 ± 0.04	0.12 ± 0.09
Campesterol	14.71 ± 0.91	14.71 ± 0.56	15.11 ± 0.63
Stigmasterol	3.76 ± 0.55	3.95 ± 1.04	4.57 ± 0.2
β-Sitosterol	75.78 ± 1.04	76.45 ± 1.19	75.19 ± 0.96
Isofucosterol	2.67 ± 1.73	2.09 ± 0.23	2.18 ± 0.29
alpha-Amyrin	0.99 ± 0.18	1.05 ± 0.14	0.89 ± 0.11
Cycloartenol	0.73 ± 0.61	0.24 ± 0.16	0.74 ± 0.57
24-Methylenecycloartanol	0.01 ± 0.01	0.04 ± 0.03	0 ± 0.01
Citrostadienol	0.06 ± 0.03	0.08 ± 0.04	0.06 ± 0.03
Total (μg/mg)	5.4 ± 0.62	5.51 ± 0.89	5.71 ± 0.86

Sterol and isoprenoid content is AVE %; STD was calculated on each sterol for the sample; $n = 5$. Units are μg (sterol and isoprenoids)/mg (dry tissue).

Table 5. The p-values of the pair-wise Student's t-test among three developmental stages.

Sterols and Isoprenoids	Heart			Kernel			Seed Coat		
	BL-BR [1]	BR-OR	BL-OR	BL-BR	BR-OR	BL-OR	BL-BR	BR-OR	BL-OR
α-Tocopherol	**0.041** [2]	**0.019**	**0.003**	0.367	0.285	0.327	0.206	0.300	0.395
Cholesterol	0.079	0.160	0.433	0.471	0.281	0.325	0.383	0.231	0.255
24-Methylenecholesterol	0.445	0.186	0.175	0.088	0.069	0.294	0.446	0.444	0.489
Campesterol	**0.028**	0.368	**0.039**	0.076	0.437	0.062	0.499	0.156	0.219
Stigmasterol	0.380	**0.017**	**0.025**	0.480	**0.027**	**0.013**	0.363	0.117	**0.008**
β-Sitosterol	0.460	0.389	0.369	**0.015**	**0.004**	0.148	0.183	0.051	0.191
Isofucosterol	**0.006**	0.078	**0.015**	**0.002**	**0.030**	0.127	0.241	0.301	0.276
alpha-Amyrin	0.409	0.442	0.408	0.442	0.153	0.174	0.282	**0.040**	0.161
Cycloartenol	0.169	0.118	0.148	0.154	0.338	0.090	0.061	**0.047**	0.484
24-Methylenecycloartanol	0.184	0.115	0.181	**0.032**	**0.018**	0.179	**0.047**	**0.032**	0.382
Citrostadienol	**0.023**	0.416	0.123	0.391	0.497	0.356	0.183	0.269	0.368
Total (μg/mg)	0.338	**0.040**	0.050	**0.021**	0.414	**0.030**	0.409	0.362	0.260

[1] BL: black; BR: brown; and OR: orange. [2] The p-values less than 0.05 are in bold. Units are μg (sterol and isoprenoid)/mg (dry tissue).

When comparing sterol composition in the different seed components, the relative percentage of stigmasterol should be noted (Tables 2–4), heart 2.80%, kernel 11.31% and seed coat 4.09%. This result differs from the reported value of stigmasterol in Runner peanuts as 11.0% [18]. However, the kernel is the largest component of the seed by mass, and as a result, would more closely reflect data on the entire seed.

2.3. Isoprenoid and Phytosterol Composition in Different Maturity Classes

It is critical for farmers to harvest peanut at the optimized maturity to maximize the crop value, and a suitable biomarker could help farmers to harvest at the right time. We compared the profiles of the sterols and isoprenoids from the three stages, and there are no statistically significant differences among the major sterols. However, we found that α-tocopherol in the peanut hearts changed dramatically crossing the maturation stages. The absolute amount of this important metabolite was increased correspondingly from 0.92 μg/mg in orange to 1.61 μg/mg in brown, and peaked at 2.08 μg/mg in black. Given the fact that vitamin E is a very important metabolite to plant biology and a valuable compound to human health, we think that the content of vitamin E could be used as a biomarker for peanut harvesting. We plan to develop a user-friendly and portable method to determine vitamin E content, which may have important practical value to peanut farmers.

3. Discussion

Peanuts are a nutrient rich plant-based protein source that contain vitamins, minerals, antioxidants and bioactive phytochemicals, leading to the perception that peanuts are a "super food" [19]. Peanut phytosterols have been shown to help lower LDL cholesterol by competing in the digestive tract with cholesterol and preventing absorption [20]. One of the major phytosterols, stigmasterol, has been investigated for its pharmacological importance as an antihypercholesterolemic, anti-inflammatory, antioxidant, hypoglycemic and antitumor effector [21,22]. In this study, it is reported that the kernel contains a higher percentage of stigmasterol as compared to the heart and seed coat. Different manufacturing processes may remove the seed coat or heart, so it is beneficial to the health of the consumer that the kernel possess relatively high concentrations of stigmasterol.

In addition to potential health benefits, stigmasterol is thought to play a role in temperature stress tolerance in plants [23]. Drought and extreme heat in the growing season can increase the sensitivity of the plant to opportunistic organisms [24]. Accumulation of critical phytosterols, such as stigmasterol, during pod development may set the foundation for physiological maturation processes and resistance to stress.

Tocopherol content can vary with environmental stress and growing location in addition to other factors [25]. In this study, hearts contained the lowest percentage of α-tocopherol as compared to the other seed components, but the amount increased during maturation. During germination, the stability of the peanut heart, or embryo, is critically important to the development of the seedling and stand establishment under adverse environmental conditions [17]. α-tocopherol deactivates ROS generated during photosynthesis and is upregulated during stressful events [17]. Immature peanut seed is less resistant to stress, and as a result, is more likely to be adversely affected during germination. The results suggest that the synthesis and accumulation of α-tocopherol in developing peanut hearts may be vitally important to seedling vigor upon germination.

4. Materials and Methods

4.1. Materials and Reagents

Epicoprostanol, 5β-cholestan-3α-ol, heptane (99%), anhydride pyridine (99.9%), and *N,O*-bis(trimethylsilyl)trifluoroacetamide (BSTFA) (99.9%) were purchased from Sigma-Aldrich (St. Louis, MO, USA). HPLC-grade *n*-hexane, HPLC-grade methanol, potassium hydroxide (85%) (KOH), acetone, and HPLC-grade dichloromethane were purchased from Thermo Fisher (Waltham, MA, USA).

4.2. Pod Blasting

Freshly harvested pods, 2017 crop year (CY), from one specific genotype of runner market-type peanuts grown under conventional cultural practices were obtained from the University of Georgia research facility under the direction of Dr. W.D. Branch. Pods were removed from the plant material, and 'pod blasted' to reveal the mesocarp. Pod blasting is a process by which in-shell peanut pods are placed in a wire basket and a residential-style pressure washer is used to spray the shell exterior with high-pressure water, removing the outer portion of the peanut hull and exposing the colored mesocarp layer underneath. The blasted pods were separated by color into three different maturity classes, orange, brown and black. After separation, the remainder of the pod outer layer was removed, and the seeds segregated for additional chemical analyses. Upon receipt at the research laboratory at LCU, samples were examined and verified to be *A. hypogaea* runner type seeds by technicians under the supervision of Dr. Julie Marshall.

4.3. Isolation of Peanut Seed Components

For each maturity class, 10–50 g of redskin kernels were weighed and dried in a forced-air oven (VWR, Radnor, PA, USA) at 130 °C for 45 min. After cooling to ambient temperature, the kernels were manually separated by removing the skin, breaking open the seed, and removing the heart to subdivide the samples into three subsections consisting of seed coats, hearts and kernels. The subsections were scaled at a specific mass for 5 replicates. Each replicate of the heart and kernel contained 50 mg while the seed coat replicates contained 20 mg. Each scaled replicate was placed in a 2 mL microcentrifuge tube with locking lid.

4.4. Fat, Protein and Sugar Analysis

To analyze the seed sample for total fat by organic solvent extraction, the exact mass of 10 g $\pm$ 0.1 g sample was recorded and the sample pre-dried in a forced-air oven at 130 °C for 45 min to remove moisture. After cooling to ambient temperature, the dried sample was quantitatively transferred to an explosion-proof blender jar (Eberbach Corporation, Belleville, MI, USA). 60 mL dichloromethane (DCM) was added and the mixture blended at high speed for 1 min. After allowing the blender jar to cool for 30 s before opening, we removed the blender lid and washed down the sides of the blender with DCM in a wash bottle. We replaced the lid and blended at high speed for an additional 1 min. We allowed the blender jar to cool for 30 s before opening and washing down the sides of the blender with DCM a second time to remove all residue. We carefully poured the blender contents into a Büchner funnel vacuum filter apparatus with a Toxicity Characteristics Leaching Procedure (TCLP) glass fiber filter, rinsing the blender jar residue with DCM into the funnel until all residue was removed. We filtered the mixture, and transferred the filtrate from the vacuum flask to a tared stainless beaker, rinsing the vacuum flask with DCM into the stainless beaker to ensure all residue was transferred. We evaporated the solvent in the stainless flask over a steam bath until all solvent had been removed. We monitored evaporation and weighed the beaker/remaining oil as needed by removing the beaker from the steam bath and allowing it to come to ambient temperature. Evaporation was complete when the mass of the oil remaining in the stainless beaker stabilized ($\leq$0.03 g change in mass over a 30-min span on the steam bath). We recorded the final weight of the beaker and oil, and calculated the percent oil using the following formula:

$$\%\ \text{Oil} = [(\text{Weight of Beaker with Oil} - \text{Empty Beaker Weight})/\text{Sample Weight}] \times 100 \quad (1)$$

The protein analysis (reference methods AOAC 992.15; AACC 46-30) was conducted by Medallion Labs (Minneapolis, MN, USA) and the sugar analysis (sugar by HPLC) was conducted by North Carolina Extension/North Carolina State University (Raleigh, NC, USA).

4.5. Preparation of Nonsaponifiable Fraction (NSF)

We prepared 10% KOH/methanolic solution by dissolving 50 g KOH in 50 mL deionized water and bringing to volume of 500 mL with methanol. We prepared internal standard by mixing the epicoprostanol 5β-cholestan-3α-ol with heptane to a final concentration of 1mg/mL. After adding 25 μL of internal standard solution and 1 mL 10% methanolic KOH to each tube, the samples were saponified at 80 °C for 2 h using a Thermomixer (Eppendorf, Hamburg, Germany) with constant shaking at 500 rpm. Once cooled to ambient temperature, the nonsaponifiable fraction (NSF) containing free sterols was extracted with 1 mL of *n*-hexane. The hexane was pooled in a 1.5 mL microcentrifuge tube and the hexane was removed by evaporation in a fume hood overnight. To the residual, 20 μL of acetone was added to dissolve the sterols. After solvation, the compounds were converted to their trimethylsilyl ester by adding 10 μL of BSTFA and 10 μL of pyridine as catalyst. The derivatization mixtures were kept at room temperature for 30 min before GC/MS analysis.

4.6. GC/MS Analysis

2 μL of derivatization mixture was injected into an Agilent GC/MS (Agilent 6890 BC coupled with 5973 mass-selective detector (MSD)) (Agilent, Santa Clara, CA, USA). The GC was equipped with an Agilent DB-5Ms+DG narrow-bore capillary column (30 m × 0.25 mm × 0.25 μm with 10 m Duraguard). The injection mode was splitless, with helium carrier gas at a constant flow of 1.2 mL/min. The GC oven was initialed at 170 °C, held for 1 min, the temperature was ramped to 280 °C at 40 °C/min and held at 280 °C for 25 min. The MSD was in electron ionization (EI) mode, scan range was from 50–550 amu, temperature of the ion source was 230 °C, the quadrupole temperature was 150 °C, and the interface was 280 °C.

The GC/MS data was processed with ChemStation software (Version E.02.02.1431, Agilent, Santa Clara, CA, USA) and Automated Mass Spectral Deconvolution & Identification System (AMDIS) (National Institute of Standards and Technology, United States Department of Commerce, Washington, DC, USA). The sterol peaks were deconvoluted using AMDIS after baseline correction and identified by their relative retention time to cholesterol and comparison to the mass spectra from commercial mass database (NIST08 mass spectral library, http://nistmassspeclibrary.com/). The GC peak representing the sterol amount generated from total ion current (TIC) was integrated using the software default parameters [26,27].

5. Conclusions

Physiological maturity of *A. hypogaea* seeds is impactful to germination efficiency, and consequently, is of economic importance to the peanut industry. Immature seeds have reduced germination frequency, are more susceptible to disease, and require more financial inputs during manufacturing processes which are dependent on fat content or roasting performance. This study seeks to characterize phytosterol and isoprenoid content in seed components as a function of physiological maturation indicated by mesocarp color class.

Peanut hearts contained the highest percentage of phytosterols and isoprenoids by mass as compared to the other seed components. The stability of the peanut heart, or embryo, is critically important to the development and vigor of the seedling during germination and plant development. Tocopherols provide plants with antioxidant capacity and provide fluidity and flexibility to cell membranes. α-tocopherol concentration in the hearts changed dramatically across maturation classes, providing a possible causal agent for poor stress adaptability in immature seeds. Given the economic importance of seed efficiency and performance in all sectors of the peanut industry, α-tocopherol content in the seed embryo could be utilized as a biomarker to optimize harvest timing and profitability. In the subsequent study, we will investigate more details to evaluate using α-tocopherol as a biomarker for peanut harvesting, and will develop a sensitive and portable α-tocopherol detection method for peanut farmers.

Stigmasterol, one of the three end products in the phytosterol pathway in *A. hypogaea*, is a significant component by percentage of seeds. The seed coat, or skin, is known to be a major contributor of antioxidant capacity and other health benefits. The relatively high concentration of stigmasterol may contribute to the health benefits reported by the addition of peanut skins to various manufactured goods. Also, stigmasterol biosynthesis is important in plant physiology to deal with low and high temperature stresses, and plays a key role in innate immunity to combat biotic stresses. Since the main function of the seed coat is to protect the embryo, high stigmasterol content may be important for seed storage and germination.

Author Contributions: The author contributions are as follows: conceptualization, W.Z. and J.A.M.; methodology, W.Z., W.D.B., L.G. and J.A.M.; formal analysis, W.Z., L.G. and J.A.M.; resources, W.D.B. and L.G.; data curation, W.Z. and L.G.; writing-original draft preparation, W.Z., L.G. and J.A.M.; writing-review and editing, L.G. and J.A.M.; project administration, J.A.M.; and funding acquisition, J.A.M.

Funding: This research was partially funded by The Welch Foundation, grant number BV-0043.

Acknowledgments: The authors would like to acknowledge Chris Cobos for technical support, Joshua Thomas and Melanie McGilton for literature research and manuscript preparation.

Conflicts of Interest: The authors declare no conflict of interest.

References

1. Duranti, M.; Gius, C. Legume seeds: Protein content and nutritional value. *Field Crop. Res.* **1997**, *53*, 31–45. [CrossRef]
2. Djemel, N.; Guedon, D.; Lechevalier, A.; Salon, C.; Miquel, M.; Prosperi, J.M.; Rochat, C.; Boutin, J.P. Development and composition of the seeds of nine genotypes of the *Medicago truncatula* species complex. *Plant Physiol. Biochem.* **2005**, *43*, 557–566. [CrossRef] [PubMed]
3. Gallardo, K.; Thompson, R.; Burstin, J. Reserve accumulation in legume seeds. *Comptes. Rendus. Biol.* **2008**, *331*, 755–762. [CrossRef] [PubMed]
4. Siddique, K.H.; Johansen, C.; Turner, N.C.; Jeuffroy, M.H.; Hashem, A.; Sakar, D.; Gan, Y.; Alghamdi, S.S. Innovations in agronomy for food legumes. A review. *Agron. Sustain. Dev.* **2012**, *32*, 45–64. [CrossRef]
5. Patee, H.E.; Johns, E.B.; Singleton, J.A.; Sanders, T.H. Composition Changes of Peanut Fruit Parts During Maturation. *Peanut. Sci.* **1974**, *1*, 57–62. [CrossRef]
6. Patee, H.E.; Salunkhe, D.K.; Sathe, S.K.; Reddy, N.R.; Ory, R.L. Legume lipids. *CRC Crit. Rev. Food Sci. Nutr.* **1983**, *17*, 97–139. [CrossRef]
7. Young, C.T.; Schadel, W.E. Microstructure of Peanut Seed: A Review. *Food Struct.* **1990**, *9*, 317–328.
8. Huang, A.H.C. Oil Bodies and Oleosins in Seeds. *Annu. Rev. Plant Physiol. Plant Mol. Biol.* **1992**, *43*, 177–200. [CrossRef]
9. Tzen, J.T.C.; Cao, Y.Z.; Laurent, P.; Ratnayake, C.; Huang, A.H.C. Lipids, Proteins, and Structure of Seed Oil Bodies from Diverse Species. *Plant Physiol.* **1993**, *101*, 267–276. [CrossRef] [PubMed]
10. Dufourc, E.J. Sterols and membrane dynamics. *J. Chem. Biol.* **2008**, *1*, 63–77. [CrossRef] [PubMed]
11. Boutté, Y.; Grebe, M. Cellular processes relying on sterol function in plants. *Cur. Opin. Plant Biol.* **2009**, *12*, 705–713. [CrossRef] [PubMed]
12. Ventura, L.; Doná, M.; Macovei, A.; Carbonera, D.; Buttafava, A.; Mondoni, A.; Rossi, G.; Balestrazzi, A. Understanding the molecular pathways associated with seed vigor. *Plant Physiol. Biochem.* **2012**, *60*, 196–206. [CrossRef]
13. Williams, E.J.; Drexler, J.S. A Non-Destructive Method for Determining Peanut Pod Maturity. *Peanut Sci.* **1981**, *8*, 134–141. [CrossRef]
14. Drexler, J.E.; Williams, E.J. *A Non-Destructive Method of Peanut Pod Maturity Classification*; American Peanut Research and Education Association: Tifton, GA, USA, 1979.
15. Young, C.T.; Pattee, H.E.; Schadel, W.E.; Sanders, T.H. Microstructure of Peanut (*Arachis hypogaea* L. cv. 'NC 7') Cotyledons During Development. *LWT-Food Sci. Technol.* **2004**, *37*, 439–445. [CrossRef]
16. Carrin, M.E.; Carelli, A.A. Peanut Oil: Compositional Data. *Eur. J. Lipid Sci. Technol.* **2010**, *112*, 697–707. [CrossRef]

17. Munné-Bosch, S.; Alegre, L. The Function of Tocopherols and Tocotrienols in Plants. *Crit. Rev. Plant Sci.* **2002**, *21*, 31–57. [CrossRef]
18. Maguire, L.; O'Sullivan, S.; Galvin, K.; O'Connor, T.; O'Brien, N. Fatty acid profile, tocopherol, squalene and phytosterol content of walnuts, almonds, peanuts, hazelnuts and the macadamia nut. *Int. J. Food Sci. Nutr.* **2004**, *55*, 171–178. [CrossRef]
19. U.S. Department of Agriculture. USDA National Nutrient Database for Standard Reference. Available online: https://www.ars.usda.gov/nea/bhnrc/ndl (accessed on 8 October 2018).
20. Jang, M.; Cai, L.; Udeani, G.O.; Slowing, K.V.; Thomas, C.F.; Beecher, C.W.W.; Fong, H.H.; Farnsworth, N.R.; Kinghorn, A.D.; Mehta, R.G.; et al. Cancer Chemopreventive Activity of Resveratrol, A Natural Product Derived From Grapes. *Science* **2008**, *57*, 130–139. [CrossRef]
21. Kaur, N.; Chaudhary, J.; Jain, A.; Kishore, L. Stigmasterol: A comprehensive review. *Int. J. Pharm. Sci. Res.* **2011**, *2*, 2259. [CrossRef]
22. Burg, V.K.; Grimm, H.S.; Rothhaar, T.L.; Grösgen, S.; Hundsdörfer, B.; Haupenthal, V.J.; Zimmer, V.C.; Mett, J.; Weingärtner, O.; Laufs, U.; et al. Plant Sterols the Better Cholesterol in Alzheimer's Disease? A Mechanistical Study. *J. Neurosci.* **2013**, *33*, 16072–16087. [CrossRef] [PubMed]
23. Senthil-Kumar, M.; Wang, K.; Mysore, K.S. AtCYP710A1 gene-mediated stigmasterol production plays a role in imparting temperature stress tolerance in *Arabidopsis thaliana*. *Plant Signal. Behav.* **2013**, *8*, e23142. [CrossRef] [PubMed]
24. Wang, K.; Senthil-Kumar, M.; Ryu, C.M.; Kang, L.; Mysore, K.S. Phytosterols Play a Key Role in Plant Innate Immunity against Bacterial Pathogens by Regulating Nutrient Efflux into the Apoplast. *Plant Physiol.* **2012**. [CrossRef] [PubMed]
25. Hashim, I.; Koehler, P.; Eitenmiller, R. Tocopherols in Runner and Virginia Peanut Cultivars at Various Maturity Stages. *J. Am. Oil Chem. Soc.* **1993**, *70*, 633–635. [CrossRef]
26. Raffaella, I.; Cardenia, V.; Rodriguez-Estrada, M. Analysis of phytosterols and phytostanols in enriched dairy products by Fast gas chromatography with mass spectrometry. *J. Sep. Sci.* **2014**, *37*, 291–2919. [CrossRef]
27. Menéndez-Carreño, M.; Knol, D.; Janssen, H.G. Development and validation of methodologies for the quantification of phytosterols and phytosterol oxidation products in cooked and baked food products. *J. Chromatogr. A* **2016**, *1428*, 316–325. [CrossRef] [PubMed]

Sample Availability: Not available.

Communication

Arginine in the FARM and SARM: A Role in Chain-Length Determination for Arginine in the Aspartate-Rich Motifs of Isoprenyl Diphosphate Synthases from *Mycobacterium tuberculosis* †

Raimund Nagel [1], Jill A. Thomas [1], Faith A. Adekunle [2], Francis M. Mann [2,*] and Reuben J. Peters [1,*]

1 Roy J. Carver Department of Biochemistry, Biophysics and Molecular Biology, Iowa State University, Ames, IA 50011, USA; rnagel@iastate.edu (R.N.); thomas.jillann@gmail.com (J.A.T.)

2 Department of Chemistry, University of Wisconsin-Parkside, Kenosha, WI 53141, USA; adeku001@rangers.uwp.edu

* Correspondence: mannf@uwp.edu (F.M.M.); rjpeters@iastate.edu (R.J.P.)

† This study is dedicated to Prof. W. David Nes on the happy occasion of his 65th birthday.

Academic Editor: Wenxu Zhou
Received: 14 September 2018; Accepted: 5 October 2018; Published: 6 October 2018

Abstract: Isoprenyl chains are found in many important metabolites. These are derived from precursors of the appropriate length produced by isoprenyl diphosphate synthases (IDSs). The human pathogen *Mycobacterium tuberculosis* makes various isoprenoids/terpenoids, with important roles in their biosynthesis played by two closely related IDSs, encoded by *grcC1* (Rv0562) and *grcC2* (Rv0989c), with Rv0989c generating the 10-carbon precursor (*E*)-geranyl diphosphate (GPP), and Rv0562 the 20-carbon precursor (*E,E,E*)-geranylgeranyl diphosphate (GGPP). Intriguingly, while Rv0562 contains the prototypical *trans*-IDS first and second aspartate-rich (DDxxD) motifs (FARM and SARM, respectively), Rv0989c uniquely contains arginine in place of the second Asp in the FARM and first Asp in the SARM. Here site-directed mutagenesis of the corresponding residues in both Rv0562 and Rv0989c reveals that these play a role in determination of product chain length. Specifically, substitution of Asp for the Arg in the FARM and SARM of Rv0989c leads to increased production of the longer 15-carbon farnesyl diphosphate (FPP), while substitution of Arg for the corresponding Asp in Rv0562 leads to increased release of shorter products, both FPP and GPP. Accordingly, while the primary role of the FARM and SARM is known to be chelation of the divalent magnesium ion co-factors that assist substrate binding and catalysis, the Arg substitutions found in Rv0989c seem to provide a novel means by which product chain length is moderated, at least in these *M. tuberculosis* IDSs.

Keywords: terpene; isoprenoid; divalent metal co-factor ligation

1. Introduction

Isoprenoids play essential roles in many metabolic pathways. In large part, due to the hydrophobic nature of polyprenyl chains, these are used to anchor proteins to the membrane or, after cyclization, form the sterols or hopanoids that are integral membrane molecules. Polyprenol-derived molecules such as dolichol are essential in the transfer of glycosyl chains from the cytoplasmatic face of the Endoplasmatic Reticulum (ER) to the ER-lumen where they are attached to proteins, while others such as quinones are involved in electron transport chains. In addition, short-chain isoprenoids ($\leq$20-carbons) can be formed into natural products, termed terpenoids, which directly exert various biological activities [1–3].

Polymerization of the appropriate length chains from the universal precursors isopentenyl diphosphate (IPP) and dimethylallyl diphosphate (DMAPP) is catalyzed by isoprenyl diphosphate synthases (IDSs). IDS can be divided on the basis of the double bond configuration, either *trans*/(*E*)- or *cis*/(Z)-, formed in their prenyl diphosphate products [4]. Shorter chain length prenyl diphosphates are generally produced by *trans*-IDS, although there are a few exceptions [5–8]. These short prenyl diphosphates with *trans* carbon-carbon double bonds can then be the allylic substrates for *cis*-IDS that catalyze further polymerization with IPP and produce long chain prenyl diphosphates such as decaprenyl diphosphate or undecaprenyl diphosphate, in which the newly added carbon–carbon double bonds are then in the *cis* conformation. Although the *trans*- and *cis*- IDSs both catalyze similar reactions, utilizing divalent metal cation co-factors/substrates (usually magnesium, Mg^{2+}) to initiate ionization of the diphosphate ester bond in their allylic substrate (e.g., DMAPP) that then condenses with the terminal alkene of IPP followed by proton elimination [9,10], these form two structurally distinct enzymatic families [4]. For example, while the *cis*-IDS utilize scattered acidic residues to coordinate Mg^{2+}, the *trans*-IDS employ two aspartate-rich DDxxD motifs, simply termed the first (FARM) and second (SARM), to bind a trio of Mg^{2+} (Figure 1) [11].

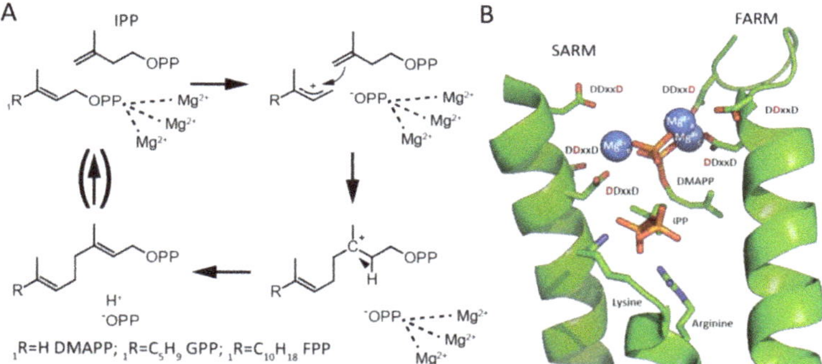

Figure 1. *Trans*-isoprenyl diphosphate synthase (IDS) reaction mechanism and coordination of substrate diphosphate moieties in the active site of *trans*-IDS. (**A**) The allylic substrate is coordinated by three Mg^{2+} co-factors that assist initiating ionization of the diphosphate ester. The resulting carbocation is attacked by the double bond of isopentenyl diphosphate (IPP), leading to formation of a bond between the allylic substrate and IPP and a shift of the carbocation. The resulting carbocation is quenched by deprotonation and formation of a new *trans*/(*E*)- double bond in the elongated product. This product can act as the allylic substrate for further elongation with another IPP molecule. (**B**) The active site of the (*E*,*E*)-farnesyl diphosphate synthase from *E. coli* complexed with its substrates IPP and dimethylallyl diphosphate (DMAPP) (Protein Data Bank (PDB) ID code: 1RIQ). The diphosphate moiety of the allylic DMAPP is activated by coordination to three Mg^{2+} that are, in turn, bound by the characteristic first and second aspartate-rich DDxxD motifs (FARM and SARM), as shown. More specifically, $Mg^{2+}{}_A$ and $Mg^{2+}{}_C$ are coordinated by the first and last aspartate of the FARM, while $Mg^{2+}{}_B$ is largely coordinated by the first aspartate of the SARM. By contrast, the diphosphate moiety of IPP is more directly bound by the basic residues shown. Note that, while the structure contains the thiolo analog of DMAPP, for illustrative purposes the color of the sulfur was changed to red to resemble that of the oxygen found in DMAPP.

The human pathogen *Mycobacterium tuberculosis* encodes an interesting set of IDSs that play various roles in construction of an array of isoprenoids/terpenoids. These serve key roles in construction of the complex mycobacterial cell wall. For example, in production of the polyprenyl phosphate serving as a glyco-carrier required for cell wall biosynthesis [6]. In particular, this 10 isoprenyl-unit carrier is produced via an unusual (*Z*,*E*)-farnesyl diphosphate (FPP) intermediate, which serves as the initial allylic substrate for the long-chain *cis*-IDS Rv2361c [12]. In turn, the 15-carbon

(*Z*,*E*)-FPP is produced by the short-chain *cis*-IDS Rv1086, which uses (*E*)-geranyl diphosphate (GPP) as its allylic substrate [13]. This unusual bacterial production of the 10-carbon GPP is mediated by the *trans*-IDS Rv0989c (*grcC2*) [14].

An intriguing cell wall lipid that appears to be unique to *M. tuberculosis* is the diterpenoid (20-carbon) nucleoside tuberculosinyl-adenosine [15]. This is generated by cyclization of (*E*,*E*,*E*)-geranylgeranyl diphosphate (GGPP) to tuberculosinyl/halima-5,13*E*-dienyl diphosphate catalyzed by Rv3377c [16,17], and subsequent addition of adenosine catalyzed by Rv3378c [18]. A role for this diterpenoid in suppressing acidification of the phagosomal compartments into which *M. tuberculosis* are taken up, which assists infiltration of the engulfing macrophage that then serve as host cells, was indicated by a genetic screen [19]. Indeed, the diterpenoid itself appears to suppress phagosomal acidification [20]. However, the genetic screen only identified Rv3377c and Rv3378c as essential for this biosynthetic process [19], and not the *trans*-IDS GGPP synthase Rv3383c (idsB) found in the same operon [21]. This is presumably due to the presence of an additional *trans*-IDS that produces the 20-carbon GGPP in *M. tuberculosis*, Rv0562 (*grcC1*) [21].

2. Results

As indicated by their shared genetic nomenclature (i.e., *grcC1* and *grcC2*), Rv0562 and Rv0989c are closely related (e.g., particularly among the *M. tuberculosis* IDSs; Figure S1), sharing >56% amino acid sequence identity, despite their divergent product lengths. A common determinant of product chain length in *trans*-IDSs is the identity of the amino acid residue in the fifth position before the FARM [22]. In particular, with smaller residues typically found in GGPP synthases and larger (aromatic) residues in those *trans*-IDS with shorter products. However, Rv0562 and Rv0989 both have small residues at this position (i.e., alanine and glycine, respectively; Figure 2). Thus, there must be an alternative determinant for the difference in product chain length between these two *trans*-IDSs.

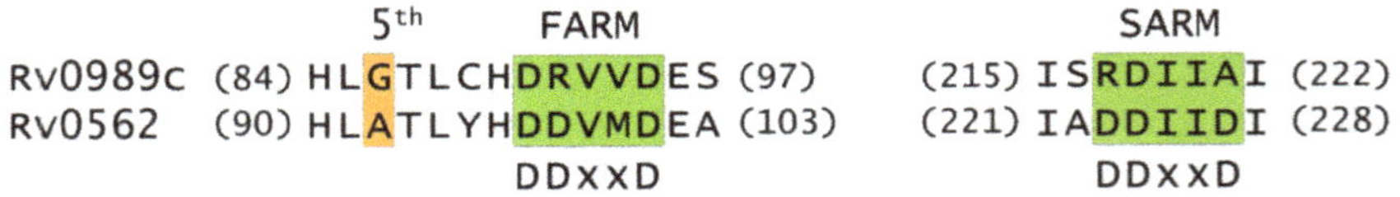

```
                    5th     FARM                          SARM
Rv0989c  (84) HLGTLCHDRVVDES (97)       (215) ISRDIIAI (222)
Rv0562   (90) HLATLYHDDVMDEA (103)      (221) IADDIIDI (228)
                      DDXXD                      DDXXD
```

Figure 2. Partial protein sequence alignment of Rv0989c and Rv0562. Rv0562 contains the canonical *trans*-IDS first and second aspartate-rich DDxxD motifs (FARM and SARM; highlighted in green), but these are both disrupted by Arg substitutions in Rv0989c. The fifth residue upstream of the FARM, which is typically responsible for product length determination in *trans*-IDS, is also highlighted here (orange).

Notably, while Rv0562 contains the canonical DDxxD sequence in both its FARM and SARM, Rv0989c instead contains arginine in place of the second Asp in its FARM and first Asp in its SARM (Figure 2). Although Rv0989c also contains an Ala in place of the last Asp of its SARM, this Asp has been indicated to be less important for *trans*-IDS activity [23]. Indeed, the *M. tuberculosis trans*-IDS that produces (*E*,*E*)-FPP, Rv3398c (idsA1) [24], has a Gly at this position. By contrast, all the other Asp in these motifs have been shown to be important for catalytic activity [4]. Perhaps not surprisingly then, Arg has not been reported in the FARM or SARM of other *trans*-IDS. Accordingly, it was hypothesized that these unique substitutions in Rv0989c might influence product chain length. This was investigated by iteratively swapping the corresponding active site residues via site-directed mutagenesis between Rv0989c and the closely related Rv0562.

The resulting mutants were purified and assayed in vitro with DMAPP and IPP. Products were observed during 30-min assays, indicating that these mutants retain reasonable amounts of catalytic activity. However, to ensure thorough representation of all synthesized products, the product ratios reported here are from overnight assays (Table 1). With such extended incubation, wild-type (WT) Rv0989c produces some (*E*,*E*)-FPP unless IPP concentrations are reduced (Table S1). Thus, the product ratio quantification assays were run with 100 μM DMAPP and 10 μM IPP overnight (Table 1). Under these conditions, WT Rv0989c produces only GPP, and WT Rv0562 still produces only GGPP. By contrast, all of the mutants yielded a variety of products. In general, substitution of Asp for the Arg in Rv0989c led to the production of longer chain product, specifically (*E,E*)-FPP (but not GGPP), in addition to GPP. Conversely, substitution of Arg for the corresponding Asp in Rv0562 led to the appearance of the shorter chain products, both (*E,E*)-FPP and GPP, with only small amounts of GGPP produced. With both of these *M. tuberculosis trans*-IDSs, while the single mutants certainly affect product outcome, the double-mutant led to the greatest change in product outcome i.e., these exhibit additive effects.

Table 1. Product profiles of Rv0562, Rv0989c, and associated mutants. Assays were completed with purified enzyme in the presence of 100 μM DMAPP and 10 μM IPP for 12 hours prior to dephosphorylation and extraction with organic solvent. Organic extracts were concentrated and analyzed via GC-FID. Product identity was confirmed via comparison to dephosphorylated authentic standards prior to integration of peak area. Product profile is represented as percentage of total isoprenoid peak areas.

Enzyme	GPP	FPP	GGPP
Rv0989c	100	0	0
Rv0989c:R92D	76	24	0
Rv0989c:R217D	40	60	0
Rv0989c:R92D/R217D	30	70	0
Rv0562	0	0	100
Rv0562:D98R	66	34	0
Rv0562:D223R	15	51	34
Rv0562:D98R/D223R	9	90	1

These results indicate that the unique Arg observed in the FARM and SARM of Rv0989c play a role in the unusual production of the 10-carbon GPP by this *M. tuberculosis trans*-IDS. This hypothesis is supported not only by the ability of substituting Asp for these Arg to increase Rv0989c product chain length, but also the inverse decrease in product chain length observed upon substituting Arg for the corresponding Asp in the closely related Rv0562 (Table 1). While the second Asp in the FARM does not typically engage the Mg^{2+} co-factors, the first Asp in the SARM usually does interact with one of this trio of divalent metal ions, specifically $Mg^{2+}{}_B$, to which this Asp provides the major enzymatic contact [11]. Thus, it seems likely that this $Mg^{2+}{}_B$ is not present in the reactions catalyzed by Rv0989c. However, the Arg found in the SARM potentially replaces this specific co-factor by directly interacting with the diphosphate moiety of the allylic substrate (e.g., much like the binding of the diphosphate moiety of IPP shown in Figure 1). In addition, the Arg found in the FARM presumably shifts this positioning of this motif, bound Mg^{2+} and, consequently, also the allylic substrate. Accordingly, it seems likely that these Arg shift the position of the allylic substrate in a manner that decreases the available space for the appended isoprenyl chain and, hence, limit the size of the final product. Regardless, although determination of the exact mechanism will require more detailed structural analysis, the ability of Arg in the FARM and SARM to affect product chain length, at least in these *M. tuberculosis trans*-IDSs, represents a novel means by which such a final product outcome can be controlled.

3. Materials and Methods

3.1. General Reagents

Unless otherwise indicated, reagents were obtained from Sigma-Aldrich (St. Louis, MO, USA). Isoprenoid substrates and standards were purchased from Isoprenoids.com (Tampa, FL, USA). Primers were synthesized by and purchased from Integrated DNA Technologies (Coralville, IA, USA).

3.2. Sequence Analysis and Alignment

Protein sequences encoded by *M. tuberculosis* H37Rv loci Rv0562, Rv0989c, Rv1086, Rv2173, Rv2361c, Rv3383c, and Rv3398c were aligned using the MUSCLE algorithm, with subsequent phylogenetic analysis using the maximum likelihood method with the JJT frequencies model with inclusion of a gamma distribution, use of all sites, and 1000 replicates for the bootstrap test, in the MEGA7 software package [25].

3.3. Site-Directed Mutagenesis

Rv0989c and Rv0562 have previously been cloned, purified, and characterized [14,21]. Site-directed mutagenesis was performed using overlapping mutagenic primers (Figure S2) on pENTR/SD/D-TOPO constructs of each gene. Final constructs were confirmed via complete sequencing prior to transfer to pDEST17 vector for protein expression and purification (Invitrogen, Carlsbad, CA, USA).

3.4. Protein Expression and Assay

Constructs were transformed into *Escherichia coli* BL21-Star (Invitrogen, Carlsbad, CA, USA) or C41 OverExpress cells (Lucigen, Middleton, WI, USA). Starter cultures were inoculated into 10 mL NZY media (10 g/L NaCl, 10 g/L casein, 5 g/L yeast extract, 1 g/L MgSO4 (anhydrous), pH 7.0) with 50 µg/mL carbenicillin and incubated at 200 rpm and 18 °C for 3 days. Starter cultures (5 mL) were used to inoculate 1000 mL fresh NZY media, also with 50 µg/mL carbenicillin, and after reaching an OD_{600} of 0.6 they were induced with 1 Mm Isopropyl β-D-1-thiogalactopyranoside (IPTG) and incubated under continuous shaking at 200 rpm at 18 °C for 12 hours. Cells were harvested by centrifugation at 5000× *g* for 10 min. The cell pellet was re-suspended in 5 mL 25 mM 3-(N-morpholino)-2-hydroxypropanesulfonic acid buffer (MOPSO), pH 7.2, 10 mM $MgCl_2$, 10% (*v*/*v*) glycerol with 10 mM imidazole, and homogenized using an EmulsiFlex C-5 (Avestin, Canada). The homogenized suspension was centrifuged at 16,000× *g* for 60 min. The supernatant was passed over 1 mL Ni-NTA agarose (Qiagen, Hilden, Germany) washed with 5 mL buffer containing 10 mM imidazole and an additional 5 mL with 50 mM imidazole. Proteins were eluted with 2 mL buffer containing 250 mM imidazole.

Enzyme assays were carried out with 300 µg purified protein in 2 mL 25 mM MOPSO, pH 7.2, 10 mM MgCl2, 10% (*v*/*v*) glycerol with the addition of 100 µM DMAPP and 10 or 100 µM IPP, the assays were incubated for either 30 minutes or 24 hours at 30 °C before addition of 200 units of calf intestinal alkaline phosphatase (Promega, Madison, WI) and the supplied buffer. The assay was dephosphorylated for 24 hours at 30 °C and extracted three times with 2 mL pentane. The organic phase was combined and concentrated under a gentle stream of N_2 until a volume of 200 µL solvent remained. The concentrated organic extracts (1 µL) were injected into 3900 Saturn GC (Varian, Palo Alto, CA) with an injector temperature of 250 °C coupled to a Saturn 2100T ion trap mass spectrometer detector for product identification. Separation was achieved using a HP-5MS column (30 m × 250 µm × 0.25 µm) (Agilent, Santa Clara, CA, USA). The chromatographic program was as follows: 50 °C for 3 min, increasing to 300 °C by 15 °C/min and held at 300 °C 3 min. Electrospray ionization scanning from 60–650 m/z was used to obtain product spectra (Figure S3). Quantification of products was completed using a Shimadzu GC02014 with flame ionization detection (GC-FID) over an SH-Rxi-5ms column (15 m × 250 µm × 0.25 µm; Shimadzu, Kyoto, Japan) using the previously described method. Products were confirmed by comparison, of both retention time and mass spectra (Figure S3), to isoprenoid diphosphate authentic standards dephosphorylated as above.

Supplementary Materials: Supplementary materials are available on line.

Author Contributions: F.M.M. and R.J.P. conceived the research and wrote the manuscript; R.N., F.M.M., J.A.T. and F.A.A. performed research, analyzed data and revised the manuscript.

Funding: This work was supported by a grant from the NIH (GM076324) to R.J.P. and a postdoctoral fellowship to R.N. from the Deutsche Forschungsgemeinschaft (DFG) NA 1261/1-2.

Acknowledgments: The authors thank Meimei Xu for assistance with constructing mutants and expression vectors.

Conflicts of Interest: The authors declare no conflict of interest.

References

1. Gershenzon, J.; Dudareva, N. The function of terpene natural products in the natural world. *Nat. Chem. Biol.* **2007**, *3*, 408–414. [CrossRef] [PubMed]
2. Pichersky, E.; Raguso, R.A. Why do plants produce so many terpenoid compounds? *New Phytol.* **2016**. [CrossRef] [PubMed]
3. Buckingham, J. *Dictionary of Natural Products (on-line web edition)*; Chapman & Hall/CRC Press: London, UK, 2007.
4. Liang, P.-H.; Ko, T.-P.; Wang, A.H.-J. Structure, mechanism and function of prenyltransferases. *Eur. J. Biochem.* **2002**, *269*, 3339–3354. [CrossRef] [PubMed]
5. Chen, A.P.; Chang, S.Y.; Lin, Y.C.; Sun, Y.S.; Chen, C.T.; Wang, A.H.; Liang, P.H. Substrate and product specificities of cis-type undecaprenyl pyrophosphate synthase. *Biochem. J.* **2005**, *386*, 169–176. [CrossRef] [PubMed]
6. Kaur, D.; Brennan, P.J.; Crick, D.C. Decaprenyl diphosphate synthesis in Mycobacterium tuberculosis. *J. Bacteriol.* **2004**, *186*, 7564–7570. [CrossRef] [PubMed]
7. Swiezewska, E.; Danikiewicz, W. Polyisoprenoids: Structure, biosynthesis and function. *Prog. Lipid Res.* **2005**, *44*, 235–258. [CrossRef] [PubMed]
8. Akhtar, T.A.; Matsuba, Y.; Schauvinhold, I.; Yu, G.; Lees, H.A.; Klein, S.E.; Pichersky, E. The tomato cisprenyltransferase gene family. *Plant J.* **2013**, *73*, 640–652. [CrossRef] [PubMed]
9. Poulter, C.D.; Rilling, H.C. Prenyltransferase: the mechanism of the reaction. *Biochemistry* **1976**, *15*, 1079–1083. [CrossRef] [PubMed]
10. Poulter, C.D.; Satterwhite, D.M.; Rilling, H.C. Prenyltransferase—Mechanism Of Reaction. *J. Am. Chem. Soc.* **1976**, *98*, 3376–3377. [CrossRef] [PubMed]
11. Aaron, J.A.; Christianson, D.W. Trinuclear Metal Clusters in Catalysis by Terpenoid Synthases. *Pure Appl. Chem.* **2010**, *82*, 1585–1597. [CrossRef] [PubMed]
12. Schulbach, M.C.; Brennan, P.J.; Crick, D.C. Identification of a Short (C15) ChainZ-Isoprenyl Diphosphate Synthase and a Homologous Long (C50) Chain Isoprenyl Diphosphate Synthase inMycobacterium tuberculosis. *J. Biol. Chem.* **2000**, *275*, 22876–22881. [CrossRef] [PubMed]
13. Schulbach, M.C.; Mahapatra, S.; Macchia, M.; Barontini, S.; Papi, C.; Minutolo, F.; Bertini, S.; Brennan, P.J.; Crick, D.C. Purification, Enzymatic Characterization, and Inhibition of theZ-Farnesyl Diphosphate Synthase from Mycobacterium tuberculosis. *J. Biol. Chem.* **2001**, *276*, 11624–11630. [CrossRef] [PubMed]
14. Mann, F.M.; Thomas, J.A.; Peters, R.J. Rv0989c encodes a novel (E)-geranyl diphosphate synthase facilitating decaprenyl diphosphate biosynthesis in Mycobacterium tuberculosis. *FEBS Lett.* **2011**, *585*, 549–554. [CrossRef] [PubMed]
15. Young, D.; Moody, D. In vivo biosynthesis of terpene nucleosides provides unique chemical markers of Mycobacterium tuberculosis infection. *Chem. Bio.* **2015**, *22*, 516–526. [CrossRef] [PubMed]
16. Mann, F.M.; Prisic, S.; Hu, H.; Xu, M.; Coates, R.M.; Peters, R.J. Characterization and inhibition of a class II diterpene cyclase from Mycobacterium tuberculosis: implications for tuberculosis. *J. Biol. Chem.* **2009**, *284*, 23574–23579. [CrossRef] [PubMed]
17. Nakano, C.; Hoshino, T. Characterization of the Rv3377c gene product, a type-B diterpene cyclase, from the Mycobacterium tuberculosis H37 genome. *Chembiochem* **2009**, *10*, 2060–2071. [CrossRef] [PubMed]
18. Layre, E.; Lee, H.J.; Young, D.C.; Martinot, A.J.; Buter, J.; Minnaard, A.J.; Annand, J.W.; Fortune, S.M.; Snider, B.B. Molecular profiling of Mycobacterium tuberculosis identifies tuberculosinyl nucleoside products of the virulence-associated enzyme Rv3378c. *Proc. National Acad. Sci.* **2014**, *111*, 2978–2983. [CrossRef] [PubMed]

19. Pethe, K.; Swenson, D.L.; Alonso, S.; Anderson, J.; Wang, C.; Russell, D.G. Isolation of Mycobacterium tuberculosis mutants defective in the arrest of phagosome maturation. *Proc. National Acad. Sci.* **2004**, *101*, 13642–13647. [CrossRef] [PubMed]
20. Mann, F.M.; Xu, M.; Chen, X.; Fulton, D.B.; Russell, D.G.; Peters, R.J. Edaxadiene: A new bioactive diterpene from Mycobacterium tuberculosis. *J. Am. Chem. Soc.* **2009**, *131*, 15726–15727. [CrossRef] [PubMed]
21. Mann, F.M.; Xu, M.; Davenport, E.K.; Peters, R.J. Functional characterization and evolution of the isotuberculosinol operon in Mycobacterium tuberculosis and related Mycobacteria. *Front Microbiol.* **2012**, *3*, 368. [CrossRef] [PubMed]
22. Ohnuma, S.; Narita, K.; Nakazawa, T.; Ishida, C.; Takeuchi, C.; Ohto, C.; Nishino, T. A role of the amino acid residue located on the fifth position before the first aspartate-rich motif of farnesyl diphosphate synthase on determination of the final product. *J. Biol. Chem.* **1996**, *271*, 30748–30754. [CrossRef] [PubMed]
23. Joly, A.; Edwards, P.A. Effect of site-directed mutagenesis of conserved aspartate and arginine residues upon farnesyl diphosphate synthase activity. *J. Biol. Chem.* **1993**, *268*, 26983–26989. [PubMed]
24. Dhiman, R.K.; Schulbach, M.C.; Mahapatra, S.; Baulard, A.R.; Vissa, V.; Brennan, P.J.; Crick, D.C. Identification of a novel class of omega,E,E-farnesyl diphosphate synthase from Mycobacterium tuberculosis. *J. Lipid Res.* **2004**, *45*, 1140–1147. [CrossRef] [PubMed]
25. Kumar, S.; Stecher, G.; Tamura, K. MEGA7: Molecular Evolutionary Genetics Analysis Version 7.0 for Bigger Datasets. *Mol. Biol. Evol.* **2016**, *33*, 1870–1874. [CrossRef] [PubMed]

Sample Availability: Samples of the compounds (plasmids) are available from the authors, while the isoprenoids are commercially available.

MDPI

Article

Ursolic Acid Attenuates Atherosclerosis in ApoE$^{-/-}$ Mice: Role of LOX-1 Mediated by ROS/NF-κB Pathway

Qiu Li [1], Wenwen Zhao [2], Xi Zeng [2] and Zhihui Hao [3,*]

1 State Key Laboratory of Quality Research in Chinese Medicine, Institute of Chinese Medical Sciences, University of Macau, Macau SAR 999078, China; liqiu370725@126.com
2 Department of Pharmacology, College of basic Medicine, Qingdao University, 308 Ningxia Road, Qingdao 266000, China; wenwenzhao0313@163.com (W.Z.); hf9079@163.com (X.Z.)
3 Agricultural Bio-Pharmaceutical Laboratory, Qingdao Agricultural University, Qingdao 266000, China
* Correspondence: abplab@126.com; Tel.: +86-532-8803-0364

Received: 28 March 2018; Accepted: 28 April 2018; Published: 7 May 2018

Abstract: Atherosclerosis, a chronic inflammatory disease, is a major contributor to cardiovascular diseases. Ursolic acid (UA) is a phytonutrient with widely biological effects including anti-oxidative, anti-inflammatory, and so on. At present, the effect of UA on atherosclerosis and the mechanism of action are still obscure. This study focused on investigating the effects of UA on atherosclerosis both in vivo and in vitro. We first selected LOX-1 as our target, which was reckoned as a new promising receptor for treating atherosclerosis. The evaluation in vitro suggested that UA significantly decreased endothelial LOX-1 expression induced by LPS both in mRNA and protein levels. Pre-treatment of UA also inhibited TLR4/MyD88 signaling activated by LPS. Moreover, UA reduced ROS production and suppressed the activation of NF-κB stimulated by LPS. Particularly, the evaluation in vivo further verified the conclusion obtained in vitro. In ApoE$^{-/-}$ mice fed with an atherogenic diet, both UA (100 mg/kg/day) and simvastatin significantly attenuated atherosclerotic plaque formation and shrunk necrotic core areas. The enhanced expression of LOX-1 in atherosclerotic aorta was also dramatically decreased by administration of UA. Taken together, these results suggested that UA, with anti-atherosclerotic activity through inhibition of LOX-1 mediated by ROS/NF-κB signaling pathways, may become a valuable vascular protective candidate for the treatment of atherosclerosis.

Keywords: atherosclerosis; ROS; HUVECs; LOX-1

---

1. Introduction

Atherosclerosis, a chronic inflammatory disease of the arterial wall, is a major cause of morbidity and mortality worldwide and is marked by the formation of atherosclerotic plaques [1]. One critical event in the initiation of atherosclerotic plaques is the uptake of ox-LDL [2]. LDL uptake is principally mediated by a variety of specific receptors including SR-A I/II, CD36, SR-BI, etc. [3]. Recently, LOX-1 has been identified as the main endothelial receptor for ox-LDL [4,5]. LOX-1 is a 50 KD transmembrane protein highly expressed on macrophages, vascular smooth muscle cells, and, especially, endothelial cells. It plays vital roles in the pathogenesis of atherosclerosis [6]. Disturbance of LOX-1-mediated signaling pathways has been proposed as a potential strategy for anti-atherosclerotic drug discovery.

TLR4-MyD88 pathway leads to activation of the transcription factor NF-κB, thereby influencing inflammatory responses. Oxidative stress is characterized by an increased production of free oxygen radicals [7]. Numerous clinical studies have demonstrated that oxidant stress is closely related to different risk factors of atherosclerosis such as hypercholesterolemia [8], hypertension [9], diabetes [10], and smoking [11]. Oxidative stress can promote conversion of LDL to atherogenic ox-LDL, contributing to the atherosclerotic plaque formation [12]. Besides, oxidant stress is closely correlated with endothelial dysfunction [13] and promotes vascular inflammatory response [14]. Furthermore, TLRs-NF-κB pathway has been reported to participate in the anti-atherosclerotic effect of several natural products such as quercetin, pycnogenol, and procyanidins [15–17].

Ursolic acid (UA) is a pentacyclic triterpenoid found in many herbs and spices like rosemary and thyme, especially in valuable Chinese medicinal herbs such as *Fructus Ligustrum lucidum* [18] and *Forsythiae fructus* [19]. It shows potentially beneficial activities in treating cardiovascular disease due to its anti-oxidative [14], anti-inflammatory effects [20], and other biological activities [21,22]. However, it also demonstrates potential adverse effects. Messner et al. [1] reported the pro-atherogenic effects of UA. There is a controversy around the effect of UA on atherosclerosis. This study is to investigate the effect of UA on cells in vitro; high-fat, diet-induced ApoE$^{-/-}$ mice in vivo; and related mechanisms such as LOX-1 expression in endothelial cells. Our findings may supply a new sight for illustrating role of UA, which benefits the development of the anti-athrogenic drug.

2. Results

2.1. UA Decreased LPS-Induced LOX-1 Expression in HUVECs

Firstly, the cytotoxic effect of UA on HUVECs was determined by MTT assay. Results showed that UA was cytotoxic to HUVECs at 50 μM (Figure S1). To minimize the cytotoxic effect of UA on cell viability, 1 μM UA was chosen for further study. Compared with untreated HUVECs, 24 h stimulation with LPS increased LOX-1 expression at both mRNA and protein levels, which were dramatically inhibited by UA pretreatment (Figure 1B,C). Furthermore, immunofluorescence results showed that UA blocked LPS-induced LOX-1 expression localizing on the cell membrane (Figure 1D).

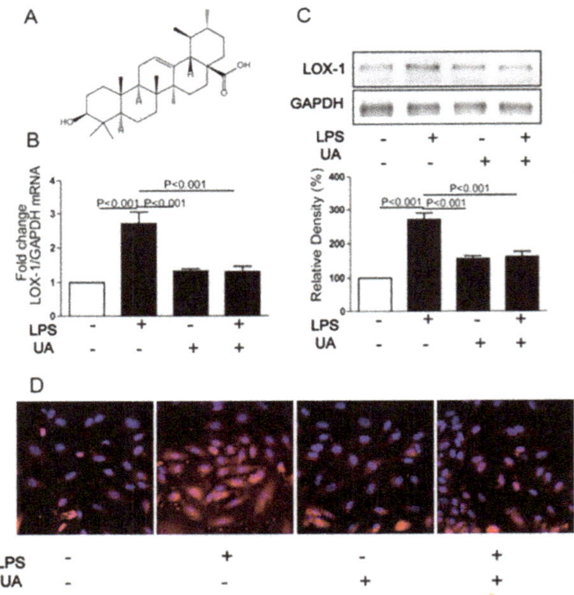

Figure 1. The structure of UA (**A**); Cells were pretreated with UA (1 μM) for 1 h and then stimulated with LPS (5 mg/mL) for 24 h; LOX-1 mRNA (**B**), protein (**C**), and localization on the membranes (**D**) were detected by real-time PCR, western blotting, and immunofluorescence (60×), respectively. UA, ursolic acid.

2.2. UA Inhibited LPS-Induced LOX-1 Expression via TLR4/MyD88 Pathway

TLR4/MyD88 signal pathway is involved in LPS-induced inflammation [23]. As expected, UA reversed LPS-induced TLR4 and MyD88 protein expressions (Figure 2A,B). Furthermore, silence of either TLR4 or MyD88 significantly decreased LOX-1 expression (Figure 2C,E).

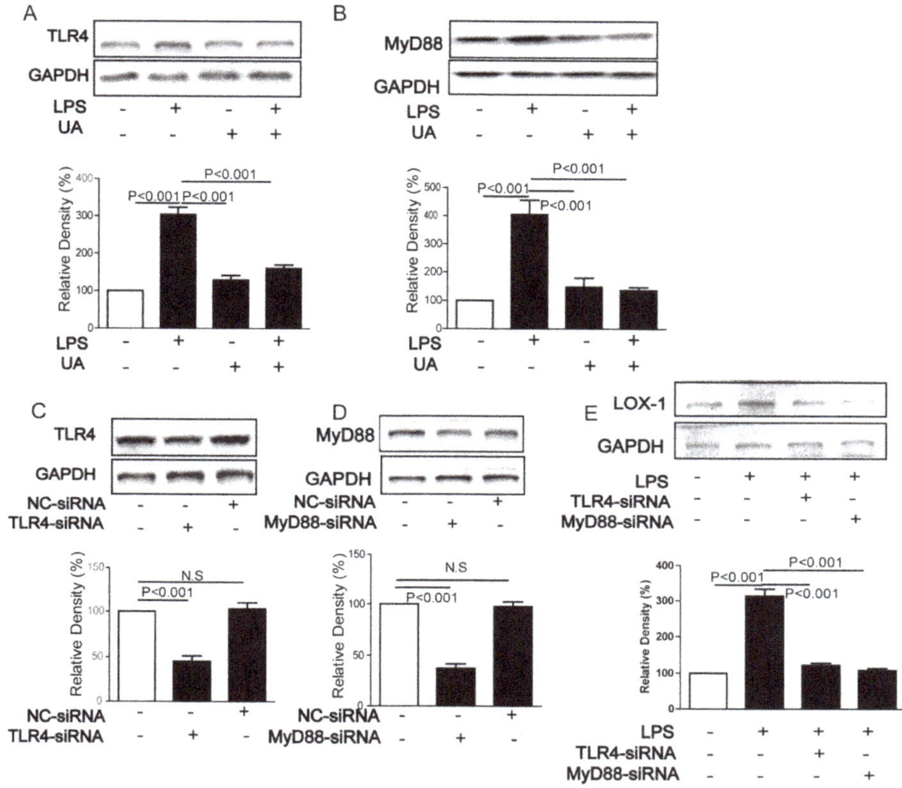

Figure 2. Cells were pretreated with UA (1 μM) for 1 h and then stimulated by LPS (5 mg/mL) for 24 h; expressions of TLR4 and MyD88 were determined by western blotting (**A**,**B**). Cells were transfected with siRNAs for TLR4 (**C**) and MyD88 (**D**) and then stimulated with LPS (5 mg/mL) for 24 h and the LOX-1 expression were determined by western blotting (**E**). Cont, control; NC-siRNA, negative control siRNA. UA, ursolic acid; N.S, no significant differences.

2.3. UA Reduced ROS Generation and Decreased NF-κB Activity to Block LOX-1 Expression

ROS is one key signaling molecule involved in inflammation, and NF-κB pathway plays an essential role in LPS-induced LOX-1 expression in HUVECs [23]. In this study, LPS induced ROS production and NF-κB activity, while UA pretreatment obviously reduced ROS generation and blocked translocation of p65 NF-κB into nucleus (Figure 3A–C). Furthermore, all NAC (ROS scavenger) and BAY (NF-κB inhibitor) inhibited LOX-1 expression (Figure 3D), hinting that UA blocked ROS/ NF-κB pathway to regulate LOX-1 expression.

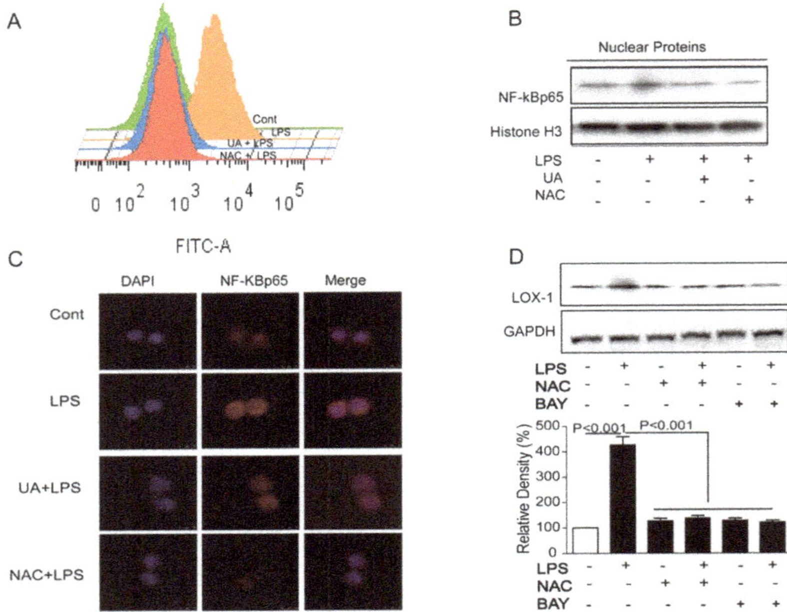

Figure 3. Cells were treated with LPS (5 mg/mL) for 4 h after pretreatment with UA (1 μM) or NAC (5 mM) for 1 h; ROS was determined by DCFH$_2$-DA (**A**); Cells were treated with LPS (5 mg/mL) for 24 h with or without pretreatment with UA (1 μM), NAC (5 mM), or BAY (10 μM) for 1 h, and expression of p65 (**B**) and LOX-1 (**D**) was detected by western blotting; Cells were treated with LPS (5 mg/mL) with or without 1 h pretreatment with UA (1 μM) and NAC (5 mM), and p65 NF-κB localization was determined by immunofluorescence (**C**) (60×). UA, ursolic acid; BAY, BAY11-7082; NAC, N-acetyl cysteine.

2.4. UA Reduced Atherosclerotic Plaque Development in ApoE$^{-/-}$ Mice

Plaques formation and necrotic core areas occurrence are two key characteristics of atherosclerosis. In this study, both plaques and necrotic core areas were observed in artery of ApoE$^{-/-}$ mice. UA administration significantly reduced plaque sizes as well as shrank necrotic core areas (Figure 4A,B).

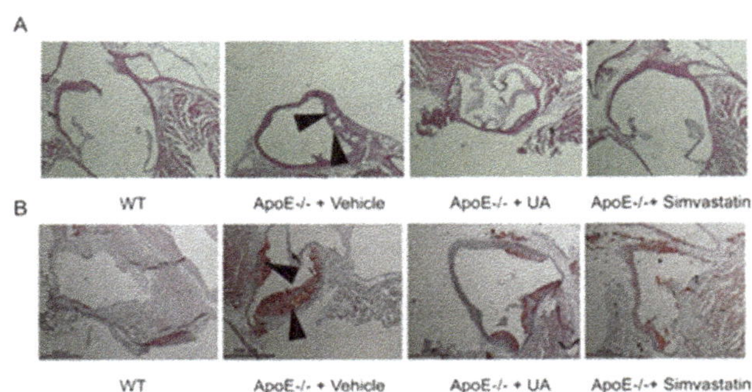

Figure 4. Aortic sinus sections were stained with Oil Red O and H&E to detect plaque sizes (40×) (**B**) and necrotic core areas (40×) (**A**), respectively. UA, ursolic acid; WT, wild type; ApoE$^{-/-}$, apolipoprotein E-deficient.

2.5. UA Inhibited LOX-1 Expression in Thoracic Aorta of ApoE$^{-/-}$ Mice

LOX-1 was expressed at low level in aorta of WT mice but significantly increased in ApoE$^{-/-}$ mice, which was dramatically decreased in UA pretreated mice (Figure 5).

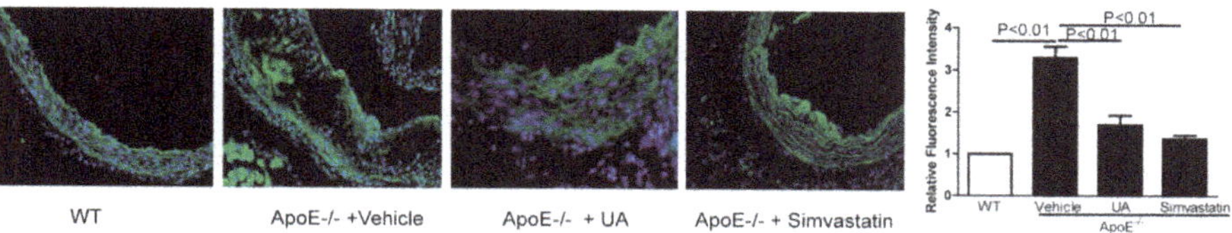

Figure 5. Expression of LOX-1 in the aorta was detected by immunofluorescence (4×). WT, wild type; ApoE$^{-/-}$, apolipoprotein E-deficient; UA, ursolic acid.

3. Discussion

Due to the high life-threatening risk and severe financial burden of surgical treatment, pharmacological options for atherosclerosis therapy have always been discussed. UA, a natural pentacyclic triterpenoid carboxylic acid, is the major component of some traditional medicine herbs, although several studies have shown that UA has anti-inflammatory, anti-oxidative, and anti-diabetes functions [24,25]. There are conflicting studies around its anti-atherosclerosis effect. LOX-1 is the primary receptor for ox-LDL uptake, which promotes key steps involved in atherosclerosis [26–29]. In this study, LPS was chosen to induce endothelial LOX-1expression. UA decreased LOX-1 expression both in mRNA and protein levels, hinting that UA has potential anti-atherosclerosis activity.

Several lines of evidence support a role of oxidative stress in the pathogenesis of many diseases including atherosclerosis, tumor, Parkinson's disease, and so on [30,31]. UA has been proved to own anti-oxidative effects. Experiment data showed that UA significantly reduced ROS generation, and ROS scavenger obviously decreased LOX-1 expression.

NF-κB plays a critical role in regulating inflammation and performs an important function in LPS-induced LOX-1 expression [23]. UA has been proved to own anti-inflammatory effects [32]. In this study, UA blocked p65NF-κB translocated into nucleus. Furthermore, NF-κB inhibitor decreased endothelial LOX-1 expression, suggesting the importance of NF-κB for regulating LOX-1.

The present findings in vitro allowed us to postulate that UA exhibited anti-atherosclerosis action in vivo. To verify this hypothesis, we detected the effects of UA on high-fat-diet-fed ApoE$^{-/-}$ mice. The experiments showed that oral administration of UA for 12 weeks dramatically reduced plaque sizes and the shrinkage of necrotic core areas occurred in model ApoE$^{-/-}$ mice providing the direct evidence for its anti-atherosclerotic effect. Besides, consistent with the in vitro results, expression of LOX-1 in the aorta was also decreased by UA, suggesting that LOX-1 is a promising therapeutic target in atherosclerosis for UA in future.

4. Materials and Methods

4.1. Materials and Reagents

UA (purity > 98%) was purchased from Chengdu Herbpurify Co. Ltd. (Chendu, China). Hoechst 33342, *N*-acetyl cysteine (NAC), LPS (Escherichia coli serotype 055:B5, LPS), 5-(6)-carboxy-2′, 7′-dichlorodihydrofluorescein diacetate (DCFH$_2$-DA), and BAY11-7082 were purchased from Sigma Aldrich (St. Louis, MO, USA). Antibodies for TLR4, MyD88 were purchased from Santa Cruz Biotechnology (Santa Cruz, CA, USA). Antibodies for NF-κB p65, GAPDH, and Histone H3 were purchased from Cell Signaling Technology (Beverly, MA, USA). Anti-LOX-1 antibody was obtained from R&D (Minneapolis, MN, USA). The BCA protein kits were purchased from Thermo Fisher (Suwanee, GA, USA). SiRNAs for TLR4, MyD88 were purchased from Gene Pharma Company

(Shanghai, China). Oil Red O Staining Kit and HE Staining Kit were obtained from Nanjing Jiancheng Bioengineering Research Institute (Nanjing, China).

4.2. Cell Culture

Human umbilical vein endothelial cells (HUVECs) (Gibco, Life Technologies Corp., Carlsbad, CA, USA) were cultured in Vascular Cell Basal Medium with Endothelial Cell Growth Kit-BBE at 37 °C in a humidified atmosphere of 5% CO_2. Before passaging cells, issue culture flasks, 96-well plates, and 6-well plates were pre-coated with 0.1% gelatin. All assays were conducted using low cell passage cells (2–5 passages).

4.3. Animal Experiment

Male ApoE$^{-/-}$ mice (6–8 weeks old) on C57BL/6J background and age-matched wild-type C57BL/6J controls were purchased from Beijing HFK Bioscience Co., Ltd. (Beijing, China). Mice were housed in SPF-grade animal facilities with a 12 h light/dark cycle at 23 °C (±2 °C). All animal procedures follow the NIH guide for the Care and Use of Laboratory Animals (NIH Publications No. 80-23, revised 1978). Starting from 6 weeks, the mice were fed with a HCD (54.35% raw grain, 20% lard, 0.15% cholesterol, 15% sucrose, 0.5% Sodium Cholate, and 10% yolk powder) for 12 weeks. All ApoE$^{-/-}$ mice were dosed daily via intragastric gavage with 100 mg·kg^{-1}·day^{-1} UA and 25 mg·kg^{-1}·day^{-1} simvastatin dissolved in 0.5% CMC-Na or administered 0.5% CMC-Na alone (vehicle control) (n = 8 per group).

4.4. Immunofluorescence Assay

Cells (1×10^4 cells/well) were seeded on glass slides in 96-well plates. After LPS treatment (with or without UA pretreatment), the slides were fixed with 4% PFA for 30 min. Then, the slides were permeabilized with PBS-T (containing 0.1% Triton x-100 in PBS solution) and blocked with PBS-B (containing 4% BSA in PBS solution). After being incubated with primary antibody (1:1000) and secondary antibody (1:5000), cells were stained with Hoechst 33342 in dark for 5 min. The protein location and expression were observed with IN Cell Analyzer 2000 (GE Healthcare).

4.5. Western Blotting

Treated HUVECs were washed twice with ice-cold PBS and lysed with RIPA buffer supplemented with a protease cocktail and phosphatase inhibitors. The cell lysates were separated using 8–10% SDS-PAGE and transferred onto PVDF membranes. After being blocked with 5% non-fat milk in TBST (20 mM Tris-HCl, 500 mM NaCl, and 0.1% Tween 20) at room temperature for 2 h, membranes were incubated with specific primary antibodies and secondary antibodies. The protein-antibody complexes were detected by ECL Advanced Western Blot Detection Kit. The intensity of the band was quantitated with Quantity One software (Bio-Rad).

4.6. SiRNA Transfection

Gene silencing experiment of TLR4, MyD88 with siRNA was performed according to our previous report [23]. Briefly, cells (1.0×10^6/well) were seeded in 6-well plate overnight. 100 pM siRNA was diluted in 100 μL Opti-MEM reduced serum medium in each well and then mixed gently. Similarly, 5 μL of Lipofectamine TM 2000 was diluted in 100 μL of Opti-MEM reduced serum medium and mixed gently. After incubation for 5 min at room temperature, the diluted siRNA and diluted lipofectamine (total volume 200 μL) were mixed gently and incubated for another 20 min at room temperature. Further, siRNA-lipofectamine complex (200 μL) was added to each well. After incubation for 4 h, the complexes were discarded and cells were cultured with completed medium.

4.7. Real-Time RT-PCR

The mRNA expression of LOX-1 was determined with real-time PCR as our previous report [23]. Briefly, total RNA was extracted using TRIzol reagent (Life Technologies, USA). cDNA was synthesized by reverse transcription (RT) using GoScript™ Reverse Transcription System (Promega, Madison, WY, USA). Quantitative real-time PCR was carried out on Stratagene Mx3005P (Agilent Technologies, Santa Clara, CA, USA) using GoTaq® qPCR Master Mix (Promega, USA). The primer sequences used in this experiment are listed for LOX-1: 5′-TGGGAAAAGAGCCAAGAGAA-3′ (forward), 5′- TAAGTGGGGCATCAAAGGA′ (reverse); anf for GAPDH: 5′-AGAAGGCTGGGGCTCATTTG-3′ (forward), 5′-AGGGGCCATCCACAGTCTTC-3′ (reverse). The levels of LOX-1 expression were determined by normalizing to GAPDH expression.

4.8. Aorta Collection and Lesion Size Evaluation

To evaluate plaque extension, frozen sections of the aortic sinus (8 mm) were stained using Oil Red O, hematoxylin, and eosin (H&E), respectively. Related experiments were performed following the method of Paigen et al. [33].

4.9. Statistical Analysis

Twenty four ApoE$^{-/-}$ mice were randomly allocated to groups and equal group sizes were obtained (n = 8 per group). Data were expressed as the means ± SD. The differences between groups were analyzed using Prism 5.0 (Graph Pad Software Inc., San Diego, CA, USA), and the statistical analysis was performed by analysis of variance (one-way ANOVA) followed by Student Newman–Keuls test.

5. Conclusions

In summary, this study showed that natural product-UA inhibited NF-κB-mediated LOX-1 expression both in vivo and in vitro through ROS generation. In view of the key roles of LOX-1 in the pathogenesis of atherosclerosis, inhibition of LOX-1 contributed to the anti-atherosclerotic effect of UA. Our study demonstrates that UA showed the effect of ameliorating atherosclerosis and put forward the potential mechanism. It supplies a new sight to illustrate the mechanism of action of UA for attenuates atherosclerosis in mice.

Supplementary Materials: The Supplementary Materials are available online, Figure S1: the cytotoxic effect of UA on HUVECs.

Author Contributions: Z.H. and Q.L. designed, performed the experiments, and wrote the paper; W.W. and X.Z. contributed to data analysis and paper writing.

Acknowledgments: This work was financially supported by National Natural Science Foundation of China (31402256); Special Fund for Agro-Scientific Research in the Public Interest, China (201303038-8); and High-Level Talent Research Foundation of Qingdao Agricultural University, China (631206).

Conflicts of Interest: The authors declare no conflict of interest.

References

1. Gaziano, T.A.; Bitton, A.; Anand, S.; Abrahams-Gessel, S.; Murphy, A. Growing epidemic of coronary heart disease in low- and middle-income countries. *Curr. Probl. Cardiol.* **2010**, *35*, 72–115. [CrossRef] [PubMed]
2. Di Pietro, N.; Formoso, G.; Pandolfi, A. Physiology and pathophysiology of oxLDL uptake by vascular wall cells in atherosclerosis. *Vasc. Pharmacol.* **2016**, *84*, 1–7. [CrossRef] [PubMed]
3. Li, A.C.; Glass, C.K. The macrophage foam cell as a target for therapeutic intervention. *Nat. Med.* **2002**, *8*, 1235–1242. [CrossRef] [PubMed]
4. Chen, M.; Masaki, T.; Sawamura, T. LOX-1, the receptor for oxidized low-density lipoprotein identified from endothelial cells: Implications in endothelial dysfunction and atherosclerosis. *Pharmacol. Ther.* **2002**, *95*, 89–100. [CrossRef]

5. Lu, J.; Mehta, J.L. LOX-1: A critical player in the genesis and progression of myocardial ischemia. *Cardiovasc. Drugs Ther.* **2011**, *25*, 431–440. [CrossRef] [PubMed]
6. Xu, S.; Ogura, S.; Chen, J.; Little, P.J.; Moss, J.; Liu, P. LOX-1 in atherosclerosis: Biological functions and pharmacological modifiers. *Cell. Mol. Life Sci.* **2013**, *70*, 2859–2872. [CrossRef] [PubMed]
7. Lushchak, V.I. Free radicals, reactive oxygen species, oxidative stress and its classification. *Chem. Biol. Interact.* **2014**, *224*, 164–175. [CrossRef] [PubMed]
8. Tie, G.; Messina, K.E.; Yan, J.; Messina, J.A.; Messina, L.M. Hypercholesterolemia induces oxidant stress that accelerates the ageing of hematopoietic stem cells. *J. Am. Heart Assoc.* **2014**, *3*, e000241. [CrossRef] [PubMed]
9. Baradaran, A.; Nasri, H.; Rafieian-Kopaei, M. Oxidative stress and hypertension: Possibility of hypertension therapy with antioxidants. *J. Res. Med. Sci. Off. J. Isfahan Univ. Med. Sci.* **2014**, *19*, 358–367.
10. Tangvarasittichai, S. Oxidative stress, insulin resistance, dyslipidemia and type 2 diabetes mellitus. *World J. Diabetes* **2015**, *6*, 456–480. [CrossRef] [PubMed]
11. Guzik, B.; Sagan, A.; Ludew, D.; Mrowiecki, W.; Chwala, M.; Bujak-Gizycka, B.; Filip, G.; Grudzien, G.; Kapelak, B.; Zmudka, K.; et al. Mechanisms of oxidative stress in human aortic aneurysms–association with clinical risk factors for atherosclerosis and disease severity. *Int. J. Cardiol.* **2013**, *168*, 2389–2396. [CrossRef] [PubMed]
12. Gokce, G.; Ozsarlak-Sozer, G.; Oran, I.; Oktay, G.; Ozkal, S.; Kerry, Z. Taurine suppresses oxidative stress-potentiated expression of lectin-like oxidized low-density lipoprotein receptor and restenosis in balloon-injured rabbit iliac artery. *Clin. Exp. Pharmacol. Physiol.* **2011**, *38*, 811–818. [CrossRef] [PubMed]
13. Odegaard, A.O.; Jacobs, D.R.; Sanchez, O.A.; Goff, D.C.; Reiner, A.P.; Gross, M.D. Oxidative stress, inflammation, endothelial dysfunction and incidence of type 2 diabetes. *Cardiovasc. Diabetol.* **2016**, *15*, 51. [CrossRef] [PubMed]
14. Siti, H.N.; Kamisah, Y.; Kamsiah, J. The role of oxidative stress, antioxidants and vascular inflammation in cardiovascular disease (a review). *Vasc. Pharmacol.* **2015**, *71*, 40–56. [CrossRef] [PubMed]
15. Martinez-Micaelo, N.; Gonzalez-Abuin, N.; Terra, X.; Richart, C.; Ardevol, A.; Pinent, M.; Blay, M. Omega-3 docosahexaenoic acid and procyanidins inhibit cyclo-oxygenase activity and attenuate NF-kappaB activation through a p105/p50 regulatory mechanism in macrophage inflammation. *Biochem. J.* **2012**, *441*, 653–663. [CrossRef] [PubMed]
16. Luo, H.; Wang, J.; Qiao, C.; Ma, N.; Liu, D.; Zhang, W. Pycnogenol attenuates atherosclerosis by regulating lipid metabolism through the TLR4-NF-kappaB pathway. *Exp. Mol. Med.* **2015**, *47*, e191. [CrossRef] [PubMed]
17. Bhaskar, S.; Sudhakaran, P.R.; Helen, A. Quercetin attenuates atherosclerotic inflammation and adhesion molecule expression by modulating TLR-NF-kappaB signaling pathway. *Cell Immunol.* **2016**, *310*, 131–140. [CrossRef] [PubMed]
18. Xia, E.Q.; Wang, B.W.; Xu, X.R.; Zhu, L.; Song, Y.; Li, H.B. Microwave-assisted extraction of oleanolic acid and ursolic acid from *Ligustrum lucidum* Ait. *Int. J. Mol. Sci.* **2011**, *12*, 5319–5329. [CrossRef] [PubMed]
19. Lee, J.Y.; Moon, H.; Kim, C.J. Effects of hydroxy pentacyclic triterpene acids from *Forsythia viridissima* on asthmatic responses to ovalbumin challenge in conscious guinea pigs. *Biol. Pharm. Bull.* **2010**, *33*, 230–237. [CrossRef] [PubMed]
20. Zhang, C.; Wang, C.; Li, W.; Wu, R.; Guo, Y.; Cheng, D.; Yang, Y.; Androulakis, I.P.; Kong, A.-N. Pharmacokinetics and Pharmacodynamics of the Triterpenoid Ursolic Acid in Regulating the Antioxidant, Anti-inflammatory, and Epigenetic Gene Responses in Rat Leukocytes. *Mol. Pharm.* **2017**, *14*, 3709–3717. [CrossRef] [PubMed]
21. Dar, B.A.; Lone, A.M.; Shah, W.A.; Qurishi, M.A. Synthesis and screening of ursolic acid-benzylidine derivatives as potential anti-cancer agents. *Eur. J. Med. Chem.* **2016**, *111*, 26–32. [CrossRef] [PubMed]
22. Do Nascimento, P.G.; Lemos, T.L.; Bizerra, A.; Arriaga, Â.; Ferreira, D.A.; Santiago, G.M.; Braz-Filho, R.; Costa, J.G.M. Antibacterial and antioxidant activities of ursolic acid and derivatives. *Molecules* **2014**, *19*, 1317–1327. [CrossRef] [PubMed]
23. Zhao, W.; Ma, G.; Chen, X. Lipopolysaccharide induced LOX-1 expression via TLR4/MyD88/ROS activated p38MAPK-NF-kappaB pathway. *Vasc. Pharmacol.* **2014**, *63*, 162–172. [CrossRef] [PubMed]
24. Vasconcelos, M.A.; Royo, V.A.; Ferreira, D.S.; Crotti, A.E.; Andrade e Silva, M.L.; Carvalho, J.C.; Bastos, J.K.; Cunha, W.R. In vivo analgesic and anti-inflammatory activities of ursolic acid and oleanoic acid from *Miconia albicans* (Melastomataceae). *Z. Naturforsch. C* **2006**, *61*, 477–482. [CrossRef] [PubMed]

25. Kurek, A.; Grudniak, A.M.; Szwed, M.; Klicka, A.; Samluk, L.; Wolska, K.I.; Janiszowska, W.; Popowska, M. Oleanolic acid and ursolic acid affect peptidoglycan metabolism in *Listeria monocytogenes*. *Antonie Van Leeuwenhoek* **2010**, *97*, 61–68. [CrossRef] [PubMed]
26. Mehta, J.L.; Chen, J.; Hermonat, P.L.; Romeo, F.; Novelli, G. Lectin-like, oxidized low-density lipoprotein receptor-1 (LOX-1): A critical player in the development of atherosclerosis and related disorders. *Cardiovasc. Res.* **2006**, *69*, 36–45. [CrossRef] [PubMed]
27. Ku, D.N.; Giddens, D.P.; Zarins, C.K.; Glagov, S. Pulsatile flow and atherosclerosis in the human carotid bifurcation. Positive correlation between plaque location and low oscillating shear stress. *Arteriosclerosis* **1985**, *5*, 293–302. [CrossRef] [PubMed]
28. Cai, J.; Hatsukami, T.S.; Ferguson, M.S.; Kerwin, W.S.; Saam, T.; Chu, B.; Takaya, N.; Polissar, N.L.; Yuan, C. In vivo quantitative measurement of intact fibrous cap and lipid-rich necrotic core size in atherosclerotic carotid plaque: Comparison of high-resolution, contrast-enhanced magnetic resonance imaging and histology. *Circulation* **2005**, *112*, 3437–3444. [CrossRef] [PubMed]
29. Zhang, T.; Su, J.; Wang, K.; Zhu, T.; Li, X. Ursolic acid reduces oxidative stress to alleviate early brain injury following experimental subarachnoid hemorrhage. *Neurosci. Lett.* **2014**, *579*, 12–17. [CrossRef] [PubMed]
30. Singh, U.; Jialal, I. Oxidative stress and atherosclerosis. *Pathophysiology* **2006**, *13*, 129–142. [CrossRef] [PubMed]
31. Hubackova, S.; Kucerova, A.; Michlits, G.; Kyjacova, L.; Reinis, M.; Korolov, O.; Bartek, J.; Hodny, Z. IFNgamma induces oxidative stress, DNA damage and tumor cell senescence via TGFbeta/SMAD signaling-dependent induction of Nox4 and suppression of ANT2. *Oncogene* **2016**, *35*, 1236–1249. [CrossRef] [PubMed]
32. Checker, R.; Sandur, S.K.; Sharma, D.; Patwardhan, R.S.; Jayakumar, S.; Kohli, V.; Sethi, G.; Aggarwal, B.B.; Sainis, K.B. Potent anti-inflammatory activity of ursolic acid, a triterpenoid antioxidant, is mediated through suppression of NF-kappaB, AP-1 and NF-AT. *PLoS ONE* **2012**, *7*, e31318. [CrossRef] [PubMed]
33. Paigen, B.; Holmes, P.A.; Mitchell, D.; Albee, D. Comparison of atherosclerotic lesions and HDL-lipid levels in male, female, and testosterone-treated female mice from strains C57BL/6, BALB/c, and C3H. *Atherosclerosis* **1987**, *64*, 215–221. [CrossRef]

Sample Availability: Samples of the compounds are not available from the authors.

Article

The Effects of Plant-Derived Oleanolic Acid on Selected Parameters of Glucose Homeostasis in a Diet-Induced Pre-Diabetic Rat Model

Mlindeli Gamede, Lindokuhle Mabuza, Phikelelani Ngubane and Andile Khathi *

Schools of Laboratory Medicine and Medical Sciences, College of Health Sciences, University of KwaZulu-Natal, Durban 4004, South Africa; 213571877@stu.ukzn.ac.za (M.G.); 211509843@stu.ukzn.ac.za (L.M.); Ngubanep1@ukzn.ac.za (P.N.)

* Correspondence: khathia@ukzn.ac.za

Academic Editor: Wenxu Zhou

Received: 22 February 2018; Accepted: 27 March 2018; Published: 29 March 2018

Abstract: Prolonged exposure to high energy diets has been implicated in the development of pre-diabetes, a long-lasting condition that precedes type 2 diabetes mellitus (T2DM). A combination of pharmacological and dietary interventions is used to prevent the progression of pre-diabetes to T2DM. However, poor patient compliance leads to negligence of the dietary intervention and thus reduced drug efficiency. Oleanolic acid (OA) has been reported to possess anti-diabetic effects in type 1 diabetic rats. However, the effects of this compound on pre-diabetes have not yet been established. Consequently, this study sought to evaluate the effects OA on a diet-induced pre-diabetes rat model. Pre-diabetic male Sprague Dawley rats were treated with OA in both the presence and absence of dietary intervention for a period of 12 weeks. The administration of OA with and without dietary intervention resulted in significantly improved glucose homeostasis through reduced caloric intake, body weights, plasma ghrelin concentration and glycated haemoglobin by comparison to the pre-diabetic control. These results suggest that OA may be used to manage pre-diabetes as it was able to restore glucose homeostasis and prevented the progression to overt type 2 diabetes.

Keywords: oleanolic acid; high-fat high-carbohydrate diet; pre-diabetes; glucose homeostasis; insulin resistance

1. Introduction

The global prevalence of type two diabetes mellitus (T2DM) has increased rapidly and has been implicated in the increased prevalence of metabolic syndrome [1,2]. One of the predisposing risk factors that are associated with T2DM include an unhealthy lifestyle such as prolonged exposure to high energy diets [3]. However the progression from normoglycemia to the diabetic state encompasses a long-lasting intermediate period of moderate hyperglycaemia and increased levels of glycated haemoglobin known as pre-diabetic state [4]. These physiological derangements have been directly linked to the pathogenesis of T2DM [5]. Disruptions in the homeostasis of appetite-regulating hormones that lead to increased food intake have also been reported in pre-diabetes [6,7]. A recent study in our laboratory showed that animals fed a high-fat high-carbohydrate (HFHC) diet for 20 weeks develop pre-diabetes and various metabolic derangements including increased HbA1c, impaired glucose tolerance and elevated levels of plasma ghrelin [8]. Literature evidence indicates that the reported changes in these markers eventually lead to diabetes associated cardiovascular and renal complications [9]. Once diagnosed, pre-diabetes management relies on the combination of diet modification and administration of insulin sensitizers such as metformin [10,11]. However, over-reliance by patients on the pharmacological interventions has resulted in diet intervention being neglected and thus reducing the efficacy of the drugs [12,13]. Therefore, new compounds that will have the desired therapeutic effect even without the need for dietary intervention are required. OA and its derivatives have also been reported to upregulate the expression of GLUT4 which increases the glucose uptake in adipose and muscle cell lines [14,15] Previous studies in our laboratory have indicated that the plant-derived triterpene oleanolic acid (OA) possesses anti-hyperglycaemic properties and consequently suppress postprandial hyperglycaemia in a streptozotocin (STZ)-induced type 1 diabetic rat model [16,17]. In addition, OA has also been reported to work synergistically with insulin therapy in lowering blood glucose as well as ameliorating renal and hepatic dysfunction in STZ-induced type 1 diabetic rats [17,18]. However, the effects of OA on diet-induced pre-diabetes are yet to be investigated. Therefore, the current study sought to evaluate the effects of administering plant-derived OA with and without diet intervention on glucose handling as well as the metabolic hormones involved in glucose homeostasis in a diet-induced pre-diabetes animal model.

2. Results

2.1. Caloric Intake

Food consumption of all experimental groups animals was determined every fourth week of the 12-week treatment period. The results showed that from the start of the treatment period (week 0), the PC group had a significantly higher caloric intake in comparison to NC (PC vs. NC) ($p < 0.05$). However, the administration of OA with and without dietary intervention resulted in a significant progressive decrease in caloric intake over the 12-week period by comparison to PC (Table 1).

Table 1. Effects of OA on food intake of rats that continued with HFHC diet during treatment and those that changed the diet. Values are presented as standard error of mean ± SEM and in percentage increase (↑ = increase and ↓ = decrease).

	Caloric Intake (kcal/g)			
Experimental Groups	**Week 0**	**Week 4**	**Week 8**	**Week 12**
NC	109.18 ± 1.9 **(100%)**	125.042 ± 2.34 ↑ **(14.52%)**	165.04 ± 1.61 ↑ **(51.16%)**	178.40 ± 0.87 ↑ **(63.34%)**
PC	121.47 ± 1.01 * **(100%)**	120.90 ± 0.64 * ↓ **(0.47%)**	206.58 ± 0.84 * ↑ **(70.07%)**	230.01 ± 0.85 * ↑ **(89.36%)**
Met	118.09 ± 0.51 * **(100%)**	100.54 ± 0.98 * α↓ **(14.86%)**	99.51 ± 1.52 * α↓ **(15.73%)**	151.66 ± 0.69 * α↑ **(28.43%)**
Met + DI	115.02 ± 0.67 * **(100%)**	102.69 ± 1.17 * α↓ **(10.72%)**	120.51 ± 0.75 * α↑ **(4.77%)**	144.72 ± 1.64 * α↑ **(25.82%)**
OA	130.35 ± 0.03 * **(100%)**	103.94 ± 2.02 * α↓ **(20.26%)**	156.09 ± 1.63 * α↑ **(19.75%)**	194.26 ± 1.85 * α↑ **(49.03%)**
OA + DI	117.58 ± 0.51 * **(100%)**	118.04 ± 0.85 *α ↑ **(0.40%)**	147.59 ± 2.74 *α ↑ **(25.52%)**	168.82 ± 2.22α ↑ **(43.58%)**

* $p < 0.05$ denotes comparison with NC; α $p < 0.05$ denotes comparison with PC.

2.2. Body Weights

Body weights of all experimental groups were monitored every fourth week of the treatment period which lasted for 12 weeks. The results showed that from the start of the treatment period (week 0), the PC group had a significantly increased body weights in comparison to NC (PC vs. NC) ($p < 0.05$). However, the administration of OA with and without dietary intervention showed a significant decrease in body weights when compared to PC ($p < 0.05$) (Figure 1).

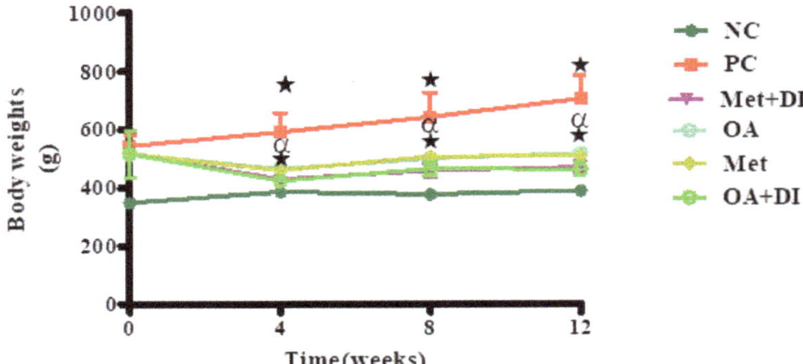

Figure 1. Effects of OA on body weights on rats that were treated OA with and without diet intervention during treatment period. Values are presented as standard error of mean ± SEM. ★ = $p < 0.05$ denotes comparison with NC; α = $p < 0.05$ denotes comparison with PC.

2.3. *Oral Glucose Tolerance (OGTT)*

The OGTT was conducted at the end of the treatment period (week 12) in all experimental groups. The results showed that at time 0, the PC had a significantly high blood glucose concentration when compared to NC and the same trend was observed for the duration of the experiment ($p < 0.05$). By the end of the experimental period, both OA-treated groups had a significantly lower blood glucose concentration when compared to PC. ($p < 0.05$) (Figure 2).

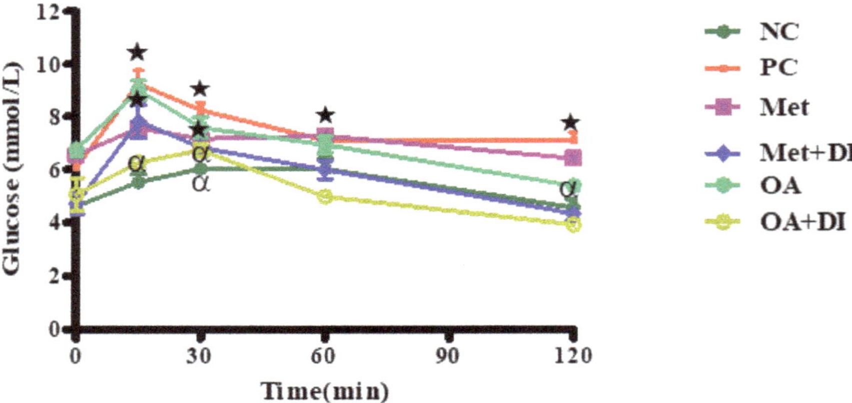

Figure 2. Effects of OA on OGTT of rats that were treated OA with and without diet intervention during treatment period. Values are presented as standard error of mean ± SEM. ★ = $p < 0.05$ denotes comparison with NC; α = $p < 0.05$ denotes comparison with PC.

2.4. *HOMAR2-IR Index*

At the end of treatment period (week 12), the HOMAR2-IR index of all animals was calculated using plasma glucose and insulin. The results showed that PC had a significantly higher HOMAR2-IR index when compared to NC ($p < 0.05$). Both OA-treated groups had a significantly lower HOMAR2-IR index in comparison to PC ($p < 0.05$) (Table 2).

Table 2. Effects of OA with and without diet intervention on HOMAR index after the treatment period. Values are presented as standard error of mean ± SEM.

HOMAR2-IR Index			
Experimental Group	**Plasma Glucose (mmol/L)**	**Plasma Insulin (mU/L)**	**HOMAR2-IR Values**
NC	4.60 ± 0.09	61.96 ± 0.90	12.670 ± 0.61
PC	5.87 ± 0.32 *	491.64 ± 3.45 *	128.26 ± 2.98 *
Met	6.55 ± 0.81 *α	169.16 ± 3.12 *α	15.93 ± 1.02 *α
Met + DI	4.52 ± 0.90α	142.94 ± *α	49.25 ± 1.15 *α
OA	6.77 ± 1.59 *α	200.58 ± 2.85 *α	60.35 ± 2.05 *α
OA + DI	4.47 ± 0.12α	68.59 ± 2.01α	13.63 ± 0.95α

* $p < 0.05$ denotes comparison with NC; α $p < 0.05$ denotes comparison with PC.

2.5. Glycated Haemoglobin Concentration (HbA1c)

All experimental groups were analyzed for HbA1c concentration at week 12. The results showed that whole blood HbA1c concentration of PC was significantly higher in comparison with NC ($p < 0.05$). The administration of OA with and without diet intervention resulted in a significant decrease in HbA1c concentration when compared to PC ($p < 0.05$) and these levels were comparable with the results of NC (Figure 3).

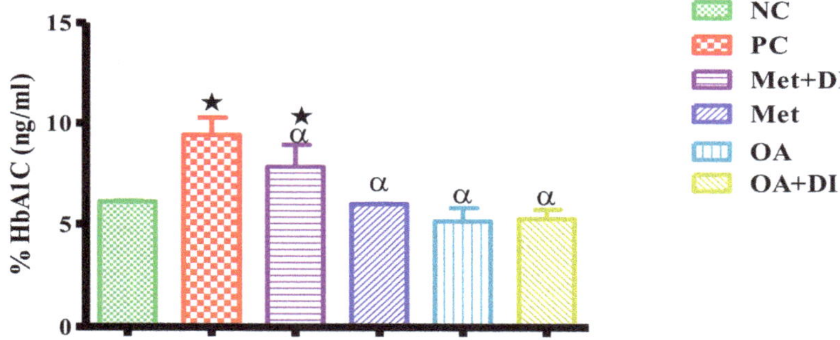

Figure 3. Effects of OA on the percentage of glycated haemoglobin concentrations of rats that continued with HFHC diet and those that changed diet or had diet intervention during treatment period. Values are presented as standard error of mean ± SEM. ★ = $p < 0.05$ denotes comparison with NC; α = $p < 0.05$ denotes comparison with PC.

2.6. Ghrelin Concentration

Terminal plasma ghrelin concentrations of all experimental groups were measured at the end of treatment period. The results showed that PC had a significantly higher plasma ghrelin concentration in comparison to NC ($p < 0.05$). However, all OA-treated animals had a significantly lower ghrelin concentration when compared to PC ($p < 0.05$) (Figure 4).

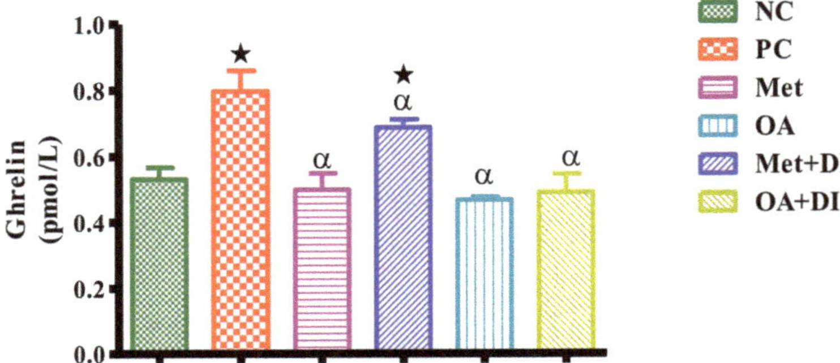

Figure 4. Effects of OA on plasma ghrelin concentrations of rats that continued with HFHC diet and those that changed diet or had diet intervention during treatment period. Values are presented as standard error of mean ± SEM. ★ = $p < 0.05$ denotes comparison with NC; α = $p < 0.05$ denotes comparison with PC.

2.7. Skeletal Muscle and Liver Glycogen Concentration

The glycogen concentration in the liver and skeletal muscle of all experimental groups was measured at the end of treatment period (week 12). The results showed that PC had significantly higher skeletal muscle and liver glycogen in comparison to NC ($p < 0.05$). However, treatment with OA resulted in a significant decrease in both skeletal muscle and liver glycogen concentrations when compared to PC ($p < 0.05$) Figures 5 and 6.

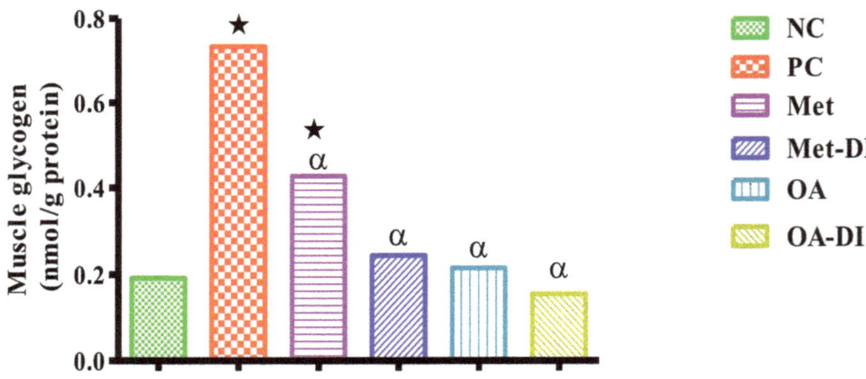

Figure 5. Effects of OA on muscle glycogen concentrations of rats that continued with HFHC diet and those that changed diet or had diet intervention during treatment period. Values are presented as mean. ★ = $p < 0.05$ denotes comparison with NC; α = $p < 0.05$ denotes comparison with PC.

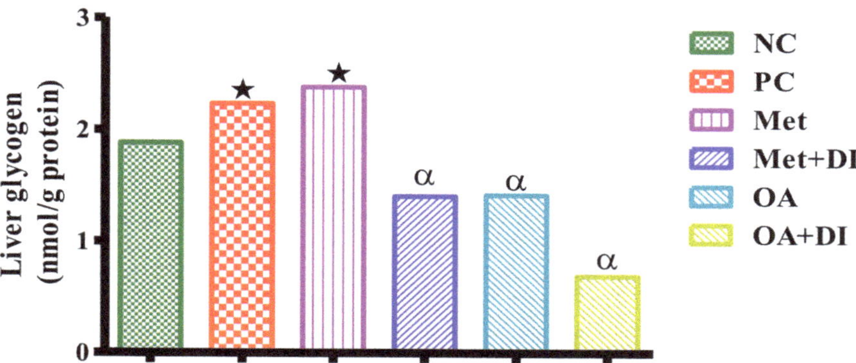

Figure 6. Effects of OA on liver glycogen concentrations of rats that continued with HFHC diet and those that changed diet or had diet intervention during treatment period. Values are presented as mean. ★ = $p < 0.05$ denotes comparison with NC; $\alpha = p < 0.05$ denotes comparison with PC.

3. Discussion

The present study investigated the effects of *Syzygium aromaticum*-derived OA on glucose homeostasis in a diet-induced pre-diabetes rat model. A recent study from our laboratory reported that prolonged exposure of rats to a HFHC diet results in insulin resistance [19]. Previous studies have reported that that OA possesses synergistic pharmacological effects with insulin in STZ-induced type 1 diabetic rats as they were found to have anti-hyperglyaecemic, antiglycation, antioxidant and hepatoprotective properties [20]. To advance from these studies, we sought to investigate the possibility of reversing pre-diabetes primarily using oleanolic acid. Furthermore, dietary interventions are clinical recommendations that accompany the use of anti-diabetic drugs. This study therefore investigated the administration of OA in both the absence and presence of diet intervention.

Under normal physiological conditions, ghrelin regulates energy metabolism through the regulation of food intake [21,22]. Ghrelin exerts orexigenic effects through activating hypothalamus receptors such as neuropeptide Y1 and Y5 while counteracting MC4R producing neurons [21]. In addition, this hormone also increases gamma aminobutyric acid (GABA) inhibitory postsynaptic currents and inhibitory synaptic contacts pre-opiomelano-cortin (POMC) neurons which counteract the effects of alpha-melanocyte stimulating hormone (αMSH) and further exacerbate orexigenic drive [23]. However, the administration of OA led to the reduction of plasma ghrelin concentration and the consequent reduction in caloric intake. Despite the unclear mechanism by which OA reduces circulating ghrelin concentration, we postulate that OA may restore ghrelin regulation through sensitizing the peripheral cells for insulin and suppress ghrelin secretion [20]. These findings of this study agree with the observations of an earlier study that showed that the administration of OA decreased ghrelin concentrations in STZ-induced type 1 diabetic rats [21]. Previous studies have shown that ghrelin also plays an important role in the regulation of glucose homeostasis through stimulating insulin release and sensitizing cells for insulin [22,24]. Ghrelin is also implicated in the inhibition of insulin stimulated glucose uptake during obesity [23]. Obeity is one of the predisposing factors for pre-diabetes as well as T2DM due to an increase in accumulation of ectopic fat deposition which is associated with insulin resistance [25]. However, several studies have shown that weight loss improves insulin sensitivity on obese individuals [26–28]. The regulation of caloric intake plays a crucial role in the management of metabolic disorders such as pre-diabetes and T2DM [29,30]. Furthermore, studies have shown that reducing caloric intake may play an important role in improving insulin sensitivity and subsequently the glucose homeostasis in pre-diabetic patients [31]. However, the modification of diets is often seen by patients as being of less significance and they tend to rely heavily on the medication [32,33]. In this study, OA mediated the reduction of caloric intake seen in this study in both the presence and absence of diet intervention resulting in decreased body weights and improvements in glucose tolerance.

Impaired fasting glucose and glucose intolerance are some of the remarkable diagnostic features of pre-diabetes [34]. In this study, we observed that prolonged exposure to HFHC diet led to both impaired fasting glucose and impaired glucose tolerance. However, with the administration of OA we observed a decrease in fasting blood glucose concentrations and 2-h postprandial glucose concentrations to within the normal range. Previously, OA has been found to increase the expression of glucose transporters such as GLUT 4 in skeletal muscle of STZ-induced type 1 diabetic rats [35]. Another study reported that OA stimulates phosphoinositol-3-kinase which phosphorylates Akt and downregulates p-mTOR to improve insulin resistance [36]. Indeed, this study additionally showed that the administration of OA in both the presence and absence of diet intervention resulted in the reduction of plasma insulin concentrations. This was further evidenced by decreases in the HOMAR2-IR index which is used to quantify insulin resistance and beta-cell function [37].

Increased glycated haemoglobin (HbA1c) is also accepted as one of the diagnostic feature of pre-diabetes [38]. Increases in HbA1c develop because of sustained high plasma glucose concentrations which lead to glucose-mediated non-enzymatic glycation of haemoglobin [39]. In the present study, we observed that HFHC diet led to increased HbA1c indicating the moderate hyperglycaemia found in pre-diabetes [40]. However, the administration of OA decreased HbA1c and this may be due to OA sensitizing skeletal myocytes for insulin which led to translocation of GLUT 4 and subsequently the uptake of glucose by the cells [41]. Furthermore, these results suggest that OA started exerting these effects at the beginning of the study as HbA1c reflects average plasma glucose over the previous eight to 12 weeks [42]. Previous studies have reported that lowering HbA1c can be a sign that there is a sustained regulation of glucose homeostasis [42]. Taken together, the results of the study suggest that OA improves insulin sensitivity in diet-induced pre-diabetes even in the absence of diet intervention.

4. Materials and Methods

4.1. Drugs and Chemicals

All chemicals and reagents were sourced from the standard pharmaceutical suppliers and were of analytical grade.

4.2. Extraction Method and Administration

OA was extracted from *Syzygium aromaticum* ((Linnaeus) Merrill & Perry) (cloves) using an established protocol from [16]. Briefly: air-dried *S. aromaticum* flower buds (500 g) were milled and sequentially extracted twice at 24 h intervals at room temperature using 1 L dichloromethane (DCM), and ethyl acetate (720 mL) on each occasion. Subsequently, the extract was concentrated under reduced pressure at 55 ± 1 °C using a rotary evaporator to yield dichloromethane solubles (DCMS) and ethyl acetate solubles (EAS). The EAS containing mixtures of oleanolic/ursolic acid and methyl maslinate/methyl corosolate were purified by silica gel 60 column chromatography with hexane: ethyl acetate solvent systems of 7:3. This yielded OA which was further purified by recrystallization from chloroform-methanol (1:1, v/v). The structure of OA was confirmed by spectroscopic analysis using 1D and 2D, 1H and 13C nuclear magnetic resonance (NMR) spectroscopic experiments. The purity of the extracted OA used in the study was greater than 95%.

4.3. Animals

Male Sprague-Dawley rats (130–160 g), bred and housed in the Biomedical Research Unit (BRU) of the University of KwaZulu-Natal were used in the study. All animal procedures and housing conditions was approved by the Animal Research Ethics Committee of the University of KwaZulu-Natal (ethics No.: AREC/035/016M). The animals were allowed access to food and fluids *ad libitum*.

4.3.1. Induction of Pre-Diabetes

Experimental pre-diabetes was induced in male Sprague-Dawley rats using a previously described protocol [19]. Briefly, the experimental animals were fed a high fat high carbohydrate (HFHC) diet supplemented with 15% fructose for 20 weeks while the control animals were exposed to standard chow for the equal number of weeks. After 20 weeks, the American Diabetes Federation criteria was used to diagnose pre-diabetes. The animals that were fed normal diet were also tested and were found to be normoglycemic and without pre-diabetes.

4.3.2. Experimental Design

The study had two major groups which was the normal group and pre-diabetic group. Pre-diabetic group was further sub-divided into six groups each group having six rats (n = 6). The groups were categorized as follows: pre-diabetic control group (PC) which are the pre-diabetic animals continued with the experimental diet throughout the study period; metformin group (Met) which are the pre-diabetic animals that continued with the experimental diet but received metformin during treatment period; metformin and diet intervention group (Met + DI) which are the pre-diabetic animals that changed to a normal diet and received metformin during treatment period; oleanolic acid group (OA) which are the pre-diabetic animals that continued with experimental diet but received oleanolic acid during treatment period as well as OA and diet intervention group (OA + DI) which are the pre-diabetic animals that changed to a normal diet and received oleanolic acid during experimental period. The normal control group was the group of animals that were fed normal diet and diagnosed as without pre-diabetes.

4.3.3. Treatment of Pre-Diabetic Animals

The experimental period lasted for 12 weeks. The animals were treated every third day where the Met and Met + DI groups received (500 mg/kg) of metformin and the OA and OA + DI groups was given (80 mg/kg) of oleanolic acid. Parameters including food intake, body weights, fasting blood glucose were monitored at week 0, 4, 8 and 12. Glucose tolerance was also evaluated at week 12 with oral glucose tolerance test (OGTT) using OGTT protocol from previous studies in our laboratory [19].

4.3.4. Blood Collection and Tissue Harvesting

After the experimental period, the animals were anaesthetized with Isofor (100 mg/kg) (Safeline Pharmaceuticals (Pty) Ltd., Roodeport, South Africa) via a gas anesthetic chamber (Biomedical Resource Unit, UKZN, Durban, South Africa) for 3 min. Blood was collected by cardiac puncture and then injected into individual pre-cooled heparinized containers. The blood was then centrifuged (Eppendorf centrifuge 5403, LGBW Germany) at 4 °C, 503× g for 15 min. Plasma was collected and stored at −80 °C in a Bio Ultra freezer (Snijers Scientific, Tilburg, NB, Netherlands) until ready for biochemical analysis. The harvested livers and skeletal muscle were rinsed with cold normal saline solution and snap frozen in liquid nitrogen before storage in a Bio Ultra freezer (Snijers Scientific) at −80 °C.

4.4. Biochemical Analysis

HbA1c and ghrelin concentrations were measured using their respective rat ELISA kits (Elabscience Biotechnology Co., Ltd., Houston, TX, USA) according to the manufacturer's instructions. Plasma insulin concentration was measured using an ultra-sensitive rat insulin ELISA kit (Mercodia AB, Sylveniusgatan 8A, SE-754 50, Uppsala, Sweden) according to the manufacturer's instructions. HOMAR-IR index was further calculated from insulin concentrations and fasting glucose.

4.5. Glycogen Assay

Glycogen analysis was performed in muscle and liver tissues using a well-established laboratory protocol [16,35,43]. The harvested liver and muscle tissues were weighed and heated with potassium hydroxide (KOH) (30%, 2 mL) at 100 °C for 30 min. Then immediately disodium sulphite (Na_2SO_4) (10%, 0.194 mL) was added in to the mixture to stop the reaction. The mixture was then allowed to cool, and the glycogen precipitate was formed. The cooled mixture with precipitate was aspirated (200 μL) and mixed with ethanol (95%, 200 μL). The precipitated glycogen was pelleted, washed and resolubilized in H_2O (1 mL). Thereafter, anthrone (0.5 g dissolve in 250 mL of sulphuric acid, 4 mL) was added and boiled for 10 min. After cooling the absorbance was read using the Spectrostar Nano spectrophotometer (BMG Labtech, Ortenburg, LGBW Germany) at 620 nm. The glycogen concentrations were calculated from the glycogen standard curve. The standard curve ranges from 200 to 1000 mg/L.

4.6. Statistical Analysis

All data was expressed as means ± S.E.M. Statistical comparisons were performed with GraphPadInStat Software (version 5.00, Graph Pad Software, Inc., San Diego, CA, USA) using one-way analysis of variance (ANOVA) followed by Steel-Dwass-Critchlow-Fligner multiple comparison test to simultaneously determine statistical differences between the means of two independent groups. A value of $p < 0.05$ was considered statistically significant.

5. Conclusions

The findings of this study confirm the previous findings that reported that OA has potential in managing diabetes. However, the findings of this study have revealed that OA can restore the regulation of glucose homeostasis in a diet-induced pre-diabetic rat model with and without the use of diet intervention. Therefore, this natural bioactive compound has shown potential in the prevention of the progression of pre-diabetes to T2DM.

Acknowledgments: The authors are grateful to the Biomedical Resource Unit, University of KwaZulu-Natal for the supply of animals and National Research Foundation (NRF) South Africa.

Author Contributions: Mlindeli Gamede and Lindokuhle Mabuza carried out experiments, study design, analysis of data, writing of manuscript. Andile Khathi and Phikelelani Ngubane were involved in conceptualization, carried out experiments, study design, analysis of data, were involved in the writing of the manuscript, and provided funding.

Conflicts of Interest: The authors declare no conflicts of interest.

References

1. Wang, C.; Zhang, Y.; Zhang, L.; Hou, X.; Lu, H.; Shen, Y.; Chen, R.; Fang, P.; Yu, H.; Li, M. Prevalence of type 2 diabetes among high-risk adults in shanghai from 2002 to 2012. *PLoS ONE* **2014**, *9*, e102926. [CrossRef] [PubMed]
2. Yoon, K.-H.; Lee, J.-H.; Kim, J.-W.; Cho, J.H.; Choi, Y.-H.; Ko, S.-H.; Zimmet, P.; Son, H.-Y. Epidemic obesity and type 2 diabetes in asia. *Lancet* **2006**, *368*, 1681–1688. [CrossRef]
3. Panchal, S.; Poudyal, H.; Iyer, A.; Nazer, R.; Alam, A.; Diwan, V.; Kauter, K.; Sernia, C.; Campbell, F.; Ward, P.; et al. High-carbohydrate, high-fat diet–induced metabolic syndrome and cardiovascular remodeling in rats. *J. Cardiovasc. Pharm.* **2011**, *57*, 611–624. [CrossRef] [PubMed]
4. Tuomilehto, J.; Lindström, J.; Eriksson, J.G.; Valle, T.T.; Hämäläinen, H.; Ilanne-Parikka, P.; Keinänen-Kiukaanniemi, S.; Laakso, M.; Louheranta, A.; Rastas, M.; et al. Prevention of type 2 diabetes mellitus by changes in lifestyle among subjects with impaired glucose tolerance. *N. Engl. J. Med.* **2001**, *344*, 1343–1350. [CrossRef] [PubMed]
5. Rebnord, E.W.; Pedersen, E.R.; Strand, E.; Svingen, G.F.T.; Meyer, K.; Schartum-Hansen, H.; Løland, K.H.; Seifert, R.; Ueland, P.M.; Nilsen, D.W.T.; et al. Glycated hemoglobin and long-term prognosis in patients with suspected stable angina pectoris without diabetes mellitus: A prospective cohort study. *Atherosclerosis* **2015**, *240*, 115–120. [CrossRef] [PubMed]

6. D'Souza, A.M.; Johnson, J.D.; Clee, S.M.; Kieffer, T.J. Suppressing hyperinsulinemia prevents obesity but causes rapid onset of diabetes in leptin-deficient lepob/ob mice. *Mol. Metab.* **2016**, *5*, 1103–1112. [CrossRef] [PubMed]
7. Walsh, J.M.; Byrne, J.; Mahony, R.M.; Foley, M.E.; McAuliffe, F.M. Leptin, fetal growth and insulin resistance in non-diabetic pregnancies. *Early Hum. Dev.* **2014**, *90*, 271–274. [CrossRef] [PubMed]
8. Man, R.E.K.; Charumathi, S.; Gan, A.T.L.; Fenwick, E.K.; Tey, C.S.; Chua, J.; Wong, T.-Y.; Cheng, C.-Y.; Lamoureux, E.L. Cumulative incidence and risk factors of prediabetes and type 2 diabetes in a singaporean malay cohort. *Diabetes Res. Clin. Pract.* **2017**, *127*, 163–171. [CrossRef] [PubMed]
9. Mosa, Z.M.; El Badry, Y.A.; Fattah, H.S.; Mohamed, E.G. Comparative study between the effects of some dietary sources and metformin drug on weight reduction in obese rats. *Ann. Agric. Sci.* **2015**, *60*, 381–388. [CrossRef]
10. Watson, C.S. Prediabetes: Screening, diagnosis, and intervention. *J. Nurse Pract.* **2017**, *13*, 216–221.e1. [CrossRef]
11. Chaudhari, P.; Vallarino, C.; Law, E.H.; Seifeldin, R. Evaluation of patients with type 2 diabetes mellitus receiving treatment during the pre-diabetes period: Is early treatment associated with improved outcomes? *Diabetes Res. Clin. Pract.* **2016**, *122*, 162–169. [CrossRef] [PubMed]
12. Courcoulas, A.P.; Belle, S.H.; Neiberg, R.H.; Pierson, S.K.; Eagleton, J.K.; Kalarchian, M.A.; DeLany, J.P.; Lang, W.; Jakicic, J.M. Three-year outcomes of bariatric surgery vs lifestyle intervention for type 2 diabetes mellitus treatment: A randomized clinical trial. *JAMA Surg.* **2015**, *150*, 931–940. [CrossRef] [PubMed]
13. Inzucchi, S.E.; Bergenstal, R.M.; Buse, J.B.; Diamant, M.; Ferrannini, E.; Nauck, M.; Peters, A.L.; Tsapas, A.; Wender, R.; Matthews, D.R. Management of hyperglycaemia in type 2 diabetes, 2015: A patient-centred approach. Update to a position statement of the american diabetes association and the european association for the study of diabetes. *Diabetologia* **2015**, *58*, 429–442. [CrossRef] [PubMed]
14. Katashima, C.K.; Silva, V.R.; Gomes, T.L.; Pichard, C.; Pimentel, G.D. Ursolic acid and mechanisms of actions on adipose and muscle tissue: A systematic review. *Obes. Rev.* **2017**, *18*, 700–711. [CrossRef] [PubMed]
15. Gajęcka, M.; Przybylska-Gornowicz, B.; Zakłos-Szyda, M.; Dąbrowski, M.; Michalczuk, L.; Koziołkiewicz, M.; Babuchowski, A.; Zielonka, Ł.; Lewczuk, B.; Gajęcki, M.T. The influence of a natural triterpene preparation on the gastrointestinal tract of gilts with streptozocin-induced diabetes and on cell metabolic activity. *J. Funct. Foods* **2017**, *33*, 11–20. [CrossRef]
16. Khathi, A.; Serumula, M.R.; Myburg, R.B.; Van Heerden, F.R.; Musabayane, C.T. Effects of syzygium aromaticum-derived triterpenes on postprandial blood glucose in streptozotocin-induced diabetic rats following carbohydrate challenge. *PLoS ONE* **2013**, *8*, e81632. [CrossRef] [PubMed]
17. Mkhwanazi, B.N.; Serumula, M.R.; Myburg, R.B.; Van Heerden, F.R.; Musabayane, C.T. Antioxidant effects of maslinic acid in livers, hearts and kidneys of streptozotocin-induced diabetic rats: Effects on kidney function. *Ren. Fail.* **2014**, *36*, 419–431. [CrossRef] [PubMed]
18. Mukundwa, A.; Langa, S.O.; Mukaratirwa, S.; Masola, B. In Vivo effects of diabetes, insulin and oleanolic acid on enzymes of glycogen metabolism in the skin of streptozotocin-induced diabetic male sprague-dawley rats. *Biochem. Biophys. Res. Commun.* **2016**, *471*, 315–319. [CrossRef] [PubMed]
19. Luvuno, M.; Kathi, A.; Mabandla, M.V. Voluntary Ingestion of a High-Fat High-Carbohydrate Diet: A Model for Prediabetes. Master's Dissertation, University of KwaZulu-Natal, KwaZulu-Natal, South Africa, 2017.
20. Chen, P.; Zeng, H.; Wang, Y.; Fan, X.; Xu, C.; Deng, R.; Zhou, X.; Bi, H.; Huang, M. Low dose of oleanolic acid protects against lithocholic acid-induced cholestasis in mice: Potential involvement of nuclear factor-e2-related factor 2-mediated upregulation of multidrug resistance-associated proteins. *Drug Metab. Dispos.* **2014**, *42*, 844–852. [CrossRef] [PubMed]
21. Luvuno, M.; Mbongwa, H.P.; Khathi, A. The effects of syzygium aromaticum-derived triterpenes on gastrointestinal ghrelin expression in streptozotocin-induced diabetic rats. *Afr. J. Tradit. Complement. Altern. Med.* **2016**, *13*, 8–14. [CrossRef] [PubMed]
22. Alamri, B.N.; Shin, K.; Chappe, V.; Anini, Y. The role of ghrelin in the regulation of glucose homeostasis. *Horm. Mol. Biol. Clin. Investig.* **2016**, *26*, 3–11. [CrossRef] [PubMed]
23. Theander-Carrillo, C.; Wiedmer, P.; Cettour-Rose, P.; Nogueiras, R.; Perez-Tilve, D.; Pfluger, P.; Castaneda, T.R.; Muzzin, P.; Schürmann, A.; Szanto, I. Ghrelin action in the brain controls adipocyte metabolism. *J. Clin. Investig.* **2006**, *116*, 1983. [CrossRef] [PubMed]

24. Jung, S.H.; Ha, Y.J.; Shim, E.K.; Choi, S.Y.; Jin, J.L.; Yun-Choi, H.S.; Lee, J.R. Insulin-mimetic and insulin-sensitizing activities of a pentacyclic triterpenoid insulin receptor activator. *Biochem. J.* **2007**, *403*, 243–250. [CrossRef] [PubMed]
25. Jung, U.J.; Choi, M.-S. Obesity and its metabolic complications: The role of adipokines and the relationship between obesity, inflammation, insulin resistance, dyslipidemia and nonalcoholic fatty liver disease. *Int. J. Mol. Sci.* **2014**, *15*, 6184–6223. [CrossRef] [PubMed]
26. Magkos, F.; Fraterrigo, G.; Yoshino, J.; Luecking, C.; Kirbach, K.; Kelly, S.C.; de las Fuentes, L.; He, S.; Okunade, A.L.; Patterson, B.W. Effects of moderate and subsequent progressive weight loss on metabolic function and adipose tissue biology in humans with obesity. *Cell Metab.* **2016**, *23*, 591–601. [CrossRef] [PubMed]
27. Canfora, E.E.; Jocken, J.W.; Blaak, E.E. Short-chain fatty acids in control of body weight and insulin sensitivity. *Nat. Rev. Endocrinol.* **2015**, *11*, 577. [CrossRef] [PubMed]
28. Merovci, A.; Solis-Herrera, C.; Daniele, G.; Eldor, R.; Fiorentino, T.V.; Tripathy, D.; Xiong, J.; Perez, Z.; Norton, L.; Abdul-Ghani, M.A. Dapagliflozin improves muscle insulin sensitivity but enhances endogenous glucose production. *J. Clin. Investig.* **2014**, *124*, 509–514. [CrossRef] [PubMed]
29. Noakes, T.D.; Windt, J. Evidence that supports the prescription of low-carbohydrate high-fat diets: A narrative review. *Br. J. Sports Med.* **2017**, *51*, 133–139. [CrossRef] [PubMed]
30. Mahapatra, D.K.; Asati, V.; Bharti, S.K. Chalcones and their therapeutic targets for the management of diabetes: Structural and pharmacological perspectives. *Eur. J. Med. Chem.* **2015**, *92*, 839–865. [CrossRef] [PubMed]
31. Evert, A.B.; Boucher, J.L.; Cypress, M.; Dunbar, S.A.; Franz, M.J.; Mayer-Davis, E.J.; Neumiller, J.J.; Nwankwo, R.; Verdi, C.L.; Urbanski, P. Nutrition therapy recommendations for the management of adults with diabetes. *Diabetes Care* **2014**, *37*, S120–S143. [CrossRef] [PubMed]
32. Colhoun, H.M.; Betteridge, D.J.; Durrington, P.N.; Hitman, G.A.; Neil, H.A.W.; Livingstone, S.J.; Thomason, M.J.; Mackness, M.I.; Charlton-Menys, V.; Fuller, J.H. Primary prevention of cardiovascular disease with atorvastatin in type 2 diabetes in the collaborative atorvastatin diabetes study (cards): Multicentre randomised placebo-controlled trial. *Lancet* **2004**, *364*, 685–696. [CrossRef]
33. Donnan, P.T.; MacDonald, T.M.; Morris, A.D. Adherence to prescribed oral hypoglycaemic medication in a population of patients with type 2 diabetes: A retrospective cohort study. *Diabet. Med.* **2002**, *19*, 279–284. [CrossRef] [PubMed]
34. Hanssen, M.J.; Wierts, R.; Hoeks, J.; Gemmink, A.; Brans, B.; Mottaghy, F.M.; Schrauwen, P.; van Marken Lichtenbelt, W.D. Glucose uptake in human brown adipose tissue is impaired upon fasting-induced insulin resistance. *Diabetologia* **2015**, *58*, 586–595. [CrossRef] [PubMed]
35. Ngubane, P.S.; Masola, B.; Musabayane, C.T. The effects of syzygium aromaticum-derived oleanolic acid on glycogenic enzymes in streptozotocin-induced diabetic rats. *Ren. Fail.* **2011**, *33*, 434–439. [CrossRef] [PubMed]
36. Wang, H.; Shi, G.; Zhang, X.; Gong, S.; Tan, S.; Chen, B.; Che, H.; Li, T. Mesoscale modelling study of the interactions between aerosols and pbl meteorology during a haze episode in china jing–jin–ji and its near surrounding region—Part 2: Aerosols' radiative feedback effects. *Atmos. Chem. Phys.* **2015**, *15*, 3277–3287. [CrossRef]
37. Wallace, T.M.; Levy, J.C.; Matthews, D.R. Use and abuse of homa modeling. *Diabetes Care* **2004**, *27*, 1487–1495. [CrossRef] [PubMed]
38. Monje, A.; Catena, A.; Borgnakke, W.S. Association between diabetes mellitus/hyperglycemia and peri-implant diseases: Systematic review and meta-analysis. *J. Clin. Periodontol.* **2017**. [CrossRef] [PubMed]
39. Richter, B.; Hemmingsen, B.; Metzendorf, M.I.; Takwoingi, Y. Intermediate hyperglycaemia as a predictor for the development of type 2 diabetes: Prognostic factor exemplar review. *Cochrane Libr.* **2017**. [CrossRef]
40. Myers, R.W.; Guan, H.-P.; Ehrhart, J.; Petrov, A.; Prahalada, S.; Tozzo, E.; Yang, X.; Kurtz, M.M.; Trujillo, M.; Trotter, D.G. Systemic pan-ampk activator mk-8722 improves glucose homeostasis but induces cardiac hypertrophy. *Science* **2017**, *357*, 507–511. [CrossRef] [PubMed]
41. Perumal, V.; Narayanan, N.; Rangarajan, J.; Palanisamy, E.; Kalifa, M.; Maheshwari, U.; Perincheri, P.S.S.; Chinnathambi, S. A study to correlate hba1c levels and left ventricular diastolic dysfunction in newly diagnosed type ii diabetes mellitus. *J. Evol. Med. Dent. Sci.* **2016**, *5*, 3412–3417. [CrossRef] [PubMed]

42. Khan, S.; Jena, G. Protective role of sodium butyrate, a hdac inhibitor on beta-cell proliferation, function and glucose homeostasis through modulation of p38/erk mapk and apoptotic pathways: Study in juvenile diabetic rat. *Chem.-Biol. Interact.* **2014**, *213*, 1–12. [CrossRef] [PubMed]
43. Musabayane, C.; Mahlalela, N.; Shode, F.; Ojewole, J. Effects of *Syzygium cordatum* (hochst.) [myrtaccae] leaf extract on plasma glucose and hepatic glycogen in streptozotocin-induce diabetic rats. *J. Ethnopharmacol.* **2005**, *97*, 485–490. [CrossRef] [PubMed]

Sample Availability: Samples of the compounds are not available from the authors.

Review

Oxysterols and Retinal Degeneration in a Rat Model of Smith-Lemli-Opitz Syndrome: Implications for an Improved Therapeutic Intervention

Steven J. Fliesler [1,2,*] and Libin Xu [3,*]

1 Departments of Ophthalmology and Biochemistry and Neuroscience Program, Jacobs School of Medicine and Biomedical Sciences, University at Buffalo, The State University of New York, Buffalo, NY 14260, USA
2 Research Service, VA Western NY Healthcare System, Buffalo, NY 14260, USA
3 Department of Medicinal Chemistry, School of Pharmacy, University of Washington, Seattle, WA 98195, USA
* Correspondence: fliesler@buffalo.edu (S.J.F.); libinxu@uw.edu (L.X.); Tel.: +1-716-862-6538 (S.J.F.); +1-206-543-1080 (L.X.)

Received: 12 September 2018; Accepted: 19 October 2018; Published: 22 October 2018

Abstract: Smith-Lemli-Opitz syndrome (SLOS) is an autosomal recessive human disease caused by mutations in the gene encoding 7-dehydrocholesterol (7DHC) reductase (DHCR7), resulting in abnormal accumulation of 7DHC and reduced levels of cholesterol in bodily tissues and fluids. A rat model of the disease has been created by treating normal rats with the DHCR7 inhibitor, AY9944, which causes progressive, irreversible retinal degeneration. Herein, we review the features of this disease model and the evidence linking 7DHC-derived oxysterols to the pathobiology of the disease, with particular emphasis on the associated retinal degeneration. A recent study has shown that treating the rat model with cholesterol plus suitable antioxidants completely prevents the retinal degeneration. These findings are discussed with regard to their translational implications for developing an improved therapeutic intervention for SLOS over the current standard of care.

Keywords: antioxidant; cholesterol; degeneration; oxysterol; retina; Smith-Lemli-Opitz syndrome

1. Introduction

Since the discovery of cholesterol (Chol) by François Poulletier de la Salle in 1796 and its subsequent naming as "cholesterine" by Michel Eugène Chevreul in 1816 [1], the structure, biosynthesis, and biological functions of sterols have been the subject of extensive investigations [2–6]. Similarly, the study of oxysterols—oxidative derivatives of sterols—has emerged and evolved as a distinct research area of interest in chemistry, biology, and medicine [7–13]. Here, we present an overview of the formation, presence, and possible pathophysiological role of oxysterols in the retina in an animal model of a human cholesterol deficiency syndrome (see below), and the implications of these findings with regard to development of a new therapeutic intervention for this disease, based upon blocking the formation of cytotoxic oxysterols.

2. The RSH/Smith-Lemli-Opitz Syndrome (SLOS)

RSH/Smith-Lemli-Opitz syndrome (SLOS) [14] is an autosomal recessive human genetic disease caused by a mutation-induced enzymatic defect in the last step in Chol synthesis, i.e., reduction of the Δ^7-double bond of 7-dehydrocholesterol (7DHC) to form Chol, catalyzed by the enzyme, DHCR7 (7-dehydrocholesterol reductase; 3β-hydroxysterol-Δ^7-reductase, EC1.3.1.21) (Figure 1) [15–19]. This results in abnormally high steady-state levels of 7DHC and abnormally low levels of Chol in bodily tissues and fluids [15,20,21]; this biochemical phenotype is unique to SLOS and, hence, 7DHC is a signature biomarker for this disease. SLOS was the first characterized "multiple congenital anomalies (MCA)" syndrome, the first of several subsequently discovered inborn errors of Chol synthesis to be described over the past nearly six decades, and the most frequent Chol biosynthesis disorder with high carrier frequency (more than 1%) [16,19,22]. Features of this disease include multiple dysmorphologies involving craniofacial and musculoskeletal abnormalities (notably 2,3-toe syndactyly), brain malformation, impaired cognitive functions, autistic and other behavioral problems, developmental delay, and failure to thrive, among other defects [16–19,22]. Visual system dysfunction is also associated with SLOS [23–25].

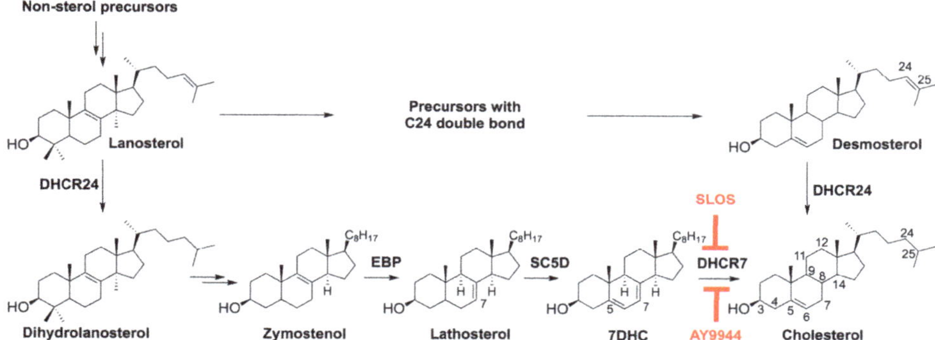

Figure 1. Post-squalene cholesterol biosynthetic pathway. Site of enzymatic inhibition by AY9944 is indicated as well as the locus of the defect in Smith-Lemli-Opitz syndrome (SLOS) at the level of DHCR7. Structures of cholesterol (Chol), 7-dehydrocholesterol (7DHC), and numbering for nomenclature are shown. EBP, 3β-hydroxysterol- Δ^8,Δ^7-isomerase; DHCR24, 3β-hydroxysterol-Δ^{24}-reductase; SC5D, sterol-C5-desaturase; DHCR7, 3β-hydroxysterol-Δ^7-reductase.

3. AY9944 and the Development of a Rat Model of SLOS

The causative link between excessive levels of blood-borne Chol, high-fat/high-Chol diets, and cardiovascular disease has been recognized since the 1950s (see ref. [3–5]). Hence, several pharmaceutical companies (e.g., Wyeth-Ayerst Laboratories, Merck, Eli Lily, Boehringer-Mannheim, etc.) have directed substantial efforts over the years toward developing Chol-lowering drugs, especially those that block the de novo synthesis of Chol—most notably, the statins, which block Chol synthesis at the level of HMG-CoA reductase, the main rate-limiting enzyme of the de novo pathway [26,27]. In addition, inhibitors that specifically target more distal enzymatic steps in Chol biosynthesis have been discovered, among them "AY9944" (*trans*-1,4-bis(2-chlorobenzylaminoethyl) cyclohexane dihydrochloride), which inhibits DHCR7—the same enzyme that is genetically abnormal in SLOS [28,29] (Figure 1). AY9944 was not successful as a cholesterol-lowering drug as it is teratogenic [30,31]. However, treating experimental animals, such as rats, with AY9944 has been employed successfully as a pharmacological approach for developing an animal model of SLOS [32]. We made modifications to the protocol originally developed by Kolf-Clauw et al. [32] to create an improved SLOS rat model that is viable for up to three postnatal months and exhibits profound elevation in 7DHC and reduction in Chol levels in the serum, liver, brain, and retina [33,34].

3.1. Retinal Degeneration in the AY9944-Induced SLOS Rat Model

The retina is rich in oxygen, cholesterol, light-absorbing retinoids, and polyunsaturated fatty acids (PUFAs), such as docosahexaenoic acid (DHA; 22:6), and is constantly exposed to light. Together, these factors make the retina susceptible to photo- or free radical-induced lipid oxidation and subsequent oxidative damage [35–38]. This is particularly relevant to this SLOS rat model because the accumulated cholesterol precursor, 7DHC, is highly prone to free radical oxidation [39] (next Section), which makes the retina even more susceptible to oxidative damage. Using the AY9944-induced SLOS rat model, we observed a progressive, irreversible, and profound retinal degeneration [34]. Remarkably, even though retina 7DHC/Chol mole ratios were >4:1 by one postnatal month, there were no appreciable structural abnormalities observed in the retina at that time point [33], although the retinal pigment epithelium (RPE) exhibited noticeable accumulation of phagosomes and membrane/lipid inclusions (Figure 2). However, by two postnatal months of treatment with AY9944, the retina exhibited marked pyknosis of the outer nuclear layer (ONL; the histological layer of the retina containing nuclei of rod and cone photoreceptor cells), degeneration of rod photoreceptor cells, and thinning of the neural retina. By three postnatal months, the severity of the retinal degeneration was more pronounced, including a loss of >25% of the photoreceptors (Figure 2). The degeneration appears to impact the photoreceptor layer almost exclusively. In fact, photoreceptor-specific cell death and dropout in this rat model has been confirmed independently, by TUNEL (terminal deoxynucleotidyl transferase dUTP nick end labeling) assay [40].

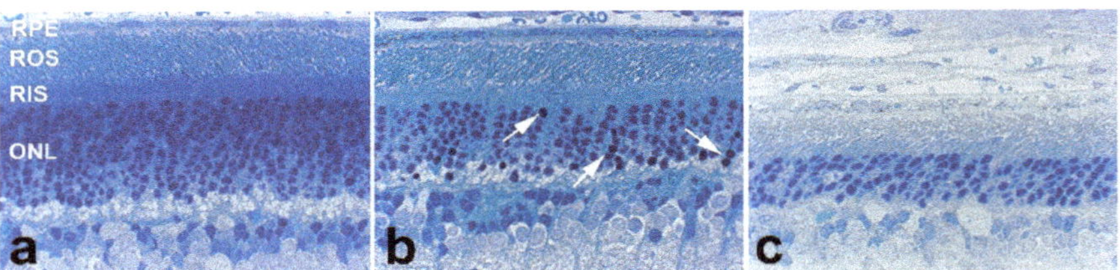

Figure 2. Histological degeneration of the retina observed in the AY9944-induced rat model of SLOS as a function of postnatal age at postnatal (**a**) one month, (**b**) two months, and (**c**) three months. Note progressive thinning of the outer nuclear layer (ONL), the appearance of pyknotic photoreceptor nuclei by two postnatal months (*arrows*, panel **b**), and the progressive shortening of rod outer segments (ROS). Epon embedment; toluidine blue stain.

Consistent with this degenerative phenotype, the electrophysiological function of the retina was also progressively and markedly compromised in this rat model of SLOS, as determined by electroretinography (ERG) [34]. Although the amplitudes of rod- and cone-driven responses to light stimulation were robust and comparable to those of untreated control rats up to one postnatal month, response amplitudes were diminished, relative to controls, at two and three postnatal months, approximately proportional to the magnitude of retinal thinning and photoreceptor loss observed histologically. In addition, the timing of the photoresponses was much slower than normal.

3.2. Sterols and Oxysterols in the Retina in the AY9944-Induced Rat Model of SLOS

As mentioned above, the 7DHC/Chol mole ratio of retinas in this rat model become markedly elevated, relative to controls, with the ratio being about 4:1 at one postnatal month and increasing to >5:1 by three postnatal months [34]. Despite the apparent correlation between an increasing 7DHC/Chol mole ratio and severity of the observed retinal degeneration, however, there is no biological evidence to suggest that 7DHC itself is cytotoxic. In fact, 7DHC is almost identical to Chol in its physical properties: Its molecular mass, number of carbons, geometric shape, and molar volume are similar to those of Chol. Studies with artificial membrane bilayers, varying the sterol/phospholipid mole ratio as well as the relative proportions of 7DHC and Chol, have shown that the packing of 7DHC in membrane bilayers is nearly the same as that of Chol [41,42]. Furthermore, the ability of 7DHC to form "lipid rafts"—highly-ordered membrane microdomains containing high concentrations of sterols and sphingolipids, compared to the dominant bulk phase lipid composition—is comparable to, if not better than, Chol [43,44]. That said, there is a critical chemical difference between 7DHC and Chol: 7DHC has an additional double bond, which is also conjugated. That structural feature makes 7DHC highly susceptible to oxidation; in fact, it is the most highly reactive lipid molecule toward free radical oxidation [45–47]. In-solution oxidation of 7DHC leads to over a dozen oxysterols [45,46] (Figure 3a), many of which are further metabolized in biological systems, leading to metabolically more stable oxysterols [48,49] (Figure 3b). For example, 7DHC 5α,6α-epoxide (7DHCep) can be readily metabolized into 3β,5α-dihydroxycholest-7-en-6-one (DHCEO) while compounds, 5α,6α-epoxycholest-7-en-3β,9α-diol (1) and 5,9-endoperoxy-cholest-7-en-3β,6α(β)-diol (EPCD-a or -b), can serve as precursors to 3β,5α,9α-trihydroxycholest-7-en-6-one (THCEO) and 3β,5α-dihydroxycholesta-7,9(11)-dien-6-one (DHCDO) [49]. On the other hand, 7DHC is also prone to oxidation by cytochrome P450 (CYP) as a number of such metabolites have been identified in vitro and in vivo [49–52] (Figure 3c). In particular, 7-DHC can be directly converted to 7-ketocholesterol (7-kChol) by CYP 7A1, which is a novel mechanism of formation for this known cytotoxic oxysterol that is normally derived from Chol [50]. Indeed, both free radical and enzymatic oxidation-derived oxysterols, including DHCEO, 7-ketocholesterol (7-kChol), 4α-hydroxy-7DHC (4α-OH-7DHC), 4β-OH-7DHC, and 24-OH-7DHC, have been identified in retinas from AY9944-treated rats [52] (Figure 3).

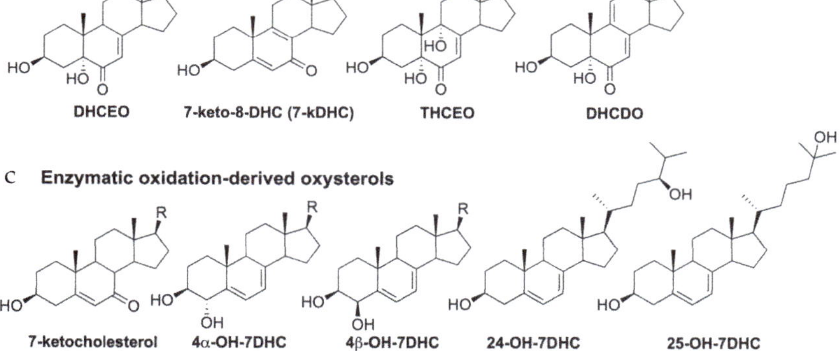

Figure 3. Chemical structure of some 7DHC-derived oxysterols: (**a**) Major oxysterols derived from free radical oxidation of 7DHC in organic solvent (benzene); (**b**) Oxysterols metabolites observed in biological systems that are derived from the primary oxysterols in solution; (**c**) Oxysterols derived from enzymatic oxidation of 7DHC by cytochrome P450 (CYP).

Several of these 7DHC-derived oxysterols have been shown to be highly toxic to cells in culture, e.g., EPCD-a, DHCEO, and 7-kChol, while others exhibit little or no cytotoxicity [53]. Importantly, cell culture studies have shown that several of these 7DHC-specific oxysterols are highly toxic to retina-derived cells, and that transformed photoreceptor-derived 661W cells are preferentially more sensitive to such oxysterol-induced cytotoxicity compared to other types of retina-derived cells (e.g., RPE and glial cells) [54]. This latter point is consistent with the finding that the retinal degeneration in the AY9944-induced SLOS rat model appears to be photoreceptor-specific (see above). The molecular basis for this relative selectivity of oxysterol toxicity remains to be elucidated. Perhaps it is due to the fact that retinal photoreceptor cells are the most highly differentiated cells in the entire body, and maintain this rather unique status at the expense of certain protective mechanisms found in other cell types. Regardless, our findings are consistent with those obtained by other investigators, which have demonstrated that different cell types and tissues exhibit different degrees of sensitivity to the cytotoxic effects of oxysterols [53,55–59]. Furthermore, intravitreal injection of a small amount of 7kChol to rats led to massive retinal degeneration within one week [52], which is also consistent with the phenotype observed in AY9944-treated rats. Lipid hydroperoxide levels are also markedly elevated in the retinas of AY9944-treated rats, compared to untreated controls, under normal ambient lighting conditions [60], and exposure of those animals to intense constant light dramatically exacerbates both the increased levels of lipid hydroperoxides and the severity of the retinal degeneration [61]. Conversely, systemic pretreatment of the SLOS rat model with an antioxidant prior to exposure to intense constant light offers significant protection against the light-induced retinal degeneration, with concomitant reduction in the levels of lipid hydroperoxides [61].

Taken together, these facts suggest that oxysterols (specifically those derived from 7DHC) may be causative in the retinal degeneration observed in the AY9944-induced SLOS rat model. If true, then blocking the formation of such oxysterols, e.g., with suitable antioxidants, should minimize or prevent the retinal degeneration from occurring [62,63]. Experiments to test that hypothesis have been performed and the results have been reported recently [64]. AY9944-treated adult rats were randomized into three groups: Group A rats were fed a Chol-free diet; Group B rats were fed a diet enriched in Chol (2 wt.%); Group C rats were fed a diet enriched in Chol (as for Group B) plus a mixture of water-soluble (vitamin C) and lipid-soluble (vitamin E) antioxidants and sodium selenite. A fourth group of rats, fed normal rat chow, served as untreated controls. At three posnatal months of age, electroretinograms were measured and then the rats were euthanized and their eyes were subjected to histological analysis. In good agreement with previous studies [34], Group A rats exhibited marked retinal degeneration and profoundly reduced ERG amplitudes, compared to untreated controls; feeding a high-cholesterol diet offered significant, but not complete, protection of retinas from degeneration, while feeding rats a diet enriched in both Chol and antioxidants provided remarkable and complete protection from retinal degeneration (no statistically significant differences between metrics of Group C and controls). The histological data from this experiment are provided in Figure 4.

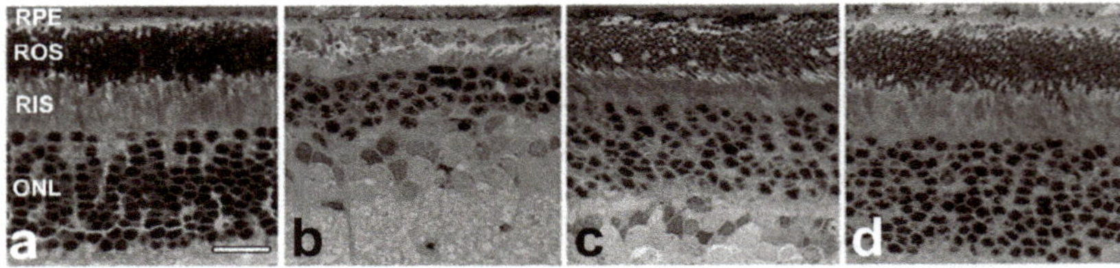

Figure 4. Retinal histology of normal control rat (**a**) and AY9944-treated (SLOS model) rats (**b**,**c**,**d**) as a function of dietary treatment (at postnatal three months of age). (**a**) Normal rat raised on standard (Chol-free) diet; (**b**) SLOS model rat raised on Chol-free diet; (**c**) SLOS model rat raised on 2%, by wt., Chol-enriched diet; (**d**) SLOS model rat raised on Chol-enriched diet supplemented with vitamins E and C and sodium selenite. Scale bar (*panel* **a**, for all panels), 20 µm. (*Adapted from Figure 1 in* [64] *with permission.*).

Consistent with these findings, and with the hypothesis that 7DHC-derived oxysterols might be causative in the retinal degeneration, the retinas of Group A rats exhibited markedly elevated oxysterol levels (compared to untreated controls), Group B rat retinas showed a statistically significant lowering of oxysterol levels (compared to Group A), and Group C rat retinas exhibited additional reduction in total oxysterol levels (*ca.* 36% compared to Group A). We note that the main 7-DHC-derived oxysterols observed in retinas of AY9944-treated rats were presumably enzymatically derived oxysterols, i.e., 7-kChol, 4α-OH-7DHC, and 4β-OH-7DHC, as many primary oxysterols derived from free radical oxidation of 7DHC are highly electrophilic and could exist in the forms of adducts with nucleophilic residues of proteins [65]. Relative quantification of the amounts of protein adducts with 7-DHC-derived oxysterols in the retinas of AY9944-treated rats has not been performed, but increased protein adduction with a common ω-6 polyunsaturated fatty acid-derived electrophile, 4-hydroxynonenal (HNE), has been reported recently [66]. Regardless, the complete rescue of the retinal degeneration by the combination of Chol and antioxidants suggest that damages caused by 7DHC-derived oxysterols, through either recepter interactions or direct adduction with proteins, are mostly prevented. Detailed assessment of such adduct formation with or without antioxidant treatment would provide further support to this conclusion.

A summary schematic, depicting a hypothetical mechanism for the AY9944-induced retinal degeneration and the potential role of antioxidants as a therapeutic intervention to protect against this degeneration, is given in Figure 5. When DHCR7 activity is compromised, as occurs in SLOS and due to inhibition by AY9944, 7DHC accumulates and Chol levels become reduced, compared to untreated controls, in bodily tissues, including the retina. A portion of the 7DHC is then oxidized to form various oxysterols, some of which are highly toxic to cells, resulting in a myriad of sequellae—including protein damage by lipid electrophiles (including some oxysterols), gene expression changes, membrane structural and functional changes, formation of reactive oxygen species (ROS) and reactive nitrogen species (RNS), etc.—that collectively lead to photoreceptor dysfunction, degeneration, death, and dropout. However, blocking oxysterol formation with suitable antioxidants can minimize or prevent these sequellae and the ensuing retinal degeneration. The model assumes that this intervention would also entail Chol supplementation, in order both to provide the required endproduct of the pathway (Chol) as well as to suppress the formation of 7DHC (feedback inhibition of *de novo* synthesis at the level of HMG-CoA reductase).

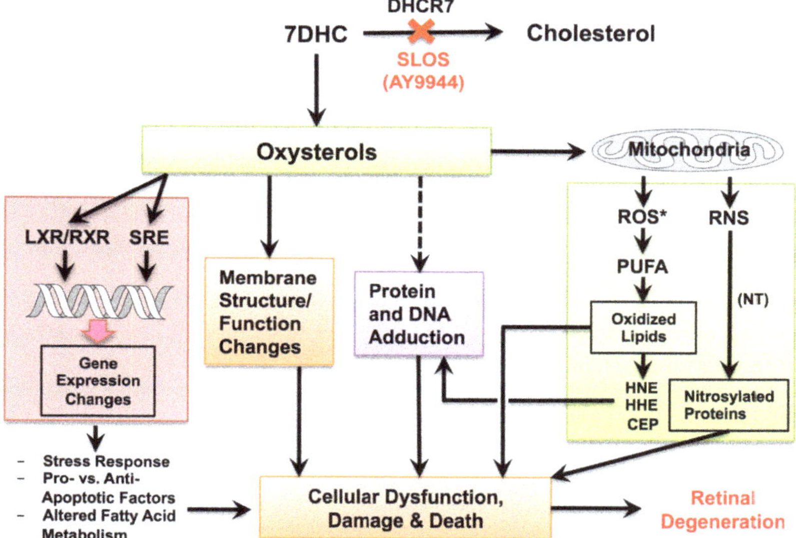

Figure 5. Schematic illustrating hypothetical mechanisms underlying retinal degeneration in the AY9944-induced rat model of SLOS and the therapeutic potential of antioxidants to prevent or minimize the degeneration. Abbreviations: LXR, liver X-receptor; RXR, retinoid X-receptor; SRE, sterol response elements; ROS*, reactive oxygen species; RNS*, reactive nitrogen species; HHE, 4-hydroxynonenal; HHE, 4-hydroxyhexenal; CEP, carboxyethylpyrrole; NT, nitrotyrosine; other abbreviations as described in text (*Adapted from Figure 29 in* [63] *with permission*.).

4. Perspective and Future Directions

While the results obtained with combined antioxidant-Chol dietary supplementation using the AY9944-induced SLOS rat model are provocative, their application to therapeutic intervention in SLOS patients, especially neonates and children, should be considered with due caution. As previously pointed out [64], the potential toxicities of antioxidants in humans must be considered as well as appropriate scaling of dosages from the rodent model to humans. Considerable additional preclinical studies as well as clinical trials are needed before these results can be adapted for human use in treating SLOS patients, including determination of optimal types, amounts, frequencey, and route of administration of antioxidants, alone and in combination.

The relevance of the retinal degeneration observed in the AY9944-induced rat SLOS model to the human disease is somewhat speculative, due to several factors. First, there is only one publication extant that describes the retinal histopathology in a human SLOS patient [67]. The findings presented in that single case report were from post-mortem eyes obtained from a one-month old male child. While the retinas exhibited essentially normal stratification of the histological cell layers and differentiation of retinal cells types, including the rods and cones, the ocular tissue sections showed signs of substantial post-mortem artifacts. However, at this early postnatal stage, retinal degeneration similar to that observed in the AY9944 rat model would not be expected. [Note also that in the rat model, the retinas are histologically and electrophysiologically normal at one postnatal month (see discussion above).] There are only two published electrophysiological (ERG) studies of human SLOS patients: One reported slower than normal rod phototransduction activation and deactivation kinetics [24], while the second [25] reported normal cone photoreceptor function in this same patient cohort. However, these findings cannot be directly compared to the AY9944 rat SLOS model because all of those patients received standard-of-care (CHOL supplementation) therapy, whereas the rat model employs a CHOL-free dietary regimen. [Note that, unlike rats, human SLOS patients are not allowed (due to medical ethics considerations) to have this standard-of-care therapy witheld.] Also, no fundus photos or OCT (optical coherence tomography) retinal imaging were provided, so the histological status of the retina in these patients cannot be ascertained. Second, the human disease is genetic and the mutations are on-board from the inception of early embyogenesis; by contrast, the rat model is pharmacological and does not introduce the inhibitor of DHCR7 (AY9944) until the second gestational week (roughly equivalent in timing to onset of the second fetal trimester in humans). The AY9944 rat SLOS model is likely an approximation of what would occur in a severely affected SLOS patient in the absence of any kind of therapeutic supplementation. That model treated with a high-CHOL diet may more faithfully approximate the cellular and electrophysiological scenario in the more mildly or moderately affected human SLOS patients' retinas.

In addition, the development of viable genetic mouse models of SLOS would be highly desirable to advance the field. Prior attempts to develop a global, homozygous *Dhcr7* knockout mouse have failed, due to early neonatal death (on postnatal day 0) [68,69]. Also, although a hypomorphic mutant *Dhcr7* knockin mouse line has been developed [70], it has been of limited value because its sterol profile progressively "self-corrects" over the first few postnatal months, for reasons that have yet to be elucidated with certainty. While treatment of normal animals with AY9944 or other DHCR7 inhibitors can mimic certain aspects of SLOS (notably the changes in tissue sterol profiles), it does not model the entire range of the SLOS phenotype and off-target pharmacological effects cannot be ruled out. Current studies are underway in our labs to develop targeted deletion of *Dhcr7* in specific retinal cell types in mice. A major obstacle to achieving this end has been the lack of a viable *Dhcr7*$^{flx/flx}$ mouse line. However, recently, that obstacle has been overcome and the targeted deletion of *Dhcr7* specifically in rod photoreceptor cells has been achieved (S.J. Fliesler, unpublished results). This breakthrough opens the way for targeted deletion of *Dhcr7*, not only in other retinal cell types (e.g., RPE cells, Müller glial cells, ganglion cells, etc.), but also in any cell type of interest throughout the body, in any tissue or organ (e.g., liver, brain, heart, kidney, etc.), using suitable available mouse lines that have targeted, promotor-driven expression of Cre recombinase in those cells, tissues, and organs of interest.

We currently are embarking on studies to evaluate Chol homeostasis in the retina, with particular emphasis on retinal photoreceptor cells vs. the contributions made by other retinal cell types as well as blood-borne Chol to the total steady-state levels of Chol in the retina.

Author Contributions: Both authors (S.J.F. and L.X.) contributed equally to Conceptualization, Methodology, Validation, Formal Analysis, Investigation, Resources, Data Curation, Writing-Original Draft Preparation, Writing-Review & Editing, Visualization, Supervision, Project Administration, and Funding Acquisition.

Funding: This research was funded, in part, by U.S.P.H.S. (NIH) grants R01 EY007361 (S.J.F.), R00 HD073270 and R01 HD092659 (L.X.); by Clinical and Translational Science Award UL1 TR001412 to the University at Buffalo-The State University of New York from the National Center for Advancing Translational Sciences (S.J.F.); by a Research to Prevent Blindness Unrestricted Grant to the Department of Ophthalmology, University at Buffalo-The State University of New York (S.J.F.); by startup funds from the Department of Medicinal Chemistry, School of Pharmacy, University of Washington (L.X.); and by facilities and resources provided by the VA Western New York Healthcare System (S.J.F.). S.J.F. is a Research Career Scientist Award recipient from the Department of Veterans Affairs, Biomedical Laboratory Research and Development Service (BLR&D). The opinions expressed herein do not necessarily reflect those of the Veterans Administration, the NIH, or the U.S. Government.

Acknowledgments: The authors congratulate W. David Nes on his lifetime of numerous accomplishments and seminal contributions to the field of sterol and steroid research. We thank the many people who contributed to the evolution and conduct of the studies reviewed herein over the years, including: Michael J. Richards, Barbara A. Nagel, Neal S. Peachey, Dana K. Vaughan, Bruce A. Pfeffer, Sriganesh Ramachandra Rao, Lara A. Skelton, Nadav I. Weinstock, Josi Herron, Kelly M. Hines, Zjelka Korade, Wei Liu, and Ned A. Porter. Portions of this work were presented at the 2018 Annual Meeting of the European Association for Eye and Vision Research (EVER 2018) in Nice, France; 4–6 October 2018 (S.J.F.).

Conflicts of Interest: The authors declare no conflict of interest. The funders had no role in the design of the study; in the collection, analyses, or interpretation of data; in the writing of the manuscript, and in the decision to publish the results.

References

1. Chevreul, M.E. Recherches chimiques sur les corps gras, et particulièrement sur leurs combinaisons avec les alcalis. Sixième mémoire. Examen des graisses d'homme, de mouton, de boeuf, de jaguar et d'oie. *Ann. Chim. Phys.* **1816**, *2*, 339–372. (In French)
2. Nes, W.D. Biosynthesis of cholesterol and other sterols. *Chem Rev.* **2011**, *111*, 6423–6451. [CrossRef] [PubMed]
3. Myant, N.B. *The Biology of Cholesterol and Related Sterols*, 1st ed.; Butterworth-Heinemann: Oxford, UK, 1981.
4. Yeagle, P.L. *Biology of Cholesterol*; CRC Press: Boca Raton, FL, USA, 1988.
5. Bittman, R. Cholesterol: Its functions and metabolism in biology and medicine. In *Subcellular Biochemistry*; Plenum Press: New York, NY, USA, 1997; Volume 28.
6. Finegold, L. (Ed.) *Cholesterol in Membrane Models*; CRC Press: Boca Raton, FL, USA, 1993.
7. Smith, L.L.; Johnson, B.H. Biological activities of oxysterols. *Free Radic Biol Med.* **1989**, *7*, 285–332. [CrossRef]
8. Luu, B.; Moog, C. Oxysterols: Biological activities and physiochemical studies. *Biochemie* **1991**, *73*, 1317–1320. [CrossRef]
9. Brown, A.J.; Jessup, W. Oxysterols and atherosclerosis. *Atherosclerosis* **1999**, *142*, 1–28. [CrossRef]
10. Schroepfer, G.J.J. Oxysterols: Modulators of cholesterol metabolism and other processes. *Physiol. Rev.* **2000**, *80*, 361–554. [CrossRef] [PubMed]
11. Griffiths, W.; Jörnvall, H. Oxysterols. *Biochim. Biophys. Res. Commun.* **2014**, *446*, 645–646. [CrossRef] [PubMed]
12. Mutemberezi, V.; Guillermot-Legris, O.; Muccioloi, G.G. Oxysterols: From cholesterol metabolites to key mediators. *Prog. Lipid Res.* **2016**, *64*, 152–169. [CrossRef] [PubMed]
13. Testa, G.; Rossin, D.; Poli, G.; Biasi, F.; Leonarduzzi, G. Implication of oxysterols in chronic inflammatory human diseases. *Biochemie* **2018**, *153*, 220–231. [CrossRef] [PubMed]
14. Smith, D.W.; Lemmli, L.; Opitz, J.M. A newly recognized syndrome of multiple congenital anomalies. *J. Pediatr.* **1964**, *64*, 210–217. [CrossRef]
15. Tint, G.S.; Batta, A.K.; Xu, G.; Shefer, S.; Honda, A.; Irons, M.; Elias, E.R.; Salen, G. The Smith-Lemli-Opitz syndrome: A potentially fatal birth defect caused by a block in the last enzymatic step in cholesterol biosynthesis. *Subcell Biochem.* **1997**, *28*, 117–144. [PubMed]

16. Porter, F.D.; Herman, G.E. Malformation syndromes caused by disorders of cholesterol synthesis. *J. Lipid Res.* **2011**, *51*, 6–34. [CrossRef] [PubMed]
17. Nowaczyk, M.J.; Irons, M.B. Smith-Lemli-Opitz syndrome: Phenotype, natural history, and epidemiology. *Am. J. Med. Genet. C Semin. Med. Genet.* **2012**, *160C*, 250–262. [CrossRef] [PubMed]
18. DeBarber, A.E.; Eroglu, Y.; Merkens, L.S.; Pappu, A.S.; Steiner, R.D. Smith-Lemli-Opitz syndrome. *Expert Rev. Mol. Med.* **2011**, *13*, e24. [CrossRef] [PubMed]
19. Kelley, R.I.; Herman, G.E. Inborn errors of sterol biosynthesis. *Annu. Rev. Genomics Hum. Genet.* **2001**, *2*, 299–341. [CrossRef] [PubMed]
20. Tint, G.S.; Irons, M.E.; Elias, E.R.; Batta, A.K.; Frieden, R.; Chen, T.S.; Salen, G. Defective cholesterol biosynthesis associated with the Smith-Lemli-Opitz syndrome. *N. Engl. J. Med.* **1994**, *330*, 107–113. [CrossRef] [PubMed]
21. Tint, G.S.; Seller, M.; Hughes-Benzie, R.; Batta, A.K.; Shefer, S.; Genest, D.; Irons, M.E.; Elias, E.R.; Salen, G. Markedly increased tissue concentrations of 7-dehydrocholesterol combined with low levels of cholesterol are characteristic of the Smith-Lemli-Opitz syndrome. *J. Lipid Res.* **1995**, *36*, 89–95. [PubMed]
22. Herman, G.E.; Kratz, L. Disorders of sterol synthesis: Beyond Smith-Lemli-Opitz syndrome. *Am. J. Med. Genet. C Semin. Med. Genet.* **2012**, *160C*, 301–321. [CrossRef] [PubMed]
23. Fierro, M.; Martinez, A.J.; Harbison, J.W.; Hay, S.H. Smith-Lemli-Opitz syndrome: Neuropathological and ophthalmological observations. *Dev. Med. Child. Neurol.* **1977**, *19*, 57–62. [CrossRef] [PubMed]
24. Elias, E.R.; Hansen, R.M.; Irons, M.; Quinn, N.B.; Fulton, A.B. Rod photoreceptor responses in children with Smith-Lemli-Opitz syndrome. *Arch. Ophthalmol.* **2003**, *121*, 1738–1743. [CrossRef] [PubMed]
25. Garry, D.; Hansen, R.M.; Moskowitz, A.; Elias, E.R.; Irons, M.; Fulton, A.B. Cone ERG responses in patients with Smith-Lemli-Opitz syndrome (SLOS). *Doc. Ophthalmol.* **2010**, *121*, 85–91. [CrossRef] [PubMed]
26. Sirtori, C.R. The pharmacology of statins. *Pharmacol. Res.* **2014**, *88*, 3–11. [CrossRef] [PubMed]
27. Davignon, J.; Montigny, M.; Dufour, R. HMG-CoA reductase inhibitors: A look back and a look ahead. *Can. J. Cardiol.* **1992**, *8*, 843–864. [PubMed]
28. Givner, M.L.; Dvornik, D. Agents affecting lipid metabolism—XV. Biochemical studies with the cholesterol biosynthesis inhibitor AY-9944 in young and mature rats. *Biochem. Pharmacol.* **1965**, *14*, 611–619. [CrossRef]
29. Kraml, M.; Marton, A.V.; Dvornik, D. Agents affecting lipid metabolism. XXVIII: A 7-dehydrocholesterol delat-7-reductase inhibitor (AY-9944) as tool in studies of delta-7-sterol metabolism. *Biochemistry* **1966**, *5*, 1060–1064.
30. Roux, C.; Horvath, C.; Dupuis, R. Teratogenic action and embryo lethality of AY 9944R: Prevention by a hypercholesterolemia-provoking diet. *Teratology* **1979**, *19*, 35–38. [CrossRef] [PubMed]
31. Roux, C.; Dupuis, R.; Horvath, C.; Talbot, J.N. Teratogenic effect of an inhibitor of cholesterol synthesis (AY 9944) in rats: Correlation with maternal cholesterolemia. *J. Nutrition.* **1980**, *110*, 2310–2312. [CrossRef] [PubMed]
32. Kolf Clauw, M.; Chevy, F.; Wolf, C.; Siliart, B.; Citadelle, D.; Roux, C. Inhibition of 7-dehydrocholesterol reductase by the teratogen AY9944: A rat model for Smith-Lemli-Opitz syndrome. *Teratology* **1996**, *54*, 115–125. [CrossRef]
33. Fliesler, S.J.; Richards, M.J.; Miller, C.-Y.; Peachey, N.S. Marked alteration of sterol metabolism and composition without compromising retinal development or function. *Invest. Ophthalmol. Vis. Sci.* **1999**, *40*, 1792–1801. [PubMed]
34. Fliesler, S.J.; Peachey, N.S.; Richards, M.J.; Nagel, B.A.; Vaughan, D.K. Retinal degeneration in a rodent model of Smith-Lemli-Opitz syndrome: Electrophysiological, biochemical, and morphological features. *Arch. Ophthalmol.* **2004**, *122*, 1190–2000. [CrossRef] [PubMed]
35. Rodriguez, I.R.; Fliesler, S.J. Photodamage generates 7-keto- and 7-hydroxycholesterol in the rat retina via a free radical-mediated mechanism. *Photochem. Photobiol.* **2009**, *85*, 1116–1125. [CrossRef] [PubMed]
36. Boulton, M.; Rózanowska, M.; Rózanowski, B. Retinal photodamage. *J. Photochem. Photobiol. B* **2001**, *64*, 144–161. [CrossRef]
37. Organisciak, D.T.; Vaughan, D.K. Retinal light damage: mechanisms and protection. *Prog. Retin. Eye Res.* **2010**, *29*, 113–134. [CrossRef] [PubMed]
38. Hunter, J.J.; Morgan, J.I.; Merigan, W.H.; Sliney, D.H.; Sparrow, J.R.; Williams, D.R. The susceptibility of the retina to photochemical damage from visible light. *Prog. Retin. Eye Res.* **2012**, *31*, 28–42. [CrossRef] [PubMed]

39. Xu, L.; Davis, T.A.; Porter, N.A. Rate constants for peroxidation of polyunsaturated fatty acids and sterols in solution and in liposomes. *J. Am. Chem. Soc.* **2009**, *131*, 13037–13044. [CrossRef] [PubMed]
40. Tu, C.; Li, J.; Sheflin, L.G.; Pfeffer, B.A.; Behringer, M.; Fliesler, S.J.; Qu, J. Ion-current-based proteomic profiling of the retina in a rat model of Smith-Lemli-Opitz syndrome. *Mol. Cell Proteomics* **2013**, *12*, 3583–3598. [CrossRef] [PubMed]
41. Serfis, A.B.; Brancato, S.; Fliesler, S.J. Comparative behavior of sterols in phosphatidylcholine-sterol monolayer films. *Biochim. Biophys. Acta* **2001**, *1511*, 341–348. [CrossRef]
42. Lintker, K.B.; Kpere-Daibo, P.; Fliesler, S.J.; Serfis, A.B. A comparison of the packing behavior of egg phosphatidylcholine with cholesterol and biogenically related sterols in Langmuir monolayer films. *Chem. Phys. Lipids* **2009**, *161*, 22–31. [CrossRef] [PubMed]
43. Keller, R.K.; Arnold, T.P.; Fliesler, S.J. Formation of 7-dehydrocholesterol-containing membrane rafts in vitro and in vivo, with relevance to the Smith-Lemli-Opitz syndrome. *J. Lipid Res.* **2004**, *45*, 347–355. [CrossRef] [PubMed]
44. Xu, X.; Bittman, R.; Duportail, G.; Heissler, D.; Vilcheze, C.; London, E. Effect of the structure of natural sterols and sphingolipids on the formation of ordered sphingolipid/sterol domains (rafts): Comparison of cholesterol to plant, fungal, and disease-associated sterols and comparison of sphingomyelin, cerebrosides, and ceramide. *J. Biol. Chem.* **2001**, *276*, 33540–33546. [PubMed]
45. Xu, L.; Korade, Z.; Porter, N.A. Oxysterols from free radical chain oxidation of 7-dehydrocholesterol: Product and mechanistic studies. *J. Am. Chem. Soc.* **2010**, *132*, 2222–2232. [CrossRef] [PubMed]
46. Xu, L.; Porter, N.A. Free radical oxidation of cholesterol and its precursors: Implications in cholesterol biosynthesis disorders. *Free Radic. Res.* **2015**, *49*, 835–849. [CrossRef] [PubMed]
47. Xu, L.; Korade, Z.; Rosado, D.A., Jr.; Liu, W.; Lamberson, C.R.; Porter, N.A. An oxysterol biomarker for 7-dehydrocholesterol oxidation in cell/mouse models for Smith-Lemli-Opitz syndrome. *J. Lipid Res.* **2011**, *52*, 1222–1233. [CrossRef] [PubMed]
48. Xu, L.; Korade, Z.; Rosdado, D.A., Jr.; Mirnics, K.; Porter, N.A. Metabolism of oxysterols derived from nonenzymatic oxidation of 7-dehydrocholesterol in cells. *J. Lipid Res.* **2013**, *54*, 1135–1143. [CrossRef] [PubMed]
49. Xu, L.; Liu, W.; Sheflin, L.G.; Fliesler, S.J.; Porter, N.A. Novel oxysterols observed in tissues and fluids of AY9944-treated rats: a model for Smith-Lemli-Opitz syndrome. *J. Lipid Res.* **2011**, *52*, 1810–1820. [CrossRef] [PubMed]
50. Shinkyo, R.; Xu, L.; Tallman, K.A.; Cheng, Q.; Porter, N.A.; Guengerich, F.P. Conversion of 7-dehydrocholesterol to 7-ketocholesterol is catalyzed by human cytochrome P450 7A1 and occurs by direct oxidation without an epoxide intermediate. *J. Biol. Chem.* **2011**, *286*, 33021–33028. [CrossRef] [PubMed]
51. Xu, L.; Mirnics, K.; Bowman, A.B.; Liu, W.; Da, J.; Porter, N.A.; Korade, Z. DHCEO accumulation is a critical mediator of pathophysiology in a Smith-Lemli-Opitz syndrome model. *Neurobiol. Dis.* **2012**, *45*, 923–929. [CrossRef] [PubMed]
52. Panini, S.R.; Sinensky, M.S. Mechanisms of oxysterol-induced apoptosis. *Curr. Opin. Lipidol.* **2001**, *12*, 529–533. [CrossRef] [PubMed]
53. Vejux, A.; Malvitte, L.; Lizard, G. Side effects of oxysterols: Cytotoxicity, oxidation, inflammation, and phospholipidosis. *Braz. J. Med. Biol. Res.* **2008**, *41*, 545–556. [CrossRef]
54. Lordan, S.; Mackrill, J.J.; O'Brien, N.M. Oxysterols and mechanisms of apoptotic signaling: implications in the pathology of degenerative diseases. *J. Nutr. Biochem.* **2009**, *20*, 321–336. [CrossRef] [PubMed]
55. Cilla, A.; Alegría, A.; Attanzio, A.; Garcia-Llatas, G.; Tesoriere, L.; Livrea, M.A. Dietary phytochemicals in the protection against oxysterol-induced damage. *Chem. Phys. Lipids.* **2017**, *207 (Pt B)*, 192–205. [CrossRef]
56. Goyal, S.; Xiao, Y.; Porter, N.A.; Xu, L.; Guengerich, F.P. Oxidation of 7-dehydrocholesterol and desmosterol by human cytochrome P450 46A1. *J. Lipid Res.* **2014**, *55*, 1933–1943. [CrossRef] [PubMed]
57. Xu, L.; Sheflin, L.G.; Porter, N.A.; Fliesler, S.J. 7-Dehydrocholesterol-derived oxysterols and retinal degeneration in a rat model of Smith-Lemli-Opitz syndrome. *Biochim. Biophys. Acta* **2012**, *1821*, 877–883. [CrossRef] [PubMed]
58. Korade, Z.; Xu, L.; Shelton, R.; Porter, N.A. Biological activities of 7-dehydrocholesterol-derived oxysterols: Implications for Smith-Lemli-Opitz syndrome. *J. Lipid Res.* **2010**, *51*, 3259–3269. [CrossRef] [PubMed]

59. Pfeffer, B.A.; Xu, L.; Porter, N.A.; Rao, S.R.; Fliesler, S.J. Differential cytotoxic effects of 7-dehydrocholesterol-derived oxysterols on cultured retina-derived cells: Dependence on sterol structure, cell type, and density. *Exp. Eye Res.* **2016**, *145*, 297–316. [CrossRef] [PubMed]
60. Richards, M.J.; Nagel, B.A.; Fliesler, S.J. Lipid hydroperoxide formation in the retina: Correlation with retinal degeneration in a rat model of Smith-Lemli-Opitz syndrome. *Exp. Eye Res.* **2006**, *82*, 538–541. [CrossRef] [PubMed]
61. Vaughan, D.K.; Peachey, N.S.; Richards, M.J.; Buchan, B.; Fliesler, S.J. Light-induced exacerbation of retinal degeneration in a rat model of Smith-Lemli-Opitz syndrome. *Exp. Eye Res.* **2006**, *82*, 496–504. [CrossRef] [PubMed]
62. Fliesler, S.J. Antioxidants: The missing key to improved therapeutic intervention in Smith-Lemli-Opitz syndrome? *Hereditary Genet.* **2013**, *2*, 119. [CrossRef] [PubMed]
63. Fliesler, S.J. Retinal Degeneration and Cholesterol Deficiency. In *Handbook of Nutrition, Diet and the Eye*, 1st ed.; Preedy, V.R., Ed.; Elsevier: London, UK, 2014; pp. 287–297, ISBN 9780124017177.
64. Fliesler, S.J.; Peachey, N.S.; Herron, J.; Hines, K.M.; Weinstock, N.I.; Ramachandra Rao, S.; Xu, L. Prevention of retinal degeneration in a rat model of Smith-Lemli-Opitz syndrome. *Sci. Rep.* **2018**, *8*, 1286. [CrossRef] [PubMed]
65. Windsor, K.; Genaro-Mattos, T.C.; Kim, H.Y.; Liu, W.; Tallman, K.A.; Miyamoto, S.; Korade, Z.; Porter, N.A. Probing lipid-protein adduction with alkynyl surrogates: Application to Smith-Lemli-Opitz syndrome. *J. Lipid Res.* **2013**, *54*, 2842–2850. [CrossRef] [PubMed]
66. Kapphahn, R.J.; Richards, M.J.; Ferrington, D.A.; Fliesler, S.J. Lipid-derived and other oxidative modifications of retinal proteins in a rat model of Smith-Lemli-Opitz syndrome. *Exp Eye Res.* **2018**, in press. [CrossRef] [PubMed]
67. Kretzer, F.L.; Hittner, H.M.; Mehta, R.S. Ocular manifestations of the Smith-Lemli-Opitz syndrome. *Arch. Ophthalmol.* **1981**, *99*, 2000–2006. [CrossRef] [PubMed]
68. Wassif, C.A.; Zhu, P.; Kratz, L.; Krakowiak, P.A.; Battaile, K.P.; Weight, F.F.; Grinberg, A.; Steiner, R.D.; Nwokoro, N.A.; Kelley, R.I.; et al. Biochemical, phenotypic and neurophysiological characterization of a genetic mouse model of RSH/Smith-Lemli-Opitz syndrome. *Hum. Mol. Genet.* **2001**, *10*, 555–564. [CrossRef] [PubMed]
69. Fitzky, B.U.; Moebius, F.F.; Asaoka, H.; Waage-Baudet, H.; Xu, L.; Xu, G.; Maeda, N.; Kluckman, K.; Hiller, S.; Yu, H.; et al. 7-Dehydrocholesterol-dependent proteolysis of HMG-CoA reductase suppresses sterol biosynthesis in a mouse model of Smith-Lemli-Opitz/RSH syndrome. *J. Clin. Investig.* **2001**, *108*, 905–915. [CrossRef] [PubMed]
70. Correa-Cerro, L.S.; Wassif, C.A.; Kratz, L.; Miller, G.F.; Munasinghe, J.P.; Grinberg, A.; Fliesler, S.J.; Porter, F.D. Development and characterization of a hypomorphic Smith-Lemli-Opitz syndrome mouse model and efficacy of simvastatin therapy. *Hum. Mol. Genet.* **2006**, *15*, 839–851. [CrossRef] [PubMed]

Article

Cytotoxic and Membrane Cholesterol Effects of Ultraviolet Irradiation and Zinc Oxide Nanoparticles on Chinese Hamster Ovary Cells

Regina E. Kuebodeaux, Paul Bernazzani and Thi Thuy Minh Nguyen *

Department of Chemistry and Biochemistry, Lamar University, Beaumont, TX 77710, USA; rekuebodeaux@gmail.com (R.E.K.); paul.bernazzani@lamar.edu (P.B.)
* Correspondence: ttnguyen15@lamar.edu; Tel.: +1-409-880-7262

Received: 30 September 2018; Accepted: 13 November 2018; Published: 15 November 2018

Abstract: Zinc Oxide (ZnO) nanoparticles are suspected to produce toxic effects toward mammalian cells; however, discrepancies in the extent of this effect have been reported between different cell lines. Simultaneously, high levels of ultraviolet (UV-C) radiation can have carcinogenic effects. The mechanism of this effect is also not well understood. Due to similarities in phenotype morphology after cell exposure to ZnO nanoparticles and UV-C irradiation, we emit the hypothesis that the toxicity of both these factors is related to damage of cellular membranes and affect their sterol content. Wild-type Chinese Hamster Ovary (CHO-K1) cells were exposed to ZnO nanoparticles or UV-C radiation. The amount of absorbed ZnO was determined by UV-visible spectroscopy and the changes in sterol profiles were evaluated by gas chromatography. Cell viability after both treatments was determined by microscopy. Comparing morphology results suggested similarities in toxicology events induced by ZnO nanoparticles and UV exposure. UV-C exposure for 360 min disrupts the sterol metabolic pathway by increasing the concentration of cholesterol by 21.6-fold. This increase in cholesterol production supports the hypothesis that UV irradiation has direct consequences in initiating sterol modifications in the cell membrane.

Keywords: UV-radiation; ZnO; toxicity; sterol content; cholesterol

1. Introduction

Advances in the field of nanotechnology where nanoparticles are commonly used in catalysts, sensors, and photovoltaic devices [1–3], as well as in the biomedical field for nanovaccines, nanodrugs, and diagnostic imaging tools [4–7], may also be associated with different health issues. The potential health impact nanoparticles raise concerns the fact that they have been incorporated into products used on a daily basis, as well as in medical products. There are several factors that determine nanoparticle cytotoxicity: the size, shape, surface charge, hydrophobicity, and heavy metal content [8]. It is important to understand what happens on a cellular level after nanoparticles have entered the human body through inhalation, ingestion, or absorption through the skin. Typically, the smaller the nanoparticle, the more cytotoxic it becomes due to its high surface area to volume ratio. Some studies revealed that not only can nanoparticles pass through cellular membranes, but they can also pass through the blood-brain barrier, which can result in organ deposition and interaction with biological systems and, therefore, alter the chemical biosynthesis [7,9–11].

Zinc Oxide (ZnO) nanoparticles are widely used [12] and recent investigations discuss their cytotoxicity, in particular, the relation between Zn^{2+} cations and the generation of reactive oxidative species, which contribute significantly to the degenerescence of macromolecules and, therefore, to the toxicity of Zn^{2+} [13,14]. In addition, the interaction of zinc in mammalian cells is complex. For example, zinc has been found to bind to enzymes related to sterol regulation [15] and help regulate the structure of cell membranes [16]. Hence, the possibility exists that ZnO nanoparticles will affect the sterol biosynthetic pathway, impacting the stability of the membrane and the phagocytosis process.

Another process that may affect sterol production in cells is the exposure to ultraviolet (UV) radiation. UV radiation is categorized into three wavelength ranges: long wave UV-A (315–400 nm), medium wave UV-B (280–315 nm), and short-wave UV-C (100–280 nm) [17–20]. The most energetic region, UV-C, is typically shielded by the ozone layer of the earth's atmosphere; however, due to ozone damage in recent years, the possibility exists that these radiations may reach the surface of the earth and affect sterol production in microalgae [21]. The effects of UV irradiation on mammalian cells remains unclear. In addition, UV radiation causes many harmful effects particularly regarding skin malignancies and severe damage to nucleic acids leading to DNA mutations; however, it can also be beneficial by creating vitamin D3 and is coupled in drug therapy applications for curing skin diseases like psoriasis and vitiligo [17,21]. The high energy radiation can cause damage to living organisms either by direct absorption through cellular chromophores or by indirect means through photo-sensitization mechanisms [21]. The indirect mechanism results in the formation of a reactive oxidative species, which can damage DNA, proteins, fatty acids, and saccharides [2]. There are therefore parallels between the effect of exposure to UV-C radiation and zinc ions. Both seem to affect the sterol biosynthetic pathways, and both can cause the generation of reactive oxidative species that lead to harmful effects.

We report the comparison of the cytotoxic effect of ZnO nanoparticles and UV-C irradiation on Chinese Hamster Ovary (CHO-K1) cells. We postulate that when exposed to either ZnO nanoparticles or UV radiation, the stability of cellular membranes will be modified because of variations in the sterol profile. Several analyses were performed on CHO cells after exposure to ZnO nanoparticles and UV-C irradiation to determine the amount of ZnO nanoparticles that cells can absorb, and alterations in cell growth and viability, DNA, and sterol profiles. The amount of ZnO nanoparticles was quantified using UV-Visible spectroscopy. Cell morphology and viability was determined with the use of compound light microscopy and the trypan blue exclusion method. Changes in DNA composition were visualized using agarose gel electrophoresis and sterol profile modifications were analyzed by gas chromatography.

2. Results and Discussion

2.1. UV-Visible Spectroscopy after ZnO Treatment

The ZnO nanoparticles used in this study were less than 100 nm in size. This is on the same scale as some biomolecules; therefore, we hypothesize that they are capable of passing through the protective cellular membrane. If so, nanoparticles could potentially enter the interior environment of the cell and this effect needs to be quantified.

UV-visible spectroscopy was used to determine the amount of ZnO that CHO-K1 cells absorbed after 24 h of exposure, by first determining the wavelength corresponding to the maximum absorbance (λmax) of the ZnO nanoparticles dissolved in the F-12K medium. Figure 1 displays the UV-visible spectra of the controls used: Ham's F-12K and 0.1 mg/mL ZnO in the F-12K medium. Nanoparticles tend to scatter the light when they are analyzed by UV-Visible spectroscopy, which shifted the baseline upward. The baseline shift was taken into account when determining the absorbance at λmax. The λmax of the 0.1 mg/mL ZnO solution was determined to be at 372 nm. The absorbance at the wavelength was calculated by subtracting the absorbance at 390 nm from the absorbance at the highest peak. Ham's F-12K was also analyzed and its λmax was found to be around 550 nm.

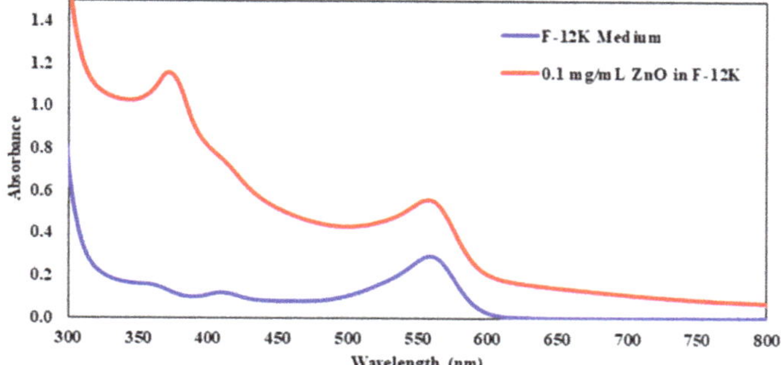

Figure 1. The UV-Visible Spectra of the controls: Ham's F-12K medium and 0.1 mg/mL of ZnO solution in an F-12K medium.

After establishing the λmax from the control sample, each ZnO exposed sample was analyzed by UV-Vis spectroscopy. This was performed to understand how the CHO cells respond to different ZnO concentration doses, and to determine at what concentration the cells reach saturation. Once the ZnO treatment was complete, the cells were pelleted by centrifugation and the clear supernatant liquid was analyzed to determine the number of unabsorbed ZnO nanoparticles. The UV-visible spectra after this first wash can be seen in Figure 2. The figure displays the unabsorbed ZnO absorbances from the wavelength range of 300–800 nm. From the lower concentrations (0–100 μg/mL), it is challenging to see a difference with the addition of ZnO. However, at 250 and 500 μg/mL, an absorbance at 372 nm can be observed clearly.

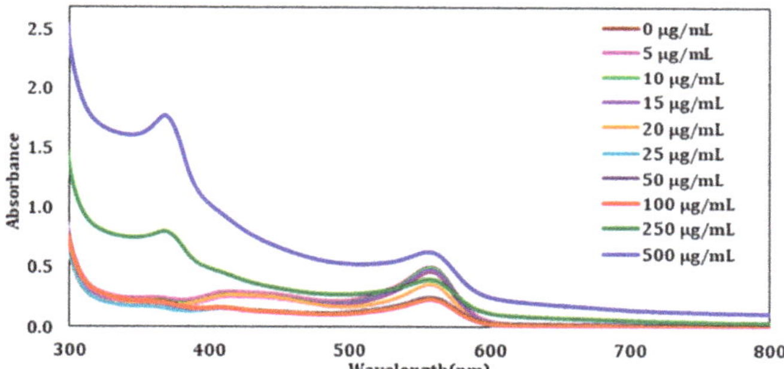

Figure 2. The UV-Visible Spectra of the supernatant of wash 1 after treatment with different ZnO concentrations.

The absorbance at 372 nm was used to determine the concentration of ZnO that CHO-K1 cells absorbed for each of the samples with the help of Equation (1).

$$C_s = C_{Initial\ [ZnO]} - \left[\frac{(C_c \times A_s)}{A_c}\right] \quad (1)$$

where C_s is the concentration of the sample and A_s is the absorbance at 372 nm for the sample. C_c and A_c correspond to the concentration and absorbance at 372 nm for the ZnO control. $C_{Initial\ [ZnO]}$ is the starting concentration of ZnO that each sample received. The concentration of each sample was calculated in mM and was plotted against the concentration of ZnO that was initially added to CHO cells (Figure 3). A steady increase in the amount of absorbed ZnO with an increase in dosage was observed. As expected, the absorption reaches a plateau when large concentrations of ZnO are present, indicating the possibility that the at least 250 μg/mL ZnO cause CHO-K1 cells to become saturated and no longer absorb or uptake the nanoparticles. The maximum amount these mammalian cells absorbed was 2.2 ± 0.2 mM ZnO. The 250 μg/mL sample had an initial concentration of 3.1 mM; therefore, around 70% of the ZnO nanoparticles were absorbed by the cells at this concentration. For the 500 μg/mL sample, 6.2 mM was the starting concentration of ZnO, which corresponds to around 35% absorption.

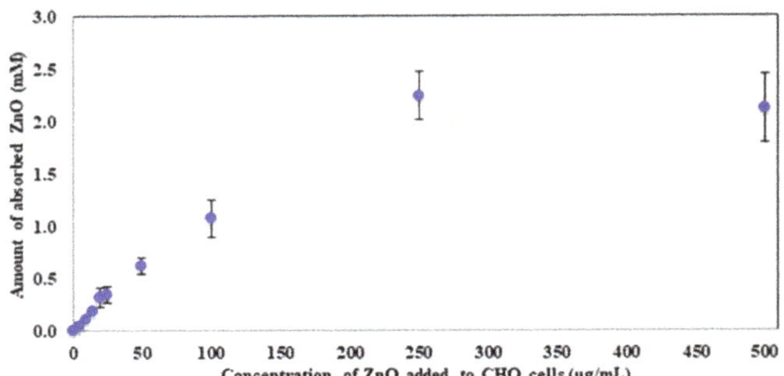

Figure 3. The average amount of absorbed ZnO (mM) after 24 h of ZnO treatment. Error bars represent the standard deviation of triplicates.

The results indicate that CHO-K1 cells are indeed capable of absorbing ZnO nanoparticles after 24 h of exposure. The amount absorbed by the cell increased with increasing ZnO dose and a saturation parameter was established at 250 μg/mL. To determine the toxicity of these nanoparticles, cell morphological changes and viability were examined by microscopy.

2.2. Cell Viability and Morphology

Cell viability and cell morphology help establish the degree of toxicity of external agents such as the addition of ZnO nanoparticles or UV-C irradiation. CHO-K1 cell viability and morphology were determined using a BioExpress GeneMate inverted microscope and trypan blue exclusion test. Trypan blue is a staining technique that allows for the observer to differentiate which cells are no longer viable. The unhealthy/dead cells have damaged cell membranes. This allows the cell to absorb the trypan blue dye and become blue, while the healthy, viable cells are not stained. Once the viable cells can be identified, the percent viability and viable cell concentration (viable cells/mL) can be calculated using a hemocytometer and Equations (2) and (3).

$$\%\ \text{Viability} = [1 - (\text{Number of blue cells} \div \text{Number of total cells})] \times 100 \quad (2)$$

$$\text{Viable cells/mL} = \text{Average number of viable cells} \times 16 \times 104 \quad (3)$$

CHO-K1 cells are adherent epithelial cells, meaning that when the cell culture matures, the cells attach to the culture flask. After the cells have adhered to the culture flask, their morphology changes to become more stretched and oblong, whereas the younger cells that have not adhered have a circular morphology. Figure 4 displays micrographs of these two different growth stages that CHO-K1 cells exhibit: the initial suspension stage (panel A) and the mature adherent stage (panels B and C). While cells in the suspension stage have a rounded morphology, mature cells in the adherent stage show significant elongation. The rounded cells still observed in panels B and C are expected and represent newly formed immature cells.

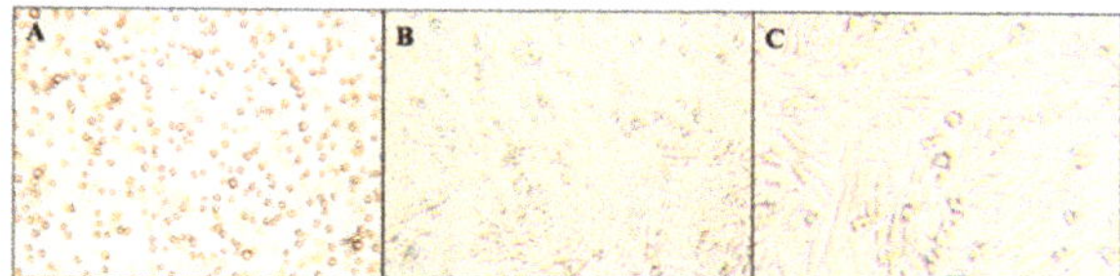

Figure 4. The wild-type CHO cell micrographs of the Initial Suspension Stage at 250× (**A**) and the Mature Adhered Phase at 250× (**B**) and at 400× (**C**).

2.3. Cell Viability and Morphology after ZnO Treatment

Figure 5 shows the percent viability curve from 0–500 μg/mL ZnO after cell incubation in the presence of different amounts of nanoparticle for 24 h at 37 °C with 5% CO_2. The cells were grown in the presence of ZnO in a complete growth medium containing Ham's F-12K media with 10% FBS and antibiotics. Cell viability after ZnO exposure was characterized with a decrease in the percent viability as the concentration of ZnO increased. From the concentrations of 15 μg/mL and lower, there was not much influence on viability. At 15 μg/mL ZnO, 91.6% of the cells were still viable. As the amount is increased to 20 μg/mL ZnO, it fell to 80.3%. A drastic drop of almost 24% in viability was observed between the ZnO concentrations of 20 to 25 μg/mL. A similar decrease in percentage was found from 25 to 50 μg/mL ZnO. Higher amounts resulted in a gradual decline in viability and, eventually, the lowest viability of 11.0% was observed at the highest concentration of 500 μg/mL ZnO.

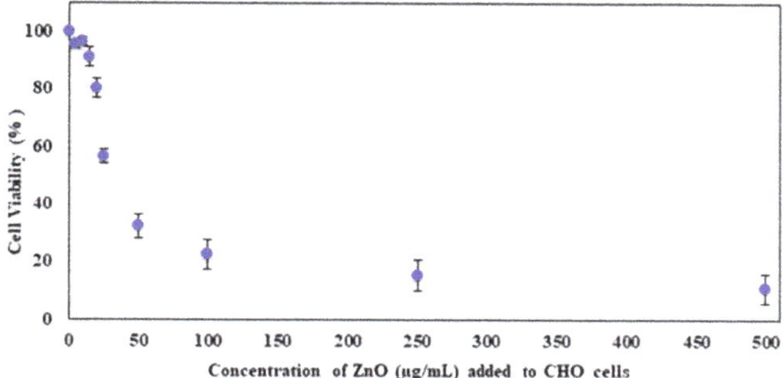

Figure 5. The average percent cell viability after 24 h of incubation with ZnO. Error bars represent the standard deviation of triplicates.

As expected, viability analysis confirmed that ZnO nanoparticles are toxic to CHO-K1 cells at higher concentrations, but morphology changes provide additional information about the toxicity of these nanoparticles. Changes in morphology could mean that the natural behavior of the cell is altered. Figure 6 shows the microscopy images of the culture flasks using an inverted microscope. The different panels represent typical behavior under different ZnO amounts. Images indicate that the morphology of CHO-K1 cells was directly impacted as the concentration of ZnO present increased. At concentrations of 20 μg/mL ZnO and below, the cells were able to maintain their natural adherent behavior. At 25 μg/mL of ZnO, some CHO cells were able to maintain the adherent morphology, but the majority of the cell culture expressed a round physiology. However, at amounts exceeding 25 μg/mL ZnO, the CHO-K1 cells stayed in the media and grew similar to suspension cells and stayed circular in form.

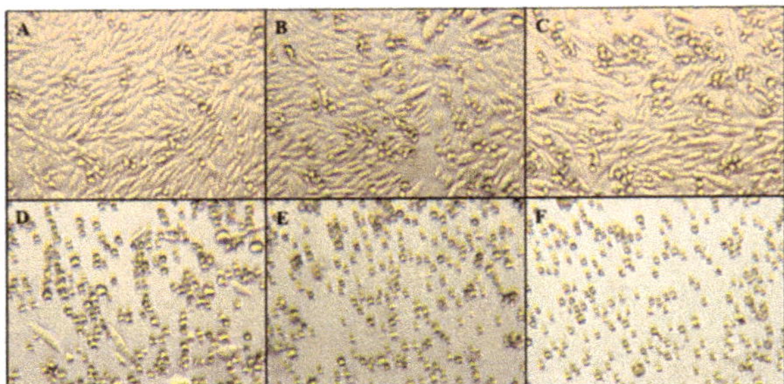

Figure 6. The CHO cell morphology micrographs (400×) of 0 μg/mL control (**A**), 10 μg/mL (**B**), 15 μg/mL (**C**), 25 μg/mL (**D**), 50 μg/mL (**E**), and 100 μg/mL ZnO (**F**) after 24 h of exposure.

The presence of Zinc oxide nanoparticles not only impacts the viability of CHO cells, but also influences their physical characteristics. When 25 μg/mL of ZnO and above were introduced, the cells were no longer able to completely adhere to the culture flask. This indicates that the mode of toxicity caused the natural function of the cell to change. These results were compared with the viability and morphological deviations after UV-C exposure.

2.4. Cell Viability and Morphology after UV-C Exposure

Figure 7 shows the viability curve of CHO cells as they were exposed to UV-C for longer times. With the increase in UV exposure time, there was a decrease in cell viability. The result provides for an almost linear relation, rather than an exponential decrease as the ZnO treatment provided. On average, cell viability dropped by 12.1%/min during the span of 360 min of UV exposure. At the maximum UV exposure time of 360 min, the number of viable cells was 27.0%.

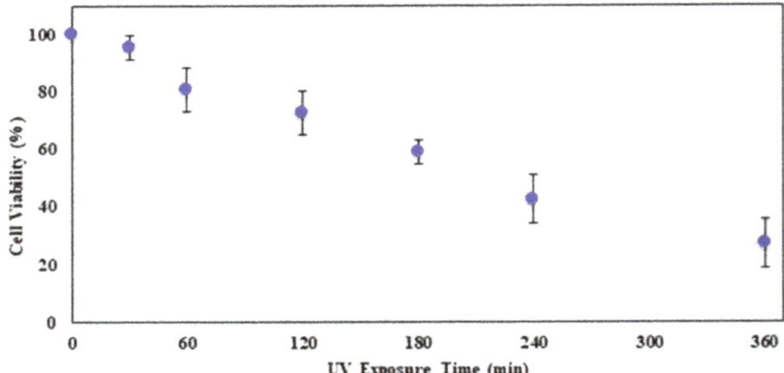

Figure 7. The average percent cell viability after different UV-C exposure times. Error bars represent the standard deviation of triplicates.

To determine the effect of UV-C irradiation on cell morphology, Figure 8 displays micrographs taken directly from the petri dish after UV exposure times of 0, 60, 180, and 360 min. All exposure times resulted in suspension-like cell cultures with the cellular structure staying in circular form. UV-C irradiation showed a similar, although stronger, impact on CHO cell morphology compared to exposure to ZnO nanoparticles. These results indicate that UV-C exposure has an extreme impact on cell morphology, but do not reveal how this radiation affects the cell membrane. As previously discussed, because the phenotype shows that cell membranes are affected, the toxicity can be attributed to either the formation of reactive oxidative species or an effect on the sterol biosynthetic pathway, which directly impacts the membrane. This latter assumption can be evaluated through gas chromatography analysis of the extracted membrane sterols after each UV exposure time.

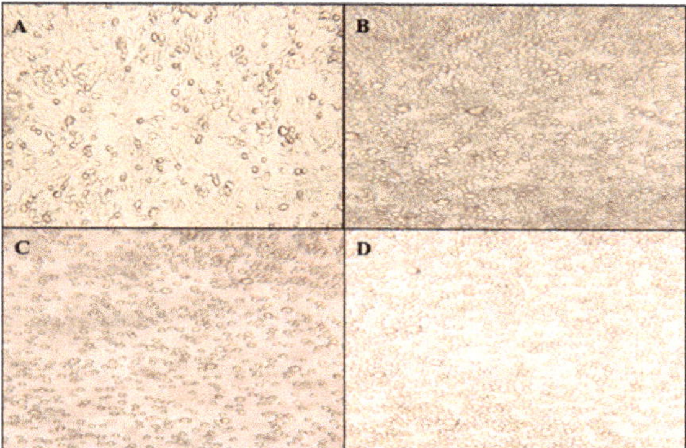

Figure 8. The CHO cell morphology micrographs (250×) of Wild-type CHO-K1 (**A**), 60 min (**B**), 180 min (**C**), and 360 min (**D**) of UV exposure.

Evaluation of the possible damage caused by reactive oxidative species on DNA can also be determined. Alterations in the DNA were first assessed before analyzing the sterol profile.

2.5. Agarose Gel Electrophoresis after ZnO Treatment

Agarose gel electrophoresis was used to evaluate possible alterations in CHO-K1 genomic DNA after incubation of different concentrations of ZnO. Electrophoresis was performed on whole genomic DNA without the use of restriction enzymes. Since restriction enzymes were not used, distinct bands were not observed in the agarose gel. Figure 9 displays the electrophoresis gel following this experiment.

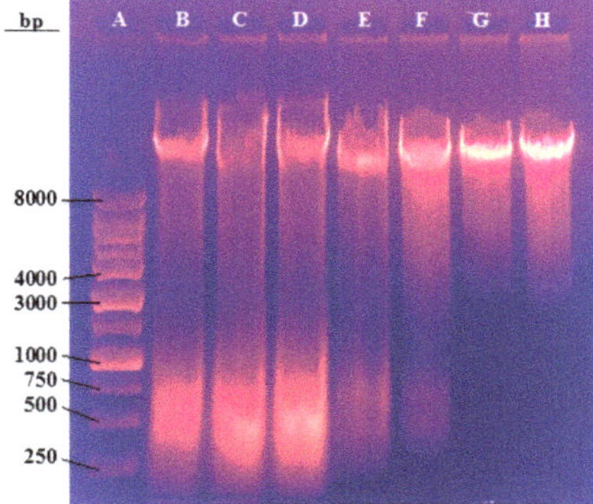

Figure 9. A total of 1% of Agarose gel of the samples exposed to different concentrations of ZnO: DNA Ladder (**A**), 0 μg/mL (**B**), 5 μg/mL (**C**), 10 μg/mL (**D**), 15 μg/mL (**E**), 25 μg/mL (**F**), 50 μg/mL (**G**), and 100 μg/mL (**H**).

The samples of 0 to 15 μg/mL ZnO (Lanes B–D) have a prominent band smear towards low molecular weight around 500 base pairs. This band corresponds to RNA molecules that were also purified during DNA extraction. Lanes E and F (20 and 25 μg/mL ZnO) displayed the bands as well, but were in trace amounts that resulted in faint bands. The higher concentrations of 50 and

100 μg/mL of ZnO (lanes G and H) had no appearance of the RNA band. This indicates that with increased ZnO dosage, CHO-K1 cell's RNA is destroyed. If RNA is no longer synthesized within the cell, then DNA replication cannot occur as well. When DNA replication is halted, the cell can no longer produce the necessary proteins for maintenance, survival, and the creation of a healthy internal environment. This explains why ZnO has such a direct effect on cell viability and morphology at the higher concentration doses.

Termination of cell replication is an extreme consequence of toxicity brought on by ZnO nanoparticles; therefore, membrane sterol modifications were not investigated. However, since the UV exposed CHO samples did not present this effect, modifications in their sterol profile was explored.

2.6. Gas Chromatography after UV Exposure

We emit the hypothesis that exposure to UV-C results in a direct effect on cell membrane integrity. Gas chromatography was utilized to analyze the sterol profile of CHO-K1 cells after UV-C exposure. Cholesterol was used as a standard at a concentration of 1.0 mg/mL. This specific sterol was chosen as a standard because cholesterol is the most abundant sterol found within mammalian cellular membranes. Figure 10 represents the typical GC trace of the cholesterol standard. The prominent peak at around 21 min corresponds to pure cholesterol, while the smaller peaks at around 25 and 26 min represent derivatives.

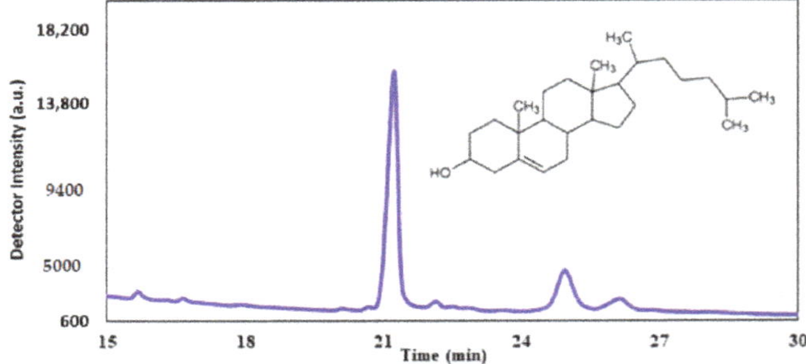

Figure 10. The GC chromatogram of standard cholesterol (1.0 mg/mL).

The sterol profiles after 0, 60, 240, and 360 min of UV-C exposure showed interesting results related to the deviation of cholesterol content. The GC chromatograms in Figure 11 show the peak related to cholesterol at a retention time of about 21 min, but do not take into account the number of cells that are viable in each sample. A decrease was observed in the amount of cholesterol after 60 min of UV exposure when compared to the sample that was not exposed. However, after 240 min of exposure, there was a slight increase in the amount of cholesterol that would generally be present under normal conditions. Intriguingly, after 360 min of exposure, the amount of cholesterol was substantially greater compared to all other samples. This provides evidence that UV-C irradiation has a direct impact on cholesterol found within the cell membrane of CHO-K1 cells, particularly after longer exposures.

Direct evaluation of the GC peak area does not demonstrate the true impact of cholesterol modification after UV exposure because each sample had a different concentration of cells that were left viable. Table 1 displays the calculated concentrations of cholesterol before and after the viability percentage was taken into account.

Figure 12 displays the calculated corrected concentration of cholesterol in μM based on the viability of each exposure time. An increase in the amount of cholesterol with increasing exposure time was found when correcting for concentration based on the number of viable cells after each exposure. These results support the hypothesis that UV-C irradiation affects cell membrane cholesterol levels.

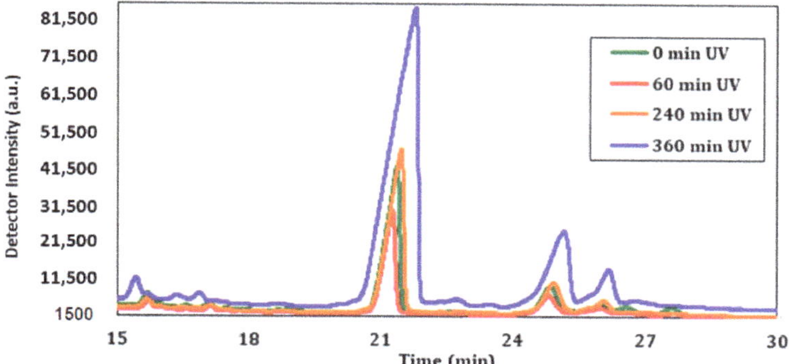

Figure 11. The GC chromatograms after CHO UV exposure for 0, 60, 240, and 360 min.

Table 1. The GC chromatogram areas, concentration (µg/µL and µM), percent viability, and calculated corrected cholesterol concentration (µg/µL and µM) based on cell viability after UV exposure starting with 1×10^7 cells per sample [a].

UV Exposure Sample (min)	Area (10^5) (a.u.)	Calc. (µg/µL)	Calc. µM (10^{-2})	Viability (%)	Corrected (µg/µL)	Corrected µM (10^{-2})
Standard	2.34	1.00	1.00	-	-	-
0	8.76	3.74	0.97	100.0	3.74	0.97
60	6.48	2.77	0.72	78.1	3.55	0.92
240	12.37	5.29	1.37	36.3	14.58	3.77
360	37.79	16.17	4.18	20.0	80.83	20.90

[a] UV exposure sterol samples were dissolved in 20 µL of methanol prior to GC injection.

The fact that viability and cholesterol are inversely proportional implies that the increase in cholesterol undermines the integrity of the cellular membrane, which results in the change in morphology and apoptosis. DNA and protein analyses need to be performed in future studies to better understand the mechanism that stimulates the spike in cholesterol production upon UV-C exposure.

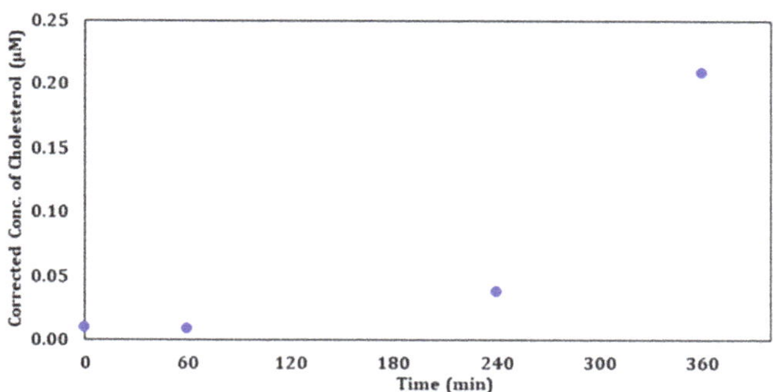

Figure 12. The corrected concentration of cholesterol (µM) after UV Exposure for 0, 60, 240, and 360 min.

3. Materials and Methods

3.1. Cell Culture

Wild-type Chinese Hamster Ovary, CCL-61, (CHO-K1) purchased from American Type Culture Collection (ATCC, Manassas, VA, USA), were cultured in Ham's F-12K with L-Glutamine medium (ATCC, 30-2004) containing 10% fetal bovine serum (FBS) (ATCC, 30-2020) and 1%

Penicillin/Streptomycin (ATCC, 30-2300). All culture media, serum, and antibiotics were purchased from ATCC. The cells were cultured at 37 °C in an atmosphere containing 5% carbon dioxide (CO_2) in a Steri-Cycle CO_2 Incubator (Thermo Scientific, Waltham, MA, USA, Model 370 Series). The cultures were grown in T-25 culture flasks until 80–90% confluency before ZnO (Sigma Aldrich, St-Louis, MO, USA, number 721077) or UV-C exposure.

3.2. Cryopreservation of CHO-K1 Cells

Cell culture was grown until 90% confluency and checked for cell viability and contamination immediately before cryopreservation. For CHO-K1 cells, the cryopreservation medium was prepared with Ham's F-12K with 20% FBS supplemented with 5% DMSO (ATCC) and 1% Penicillin/Streptomycin. Care was given to not add undiluted DMSO to the cell suspension as the dissolution of DMSO in aqueous solutions gives off heat.

The cells were collected by gentle centrifugation with a Thermo Scientific Sorvall Legend X1R centrifuge at 2500 rpm for 10 min, followed by the suspension of the pellet with the cryopreservation medium at a concentration of 1×10^6 to 5×10^6 viable cells/mL. The cryopreservation vials were labeled and 1 mL of the cell suspension was added to each vial and then sealed. The cells were allowed to equilibrate in the freeze medium at room temperature for a minimum of 15 min, but no longer than 40. This time was usually taken up in dispensing aliquots of the cell suspension into the vials. If left for longer than 40 min at room temperature, the cell viability may decline due to the DMSO. The vials were then placed on ice and stored in an −80 °C freezer for at least 24 h. Then the vials were transferred quickly to an ABS1 CryoMax Liquid Nitrogen Dewar (American BioTech Supply, Phenix Research Products, Candler, NC, USA) for long-term storage in the liquid nitrogen vapor phase. The location of vials within the Dewar was then recorded for future reference.

3.3. Subculturing CHO-K1 Cells

CHO-K1 cells were sub-cultured after they were 90% confluent in a T-25 flask. The culture medium was removed and discarded. Then the cell layer was rinsed with Ham's F-12K without serum several times and discarded. This was done to remove all traces of the serum, which contains the trypsin inhibitor. After rinsing, 2 mL of 0.25% (w/v) Trypsin-0.53 mM EDTA solution (ATCC) was added to each flask. Trypsin is an enzyme that allows the adherent cells to detach from the culture flask. To facilitate dispersal, the cells were placed at 37 °C and took about 10–15 min to complete. The cells should be monitored under an inverted microscope until the cell layer is dispersed in the solution. After the cells had completely detached from the flask, F-12K media without serum was added and the cells were aspirated by gently pipetting. Cell count and viability were determined, followed by centrifugation at 2500 rpm for 5 min. The F-12K media was discarded and fresh media was added and the centrifugation process was repeated. Then complete growth medium of F-12 K supplemented with 10% FBS with 1% Penicillin/Streptomycin was added to the cell pellet and appropriate aliquots of the cell suspension were added to new T-25 culture vessels. The cultures were then incubated at 37 °C and were examined the following days to ensure cell reattachment and active growth.

3.4. ZnO Nanoparticle Treatment and Analysis

CHO-K1 cells were exposed to 0, 5, 10, 15, 20, 25, 50, 100, 250, and 500 µg/mL ZnO for 24 h at 37 °C with 5% CO_2 in either 12 well-culture plates or T-25 culture vessels. The dose selection was obtained experimentally. Systematic trials over a large range of dosage showed a range where cells were still viable. Sigma Aldrich (Sigma Aldrich, St-Louis, MO, USA) provided ZnO nanoparticles that were less than 100 nm in size. Prior to ZnO incubation, the cells were cultured until 90% confluency and then were sub-cultured and split into additional flasks and grown until they were 90% confluent. Tables 2 and 3 provides the specific volumes and concentrations used for the 24 h ZnO treatment when a 12-well plate or a T-25 flask was used, respectively.

A Thermo Scientific Evolution 260 Bio UV-Visible Spectrophotometer was utilized to determine the concentration of ZnO nanoparticles that were unabsorbed after treatment. This was done by measuring the absorbance of three supernatants from 200–800 nm of each sample. After ZnO treatment, the cells were centrifuged at 2500 rpm for 5 min and the supernatant after centrifugation was analyzed by UV-Vis spectroscopy and classified as wash 1. Then PBS was added to the pelleted cells, resuspended, and centrifuged. This supernatant was analyzed and termed wash 2. The process was repeated one more time with PBS to yield the final wash 3. The controls used were 1 mg/mL and 0.1 mg/mL ZnO in F-12K media with 10% FBS, PBS, and F-12K media with 10% FBS. Water was used as a blank to calibrate the spectrophotometer.

Table 2. The concentration and volumes of ZnO and the culture medium used for ZnO treatment in a 12-well plate [a].

Flask	[ZnO] (μg/mL)	[ZnO] (μM)	Cell + Media (μL)	ZnO + Media (1 mg/mL) (μL)	Media (mL)	Total (mL)
1	0	0.0	650	0	4.350	5
2	5	61.5	650	25	4.325	5
3	10	123.0	650	50	4.300	5
4	15	184.5	650	75	4.275	5
5	25	307.5	650	125	4.225	5
6	50	615.0	650	250	4.100	5
7	100	1230.0	650	500	3.850	5

[a] CHO cells were seeded at 5×10^6 cells/well.

Table 3. The concentration and volumes of ZnO and the culture medium used for ZnO treatment in a T-25 culture flask [a].

Sample	[ZnO] (μg/mL)	[ZnO] (μM)	Cell + Media (μL)	ZnO + Media (0.1 mg/mL) (μL)	Media (μL)	Total (μL)
1	0	0.0	100	0	900	1000
2	5	61.5	100	50	850	1000
3	10	123.0	100	100	800	1000
4	15	184.5	100	150	750	1000
5	20	246.0	100	200	700	1000
6	25	307.5	100	250	650	1000
7	50	615.0	100	500	400	1000
8	100	1230.0	100	100 *	800	1000
9	250	3075.0	100	250 *	650	1000
10	500	6150.0	100	500 *	400	1000

[a] CHO cells were seeded at 5×10^6 cells/flask. * Taken directly from 1.0 mg/mL stock of ZnO/Ethanol.

3.5. DNA Purification and Agarose Gel Electrophoresis

Chinese Hamster Ovary cell DNA was extracted and purified after ZnO treatment for 24 h at 37 °C with PureLink Genomic DNA Mini Kit provided by Invitrogen. The procedure of preparing cell lysates, purifying, and extracting DNA was performed as the kit described for mammalian cells.

Agarose cell electrophoresis was performed with 1% agarose gel, 1× Tris/Acetic Acid/EDTA (TAE) buffer (Bio-Rad, Hercules, Berkeley, CA, USA), and ethidium bromide in a Bio-Rad Mini-Sub Cell GT apparatus. A PowerPac basic power supply from Bio-Rad was used to provide the voltage current during electrophoresis. The DNA samples were prepared with 6× loading buffer to yield a 1× sample and a DNA marker was loaded during each run as a control. The agarose gel was run at 100 V for approximately 30 min and then observed under UV for analysis.

3.6. *UV-C Irradiation Treatment*

UV irradiation treatment of 60 mm Petri dishes, prepared for sterol extraction following different exposure times: 0, 30, 60, 120, 180, 240, 360 min was performed in a Luzchem (Gloucester, ON, Canada) LZC-4× photoreactor (4 UV-C lamps, dose of 60,000 mW/m^2 with emission 235–280 nm, and peak emission at 254 nm) such that the lamps were above the samples. Prior to the addition in a petri dish, the CHO-K1 cells were cultured, sub-cultured, and then cultured until 90% confluency at 37 °C with 5% CO_2 in F-25K culture flasks. Then the cells were added to a petri dish at a concentration of 1×10^6 cells/mL with a final volume of 8 mL of Ham's F-12K supplemented with 10% FBS media with 1% penicillin-streptomycin. After each exposure time, the cell viability was analyzed using a 0.4% solution of Trypan Blue (American BioInnovations, Sparks, MD, USA).

3.7. *Sterol Extraction after UV-C Exposure*

The cell culture remaining after the allotted UV-C exposure time was washed twice with distilled water (dH_2O) and centrifuged at 2500 rpm for 10 min. Between each centrifugation, the clear supernatant fluid was discarded and the pellet was resuspended with fresh dH_2O. After the washes were complete, the cell pellet was stored at −20 °C overnight and sterols were extracted the following day.

To begin the sterol extraction process, 500 μL of dH_2O and 1 mL of 10% KOH in methanol was added to the cell pellet in a 15-mL centrifuge tube. The dH_2O/KOH/cell solution was then transferred to a small glass test tube and 100 μL of DMSO and a boiling chip was added. The contents within the test tube were then heated to 90–100 °C and allowed to boil for about 20 min. This was followed by the addition of 500 μL of dH_2O and then the solution was allowed to cool down to room temperature. This process allows for saponification to occur. The sterols within the cells are released because saponification causes hydrolysis of sterol esters and destroys the cellular tissues.

The free sterols were then extracted with 2 mL of hexane three times. The sterols travel to the clear organic hexane layer, and the polar compounds, such as phospholipids are attracted to the aqueous layer. Between each hexane addition, the test tubes were vortexed and centrifuged at 3300 rpm for a minute to allow the sterols to reach the organic layer more efficiently. The top organic layer was extracted and transferred to a smaller test tube and dried completely with N_2(g) at 60 °C.

Once the hexane was completely evaporated, acetone was added to the glass tube to dissolve any residue that might be present. The sample was then vortexed, sonicated to remove sterols from the side of the glass, and dried completely with N_2(g) at 60 °C. The same process of vortexing, sonicating, and drying with N_2(g) was performed with methanol two times after the acetone evaporated. Once completely dry, the sterol samples are ready for GC injection after the addition of 20.0 μL of methanol. The sterol samples were stored at −20 °C for future GC analysis.

3.8. *Gas Chromatography*

A Hewlett Packard 5890 series II model equipped with a flame ionization detector and non-polar capillary column (TG-SQC, 15 m, I.D.: 0.25 mm, Film: 0.25 μm) was used for the GC analysis. The initial temperature was set to 170 °C for 3 min and then ramped up at a rate of 20 °C/min to a final temperature of 280 °C, followed by an isothermal step for a total time of 30.5 min. The inlet temperature was 245 °C and the detector temperature was set to 280 °C. Cholesterol was used as a control to demonstrate retention time standards. The sterol profile of the samples isolated after UV exposure was analyzed after injecting 1 μL of sample into the GC, then plotting retention time against detector intensity. The peaks in the samples were analyzed using retention time in relation to the standard cholesterol retention time. The concentration of the cholesterol in each sample was calculated by the ratio of the area of the standard cholesterol peak compared to the cholesterol peak area of each sample.

4. Conclusions

Exposure of CHO-K1 cells to ZnO nanoparticles and UV-C irradiation proved to be damaging in terms of cell viability. UV-visible spectroscopy revealed that the 250 μg/mL of ZnO dose corresponds to the maximum absorption of ZnO nanoparticles. Similarities in cell morphology were observed when CHO cells were exposed to either UV radiation or ZnO nanoparticles. However, agarose gel electrophoresis performed after treatment demonstrated possible depletion in the ability to produce RNA after the cells were treated with ZnO concentrations of 25 μg/mL and higher, which suggests that the principal mode of toxic action of ZnO is through the generation of reactive oxidative species. Gas chromatography traces revealed that CHO-K1 irradiated with UV-C for 0, 60, 240, and 360 min had dramatic increases in cholesterol concentration present in cellular membranes compared to the wild-type CHO cells that received no UV treatment. The sharpest increase in cholesterol was observed at 360 min of exposure, which corresponded to about 20% cell viability.

Author Contributions: P.B. and T.T.M.N. conceived and designed the experiments; R.E.K. performed the experiments; P.B., R.E.K. and T.T.M.N. analyzed the data and wrote the paper.

Funding: This research was partially funded by the Welch Foundation (V-0004).

Conflicts of Interest: The authors declare no conflict of interest.

References

1. Shang, L.; Nienhaus, K.; Nienhaus, G.U. Engineered nanoparticles interactiong with cells: Size matters. *J. Nanobiotechnol.* **2014**, *12*. [CrossRef] [PubMed]
2. Saha, K.; Agasti, S.S.; Kim, C.; Li, X.N.; Rotello, V.M. Gold nanoparticles in chemical and biological sensing. *Chem. Rev.* **2012**, *112*, 2739–2779. [CrossRef] [PubMed]
3. Stratakis, E.; Kymakis, E. Nanoparticle-based plasmonic organic photovoltaic devices. *Mater. Today* **2013**, *16*, 133–146. [CrossRef]
4. Saroja, C.; Lakshmi, P.; Bhaskaran, S. Recent trends in vaccine delivery systems, a review. *Int. J. Pharm. Investig.* **2011**, *1*, 64–74.
5. Meng, F.H.; Cheng, R.; Deng, C.; Zhong, Z.Y. Intracellular drug release nanosystems. *Mater. Today* **2012**, *15*, 436–442. [CrossRef]
6. Yoo, J.W.; Irvine, D.J.; Di2scher, D.E.; Mitragotri, S. Bio-inspired, bioengineered and biomimetic drug delivery carriers. *Nat. Rev. Drug Discov.* **2011**, *10*, 521–535. [CrossRef] [PubMed]
7. Sambale, F.; Wagner, S.; Stahl, F.; Khaydarov, R.R.; Scheper, T.; Bahnemann, D. Investigations of the toxic effect of silver nanoparticles on mammalian cell lines. *J. Nanopart.* **2015**, *2015*. [CrossRef]
8. Meindl, C.; Kueznik, T.; Bosch, M.; Roblegg, E.; Frohlich, E. Intracellular calcium levels as screening tool for nanoparticle toxicity. *J. Appl. Toxicol.* **2015**, *35*, 1150–1159. [CrossRef] [PubMed]
9. Kim, J.S. Toxicity and tissue distribution of magnetic nanoparticles in mice. *Toxicol. Sci.* **2006**, *89*, 338–347. [CrossRef] [PubMed]
10. Foley, S.; Crowley, C.; Smaihi, M. Cellular localisation of a water-soluble fullerance derivative. *Biochem. Biophys. Res. Commun.* **2002**, *294*, 116–119. [CrossRef]
11. Kashiwada, S. Distribution of nanoparticles in the see-through medaka. *Environ. Health Perspect.* **2006**, *114*, 1697–1702. [CrossRef] [PubMed]
12. Zhang, Y.; Nguyen, K.C.; Lefebvre, D.E.; Shwed, P.S.; Crosthwait, J.; Bondy, G.S.; Tayabali, A.F. Critical experimental parameters related to the cytotoxicity of zinc oxide nanoparticles. *J. Nanopart. Res.* **2014**, *16*, 2440. [CrossRef] [PubMed]
13. Guo, D.; Bi, H.; Liu, B.; Wu, Q.; Wang, D.; Cui, Y. Reactive oxygen species-induced cytotoxic effects in zinc oxide nanoparticles in rat retinal ganglion cells. *Toxicol. In Vitro* **2013**, *27*, 731–738. [CrossRef] [PubMed]
14. Song, W.; Zhang, J.; Guo, J.; Zhang, J.; Ding, F.; Li, L.; Sun, Z. Role of the dissolved zinc ion and reactive oxygen species in cytotoxicity of ZnO nanoparticles. *Toxicol. Lett.* **2010**, *199*, 389–397. [CrossRef] [PubMed]
15. Zhang, L.; Fairall, L.; Goult, B.; Calkin, A.C.; Hong, C.; Millard, C.J.; Tontonoz, P.; Schwabe, J.W.R. The IDOL-UBE2D complex mediates sterol-dependent degradation of the LDL receptor. *Genes Dev.* **2011**, *25*, 1262–1274. [CrossRef] [PubMed]

16. Stohs, S.J.; Bagchi, D. Oxidative mechanisms in the toxicity of metal ions. *Free Radic. Biol. Med.* **1995**, *18*, 321–336. [CrossRef]
17. Svobodova, A.; Walterova, D.; Vostalova, J. Ultraviolet light induced alteration to the skin. *Biomed. Pap. Med. Fac. Univ. Palacky Olomouc Czech Repub.* **2006**, *150*, 25–38. [CrossRef] [PubMed]
18. Hussein, M.R. Ultraviolet radiation and skin cancer: Molecular mechanisms. *J. Cutan. Pathol.* **2005**, *32*, 191–205. [CrossRef] [PubMed]
19. Clydesdale, G.J.; Dandie, G.W.; Muller, H.K. Ultraviolet light induced injury: Immunology and inflammatory effects. *Immunol. Cell Biol.* **2001**, *79*, 547–568. [CrossRef] [PubMed]
20. Duthie, M.S.; Kimber, I.; Norval, M. The effects of ultraviolet radiation on the human immune system. *Br. J. Dermatol.* **1999**, *140*, 995–1009. [CrossRef] [PubMed]
21. Ahmed, F.; Schenk, P.M. UV-C Radiation increases sterol production in the microalga Pavlova lutheri. *Phytochemistry* **2017**, *139*, 25–32. [CrossRef] [PubMed]

Sample Availability: Samples of the compounds are not available from the authors.

Article

Nine New Gingerols from the Rhizoma of *Zingiber officinale* and Their Cytotoxic Activities

Zezhi Li [1], Yanzhi Wang [1,2,*], MeiLing Gao [1], Wanhua Cui [1], Mengnan Zeng [1], Yongxian Cheng [1,3] and Juan Li [1]

1 School of Pharmacy, Henan University of Chinese Medicine, Zhengzhou 450046, China; lzzdyq1992@163.com (Z.L.); gaoxiaomei6266@126.com (M.G.); qingyixin@163.com (W.C.); 13598851831@139.com (M.Z.); yxcheng@szu.edu.cn (Y.C.); jli_henantcm2017@163.com (J.L.)

2 Collaborative Innovation Center for Respiratory Disease Diagnosis, Treatment and New Drug Research and Development of Henan Province, Henan University of Chinese Medicine, Zhengzhou 450046, China

3 Guangdong Key Laboratory for Genome Stability & Disease Prevention, School of Pharmaceutical Sciences, School of Medicine, Shenzhen University Health Science Center, Shenzhen 518060, China

* Correspondence: wangyz@hactcm.edu.cn; Tel.: +86-136-7365-4931

Received: 9 December 2017; Accepted: 31 January 2018; Published: 2 February 2018

Abstract: Nine new gingerols, including three 6-oxo-shogaol derivatives [(*Z*)-6-oxo-[6]-shogaol (**1**), (*Z*)-6-oxo-[8]-shogaol (**2**), (*Z*)-6-oxo-[10]-shogaol (**3**)], one 6-oxoparadol derivative [6-oxo-[6]-paradol (**4**)], one isoshogaol derivative [(*E*)-[4]-isoshogaol (**5**)], and four paradoldiene derivatives [(4*E*,6*Z*)-[4]-paradoldiene (**8**), (4*E*,6*E*)-[6]-paradoldiene (**9**), (4*E*,6*E*)-[8]-paradoldiene (**10**), (4*E*,6*Z*)-[8]-paradoldiene (**11**)], together with eight known analogues, were isolated from the rhizoma of *Zingiber officinale*. Their structures were elucidated on the basis of spectroscopic data. It was noted that the isolation of 6-oxo-shogaol derivatives represents the first report of gingerols containing one 1,4-enedione motif. Their structures were elucidated on the basis of spectroscopic and HRESIMS data. All the new compounds were evaluated for their cytotoxic activities against human cancer cells (MCF-7, HepG-2, KYSE-150).

Keywords: *Zingiber officinale*; gingerols; cytotoxic activity

---

1. Introduction

Ginger, also known as white ginger, is the dry rhizome of *Zingiber officinale*, which has been a popular spice world-wide. As a Chinese medicine, it has been used in the treatment of nausea and vomiting, coughs, cold and so forth, for more than 2000 years [1–3]. In the traditional Chinese medicine theory system, *Z. officinale* is a warm medicine which can warm the spleen and stomach to dispel cold and warm the lungs, to reduce or eliminate dampness and phlegm (Huiyang Tongmai). Pharmacological studies on *Z. officinale* showed its effects on oxidative stress [4,5], tumor [6,7], degenerative diseases [8], vomiting [9], the cardiovascular system [10], and indigestion [11,12]. Chemical investigations have revealed the presence of volatile oils, gingerols, and diarylheptanoids, responsible for its pungency and other pharmacological properties [13]. Among the various activities of gingerols, numerous documents have reported anti-tumor activities, so we were interested in finding new types of gingerols with anti-tumor activity. Our efforts resulted in the isolation of nine new gingerols and eight known analogues (Figure 1), whose cytotoxic activities against human cancer cells (MCF-7, HepG-2, KYSE-150) were subsequently evaluated.

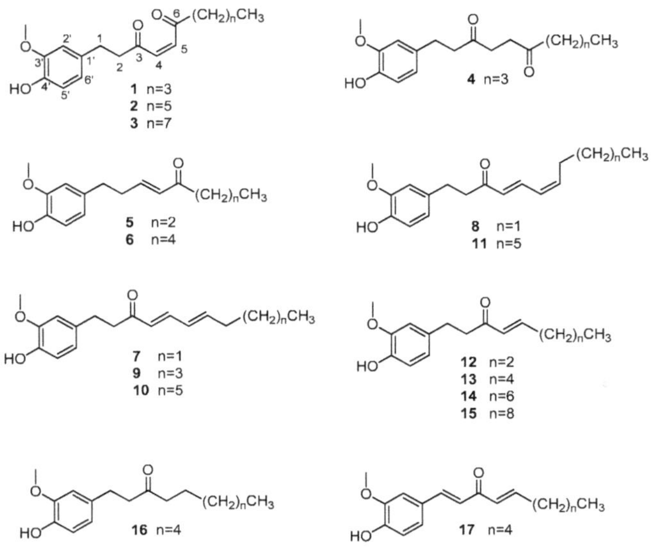

Figure 1. Structures of compounds 1–17.

2. Results and Discussion

Structure Elucidation of the Compounds

The EtOAc extract of *Z. officinale* rhizoma was submitted to a combination of column chromatography to produce compounds **1–17** (Figure 1).

Compound **2** was isolated as yellow oil and is demonstrated by a molecular formula of $C_{19}H_{26}O_4$, based on the HRESIMS ion peak at m/z 341.1727 $[M + Na]^+$ (calcd. for 341.1729).The ^{1}H-NMR (Table 1) spectrum of **2** shows the typical pattern of a coupling group of 1,3,4-trisubstituted benzene rings at δ_H 6.81 (1H, d, J = 8.0 Hz, H-5′), 6.67 (1H, d, J = 1.5 Hz, H-2′), 6.66 (1H, dd, J = 8.0, 1.5 Hz, H-6′), two olefinic protons [δ_H 6.83 (2H, s, H-4, H-5)], a methoxy group [δ_H 3.85 (3H, s, 3′-OCH_3)], seven methylene groups [δ_H 2.86 (2H, m, H-1), 2.94 (2H, m, H-2), 2.59 (2H, t, J = 7.4 Hz, H-7), 1.60 (2H, m, H-8), 1.28 (6H, m, H-9, H-10, H-11)] and a methyl group [δ_H 0.85 (3H, t, J = 6.4 Hz, H-12)]. In the ^{13}C-NMR spectrum of **2**, 19 carbon signals were observed, which included two carbonyls [δ_C 200.7 (C-3) and 199.8 (C-6)], six aromatic carbons [δ_C 146.4 (C-3′), 144.0 (C-4′), 132.4 (C-1′), 120.8 (C-6), 114.4 (C-5), 111.0 (C-2)], two olefinic carbons [δ_C 136.4 (C-4) and 136.1 (C-5)], and eight aliphatic carbons [δ_C 43.5 (C-2), 41.7 (C-7), 31.5 (C-10), 29.3 (C-1), 28.8 (C-9), 23.7 (C-8), 22.4 (C-11), 14.0 (C-12)]. The planar structure of **2** was further demonstrated by analyses of 2D NMR spectra. The HSQC showed that there is a correlation from δ_H 6.83 (2H, s, H-4, H-5) to two olefinic carbons (δ_C 136.4, C-4 and 136.1, C-5); while the HMBC presented a correlation from δ_H 6.83 (2H, s, H-4, H-5) to two carbonyl carbons [δ_C 200.7 (C-3) and 199.8 (C-6)], as well as a correlation from the methylene protons (H-1, H-2, H-7 and H-8) to two carbonyl carbons, but lacking the correlations from H-4 and H-5 to C-2 and C-7 or from H-2 and H-7 to C-4 and C-5 (Figure 2). It was found that there are two olefinic carbons between two carbonyl carbons. In addition, the methoxy group at δ_H 3.85 (3′-OCH_3) was located at C-3′ by the HMBC correlations from 3′-OCH_3 to C-3′, the substituent at C-4′was identified as a hydroxy group because of its ^{13}C-NMR chemical shift, and the side chain was located at C-1′ by HMBC correlations from H-1 to C-1′, C-2′ and C-3′ and from H-2 to C-1′.

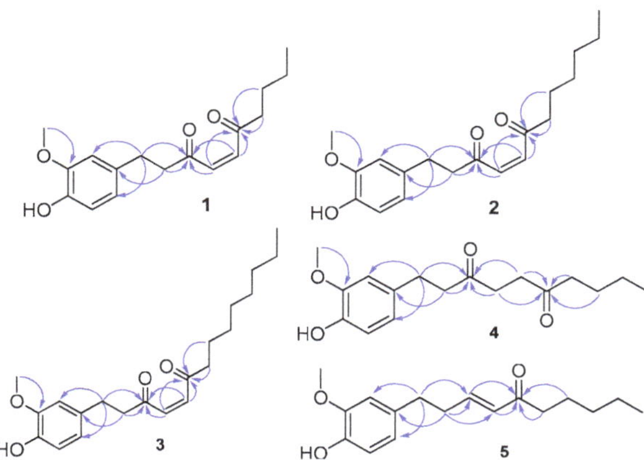

Figure 2. Key HMBC correlations of compounds **1**–**5**.

Table 1. ^{1}H- (500 MHz) and ^{13}C-NMR (125 MHz) data for compounds **1**–**4** in $CDCl_3$ (δ in ppm, J in Hz).

	1		**2**		**3**		**4**	
Position	δ_C	δ_H	δ_C	δ_H	δ_C	δ_H	δ_C	δ_H
1′	132.4		132.4		132.4		132.9	
2′	111.0	6.67, d (1.5)	111.0	6.67, d (1.5)	111.0	6.67, s	111.0	6.66, d (1.9)
3′	146.4		146.4		146.4		146.4	
4′	144.0		144.0		144.0		143.9	
5′	114.4	6.81, d (8.0)	114.4	6.81, d (8.0)	114.4	6.80, d (8.0)	114.2	6.79, d (8.0)
6′	120.8	6.65, dd (8.0,1.5)	120.8	6.66, dd (8.0,1.5)	120.8	6.65, d (8.0)	120.7	6.62, dd (8.0,1.9)
1	29.3	2.86, m	29.3	2.86, m	29.3	2.88, m	29.4	2.73, m
2	43.4	2.92, m	43.5	2.94, m	43.5	2.92, m	44.6	2.80, m
3	200.7		200.7		200.7		208.8	
4	136.4	6.82, s	136.4	6.83, s	136.4	6.83, s	36.1	2.65, m
5	136.0	6.82, s	136.1	6.83, s	136.0	6.83, s	35.9	2.62, m
6	199.8		199.8		199.8		209.7	
7	41.3	2.59, t (7.4)	41.7	2.59, t (7.4)	41.6	2.58, t (7.4)	42.5	2.42, t (7.4)
8	25.7	1.58, m	23.7	1.60, m	23.7	1.59, m	25.9	1.54, m
9	22.1	1.31, m	28.8	1.28, m	29.0	1.24, m	22.	1.27, m
10	13.7	0.88, t (7.3)	31.5	1.28, m	29.3	1.24, m	13.8	0.87, t (7.4)
11			22.4	1.28, m	29.3	1.24, m		
12			14.0	0.85, t (6.4)	31.7	1.24, m		
13					22.3	1.24, m		
14					14.0	0.85, t (6.4)		
3′-OCH_3	55.8	3.85, s	55.9	3.85, s	55.8	3.85, s	55.8	3.85, s

Finally, our focus was on determining the *cis*/*trans* configuration of double bonds. Due to the molecular symmetry, the magnetically equivalent vinylic protons of **2** behave as a singlet peak. Thus, it was impossible to directly determine the *E*/*Z*-configuration of the double bonds by the size of the vicinal coupling constants. We tested a variety of methodologies to solve this problem, including trying to cultivate single crystals, changing different deuterated reagents, and looking for more evidence on 2D NMR spectra, but none of them succeeded. After reviewing relevant literature [14], we found that olefinic protons split obviously when the compounds containing 1,4-enedione are in NMR spectra with deuterated benzene as a solvent. Subsequently, we changed the solvent to deuterated benzene, and we found that the singlet peak of olefinic protons became a quadruplet of AB system. The AB system is an advanced coupling and the coupling constant cannot be calculated directly. We calculated the actual chemical shift of the quadruplet, and the coupling constant was calculated to be 10.4 Hz by the chemical shift. Therefore, the two olefinic protons were *cis*-oriented, and the structure of **2** was determined to be (*Z*)-6-oxo-[8]-shogaol.

Compound **3** was obtained as yellow oil and shows a [M + Na]$^+$ ion at m/z 369.2042 in the HRESIMS, consistent with a molecular formula of $C_{21}H_{30}O_4$ (calcd. for 369.2042). Compound **1** was obtained as yellow oil, with the molecular formula $C_{17}H_{22}O_4$, as determined by the HREIMS at m/z 313.1414 for the [M + Na]$^+$ (calcd. for 313.1416). The ^{1}H- and ^{13}C-NMR data (Table 1) of **3** and **1** are similar to those of **2**, suggesting that they have the same carbon skeleton. The significant differences in these three compounds are the number of carbons in the aliphatic region of the ^{13}C-NMR and the ^{1}H-NMR spectra. Compound **3** has two more methylene groups in its side-chain than **2**, while **1** has two fewer methylene groups in side-chain than **2**. Similarly, the double bond moieties of **3** and **1** both behaved as a typical AB system in their ^{1}H-NMR spectrum when deuterated benzene was used as the solvent, and the coupling constants of these two compounds were calculated to be 10.4 Hz. On the basis of this evidence and literature comparison, the structures of **3** and **1** were determined to be (*Z*)-6-oxo-[10]-shogaol and (*Z*)-6-oxo-[6]-shogaol, respectively.

Compound **4** was obtained as brown oil. The molecular formula of **4** is assigned as $C_{17}H_{24}O_4$ on the basis of HRESIMS (m/z 315.1572 [M + Na]$^+$, calcd. for 315.1573). The NMR data (Table 1) of **4** closely resemble those of **1**; the analysis of the ^{1}H-NMR and HSQC data of **1** revealed that the significant differences in **4** are the absence of two olefinic protons at C-4 and C-5 and the presence of two additionally multiplet methylene groups at δ_H 2.65 and δ_H 2.62. This indicates that the 1,4-enedione-like moiety in **1** was replaced by a 1,4-dicarbonyl-like group in **4**. This deduction was supported by the HMBC correlations from the methylene protons, H-4 and H-5 (δ_H 2.65 and 2.62), to two carbonyl carbons (δ_C 209.7, C-3 and 208.7, C-6), from H-1 (δ_H 2.73) and H-2 (δ_H 2.80), to C-3 and from H-7 (δ_H 2.42), and H-8 (δ_H 1.54) to C-6 (Figure 2). Accordingly, the structure of **4** was further confirmed by the combined analyses of HSQC and HMBC data, and elucidated as 6-oxo-[6]-paradol.

Compound **5** was obtained as yellow oil. Its molecular formula, $C_{15}H_{20}O_3$, was established by HRESIMS at m/z 271.1312 for the [M + Na]$^+$ (calcd. for 271.1310). The ^{1}H-NMR spectrum (Table 2) of **5** exhibited the 1,3,4-tetrasubstituted aromatic moiety [δ_H 6.82(1H, d, J = 8.5 Hz, H-5′), 6.65 (1H, dd, J = 8.5, 1.9 Hz, H-6′), 6.64 (1H, s, H-2′)], two *trans*-conformational olefinic protons [δ_H 6.81(1H, m, H-3), 6.10 (1H, dd, J = 16.0, 1.5 Hz, H-4)], a methoxy group [δ_H 3.85 (3H, s, 3′-OCH$_3$)], four methylene groups [δ_H 2.70 (2H, t, J = 7.2 Hz, H-1), 2.48 (2H, t, J = 7.2 Hz, H-2), 2.47 (2H, t, J = 7.0 Hz, H-6) and 1.61 (2H, m, H-7)], and a methyl group [δ_H 0.83 (3H, t, J = 7.4 Hz, H-8)]. The ^{13}C-NMR and HSQC spectra of **5** revealed the presence of a carbonyl carbon [δ_C 200.7 (C-5)], two olefinic carbons [δ_C 145.9 (C-3) and δ_C 130.7 (C-4)], six aromatic carbons [δ_C 146.4 (C-3′), 143.9 (C-4′), 132.7 (C-1′), 120.9 (C-6′), 114.3 (C-5′) and 110.9 (C-2′)], a methoxy carbon [δ_C 55.85 (3′-OCH$_3$)], four methylene carbons [δ_C 42.0 (C-6), 34.47 (C-2), 34.16 (C-1), and 17.5 (C-7)], a methyl carbon [δ_C 13.8 (C-8)]. The NMR data of **5** are similar to those of [5]-shogaol [15], a typical α,β-unsaturated ketone-type structure. The significant difference in **5** is that the position of the olefinic carbons have changed. This deduction was supported by the HMBC correlations from the methylene protons, H-1 to C-2 and C-3, from H-2 to C-1, C-3 and C-4 and from H-6 and H-7 to C-5. In addition, H–H COSY correlations between H-2 and H-3 also proved this deduction (Figure 3). Accordingly, the structure of **5** was further confirmed by the combined analyses of HSQC and HMBC data. Thus, the structure of **5** was determined as (*E*)-[4]-isoshogaol.

Compound **9**, obtained as yellow oil has the molecular formula, $C_{19}H_{26}O_3$, based on the HRESIMS showing the [M + Na]$^+$ ion at m/z 325.1780 (calcd. for 325.1782). The ^{1}H-NMR data (Table 2) of **9** suggest the presence of a typical pattern of a coupling group of 1,3,4-trisubstituted benzene rings at δ_H 6.80 (1H, d, J = 8.0 Hz, H-5′), 6.68 (1H, d, J = 1.5 Hz, H-2′) and, 6.66 (1H, dd, J = 8.0, 1.5 Hz, H-6′), four olefinic protons [δ_H 7.10(1H, m, H-5), 6.14 (1H, m, H-6) at 6.13 (1H, m, H-7) and 6.05 (1H, d, J = 15.5 Hz, H-4)], a methoxy group [δ_H 3.85 (3H, s, 3′-OCH$_3$)], six methylene groups [δ_H 2.85 (2H, m, H-2) at 2.82 (2H, m, H-1), 2.15 (2H, m, H-8), 1.40 (2H, m, H-9) and 1.27 (4H, m, H-10 and H-11)], and a methyl group [δ_H 0.86 (3H, t, J = 7.0 Hz, H-12)]. The ^{13}C-NMR spectrum of **9** displays 19 signals, assigned by HSQC data to six aromatics [δ_C 146.4 (C-3′), 143.8 (C-4′), 133.2 (C-1′), 120.8 (C-6′), 114.3 (C-5′) and 111.1 (C-2′)], six methylene [δ_C 42.4 (C-2), 33.1 (C-8), 31.3 (C-10), 29.9 (C-1), 28.3 (C-9), 22.4 (C-11)], four olefinic carbons [δ_C 146.0 (C-7), 143.3 (C-5), 128.8 (C-6), 127.7 (C-4)], one ketone carbonyl [δ_C 199.9 (C-3)], and two methyls [δ_C 55.83 (3′-OCH$_3$), 14.0 (C-12)], with the former being a methoxy. The linkage of the conjugate double bond moiety to carbonyl carbon (C-3) was determined by the HMBC correlations of H-4 and H-5 to C-3. In addition, the significant chemical shift differences in ^{13}C-NMR between C-4 and C-5, C-6 and C-7 also proved this deduction (Figure 3).

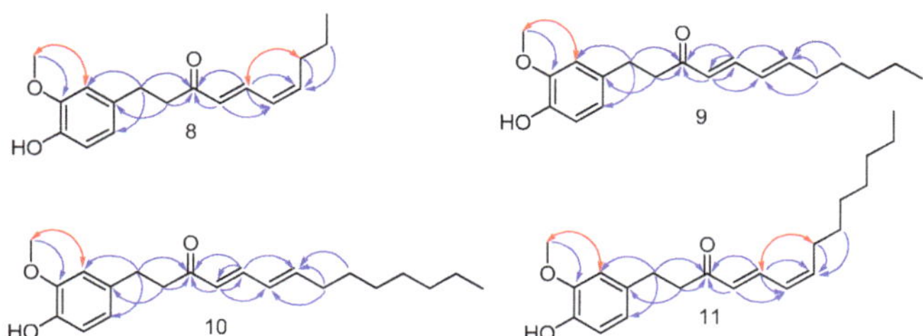

Figure 3. Key HMBC (blue) and NOESY (red) correlations of compounds **8–11**.

Table 2. ^{1}H- (500 MHz) and ^{13}C-NMR (125 MHz) data for compounds **5** and **8–11** in CDCl$_3$ (δ in ppm, J in Hz).

	5		8		9		10		11	
Position	δ_C	δ_H	δ_C	δ_H	δ_C	δ_H	δ_C	δ_H	δ_C	δ_H
1′	132.7		133.1		133.2		133.2		133.2	
2′	110.9	6.64, s	111.1	6.69, d(1.5)	111.1	6.68, d(1.5)	111.1	6.69, d(1.6)	111.1	6.68, d (1.5)
3′	146.4		146.4		146.4		146.4		146.4	
4′	143.9		143.8		143.8		143.8		143.9	
5′	114.3	6.82, d (8.5)	114.3	6.81, d (8.0)	114.3	6.80, d (8.0)	114.3	6.80, d (8.0)	114.3	6.81, d (8.0)
6′	120.9	6.65, dd (8.5, 1.9)	120.7	6.66, dd (8.0, 1.5)	120.8	6.66, dd (8.0, 1.5)	120.7	6.66, dd (8.0, 1.6)	120.8	6.66, dd (8.0, 1.5)
1	34.2	2.70, t (7.2)	29.9	2.83, m	29.9	2.82, m	29.9	2.82, m	29.9	2.83, m
2	34.5	2.48, t (7.2)	42.8	2.85, m	42.4	2.85, m	42.3	2.85, m	42.90	2.85, m
3	145.9	6.81, m	200.0		199.9		199.9		199.9	
4	130.7	6.10, dd (16.0,1.5)	129.3	6.14, d (15.3)	127.7	6.05, d (15.5)	127.7	6.05, d (15.5)	129.3	6.14, d (15.5)
5	200.7		137.4	7.48, m	143.3	7.10, m	143.4	7.10, m	137.4	7.45, m
6	42.0	2.47, t (7.0)	127.0	6.11, q (11.2)	128.8	6.14, m	128.7	6.14, m	126.8	6.09, q (11.0)
7	17.5	1.61, m	142.6	5.89, m	146.0	6.13, m	146.0	6.13, m	143.0	5.88, m
8	13.8	0.83, t (7.4)	30.3	2.25, m	33.1	2.15, m	33.1	2.15, m	29.3	2.26, q (7.5)
9			22.5	1.42, m	28.3	1.40, m	28.6	1.40, m	28.3	1.38, m
10			13.6	0.90, t (7.4)	31.3	1.27, m	29.1	1.25, m	29.1	1.25, m
11					22.4	1.27, m	29.0	1.25, m	29.0	1.25, m
12					14.0	0.86, t (7.0)	31.6	1.25, m	31.7	1.25, m
13							22.6	1.25, m	22.6	1.25, m
14							14.0	0.86, t (6.7)	14.1	0.85, t (6.1)
3′-OCH$_3$	55.85	3.85	55.81	3.85, s	55.83	3.8, s	55.8	3.85, s	55.85	3.85, s

A large coupling constant between H-4 and H-5 (J = 15.5 Hz) indicated that the two olefinic protons are *trans*-oriented. We made a bold guess to determine the orientation of the other two olefins in the conjugated double bonds. It is well known that a conjugated double bond forms a conjugated π system by resonance hybrid, where the double bond between carbon and carbon tends to become longer and the single bond tends to become shorter; thus, the bond between each carbon has the characteristics of double bonds, and each bond cannot be rotated. Therefore, a NOESY correlation should be effective in this conjugated double bond moiety. So, if H-6 and H-7 are *cis*-oriented, the NOESY correlation between H-5 and H-8 can be observed, but if they are *trans*-oriented, this correlation cannot occur. Hence, the absolute configuration of **9** was determined by detailed analysis of its NOESY spectrum, and we did not observe the NOESY interaction between H-5 and H-8 in the NOESY spectrum; in this case, we speculated that H-6 and H-7 must be *trans*-oriented, and previously reported literature [16–18] with similar structures confirms our conjecture. Thus, on the basis of this evidence and literature comparison, the structure of **9** was determined to be (4*E*,6*E*)-[6]-paradoldiene.

The HRESIMS of Compound **10** was obtained as yellow oil and showed a [M + Na]$^+$ ion at m/z 353.2091, consistent with a molecular formula of $C_{21}H_{30}O_3$ (calcd. for 353.2093). The ^{1}H- and ^{13}C-NMR data (Table 2) of **10** closely resemble those of **9**, with a slight difference in the number of carbons in the aliphatic region of the NMR spectra. From this, it can be seen that the side chain of compound **10** should have two more methylene groups than that of compound **9**. Thus, the structure of **10** was determined to be (4*E*,6*E*)-[8]-paradoldiene.

Compound **8** was obtained as yellow oil, and has the molecular formula, $C_{17}H_{22}O_3$, by HRESIMS [found *m*/*z* 297.1465 (calcd. for 297.1467) $[M + Na]^+$].The ^1H- and ^{13}C-NMR spectroscopic data (Table 2) of **8** are similar to those of **9**. The main difference is the chemical shift of the carbons and protons of the conjugated double bond. A large coupling constant between H-4 and H-5 (J = 15.3 Hz) indicated that the two protons were *trans*-oriented, and a smaller coupling constant between H-6 and H-7 (J = 11 Hz) indicated that the two protons were *cis*-oriented. In addition, the NOESY correlation of H-5 and H-8 demonstrates that H-6 and H-7 are *cis*-oriented (Figure 3). Thus, the structure of **8** was determined to be (4*E*,6*Z*)-[4]-paradoldiene.

Compound **11** was obtained as yellow oil and has the same molecular formula ($C_{21}H_{30}O_3$) as **10** by HRESIMS analysis. The ^1H- and ^{13}C-NMR data (Table 2) of **11** closely resemble those of **8**, with a slight difference in the number of carbons in the aliphatic region of the NMR spectra. From this, it can be seen that the side chain of compound **11** should have four more methylene groups than that of compound **8**. In addition, the NOESY correlation of H-2′ and 3′-OCH_3 also demonstrates that the methoxy is located at C-3′ (Figure 3). Thus, the structure of **11** was determined to be (4*E*,6*Z*)-[8]-paradoldiene.

Known metabolites were identified as (4*E*)-[4]-shogaol (**12**) [15], (4*E*)-[6]-shogaol (**13**) [15], (4*E*)-[8]-shogaol (**14**) [15], (4*E*)-[10]-shogaol (**15**) [15,19], [8]-paradol (**16**) [15], (1*E*,4*E*)-[6]-dehydroshogaol (**17**) [15] (Figure 1) by comparing their spectroscopic data (^1H-NMR, ^{13}C-NMR, ^1H-^1H COSY, HSQC, HMBC, and NOESY in Supplementary Materials) with literature values. In addition, by comparing the spectral data of the new compounds that have been identified, we have identified the following two compounds: (3*E*)-[6]-isoshogaol (**6**) and (4*E*,6*Z*)-[4]-paradoldiene (**7**).

3. Experimental Section

3.1. General Experimental Procedures

UV spectra were measured on a Thermo EVO 300 spectrophotometer (Thermo Fisher Scientific, Madison, WI, USA). IR spectra were recorded on a Thermo Nicolet IS 10 spectrometer (Thermo Fisher Scientific, Madison, WI, USA). NMR spectra were recorded on a Bruker Avance III 500 spectrometer (Bruker Biospin, Fallanden, Switzerland) with TMS as an internal standard. HRESIMS data were recorded on a Bruker maxis HD Mass Q-TOF LC/MS spectrometer (Bruker Daltonics, Billerica, MA, USA). Preparative HPLC was performed on a Sepuruisi LC-52 instrument (Beijing Sepurusi Scientific Co., Ltd., Beijing, China) with an UV200 detector (Beijing Sepurusi Scientific Co., Ltd., Beijing, China), using a YMC-Pack ODS-A column (250 mm × 20 mm, 5 μm). Column chromatography was undertaken on HP-20 macroporous resin (Mitsubishi Chemical Co., Ltd., Tokyo, Japan), SephadexLH-20 (Amersham Pharmacia, Uppsala, Sweden), ODS (50 μm, YMC Co. Ltd., Kyoto, Japan), and silica gel (200–300 mesh, 100–200 mesh, Qingdao Marine Chemical Inc., Qingdao, China). TLC was carried out with glass that was pre-coated with silica gel GF 254 (Qingdao Marine Chemical Inc.) The human heptocelluar (HepG-2) cell line, human mammary cancer (MCF-7) cell line, and human esophageal cancer (KYSE-150) cell line were purchased from the Institute of Materia Medica, Chinese Academy of Medical Sciences and Peking Union Medical College, Beijing, China.

3.2. Plant Material

The rhizoma of *Z. officinale* was collected from Bozhou herbal medicine market in Anhui province, China in November 2014, and the locality of growth was Luoping County, Yunnan Province. It was identified by Prof. Sui-Qing Chen at Henan University of Chinese Medicine. A voucher specimen

(No. 20141192A) was deposited at the Department of Natural Medicinal Chemistry, School of Pharmacy, Henan University of Chinese Medicine, Zhengzhou, China.

3.3. Extraction and Isolation

The dried rhizoma of *Z. officinale* (50 kg) was crushed into coarse powder and macerated for 1 h with 200 L EtOAc, refluxed at 75 °C (3 × 200 × 1.5 h). After removing the solvent under reduced pressure, the EtOAc extract (4 kg) was dissolved in 60% EtOH and divided into three parts (Fr.1–Fr.3) using Diaion HP-20 macroporous resin column eluted with gradient aqueous EtOH (60%, 80%, and 100%). Fr.1 (1.5 kg) was applied to a silica gel (100–200 mesh) column and eluted successively with CH_2Cl_2: MeOH (100:0, 30:1, 20:1, 10:1, 5:1, 2:1, 0:100) to yield seven subfractions (Fr.1.1–Fr.1.7). Fraction Fr.1.1 (500 g) was subjected to another silica gel (200–300 mesh) column eluting with a step-gradient of petroleum ether/acetone (from 100:0 to 0:100) to provide nine subfractions: Fr.1.1.1–Fr.1.1.9. Fr.1.1.4 (80 g) was separated by a MCI gel column (MeOH/H_2O, 10–70%) to get five fractions (Fr.1.1.4.1–Fr.1.1.4.5), of which, fraction Fr.1.1.4.3 (12 g) was purified by Sephadex LH-20 (MeOH), followed by semi-preparative HPLC (MeCN/H_2O, 65%, flow rate: 3 mL/min) to yield **17** (21.2 mg, R_t = 16.8 min), **2** (11.2 mg, R_t = 17.6 min), **13** (33.6 mg, R_t = 20.1 min), **16** (18.1 mg, R_t = 23.5 min) and **14** (78.5 mg, R_t = 43.8 min). Fr.1.1.4.2 (10 g) was chromatographed over a silica gel (200–300 mesh) column, eluted with a step-gradient of petroleum ether/EtOAc (from 100:0 to 0:100) and was further purified with semi-preparative HPLC (MeCN/H_2O, 55%, flow rate: 3 mL/min) to yield **1** (91.2 mg, R_t = 8.9 min), **4** (43.3 mg, R_t = 9.6 min), **12** (16.7 mg, R_t = 10.7 min) and **3** (27.3 mg, R_t = 35.5 min). Fr.1.1.4.5 (9 g) was subjected to a silica gel (400–500 mesh) column, eluted with a step-gradient of petroleum ether/EtOAc (from 100: 0 to 0: 100) to get six fractions (Fr.1.1.4.5.1–Fr.1.1.4.5.6). The Fr.1.1.4.5.3 (1.2 g) was further purified by semi-preparative HPLC (MeCN/H_2O, 65%, flow rate: 3 mL/min) to yield **5** (22.3 mg, R_t = 19.8 min) and **6** (17.6 mg, R_t = 14.3 min). Fr.1.1.4.5.1 (1.7 g) was further purified by semi-preparative HPLC (MeCN/H_2O, 80%, flow rate: 3 mL/min) to yield **7** (28.8 mg, R_t = 8.7 min), **8** (34.7 mg, R_t = 9.1 min), **9** (59.9 mg, R_t = 15.2 min), **10** (71.6 mg, R_t = 27.2 min), **11** (55.7 mg, R_t = 27.7 min), and **15** (59.1 mg, R_t = 35.6 min).

3.4. Spectroscopic and Physical Data

Compound **1**: yellow oil; UV (MeOH) λ_{max} (logε) 281 (2.83), 202(3.91) nm; IR (neat) ν_{max} 3391, 2957, 1682, 1516, 1272, 1033 cm^{-1}. ^{1}H (500 MHz, $CDCl_3$) and ^{13}C (125 MHz, $CDCl_3$) data are presented in Table 1. HREIMS *m/z*: 313.1414 [M + Na] $^+$ (calcd. for $C_{17}H_{22}O_4$, 313.1416).

Compound **2**: brown oil; UV (MeOH) λ_{max} (logε) 281 (3.09), 203 (4.15) (0.2) nm; IR (neat) ν_{max} 3407, 2931, 1682, 1516, 1272, 1033 cm^{-1}. ^{1}H (500 MHz, $CDCl_3$) and ^{13}C (125 MHz, $CDCl_3$) data are presented in Table 1. HREIMS *m/z*: 341.1727 [M + Na] $^+$ (calcd. for $C_{19}H_{26}O_4$, 341.1729).

Compound **3**: yellow oil; UV (MeOH) λ_{max} (logε) 281 (2.85), 203 (3.95) nm; IR (neat) ν_{max} 3372, 2922, 1681, 1521, 1278, 1030 cm^{-1}. ^{1}H (500 MHz, $CDCl_3$) and ^{13}C (125 MHz, $CDCl_3$) data are presented in Table 1. HREIMS *m/z*: 369.2042 [M + Na]$^+$ (calcd. for $C_{21}H_{30}O_4$, 369.2042).

Compound **4**: yellow oil; UV (MeOH) λ_{max} (logε) 281 (2.85), 206 (3.96) nm; IR (neat) ν_{max} 3409, 2956, 1707, 1515, 1268, 1033 cm^{-1}. ^{1}H (500 MHz, $CDCl_3$) and ^{13}C (125 MHz, $CDCl_3$) data are presented in Table 1. HREIMS *m/z*: 315.1572 [M + Na]$^+$ (calcd. for $C_{17}H_{24}O_4$, 315.1573).

Compound **5**: yellow oil; UV (MeOH) λ_{max} (logε) 281 (2.79), 203 (3.98) nm; IR (neat) ν_{max} 3356, 2963, 2940, 1664, 1516, 1273, 1035 cm^{-1}. ^{1}H (500 MHz, $CDCl_3$) and ^{13}C (125 MHz, $CDCl_3$) data are presented in Table 2. HREIMS *m/z*: 271.1312 [M + Na]$^+$ (calcd. for $C_{15}H_{20}O_3$, 271.1310).

Compound **8**: yellow oil; UV (MeOH) λ_{max} (logε) 279 (3.58), 202 (4.02) nm; IR (neat) ν_{max} 3381, 2957, 1631, 1593, 1515, 1271, 1034 cm^{-1}. ^{1}H (500 MHz, $CDCl_3$) and ^{13}C (125 MHz, $CDCl_3$) data are presented in Table 2. HREIMS *m/z*: 297.1465 [M + Na]$^+$ (calcd. for $C_{17}H_{22}O_3$, 297.1467).

Compound **9**: yellow oil; UV (MeOH) λ_{max} (logε) 279(3.44), 203 (3.93) nm; IR (neat) ν_{max} 3427, 2929, 1633, 1595, 1516, 1271, 1033 cm^{-1}. ^{1}H (500 MHz, $CDCl_3$) and ^{13}C (125 MHz, $CDCl_3$) data are presented in Table 2. HREIMS *m/z*: 325.1780 [M + Na]$^+$ (calcd. for $C_{19}H_{26}O_3$, 325.1782).

Compound **10**: yellow oil; UV (MeOH) λ_{max} (logε) 278 (3.60), 202 (4.07) nm; IR (neat) ν_{max} 3368, 2927, 1634, 1595, 1516, 1271, 1034 cm^{-1}. ^{1}H (500 MHz, CDCL$_3$) and ^{13}C (125 MHz, CDCl$_3$) data are presented in Table 2. HREIMS m/z: 353.2091 [M + Na]$^+$ (calcd. for $C_{21}H_{30}O_3$, 353.2093).

Compound **11**: yellow oil; UV (MeOH) λ_{max} (logε) 279 (3.47), 202 (4.15) nm; IR (neat) ν_{max} 3358, 2927, 1597, 1515, 1270, 1035 cm^{-1}. ^{1}H (500 MHz, CDCl$_3$) and ^{13}C (125 MHz, CDCl$_3$) data are presented in Table 2. HREIMS m/z: 353.2094 [M + Na]$^+$ (calcd. for $C_{21}H_{30}O_3$, 353.2093).

3.5. Cytotoxic Assay

Tumor cells were maintained in RPMI-1640 medium containing 10% heat-inactivated fetal bovine serum, penicillin (100 U/mL) and streptomycin (100 μg/mL), under humidified air with 5% CO_2 at 37 °C. Exponentially growing cells were seeded into 96-well tissue culture-treated plates and precultured for 24 h; the cells were treated with serum-free medium containing various concentrations of the compounds. After 48 h of incubation, 20 μL of MTT (5 mg/mL in PBS) was added to each well. The cells were incubated at 37 °C for 4 h. After removal of the medium, the cells were treated with 100 μL dimethyl sulfoxide (DMSO) for 10 min, and then the optical density was measured at 570 nm using a microplate reader (iMARK TM microplate reader, Bio-Rad, Hercules, CA, USA). The cytotoxic activities of isolated compounds were tested against MCF-7, HepG-2 and KYSE-150 cell lines. The positive control was 5-fluorouracil.

As the major effective constituent of *Z. officinale*, gingerols have attracted considerable interest in the fields of chemistry and biology. A rising number of papers have reported the biological activities of gingerols, especially on their antitumor activity [2–8]. Among these compounds, 6-oxo-shogaol derivatives, paradoldiene derivatives and 6-oxo-paradol derivatives are three rare kinds of structures and there are few reports of their biological activities.

We tested all the newly isolated compounds against HepG-2, MCF-7 and KYSE-150 tumor cells (Table 3). The 6-oxo-shogaol derivatives showed cytotoxic activities against both HepG-2 and MCF-7 cells. Among them, (*Z*)-6-oxo-[6]-shogaol (**1**) was the most prominent one, in terms of its cytotoxic activities, whereas this kind of compound has little cytotoxic effect on the KYSE-150 cell line. In addition, we also found that the cytotoxic activities of these compounds against HepG-2 and MCF-7 cells gradually decreased with an increase in the carbon chain. In contrast, among the paradoldiene derivatives, only (4*E*,6*Z*)-[4]-paradoldiene (**8**) showed better activity against HepG-2 and MCF-7 cells, while the other compounds had no obvious effects on these three kinds of cells. It indicated that with an increase in the carbon chain, the cytotoxic activities of HepG-2 and MCF-7 cells decreased dramatically. We only isolated one 6-oxo-paradol derivative, 6-oxo-[6]-paradol (**4**), which was very similar to the structure of (*Z*)-6-oxo-[6]-shogaol (**1**), with only one double bond between the carbonyl groups on the carbon chain, but it had no significant activity against these three tumor cells. We also isolated one isoshogaol derivative, [4]-isoshogaol (**5**), which was found to produce better activity on all three tested cells.

Table 3. Cytotoxicities of compounds **1–5** and **13–16** against MCF-7, HepG-2 and KYSE-150 cell lines (IC_{50}, μM).

Compound	IC_{50} (μM)		
	HepG-2	Mcf-7	KYSE-150
1	8.92 ± 0.34	6.27 ± 0.21	>50
2	45.14 ± 1.69	47.22 ± 2.31	>50
3	>50	>50	>50
4	>50	>50	>50
5	14.87 ± 0.57	>50	>50
13	21.56 ± 1.47	22.85 ± 1.01	20.41 ± 0.53
14	>50	>50	>50
15	>50	>50	>50
16	>50	>50	>50
5-Fluorouracil	8.18 ± 0.53	7.35 ± 0.37	13.26 ± 0.47

3.6. Statistical Analysis

The results of each group are expressed as mean ± SD values. Data were analyzed through the one-way ANOVA between control and sample treated groups, in Microsoft Excel. A statistically significant difference among groups was considered to be represented by $p < 0.05$.

4. Conclusions

In-depth research on ginger has been conducted by scholars across the world. In addition, gingerols, the main component of ginger, have been studied and applied in many fields. Though gingerols contain many varieties, it is not an easy task to find new gingerols. In this experiment, nine new compounds were isolated, and among them, 6-oxo-shogaol, 6-oxo-paradol and paradoldiene are rare structures in gingerol derivatives. Moreover, isolation of 6-oxo-shogaol derivatives from the rhizoma of *Z. officinale* is reported here for the first time, which also enriches our knowledge about the chemical diversity of this plant.

A wide range of gingerol derivatives, and many kinds of activities have been reported [1–10]. However, the 6-oxo-shogaol derivatives, paradoldiene derivatives and 6-oxo-paradol derivatives were rarely investigated, and the tested activity of these three derivatives against HepG-2, MCF-7 and KYSE-150 cells is reported here for the first time. To evaluate the efficacy of these nine compounds, the MTT cytotoxicity assay was performed, and significant inhibitory effects were found on the proliferation of these three kinds of cell lines. By comparing the activity of these three kinds of derivatives, we hypothesized that the α,β-unsaturated ketone structure is a key construction unit to cytotoxic activities of gingerol compounds, and the cytotoxic activity decreases rapidly with an increase in the carbon chain.

Our research enriched the number of types of gingerol derivatives, and these results will broaden the application field of gingerols.

Supplementary Materials: The following are available online.

Acknowledgments: This work was supported by following grants: Technological Innovation Team of Henan University of Chinese Medicine, Science and Technology Innovation Talent Program Funding Project (2015XCXTD02).

Author Contributions: Yanzhi Wang, Zezhi Li and Yongxian Cheng designed research; Zezhi Li, Meiling Gao, Wanhua Cui, Mengnan Zeng and Juan Li performed research and analyzed the data; Yongxian Cheng gave suggestions on paper writing, Zezhi Li wrote the paper. All authors read and approved the final manuscript.

Conflicts of Interest: The authors declare no conflicts of interest.

References

1. Chang, T.T.; Chen, K.C.; Chang, K.W.; Chen, H.Y.; Tsai, F.J.; Sun, M.F.; Chen, C.Y. In silico pharmacology suggests ginger extracts may reduce stroke risks. *Mol. Biosyst.* **2011**, *7*, 2702–2710. [CrossRef] [PubMed]
2. Li, Q.Q.; Cai, P.W.; Fu, M.W.; Ying, P. Combined administration of the mixture of honokiol and magnolol and ginger oil evokes antidepressant-like synergism in rats. *Arch. Pharm. Res.* **2009**, *32*, 1281–1292.
3. Therkleson, T. Ginger compress therapy for adults with osteoarthritis. *J. Adv. Nurs.* **2010**, *66*, 2225–2233. [CrossRef] [PubMed]
4. Masuda, Y.; Hisamoto, M.; Nakatani, N. Antioxidant properties of gingerol related compounds from ginger. *Biofactors* **2004**, *21*, 293–296. [CrossRef] [PubMed]
5. Wang, L.X.; Yang, S.W.; Li, B.; Zhang, K.; Li, Y.; Hu, Y.N. Analysis of antioxidant activity of gingerol from ginger by supercritical CO_2 fluid extraction. *China Food Addit.* **2014**, *8*, 96–101.
6. Chrubasik, S.; Pittler, M.H.; Roufogalis, B.D. Zingiber is rhizoma: A comprehensive review on the ginger effect and efficacy profiles. *Phytomedicine* **2005**, *12*, 684. [CrossRef] [PubMed]
7. Weng, C.J.; Wu, C.F.; Huang, H.W.; Ho, C.T.; Yen, G.C. Anti-invasion effects of 6-shogaol and 6-gingerol, two active components in ginger, on human hepatocarcinoma cells. *Mol. Nutr. Food Res.* **2010**, *54*, 1618. [CrossRef] [PubMed]

8. Hong, L.J.; Zheng, Z.X.; Hyun, J.K.; Steven, P.M.; Barbara, N.T.; David, R.G. Metabolic profiling and phylogenetic analysis of medicinal *Zingiber* species: Tools for authentication of ginger (*Zingiber officinale* Rosc.). *Phytochemistry* **2006**, *67*, 1673–1685.
9. Shukla, Y.; Singh, M. Cancer preventive properties of ginger: A brief review. *Food Chem. Toxicol.* **2007**, *45*, 683. [CrossRef] [PubMed]
10. Zhou, J.; Yang, W.P.; Li, Y.L.; Yang, M.; Lei, Z.X. Effects of dried ginger decoction on plasma angiotensin II, serum tumor necrosis factor alpha, malondialdehyde and nitric oxide in rats with acute myocardial ischemia. *Lishizhen Med. Mater. Med. Res.* **2014**, *25*, 288–290.
11. Nafiseh, S.M.; Reza, G.; Gholamreza, A.; Mitra, H.; Leila, D.; Mohammad, R.M. Anti-oxidative and anti-inflammatory effects of ginger in health and physical activity: Review of current evidence. *Int. J. Prev. Med.* **2013**, *4*, 36.
12. Jiang, S.Z.; Liao, K. Effect of alcohol extract from dried ginger on experimental gastric ulcer. *Chin. J. Ethnomed. Ethnopharm.* **2010**, *19*, 79–80.
13. Yu, S.J.; Qin, H.L.; Yong, Z.; Yi, Q.L.; Jian, B.L. Transcriptome analysis reveals the genetic basis underlying the biosynthesis of volatile oil, gingerols, and diarylheptanoids in ginger (*Zingiber officinale* Rosc.). *Bot. Stud.* **2017**, *58*, 41. [CrossRef]
14. Roberto, B.; Giovanna, B. A New Stereoselective Synthesis of (*E*)-α,β-Unsaturated-γ-dicarbonyl Compounds by the Henry Reaction. *J. Org. Chem.* **1994**, *59*, 5466–5467.
15. Hung, C.S.; Ching, Y.C.; Ping, C.K.; You, C.W.; Yu, Y.C. Synthesis of analogues of gingerol and shogaol, the active pungent principles from the rhizomes of *Zingiber officinale* and evaluation of their anti-platelet aggregation effects. *Int. J. Mol. Sci.* **2014**, *15*, 3926–3951.
16. Christopher, J.H.; Steven, V.L.; Edward, A. 1,5-Asymmetric induction of chirality using π-allyltricarbonyliron lactone complexes: Highly diastereoselective synthesis of π-functionalised carbonyl compounds. *Org. Biomol. Chem.* **2003**, *1*, 3208–3216.
17. Alberto, A.; Rosanna, C.; Gianluca, N.; Orso, V.P.; Sergio, Q. A new strain of streptomyces: An anthracycline containing a C-Glucoside moiety and a chiral decanol. *Phytochemistry* **1988**, *27*, 3611–3617.
18. Guillermo, T.; Carmen, S.; Pilar, R.; Marta, P.; Librada, M.C.; Carmen, C. Streptenols F-I isolated from the marine-derived *Streptomyces misionensis* BAT-10-03-023. *J. Nat. Prod.* **2017**, *80*, 1034–1038.
19. Chen, C.C.; Rosen, R.T.; Ho, C.T. Chromatographic analyses of isomeric shogaol compounds derived from isolated gingerol compounds of ginger (*Zingiber Officinale*, roscoe). *J. Chromatogr. A* **1986**, *360*, 175–184. [CrossRef]

Sample Availability: Samples of the compounds **4**, **8–15** are available from the authors.

Article

Lupeol, a Pentacyclic Triterpene, Promotes Migration, Wound Closure, and Contractile Effect In Vitro: Possible Involvement of PI3K/Akt and p38/ERK/MAPK Pathways

Fernando Pereira Beserra [1,*], Meilang Xue [2], Gabriela Lemos de Azevedo Maia [3], Ariane Leite Rozza [1], Cláudia Helena Pellizzon [1] and Christopher John Jackson [2]

1 Department of Morphology, Institute of Biosciences, São Paulo State University (UNESP), Botucatu 18618-689, São Paulo, Brazil; arianerozza@gmail.com (A.L.R.); claudia.pellizzon@gmail.com (C.H.P.)

2 Sutton Research Laboratory, Kolling Institute of Medical Research, the University of Sydney at Royal North Shore Hospital, St Leonard, NSW 2065, Australia; meilang.xue@sydney.edu.au (M.X.); chris.jackson@sydney.edu.au (C.J.J.)

3 Department of Pharmacy, Federal University of São Francisco Valley (UNIVASF), Petrolina 56304-205, Pernambuco, Brazil; gabriela.lam@gmail.com

* Correspondence: fernando.beserra@unesp.br

Academic Editors: Wenxu Zhou and De-an Guo
Received: 6 September 2018; Accepted: 26 October 2018; Published: 30 October 2018

Abstract: Skin wound healing is a dynamic and complex process involving several mediators at the cellular and molecular levels. Lupeol, a phytoconstituent belonging to the triterpenes class, is found in several fruit plants and medicinal plants that have been the object of study in the treatment of various diseases, including skin wounds. Various medicinal properties of lupeol have been reported in the literature, including anti-inflammatory, antioxidant, anti-diabetic, and anti-mutagenic effects. We investigated the effects of lupeol (0.1, 1, 10, and 20 μg/mL) on in vitro wound healing assays and signaling mechanisms in human neonatal foreskin keratinocytes and fibroblasts. Results showed that, at high concentrations, Lupeol reduced cell proliferation of both keratinocytes and fibroblasts, but increased in vitro wound healing in keratinocytes and promoted the contraction of dermal fibroblasts in the collagen gel matrix. This triterpene positively regulated matrix metalloproteinase (MMP)-2 and inhibited the NF-κB expression in keratinocytes, suggesting an anti-inflammatory effect. Lupeol also modulated the expression of keratin 16 according to the concentration tested. Additionally, in keratinocytes, lupeol treatment resulted in the activation of Akt, p38, and Tie-2, which are signaling proteins involved in cell proliferation and migration, angiogenesis, and tissue repair. These findings suggest that lupeol has therapeutic potential for accelerating wound healing.

Keywords: lupeol; keratinocytes; fibroblasts; wound healing; cell migration

1. Introduction

The skin is the largest organ of the human body limiting the organism's exterior with the external environment whose main functions are protection against external agents, thermoregulation, and perception [1]. The breakdown of skin integrity caused by injury resulting from various events, such as trauma, heat or cold exposure, burns, or blood circulation problems, may render the patient more susceptible to acute or chronic infections, electrolyte imbalance, and fluid loss [2]. In more severe cases, limb amputation may be necessary. Accelerating the wound healing process will assist in circumventing these complications.

Wound healing is a highly complex physiological process involving ordered events classified into three phases that overlap: hemostasis and inflammation (inflammatory phase), granulation tissue formation and re-epithelialization (proliferative phase), and wound contraction and tissue remodeling (extracellular matrix remodeling phase) [3]. This is a natural phenomenon that occurs through cellular and molecular responses and interactions with the main objective of reconstituting and restoring the integrity of the injured tissue [4]. During the cutaneous wound healing process, re-epithelialization via keratinocyte proliferation and migration and (myo)fibroblasts contraction result in wound closure [5].

Medicinal plants have long been reported as a therapeutic alternative for various diseases in many countries. The use of plant extracts or their isolated products for acute and chronic wounds and burns has been well described in the literature [6]. Many of these extracts or plant-derived compounds to treat wounds have pharmacological properties essential for injury repair, such as anti-inflammatory and antioxidant activities [6]. Curcumin, a polyphenol from the rhizomes of *Curcuma longa*, has shown antioxidant properties in several studies for the treatment of many diseases, including skin disorders. Topical applications of curcumin have been widely used when researching cutaneous wound healing. Manca et al. performed in vitro and in vivo studies with curcumin-loaded nanovesicles, which showed antioxidant effects on keratinocytes by reducing oxidative stress and producing an anti-inflammatory effect through decreasing myeloperoxidase activity, edema formation, and providing stimulus for cutaneous re-epithelialization [7]. Quercetin is another well-known natural compound due to its strong antioxidant properties and belongs to the class of flavonoids. A recent study developed from a multiphase system containing quercetin-loaded liposomes showed a significant reduction of lesion area in rats and confirmed its potential in the skin wound treatment [8].

In addition to phenolic compounds and flavonoids, many plant-derived secondary metabolites with antioxidant and anti-inflammatory properties, including the triterpenes, are capable of promoting these healing effects in in vivo and in vitro models [9–11]. Triterpenes chemically belong to the class of isoprenoids, which are distributed widely in various parts of the plant from the roots to the fruits. Several studies have shown that triterpenes from medicinal plants improve the quality of healing through mechanisms ranging from the regulation of pro- and anti-inflammatory mediators, chemokines, growth factors, inducing the formation of granulation, re-epithelialization, and wound contraction [12–15].

Bowdichia virgilioides (Fabaceae), is a medicinal plant widely known as "sucupira-preta", predominantly found in the Brazilian Cerrado in the north, northeast, and central regions of the country. In popular medicine, there are reports of the use of its bark and seeds in the form of an infusion to treat various diseases, such as arthritis, diabetes, bronchitis, and skin wounds [16]. Phytochemical studies revealed that the bark and roots contain a large number of alkaloids and terpenoids, such as lupeol [17,18], volatile constituents and flavonoids [19,20], and anthocyanins [21].

Lupeol is a pentacyclic triterpene found in various species in the plant kingdom, including some vegetables and fruits, such as cucumber, tomato, white cabbage, fig, and guava [22,23]. This molecule exhibits a spectrum of pharmacological activities against various acute or chronic diseases, including arthritis, renal disorders, diabetes, hepatotoxicity, cardiovascular disease, cancer, and microbial infections [24–27].

In this study, we aimed to explore the effects of the triterpene lupeol on skin wound healing in vitro by investigating proliferation, migration, and cell contraction, as well as its signaling mechanisms involved using human keratinocytes and fibroblasts.

2. Results

2.1. *High Concentrations of Lupeol Decrease Proliferation and Cause Cytotoxicity in Keratinocytes and Fibroblasts*

Human keratinocytes or fibroblasts were treated with lupeol at various concentrations ranging from 0.1 to 20 μg/mL before cell proliferation and viability were assessed by crystal violet and MTT ([3-(4,5-Dimethylthiazol-2-yl)-2,5-Diphenyltetrazolium Bromide]) assay, respectively. The results showed that lupeol at 10 μg/mL and 20 μg/mL, significantly inhibited keratinocyte proliferation after 24 h treatment by 53% and 64%, respectively (Figure 1A). Lupeol at 1 μg/mL increased fibroblast proliferation significantly by 12%, whereas the higher concentration (20 μg/mL) inhibited cell proliferation in relation to the control (19%) (Figure 1B). This triterpene did not affect keratinocyte viability but showed cytotoxicity to fibroblasts at the higher concentration (20 μg/mL) (Figure 2A,B).

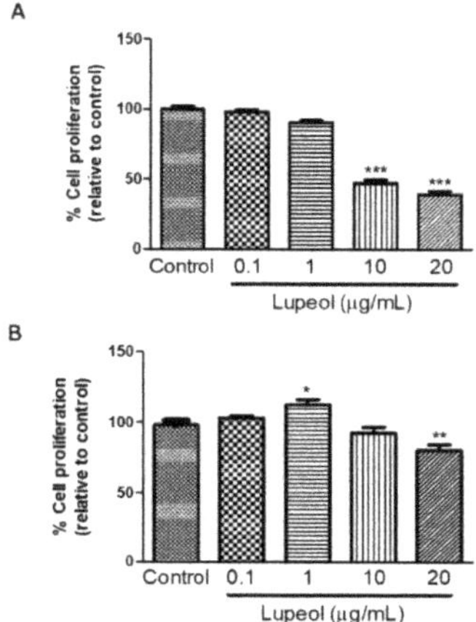

Figure 1. Cell proliferation of (**A**) epidermal keratinocytes and (**B**) dermal fibroblasts in response to lupeol. Cells were seeded in a 96-well plate. After overnight attachment, different concentrations of lupeol were added, and the cells were left for 24 h at 37 °C. Cell proliferation was measured by crystal violet assay and calculated by a comparison of the values from the lupeol treatment group with the control group. Data are expressed as mean ± standard error of the mean (SEM). * $p < 0.05$, ** $p < 0.01$, and *** $p < 0.001$ versus control group.

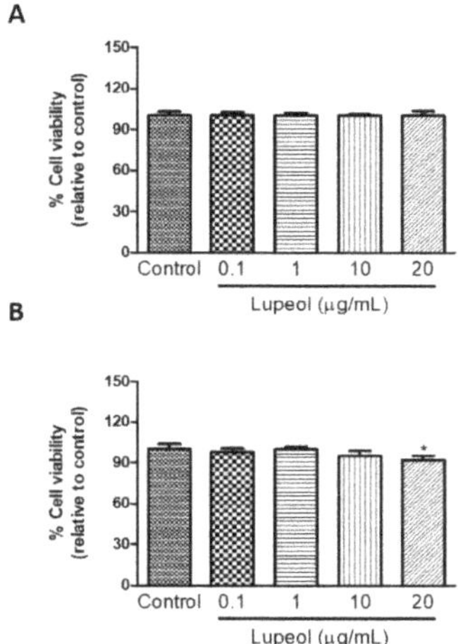

Figure 2. The viability of (**A**) epidermal keratinocytes and (**B**) dermal fibroblasts in response to lupeol. Cells were seeded in a 96-well plate and, after overnight attachment, different concentrations of lupeol were added, and the cells were left for 24 h at 37 °C. Cell viability was measured by MTT assay and calculated by a comparison of the values from the lupeol treatment group with the control group. Data are expressed as mean ± standard error of the mean (SEM). * $p < 0.05$ versus control group.

2.2. Lupeol Enhances Migration and Wound Closure in Human Epidermal Keratinocytes

The scratch wound healing assay revealed that lupeol (0.1 and 1 µg/mL) significantly increased the wound closure rate compared to the control after 24 h (Figure 3A). The cells treated with the lowest concentration of lupeol, 0.1 µg/mL, showed a 59% increase in the wound closure rate compared to the control ($p < 0.001$). Lupeol at 1 µg/mL also showed a potent wound healing effect on epidermal keratinocytes by 39% ($p < 0.05$). When used at higher concentrations, lupeol did not cause any significant change in wound closure rate.

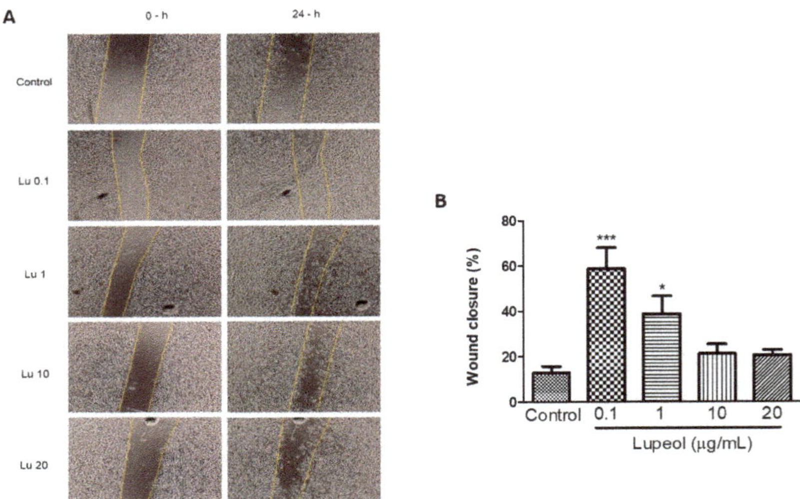

Figure 3. Wound healing effect of lupeol on human epidermal keratinocytes in the scratch assay after 24 h incubation. Cells were incubated with lupeol at concentrations of 0.1, 1, 10, and 20 µg/mL or DMSO at 10 µg/mL as a negative control. (**A**) Representative images of scratch assay at 0 and 24 h. Lu: lupeol. (**B**) Dose-response effect of lupeol on wound closure. Data are expressed as mean ± standard error of the mean (SEM). * $p < 0.05$ and *** $p < 0.001$ versus control group.

We further examined whether lupeol stimulates migration in vitro. Scratch migration of human keratinocytes treated with lupeol showed a significant increase in migration at all concentrations tested (0.1 µg/mL 93%, $p < 0.001$; 1 µg/mL 96%, $p < 0.001$; 10 µg/mL 94%, $p < 0.001$; and 20 µg/mL 83%, $p < 0.05$) compared with the control group. Although the higher concentrations were able to inhibit cell proliferation, as shown in Figure 1, these concentrations showed a potent increase in epidermal keratinocytes cell migration (Figure 4B).

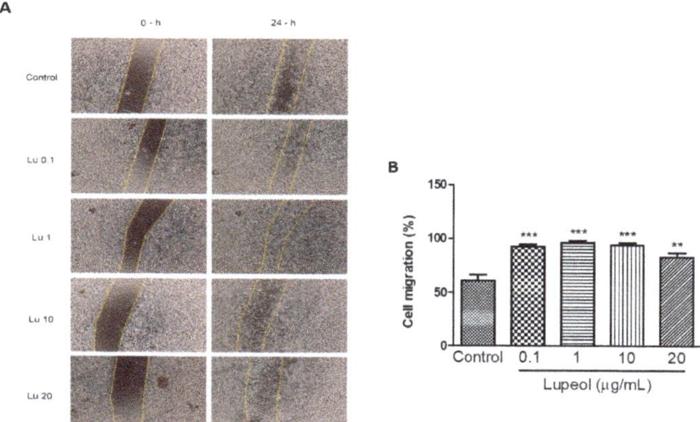

Figure 4. Keratinocyte migration using a scratch assay after 24 h of incubation with lupeol. Cells were incubated with lupeol in different concentrations at 0.1, 1, 10, or 20 μg/mL or DMSO 0.25% at 10 μg/mL as negative control. (**A**) Representative images of scratch assay at 0 and 24 h. Lu: lupeol. (**B**) Effect of lupeol on migration cell. Data are expressed as mean ± standard error of the mean (SEM). ** $p < 0.01$ and *** $p < 0.001$ versus control group.

2.3. Lupeol Promotes Contractile Effect on a Collagen Gel Matrix

To obtain additional information about wound healing mechanisms, we assessed the ability of lupeol to contract a collagen gel matrix containing dermal fibroblasts (Figure 5A). Collagen gel assays provide a convenient platform to investigate a cell's contractile ability that closely represents cell behavior in vivo. The results showed lupeol significantly increased the contractile effect on collagen gels at 1 μg/mL ($p < 0.01$), 10 ($p < 0.01$), and 20 μg/mL ($p < 0.01$) compared to the control after 48 h of treatment (Figure 5B).

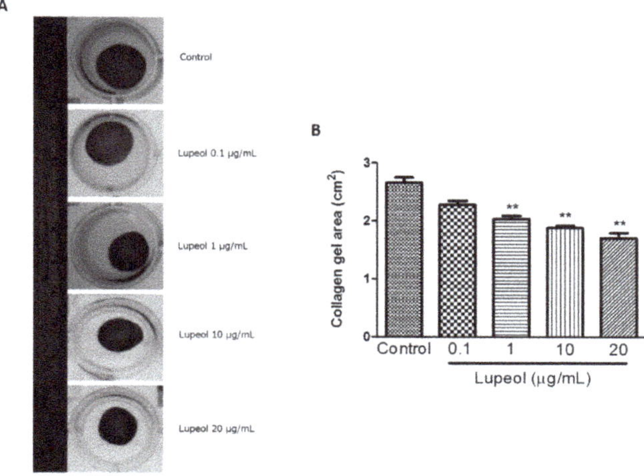

Figure 5. Contractility of lupeol on human fibroblasts. Fibroblasts were cultured within 3D-collagen matrix incubated with lupeol at concentrations of 0.1, 1, 10, or 20 μg/mL or DMSO 0.25% at 10 μg/mL as the negative control. The gels were carefully released from the side wall of the wells and the contractile activity of lupeol was quantified after 48 h of treatment. (**A**) Representative images of collagen matrix at 48 h. (**B**) Semi-quantified results of contractility of lupeol on fibroblasts, expressed as a mean ± SEM. ** $p < 0.01$ versus control group.

2.4. Lupeol Regulates via PI3K/Akt and MAPK Pathways

To identify which signaling pathways are involved in the regulation of human keratinocyte migration promoted by lupeol, we investigated the respective roles of PI3K/Akt, Tie-2, MAPK/ERK/p38, NFκB, and MMP-2 in lupeol-accelerated keratinocyte migration.

As shown in Figure 6A,B, 1 μg/mL lupeol treatment sharply increased the activity of Akt compared with the control ($p < 0.001$). Numerically smaller but significant increases in Akt were observed at all other concentrations (Figure 6B). Lupeol also increased Akt phosphorylation at all concentrations tested (Figure 6C). There was no change in the levels of ERK or p-ERK line response to lupeol (Figure 6D,E).

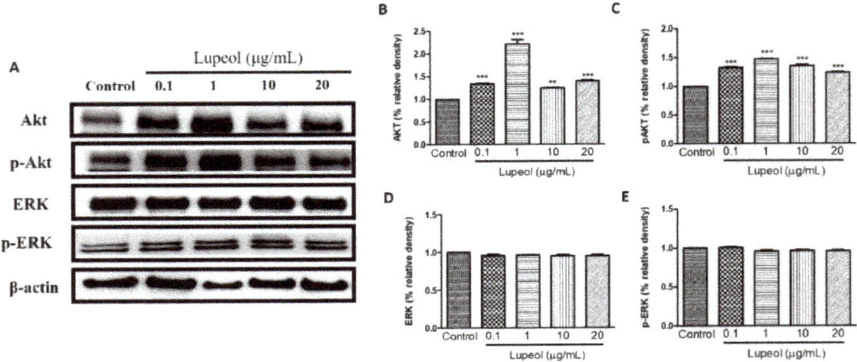

Figure 6. Lupeol regulates the expression of Akt and ERK in epidermal keratinocytes. (**A**) Western blot showing expression of Akt, p-Akt, ERK, and p-ERK in keratinocytes after 24 h treatment with lupeol in different concentrations. (**B–E**) All the proteins were semi-quantitated using ImageJ software and normalized to the levels of β-actin. Data are expressed as mean ± SEM. ** $p < 0.01$ and *** $p < 0.001$ versus control group.

The level of p-p38 and p38 proteins expressed by keratinocytes showed significant changes after treatment with lupeol. A significant increase in p38 was observed in cells treated with lupeol at 1 ($p < 0.05$), 10 ($p < 0.001$) and 20 μg/mL ($p < 0.001$) compared to the control (Figure 7B). p-p38 showed a significant dose-dependent increase in response to all concentrations of lupeol (Figure 7C). MMP-2 activity was significantly enhanced after treatment with 1, 10, and 20 μg/mL of lupeol (Figure 7E). The effect of treatment with lupeol on the expression of pro-inflammatory mediator NFκB was also evaluated. As shown in Figure 7D, lupeol at concentrations of 10 and 20 μg/mL significantly decreased the expression of NFκB compared to the control.

We examined the influence of lupeol treatment on Tie-2 expression, a protein-tyrosine kinase receptor expressed by endothelial and epithelial cells, which functions to stabilize the cell barrier. Treatment with lupeol showed a dose-dependent decrease in protein expression of Tie-2, with a concomitant increase in the levels of its phosphorylated isoform, p-Tie-2, clearly indicating that lupeol activates Tie-2 (Figure 8C).

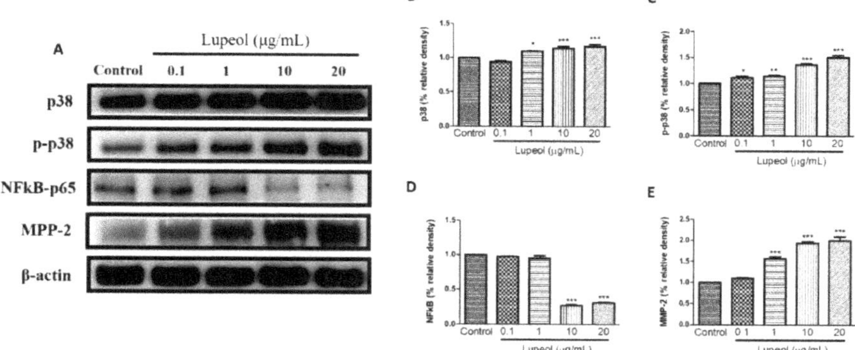

Figure 7. Lupeol regulates the expression of p-38, NF-κB-p65 and MMP-2 in epidermal keratinocytes. (**A**) Western blot showing expression of p38, p-p38, NF-κB-p65, and MMP-2 in keratinocytes after 24 h treatment with lupeol in different concentrations. (**B–E**) All the proteins were semi-quantitated using Image J software and normalized to the levels of β-actin. Data are expressed as mean ± SEM. * $p < 0.05$, ** $p < 0.01$, and *** $p < 0.001$ versus control group.

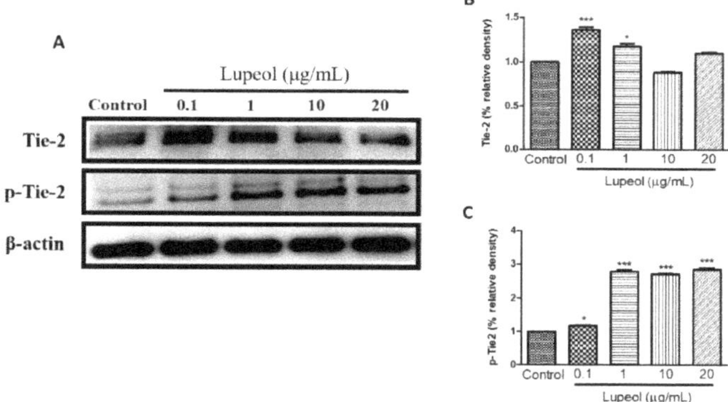

Figure 8. Lupeol regulates the expression of Tie-2 in epidermal keratinocytes. (**A**) Western blot showing expression of Tie-2 and p-Tie-2 in keratinocytes after 24 h treatment with lupeol in different concentrations. (**B**,**C**) All the proteins were semi-quantitated using Image J software and normalized to the levels of β-actin. Data are expressed as mean ± SEM. * $p < 0.05$ and *** $p < 0.001$ versus control group.

2.5. Lupeol Regulates the Differentiation of Cytokeratin 16

To understand the effect of lupeol on the differentiation of keratinocytes, we investigated keratin 16 expression in keratinocytes after 24 h treatment. As shown in Figure 9A,B, 0.1 μg/mL lupeol treatment increased keratin 16 compared to control ($p < 0.001$). Significant decreases in keratin 16 were observed with lupeol at 1 and 10 μg/mL, with no change in response to 20 μg/mL.

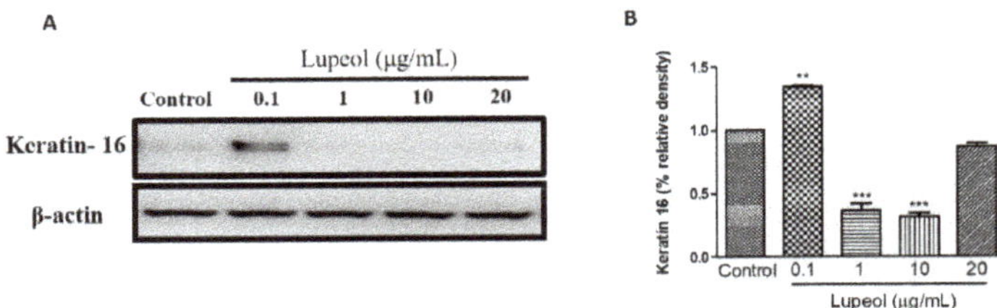

Figure 9. Lupeol regulates the expression of keratin 16 in epidermal keratinocytes. (**A**) Western blot showing expression of keratin 16 in keratinocytes after 24-h treatment with lupeol. (**B**) All the proteins were semi-quantitated using ImageJ software and normalized to the levels of β-actin. Data are expressed as mean ± SEM. ** $p < 0.01$ and *** $p < 0.001$ versus control group.

3. Discussion

Lupeol, a natural compound isolated from *B. virgiliodes*, plays protective roles in a variety of cancers, skin inflammation, pancreatitis, arthritis, diabetes, and hepatic and cardiovascular diseases [28–34], but little is known about its effects on in vitro wound healing. In this study, we demonstrated that lupeol enhances in vitro wound healing, possibly by stimulating the survival, migration, and contraction of epidermal keratinocytes and/or dermal fibroblasts (Figure 10).

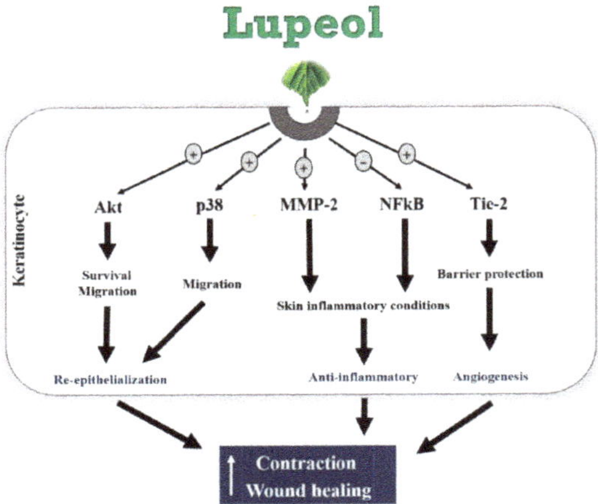

Figure 10. Hypothetical model of the regulatory mechanisms of lupeol in human keratinocytes on wound healing.

Wound healing is a highly complex biochemical and physiological process involving various cellular and molecular mediators, such as chemokines and cytokines, inflammatory infiltrate, immune system cells, growth factors, and extracellular matrix proteins [2]. Keratinocyte proliferation and migration are crucial events in wound repair. These cells migrate toward the wound surface and fill the wound area to cover the open space exposed to infections, thereby contributing to the wound healing effect [5,35]. Keratinocyte migration may occur independently of keratinocyte proliferation during re-epithelialization [36]. This cell mechanism is associated with the formation and disassembly of cell adhesion sites and cytoskeletal reorganization [35]. Using an in vitro migration assay, which did not involve cell proliferation, we found that lupeol had a strong effect at all concentrations, indicating that lupeol may accelerate cutaneous wound repair by stimulating the migration of keratinocytes. In contrast, we found that lupeol at low concentrations (0.1 and 1 μg/mL) exhibited a potent wound healing effect in vitro, a process in which both proliferation and migration are involved. Our study also demonstrated that lupeol, at high concentrations, inhibited cell proliferation in both cell types and stimulated cytotoxicity in fibroblasts. Taken together, these results suggest that cell migration, rather than cell proliferation, is the more prominent pathway for lupeol to promote cutaneous re-epithelialization and wound repair in keratinocytes.

After re-epithelialization and extracellular matrix reorganization, wound contraction is an important event during the wound healing process. Our collagen gel contraction assay showed that lupeol treatment stimulated the contractile effect of fibroblasts embedded in a collagen gel solution in a dose-dependent manner. This culture system has been widely used as a model for pro-contractile remodeling of the extracellular matrix by dermal fibroblasts for situations such as wound healing [37]. This fibroblast-embedded collagen gels system allows cells to develop endogenous tension through both mechanical and biochemical signals relevant for cell contractility, thereby triggering signal transduction cascades capable of modulating transmembrane and/or intracellular receptors to promote intracellular responses such as gene expression and protein synthesis [38].

It has been reported that activation of the PI3K/Akt pathway triggers mechanisms responsible for cell polarity capable of influencing the migration speed, thus leading to the migratory activity of various cell types, including keratinocytes and fibroblasts [39,40]. In addition, PI3K/Akt is a pathway that has been implicated as an important mediator in the control of survival/cell growth, malignant and metastatic oncogenic transformation, and the regulation of various diseases [41]. In our study, protein expression determined via Western blot clearly showed that lupeol significantly up-regulated Akt expression and activation in epidermal keratinocytes, suggesting that Akt may be a key signaling component in lupeol-induced keratinocyte migration and wound healing.

MAPK signaling is an important pathway formed by a chain of proteins in the cell, which is responsible for modulating the synthesis and release of growth factors involved in cell proliferation and migration during wound healing [42]. The MAPKs consist of extracellular signal-regulated protein kinase (ERK), c-Jun NH2-terminal kinase (JNK), and p38 mitogen-activated protein kinases [43]. ERK and p38 are crucial signaling molecules in wound healing. ERK plays a crucial role in the regulation of cell migration, proliferation, differentiation, and cell survival [44]. Although the role of the p38 MAPK signaling pathway in wound healing has not yet been fully elucidated, recent studies have suggested its involvement in the migration of keratinocytes to wounds [45]. In this study, lupeol did not significantly change ERK expression or activation, but increased the p38 activation in a concentration-dependent manner. As such, we hypothesize that lupeol-induced migration and wound closure in human keratinocytes occurs through activation of p38 MAPK.

Inflammation is a crucial event characterized by the infiltration of inflammatory cells into the injured tissue. Successful tissue regeneration is directly related to resolution of the inflammatory phase, since low recruitment of inflammatory cells is associated with delayed healing, and excessive inflammation can result in chronic wounds, complicating the repair formation of a scar. The NF-κB pathway plays a key role in controlling the expression of various inflammatory genes including TNF-α (Tumor necrosis factor-alpha), cell adhesion molecules such as E-selectin, and vascular cell adhesion molecule-1 [46]. This protein complex is located within the cytoplasm as a pair of dimers (p50/p65). Once activated by inflammatory stimuli, including surgical lesions, NF-κB dimers migrate to the nucleus of the cell binding to the molecules of DNA, thereby inducing the transcription of a variety of genes responsible for proliferation, migration, cell cycling, inhibition of apoptosis, and most importantly, inflammation [47,48]. In our study, high doses of lupeol suppressed the activation and expression of NF-κB in keratinocytes, thus contributing to the dampening of the inflammatory process in cutaneous wounds.

Another interesting finding in this study was that lupeol significantly increased the expression of MMP-2 in a concentration-dependent manner in human keratinocytes. MMPs present in keratinocytes are enzymes involved not only in the modulation of inflammation but also in the proliferative phase, in cell proliferation/migration control, and contribute to skin barrier function and extracellular matrix remodeling [49,50]. MMP-2, in particular, has strong anti-inflammatory properties [50]. Our findings, which showed that lupeol induced the expression of MMP-2 in keratinocytes, provide a possible explanation for cells' increased cell migration and reduced inflammation.

Angiogenesis is a process regulated by complex mechanisms involving the proliferation and migration of endothelial cells, maturation and formation of a new basement membrane, and the creation of new blood vessels from existing ones [51–53]. Tie-2, a protein-tyrosine kinase cell surface receptor, plays an important role during angiogenesis. Studies have shown that both Tie-2 and its activated form, p-Tie2, are present on neonatal foreskin and adult skin epidermis [54]. Major ligands include angiopoietin-1 and -2 proteins with similar binding affinity and Tie-2 activation by Ang-1 in endothelial cells is able to enhance local vascularization by promoting blood vessels resistance and stability, and mediating angiogenesis by VEGF (Vascular endothelial growth factor) activating [55]. Lupeol treatment caused an increase in Tie-2 expression in human epidermal keratinocytes. Similarly, p-Tie2 was expressed in keratinocytes treated with lupeol. These findings support Tie-2 as an essential signaling component in lupeol-induced wound healing in human keratinocytes.

The epidermal layer of the skin is composed of keratinocytes in various stages of cell differentiation. After a damaging stimulus, keratinocytes are able to express different types of keratin proteins, different from those present in the healthy epidermis [56]. Cytokeratin 16 is rarely expressed under normal skin conditions, but may be found in oral mucosa and nails. This cytokeratin has a crucial role as a stimulatory keratin in skin diseases, especially in psoriasis, hypertrophic scars, other inflammatory conditions, and several types of cancers including squamous cell carcinomas [57,58]. It is important to note that reduced expression of keratin 16 contributes to the re-organization of important organelles of the cell, such as the cytoskeleton, which can directly affect the migration of keratinocyte [56]. Our study showed that lupeol treatment is capable of up- and down-regulating keratin 16 expression according to the concentration tested.

4. Materials and Methods

4.1. Plant Material, Extraction and Isolation of Lupeol

Bowdichia virgilioides Kunth. (stem bark) was collected in December 2014 in the surroundings of Santa Rita, State of Paraíba, Brazil, a coastal area around the Atlantic Forest. A voucher specimen (*Agra et Góis 6243*) was deposited at the Herbarium Prof. Lauro Pires Xavier (JPB), and in the reference collection of the Laboratory of Pharmaceutical Technology from Federal University of Paraíba, João Pessoa, Brazil. Three kg of air-dried ground stem bark of *Bowdichia virgilioides* were exhaustively extracted with 95% alcohol solution. The extracted solution was filtered and the solvents were subjected to the evaporation method under reduced pressure with rotary evaporation (Solab, Piracicaba, São Paulo, Brazil) at 40 °C to obtain the final ethanolic extract (tHE, 250 g). The EtOHE was partitioned using solvents in increasing polarity (hexane, chloroform, and methanol). The hexane residue (49 g) was subjected to repeated washings with acetone under stirring followed by filtration. The solid obtained was recrystallized from chloroform and hexane, resulting in white crystals which were performed by analyzing ^{1}H and ^{13}C-NMR spectral data (Bruker, Billerica, MA, USA), compared with those published in the literature [59] and identified as lupeol (Figure 11) substance (3 g). The comparison of ^{1}H and ^{13}C-NMR data of lupeol isolated and ^{13}C-NMR lupeol data in the literature [60] are presented at Table A1. A stock solution of lupeol (10 mg/mL) was prepared by dilution in dimethyl sulfoxide (DMSO) and alcohol. We defined as standard protocol for all treatments the final concentration of DMSO and alcohol 0.25 and 0.075%, respectively, which did not show cytotoxicity in previous reports [61].

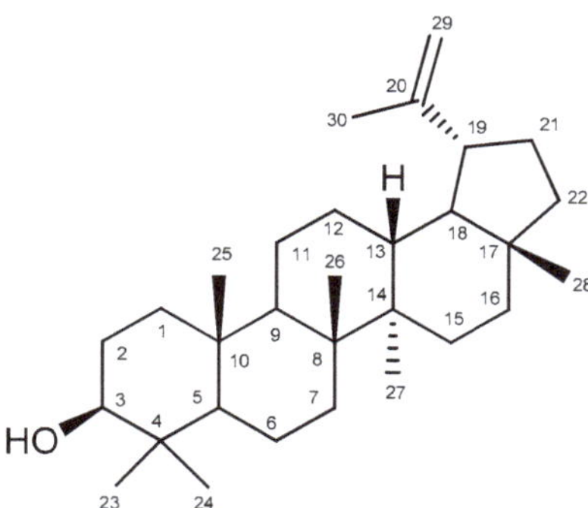

Figure 11. Chemical structure of lupeol.

4.2. Cell Isolation and Culture

Human primary epidermal keratinocytes and dermal fibroblasts were obtained from human neonatal foreskins following the standard protocols of the local ethics committee [62]. Keratinocytes were cultured in specific culture medium (Keratinocyte serum-free medium, K-SFM) and fibroblasts were cultured in Dublecco Modified Eagle's Medium (DMEM), supplemented with 10% fetal bovine serum (FBS) and 100 units penicillin/streptomycin (Gibco-Life Technologies, Grand Island, NY, USA). All cells were plated and incubated at 37 °C in a 5% CO_2 atmosphere and the culture media were changed three times per week. After reaching 70% confluency, the cells were trypsinised and seeded on specific plates for each analysis and incubated for 12 h prior to each experimental procedure.

4.3. Cell Proliferation Assay

The cell proliferation test was performed using the crystal violet assay. Cells were cultured to confluency and seeded in 96-well microplates (1×10^3 cells/well) and after 24 h of incubation, the cells were treated in the absence or presence of lupeol at different concentrations (0.1, 1, 10, or 20 μg/mL). After 24 h of lupeol treatment, the cells were stained for 15 min with a solution of crystal violet (1 μg/mL) and subsequently washed with distilled water to remove any unbound dye residue. After complete drying of each well, 0.1% sodium dodecyl sulfate (SDS) buffer diluted in PBS (Phosphate-buffered saline) was added. The absorbance was read at 570 nm using a microplate reader.

4.4. Cytotoxicity Assay

Cell viability was assessed using MTT-based metabolic assay. Cells were cultured to confluency, and switched to DMEM with 0.1% BSA (Bovine serum albumin) for fibroblasts and adult keratinocyte growth medium without supplements for 24 h. MTT was added 4 h prior to completion of experiments. The absorbance was read at 570 nm using a microplate reader. Final results are expressed as percentages of controls.

4.5. In Vitro Migration ("Scratch") Assay

Cell migration was assessed by "scratch" assay. Keratinocytes were seeded in 24-well plates and cultured to confluency. After 24 h incubation, a pretreatment with mitomycin (10 μg/mL) was performed to avoid any influence of cell proliferation [63]. Cells were scratched with a 100-μL blue plastic pipette tip after 2 h of pre-treatment with mitomycin, thereby creating a cell-free area measuring approximately 2 mm in width and photographed under phase-contrast microscopy (Olympus IX73 microscope, Olympus, Shinjuku, Tokyo, Japan). Cells were then immediately treated with lupeol at 0, 0.1, 1, 10 or 20 μg/mL concentrations and after 24 h new photographs were taken. The analysis was performed by counting cells that had moved from the initial area. The percentage of cell migration was calculated by the following formula: [number of moved cells after lupeol treatment/number of moved cells in the control condition] × 100.

4.6. In Vitro Wound Healing ("Scratch") Assay

In vitro wound healing test was also evaluated using the scratch assay. Keratinocytes were cultured to confluency using 24-well plates and incubated for 24 h. The same procedure performed for scratch induction, as shown in Section 4.5., was also followed, except that mitomycin C was not added. Photographs were taken immediately after scratch and the cells were then treated under the same conditions performed in the cell migration test. After 24 h, new photographs were taken to analyze through the images the wound area remaining. The wound closure rate was determined by the initial and final wound areas during the wounding induction and wound closure percentage calculated by the following formula: [(initial − final)/initial] × 100.

4.7. Collagen Gel Contraction Assay

The collagen gel contractility test was performed using a collagen solution prepared following manufacturer's protocols. Dermal fibroblasts were cultured to confluence before being embedded in a 3D collagen matrix to measure their long-term contractility. Fibroblasts were added to the collagen gel solution and seeded in 24-well plates at a density of 1.7×10^5 cells/mL in each well. After a 1 h incubation to allow the collagen solution to gel, DMEM was added to each well and incubated for overnight. After the incubation, three washes with PBS were performed to remove serum, and DMEM containing 0.5% of FBS was added. Cells were then treated with lupeol at 0.1, 1, 10, or 20 μg/mL concentrations and gel attachment around the side wall of the well was carefully released with a spatula immediately after treatment. The contractile activity was evaluated after 48 h of treatments for each well using ImageJ 1.47i software.

4.8. Cell Lysate Preparation and Western Blot Analysis

Keratinocytes with required confluence were pre-incubated with 0.1–20 μg/mL lupeol for 24 h. After 24 h incubation, cells were then lysed in ice-cold lysis buffer supplemented with protease inhibitors and phosphate inhibitors and centrifuged at 12,000× g for 5 min. An aliquot of cell lysate was separated by 10% SDS-PAGE (Sodium dodecyl sulfate polyacrylamide gel electrophoresis) and transferred to a polyvinylidene difluoride membrane. After blocking in a 5% skim milk powder solution, the membranes were incubated with B-actin, phosphorylated (P) forms of p38 (Tyr182) and p38 (Santa Cruz Biotechnology, Santa Cruz, CA, USA) phospho-Akt (Ser473), Akt, phospho-p44/42 MAPK (Erk1/2) (Thr202/Tyr204), p44/42 MAPK (Erk1/2), Tie2 (D9D10), phospho-Tie2 (Tyr992), NF-κB-p65 (Ser536), MMP-2 (Cell Signaling Technology, Beverly, MA, USA), and Keratin 16 (Thermo Fisher Scientific, Waltham, MA, USA) overnight at 4 °C. The next day, all membranes were washed and incubated in secondary antibody for 1 h at room temperature. Proteins were detected by the ECL (electrochemiluminescence) detection system, (Amersham Biosciences, Little Chalfont, UK) and analyzed by ImageJ software.

4.9. Statistical Analysis

All data are expressed as the mean ± standard error of the mean (SEM). Statistical significance was performed by one-way ANOVA followed by Tukey's test for all analyses. All data were examined using GraphPad Prism 6.0 software (GraphPad Software, Inc., La Jolla, CA, USA) and $p < 0.05$ was considered statistically significant.

5. Conclusions

In conclusion, the results of this study demonstrated the involvement of lupeol in the closure of skin wounds through the stimulation of the migration of keratinocytes and increased contraction of fibroblasts embedded in a collagen matrix. The underlying mechanism for the positive effect of lupeol on wound healing may involve the activation of PI3k/Akt and p38 MAPK, suppression of NF-κB signaling and Keratin 16, as well as the cyto-protective effects of MMP-2 and Tie-2. Although these results provide important information for lupeol in promoting wound healing in vitro, further in vivo and clinical studies are required to explore these and other pathways, as well as to develop lupeol as a therapeutic agent in the treatment of cutaneous wounds.

Author Contributions: F.P.B., M.X. and C.J.J. designed the experiments. F.P.B. and M.X. performed the experiments. C.J.J. and M.X. helped interpreting the results. F.P.B., M.X., C.J.J., C.H.P., A.L.R. and G.L.d.A.M. contributed to the writing and the revision of this manuscript.

Funding: This research was supported mainly by the Science without borders program, Csf/CNPq: 02513/2015-7 and Sutton Arthritis Research Laboratory, University of Sydney at Royal North Shore Hospital.

Acknowledgments: We thank Agnes Chan for her general help and Gordon Campbell, Sydney Adventist Hospital, for providing foreskin tissue.

Conflicts of Interest: The authors declare no conflict of interest.

Appendix A

Table A1. ^1H and ^{13}C-NMR (1D, 2D) spectra data for lupeol isolated from the *Bowdichia virgilioides* (400 MHz, $CDCl_3$) and ^{13}C-NMR data lupeol.

			Lupeol (Isolated)			Lupeol (Literature)
C/H	$\delta^{13}C$/ppm	DEPT	δ^1H/ppm (H Multiplicity)	HMBC	COSY	$\delta^{13}C$/ppm
1	38.75	CH_2				38.9
2	27.48	CH_2	1.89 (2H, m)		3.16 (H-3)	27.6
3	78.97	CH	3.16 (1H, m)		1.89 (H-2)	79.2
4	38.89	C				39.0
5	55.32	CH	0.66 (1H, m)		1.36 (H-6)	55.5
6	18.93	CH_2	1.36 (2 H, m)			18.5
7	34.31	CH_2				34.4
8	40.83	C				41.0
9	50.44	CH	1.24 (1H, sl)			50.6
10	37.17	C				37.3
11	20.97	CH_2				21.1
12	25.13	CH_2				25.3
13	38.05	CH				38.2
14	42.09	C				43.0
15	27.40	CH_2				27.7
16	35.62	CH_2				35.8
17	43.10	C				43.2
18	48.30	CH			2.37 (H-19)	48.5
19	48.02	CH	2.37 (1H, m)		1.36 (H-18)	
20	150.93	C				151.2
21	29.87	CH_2				30.0
22	40.05	CH_2				40.2
23	28.06	CH_3	0.94 (3H, s)	55.32 (C-5); 78.97 (C-3); 15.49 (C-24)		28.2
24	15.49	CH_3	0.74 (3H, s)	55.32 (C-5); 78.97 (C-3); 28.06 (C-23); 38.89 (C-4)		15.6
25	16.03	CH_3	0.80 (3H, s)	50.44 (C-9); 55.32 (C-5)		16.2
26	16.20	CH_3	1.01 (3H, s)	50.44 (C-9); 34.31 (C-7)		16.3
27	14.61	CH_3	0.92 (3H, s)	27.40 (C-15)		14.7
28	18.08	CH_3	0.77 (3H, s)	43.10 (C-17); 48.30 (C-18); 35.62 (C-16); 40.05 (C-22)		18.1
29	109.46	CH_2	4.67 (1H, sl) 4.55 (1H, sl)	48.02 (C-19); 19.38 (C-30); 48.02 (C-19); 19.38 (C-30)	4.67 (H-29 a) 4.55 (H-29 b)	109.5
30	19.38	CH_3	1.66 (3H, s)	109.46 (C-29); 151.93 (C-20); 48.02 (C-19)		19.5

References

1. Chen, J.C.; Lin, B.B.; Hu, H.W.; Lin, C.; Jin, W.Y.; Zhang, F.B.; Zhu, Y.A.; Lu, C.J.; Wei, X.J.; Chen, R.J. NGF accelerates cutaneous wound healing by promoting the migration of dermal fibroblasts via the PI3K/Akt-Rac1-JNK and ERK pathways. *BioMed Res. Int.* **2014**, *2014*, 547187. [CrossRef] [PubMed]
2. Gurtner, G.C.; Werner, S.; Barrandon, Y.; Longaker, M.T. Wound repair and regeneration. *Nature* **2008**, *453*, 314–321. [CrossRef] [PubMed]
3. Demidova-Rice, T.N.; Hamblin, M.R.; Herman, I.M. Acute and impaired wound healing: Pathophysiology and current methods for drug delivery, Part 2: Role of growth factors in normal and pathological wound healing: Therapeutic potential and methods of delivery. *Adv. Skin Wound Care* **2012**, *25*, 349–370. [CrossRef] [PubMed]
4. Eming, S.A.; Martin, P.; Tomic-Canic, M. Wound repair and regeneration: Mechanisms, signaling, and translation. *Sci. Transl. Med.* **2014**, *6*, 265sr6. [CrossRef] [PubMed]
5. Krafts, K.P. Tissue repair: The hidden drama. *Organogenesis* **2010**, *6*, 225–233. [CrossRef] [PubMed]
6. Süntar, I.; Akkol, E.K.; Nahar, L.; Sarker, S.D. Wound healing and antioxidant properties: Do they coexist in plants? *Free Radic. Antioxid.* **2012**, *2*, 1–7. [CrossRef]
7. Manca, M.L.; Castangia, I.; Zaru, M.; Nácher, A.; Valenti, D.; Fernàndez-Busquets, X.; Fadda, A.M.; Manconi, M. Development of curcumin loaded sodium hyaluronate immobilized vesicles (hyalurosomes) and their potential on skin inflammation and wound restoring. *Biomaterials* **2015**, *71*, 100–109. [CrossRef] [PubMed]
8. Jangde, R.; Srivastava, S.; Singh, M.R.; Singh, D. In vitro and In vivo characterization of quercetin loaded multiphase hydrogel for wound healing application. *Int. J. Biol. Macromol.* **2018**, *115*, 1211–1217. [CrossRef] [PubMed]
9. Mukherjee, H.; Ojha, D.; Bharitkar, Y.P.; Ghosh, S.; Mondal, S.; Kaity, S.; Dutta, S.; Samanta, A.; Chatterjee, T.K.; Chakrabarti, S.; et al. Evaluation of the wound healing activity of *Shorea robusta*, an Indian ethnomedicine, and its isolated constituent(s) in topical formulation. *J. Ethnopharmacol.* **2013**, *149*, 335–343. [CrossRef] [PubMed]
10. Chen, X.; Peng, L.H.; Li, N.; Li, Q.M.; Li, P.; Fung, K.P.; Leung, P.C.; Gao, J.Q. The healing and anti-scar effects of astragaloside IV on the wound repair in vitro and in vivo. *J. Ethnopharmacol.* **2012**, *139*, 721–727. [CrossRef] [PubMed]
11. Kimura, Y.; Sumiyoshi, M.; Samukawa, K.; Satake, N.; Sakanaka, M. Facilitating action of asiaticoside at low doses on burn wound repair and its mechanism. *Eur. J. Pharmacol.* **2008**, *584*, 415–423. [CrossRef] [PubMed]
12. Kim, W.K.; Song, S.Y.; Oh, W.K.; Kaewsuwan, S.; Tran, T.L.; Kim, W.S.; Sung, J.H. Wound-healing effect of ginsenoside Rd from leaves of *Panax ginseng* via cyclic AMP-dependent protein kinase pathway. *Eur. J. Pharmacol.* **2013**, *702*, 285–293. [CrossRef] [PubMed]
13. Chen, X.; Peng, L.H.; Shan, Y.H.; Li, N.; Wei, W.; Yu, L.; Li, Q.M.; Liang, W.Q.; Gao, J.Q. Astragaloside IV-loaded nanoparticle-enriched hydrogel induces wound healing and anti-scar activity through topical delivery. *Int. J. Pharm.* **2013**, *447*, 171–181. [CrossRef] [PubMed]
14. Sharath, R.; Harish, B.G.; Krishna, V.; Sathyanarayana, B.N.; Swamy, H.M. Wound healing and protease inhibition activity of Bacoside-A, isolated from *Bacopa monnieri* wettest. *Phytother. Res.* **2010**, *24*, 1217–1222. [CrossRef] [PubMed]
15. Liu, M.; Dai, Y.; Li, Y.; Luo, Y.; Huang, F.; Gong, Z.; Meng, Q. Madecassoside isolated from *Centella asiatica* herbs facilitates burn wound healing in mice. *Planta Med.* **2008**, *74*, 809–815. [CrossRef] [PubMed]
16. Barros, W.M.; Rao, V.S.N.; Silva, R.M.; Lima, J.C.S.; Martins, D.T.O. Anti-inflammatory effect of the ethanolic extract from *Bowdichia virgilioides* H.B.K stem bark. *An. Acad. Bras. Cienc.* **2010**, *82*, 609–616. [CrossRef] [PubMed]
17. Melo, F.N.; Navarro, V.R.; Silva, M.S.; Da-Cunha, E.V.; Barbosa-Filho, J.M.; Braz-Filho, R. Bowdenol, a new 2,3-dihydrobenzofuran constituent from *Bowdichia virgilioides*. *Nat. Prod. Lett.* **2001**, *15*, 261–266. [CrossRef] [PubMed]
18. Velozo, L.S.M.; Da Silva, B.P.; Da Silva, E.M.B.; Parente, J.P. Constituents from the roots of *Bowdichia virgilioides*. *Fitoterapia* **1999**, *70*, 532–535. [CrossRef]

19. Bezerra-Silva, P.C.; Santos, J.C.; Santos, G.K.; Dutra, K.A.; Santana, A.L.; Maranhão, C.A.; Nascimento, M.S.; Navarro, D.M.; Bieber, L.W. Extract of *Bowdichia virgilioides* and maackiain as larvicidal agent against Aedes aegypti mosquito. *Exp. Parasitol.* **2015**, *153*, 160–164. [CrossRef] [PubMed]
20. Sharma, B.; Balomajumder, C.; Roy, P. Hypoglycemic and hypolipidemic effects of flavonoid rich extract from *Eugenia jambolana* seeds on streptozotocin induced diabetic rats. *Food Chem. Toxicol.* **2008**, *46*, 2376–2383. [CrossRef] [PubMed]
21. Grace, M.H.; Ribnicky, D.M.; Kuhn, P.; Poulev, A.; Logendra, S.; Yousef, G.G.; Raskin, L.; Lila, M.A. Hypoglycemic activity of a novel anthocyanin-rich formulation from lowbush blueberry, *Vaccinium angustifolium* Aiton. *Phytomedicine* **2009**, *16*, 406–415. [CrossRef] [PubMed]
22. Thimmappa, R.; Geisler, K.; Louveau, T.; O'Maille, P.; Osbourn, A. Triterpene biosynthesis in plants. *Annu. Rev. Plant Biol.* **2014**, *65*, 225–257. [CrossRef] [PubMed]
23. Saleem, M. Lupeol, a novel anti-inflammatory and anti-cancer dietary triterpene. *Cancer Lett.* **2009**, *285*, 109–115. [CrossRef] [PubMed]
24. Badshah, H.; Ali, T.; Rehman, S.U.; Amin, F.U.; Ullah, F.; Kim, T.H.; Kim, M.O. Protective effect of lupeol against lipopolysaccharide-induced neuroinflammation via the p38/c-Jun N-terminal kinase pathway in the adult mouse Brain. *J. Neuroimmune Pharmacol.* **2016**, *11*, 48–60. [CrossRef] [PubMed]
25. Alqahtani, A.; Hamid, K.; Kam, A.; Wong, K.H.; Abdelhak, Z.; Razmovski-Naumovski, V.; Chan, K.; Li, K.M.; Groundwater, P.W.; Li, G.Q. The pentacyclic triterpenoids in herbal medicines and their pharmacological activities in diabetes and diabetic complications. *Curr. Med. Chem.* **2013**, *20*, 908–931. [CrossRef] [PubMed]
26. Yokoe, I.; Azuma, K.; Hata, K.; Mukaiyama, T.; Goto, T.; Tsuka, T.; Imagawa, T.; Itoh, N.; Murahata, Y.; Osaki, T.; et al. Clinical systemic lupeol administration for canine oral malignant melanoma. *Mol. Clin. Oncol.* **2015**, *3*, 89–92. [CrossRef] [PubMed]
27. Sudhahar, V.; Ashok Kumar, S.; Varalakshmi, P.; Sujatha, V. Protective effect of lupeol and lupeol linoleate in hypercholesterolemia associated renal damage. *Mol. Cell. Biochem.* **2008**, *317*, 11–20. [CrossRef] [PubMed]
28. Kwon, H.H.; Yoon, J.Y.; Park, S.Y.; Min, S.; Kim, Y.I.; Park, J.Y.; Lee, Y.S.; Thiboutot, D.M.; Suh, D.H. Activity-guided purification identifies lupeol, a pentacyclic triterpene, as a therapeutic agent multiple pathogenic factors of acne. *J. Investig. Dermatol.* **2015**, *135*, 1491–1500. [CrossRef] [PubMed]
29. Harish, B.G.; Krishna, V.; Santosh Kumar, H.S.; Khadeer Ahamed, B.M.; Sharath, R.; Kumara Swamy, H.M. Wound healing activity and docking of glycogen-synthase-kinase-3-β-protein with isolated triterpenoid lupeol in rats. *Phytomedicine* **2008**, *15*, 763–767. [CrossRef] [PubMed]
30. Deutschlander, M.S.; Lall, N.; Van de Venter, M.; Hussein, A.A. Hypoglycemic evaluation of a new triterpene and other compounds isolated from *Euclea undulata Thunb.* var. *myrtina* (Ebenaceae) root bark. *J. Ethnopharmacol.* **2011**, *133*, 1091–1095. [CrossRef] [PubMed]
31. Kim, S.J.; Cho, H.I.; Kim, S.J.; Kim, J.S.; Kwak, J.H.; Lee, D.U.; Lee, S.K.; Lee, S.M. Protective effects of lupeol against D-galactosamine and lipopolysaccharide-induced fulminant hepatic failure in mice. *J. Nat. Prod.* **2014**, *77*, 2383–2388. [CrossRef] [PubMed]
32. Kim, M.J.; Bae, G.S.; Choi, S.B.; Jo, I.J.; Kim, D.G.; Shin, J.Y.; Lee, S.K.; Kim, M.J.; Song, H.J.; Park, S.J. Lupeol protects against cerulein-induced acute pancreatitis in mice. *Phytother. Res.* **2015**, *29*, 1634–1639. [CrossRef] [PubMed]
33. Saratha, V.; Subramanian, S.P. Lupeol, a triterpenoid isolated from *Calotropis gigantea* latex ameliorates the primary and secondary complications of FCA induced adjuvant disease in experimental rats. *Inflammopharmacology* **2012**, *20*, 27–37. [CrossRef] [PubMed]
34. Ardiansyah, Y.E.; Shirakawa, H.; Hata, K.; Hiwatashi, K.; Ohinata, K.; Goto, T.; Komai, M. Lupeol supplementation improves blood pressure and lipid metabolism parameters in stroke-prone spontaneously hypertensive rats. *Biosci. Biotechnol. Biochem.* **2012**, *76*, 183–185. [CrossRef] [PubMed]
35. Andriessen, M.P.; van Bergen, B.H.; Spruijt, K.I.; Go, I.H.; Schalkwijk, J.; van de Kerkhof, P.C. Epidermal proliferation is not impaired in chronic venous ulcers. *Acta Derm. Venereol.* **1995**, *75*, 459–462. [PubMed]
36. Santoro, M.M.; Gaudino, G. Cellular and molecular facets of keratinocyte reepithelization during wound healing. *Exp. Cell Res.* **2005**, *304*, 274–286. [CrossRef] [PubMed]
37. Grinnell, F. Fibroblast-collagen-matrix contraction: Growth-factor signalling and mechanical loading. *Trends Cell Biol.* **2000**, *10*, 362–365. [CrossRef]

38. Hashimoto, K.; Kajitani, N.; Miyamoto, Y.; Matsumoto, K.I. Wound healing-related properties detected in an experimental model with a collagen gel contraction assay are affected in the absence of tenascin-X. *Exp. Cell Res.* **2018**, *1*, 102–113. [CrossRef] [PubMed]
39. Sasaki, A.T.; Chun, C.; Takeda, K.; Firtel, R.A. Localized Ras signaling at the leading edge regulates PI3K, cell polarity, and directional cell movement. *J. Cell Biol.* **2004**, *167*, 505–518. [CrossRef] [PubMed]
40. Sepe, L.; Ferrari, M.C.; Cantarella, C.; Fioretti, F.; Paolella, G. Ras activated ERK and PI3K pathways differentially affect directional movement of cultured fibroblasts. *Cell. Physiol. Biochem.* **2013**, *31*, 123–142. [CrossRef] [PubMed]
41. Yu, J.S.; Cui, W. Proliferation, survival and metabolism: The role of PI3K/AKT/mTOR signalling in pluripotency and cell fate determination. *Development* **2016**, *17*, 3050–3060. [CrossRef] [PubMed]
42. Muthusamy, V.; Piva, T.J. The UV response of the skin: A review of the MAPK, NFκB and TNFα signal transduction pathways. *Arch. Dermatol. Res.* **2010**, *302*, 5–17. [CrossRef] [PubMed]
43. Kim, E.K.; Choi, E.J. Pathological roles of MAPK signaling pathways in human diseases. *Biochim. Biophys. Acta Mol. Basis Dis.* **2010**, *1802*, 396–405. [CrossRef] [PubMed]
44. Roskoski, R., Jr. ERK1/2 MAP kinases: Structure, function, and regulation. *Pharmacol. Res.* **2012**, *2*, 105–143. [CrossRef] [PubMed]
45. Loughlin, D.T.; Artlett, C.M. Modification of collagen by 3-deoxyglucosone alters wound healing through differential regulation of p38 MAP kinase. *PLoS ONE* **2011**, *6*, e18676. [CrossRef] [PubMed]
46. Landen, N.X.; Li, D.; Ståhle, M. Transition from inflammation to proliferation: A critical step during wound healing. *Cell. Mol. Life Sci.* **2016**, *20*, 3861–3885. [CrossRef] [PubMed]
47. Monkkonen, T.; Debnath, J. Inflammatory signaling cascades and autophagy in cancer. *Autophagy* **2018**, *2*, 190–198. [CrossRef] [PubMed]
48. Hoesel, B.; Schmid, J.A. The complexity of NF-κB signaling in inflammation and cancer. *Mol. Cancer* **2013**, *12*, 86. [CrossRef] [PubMed]
49. Xue, M.; Le, N.T.; Jackson, C.J. Targeting matrix metalloproteases to improve cutaneous wound healing. *Expert Opin. Ther. Targets* **2006**, *10*, 143–155. [CrossRef] [PubMed]
50. Rossi, H.S.; Koho, N.M.; Ilves, M.; Rajamäki, M.M.; Mykkänen, A.K. Expression of extracellular matrix metalloproteinase inducer and matrix metalloproteinase-2 and -9 in horses with chronic airway inflammation. *Am. J. Vet. Res.* **2017**, *11*, 1329–1337. [CrossRef] [PubMed]
51. Oklu, R.; Walker, T.G.; Wicky, S.; Hesketh, R. Angiogenesis and current antiangiogenic strategies for the treatment of cancer. *J. Vasc. Interv. Radiol.* **2010**, *21*, 1791–1805. [CrossRef] [PubMed]
52. Kong, D.; Yamori, T.; Kobayashi, M.; Duan, H. Antiproliferative and antiangiogenic activities of smenospongine, a marine sponge sesquiterpene aminoquinone. *Mar. Drugs* **2011**, *9*, 154–161. [CrossRef] [PubMed]
53. Chapnick, D.A.; Liu, X. Leader cell positioning drives wound-directed collective migration in TGFβ-stimulated epithelial sheets. *Mol. Biol. Cell* **2014**, *25*, 1586–1593. [CrossRef] [PubMed]
54. Xue, M.; Chow, S.O.; Dervish, S.; Chan, Y.K.A.; Julovi, S.M.; Jackson, C.J. Activated protein C enhances human keratinocyte barrier integrity via sequential activation of epidermal growth factor receptor and tie2. *J. Biol. Chem.* **2011**, *286*, 6742–6750. [CrossRef] [PubMed]
55. Findley, C.M.; Cudmore, M.J.; Ahmed, A.; Kontos, C.D. VEGF induces Tie2 shedding via a phosphoinositide 3-kinase/Akt-dependent pathway to modulate Tie2 signaling. *Arterioscler. Thromb. Vasc. Biol.* **2007**, *27*, 2619–2626. [CrossRef] [PubMed]
56. Tomikawa, K.; Yamamoto, T.; Shiomi, N.; Shimoe, M.; Hongo, S.; Yamashiro, K.; Yamaguchi, T.; Maeda, H.; Takashiba, S. Smad2 decelerates re-epithelialization during gingival wound healing. *J. Dent. Res.* **2012**, *91*, 764–770. [CrossRef] [PubMed]
57. Paramio, J.M.; Casanova, M.L.; Segrelles, C.; Mittnacht, S.; Lane, E.B.; Jorcano, J.L. Modulation of cell proliferation by cytokeratins K10 and K16. *Mol. Cell. Biol.* **1999**, *4*, 3086–3094. [CrossRef] [PubMed]
58. Maruthappu, T.; Chikh, A.; Fell, B.; Delaney, P.J.; Brooke, M.A.; Levet, C.; Moncada-Pazos, A.; Ishida-Yamamoto, A.; Blaydon, D.; Waseem, A.; et al. Rhomboid family member 2 regulates cytoskeletal stress-associated Keratin 16. *Nat. Commun.* **2017**, *8*, 14174. [CrossRef] [PubMed]
59. Asha, R.; Devi, V.G.; Abraham, A. Lupeol, a pentacyclic triterpenoid isolated from *Vernonia cinerea* attenuate selenite induced cataract formation in Sprague Dawley rat pups. *Chem. Biol. Interact.* **2016**, *245*, 20–29. [CrossRef] [PubMed]

60. Suryati, S.; Nurdin, H.; Dachriyanus, D.; Lajis, M.N.H. Structure elucidation of antibacterial compound from *Ficus deltoidea* Jack leaves. *Indones. J. Chem.* **2011**, *11*, 67–70. [CrossRef]
61. Rauth, S.; Ray, S.; Bhattacharyya, S.; Mehrotra, D.G.; Alam, N.; Mondal, G.; Nath, P.; Roy, A.; Biswas, J.; Murmu, N. Lupeol evokes anticancer effects in oral squamous cell carcinoma by inhibiting oncogenic EGFR pathway. *Mol. Cell. Biochem.* **2016**, *417*, 97–110. [CrossRef] [PubMed]
62. Xue, M.; Thompson, P.; Kelso, I.; Jackson, C. Activated protein C stimulates proliferation, migration and wound closure, inhibits apoptosis and upregulates MMP-2 activity in cultured human keratinocytes. *Exp. Cell Res.* **2004**, *299*, 119–127. [CrossRef] [PubMed]
63. Grada, A.; Otero-Vinas, M.; Prieto-Castrillo, F.; Obagi, Z.; Falanga, V. Research Techniques Made Simple: Analysis of Collective Cell Migration Using the Wound Healing Assay. *J. Investig. Dermatol.* **2017**, *137*, e11–e16. [CrossRef] [PubMed]

Sample Availability: Samples of the compounds are not available from the authors.

Article

Leishmanicidal Activity of Withanolides from *Aureliana fasciculata* var. *fasciculata*

Simone Cristina de M. Lima [1,†], Juliana da Silva Pacheco [2,†], André M. Marques [3], Eduardo Raul Pereira Veltri [2], Rita de Cássia Almeida-Lafetá [4], Maria Raquel Figueiredo [3], Maria Auxiliadora Coelho Kaplan [1] and Eduardo Caio Torres-Santos [2,*]

1 Instituto de Pesquisas de Produtos Naturais (IPPN), Universidade Federal do Rio de Janeiro, Av. Carlos Chagas Filho-Cidade Universitária, 21941-902 Rio de Janeiro, Brazil; smnioe@gmail.com (S.C.d.M.L.); makaplan@nppn.ufrj.br (M.A.C.K.)

2 Laboratório de Bioquímica de Tripanosomatídeos, Instituto Oswaldo Cruz, Fundação Oswaldo Cruz (FIOCRUZ), 210360-040 Rio de Janeiro, Brazil; juspacheco@hotmail.com (J.d.S.P.); erpv1994@gmail.com (E.R.P.V.)

3 Departamento de Produtos Naturais, Farmanguinhos, Fundação Oswaldo Cruz (FIOCRUZ), 21041-250 Rio de Janeiro, Brazil; andrefarmaciarj@yahoo.com.br (A.M.M.); mraquelf6@yahoo.com.br (M.R.F.)

4 Faculdade de Educação Tecnológica do Rio de Janeiro (FAETERJ-Rio), 21351-290 Rio de Janeiro, Brazil; ritalafeta@yahoo.com.br

* Correspondence: ects@ioc.fiocruz.br; Tel.: +55-(21)-3865-8247

† These authors contributed equally to this work.

Academic Editor: Wenxu Zhou

Received: 25 October 2018; Accepted: 27 November 2018; Published: 30 November 2018

Abstract: Leishmaniasis is the generic denomination to the neglected diseases caused by more than 20 species of protozoa belonging to the genus *Leishmania*. The toxic and parenteral-delivered pentavalent antimonials remain to be the first-line treatment. However, all the current used drugs have restrictions. The species *Aureliana fasciculata* (Vell.) Sendtner var. *fasciculata* is a native Brazilian species parsimoniously studied on a chemical point of view. In this study, the antileishmanial activity of *A. fasciculata* was evaluated. Among the evaluated samples of the leaves, the dichloromethane partition (AFfDi) showed the more pronounced activity, with IC_{50} 1.85 μg/ml against promastigotes of *L. amazonensis*. From AFfDi, two active withanolides were isolated, the Aurelianolides A and B, with IC_{50} 7.61 μM and 7.94 μM, respectively. The withanolides also proved to be active against the clinically important form, the intracellular amastigote, with IC_{50} 2.25 μM and 6.43 μM for Aurelianolides A and B, respectively. Furthermore, withanolides showed results for in silico parameters of absorption, distribution, metabolism, excretion, and toxicity (ADMET) similar to miltefosine, the reference drug, and were predicted as good oral drugs, with the advantage of not being hepatotoxic. These results suggest that these compounds can be useful as scaffolds for planning drug design.

Keywords: leishmania; solanaceae; withanolides; aurelianolides

---

1. Introduction

Leishmaniasis is the generic denomination to the diseases caused by more than 20 species of protozoa belonging to the genus *Leishmania*. Cutaneous leishmaniasis (CL) is the most prevalent form, caused mainly by *L. major* and *L. tropica* in Old World and by *L. braziliensis*, *L. guyanensis*, and *L. amazonensis* in New World, with an estimated 0.6 million to 1 million new cases occurring worldwide annually [1]. The toxic and parenteral-delivered pentavalent antimonials (N-methylglucamine antimoniate and sodium stibogluconate) remain to be the first-line treatment for CL in most countries. In addition, amphotericin B (conventional and liposomal) is used as an alternative in cases of unresponsiveness. The Food and Drug Administration (FDA) authorizes the use of miltefosine, the only oral drug available, for all clinical manifestations of leishmaniasis in the United States, including CL, but its efficacy in some endemic countries in South America is variable [2]. Other drugs introduced for CL treatment include pentamidine and paromomycin. In milder cases, the use of antimony in combination with cryotherapy is recommended [3]. However, all these drugs may lead to serious side effects, high toxicity or induction of parasite resistance [4–6].

The Solanaceae Family is considered one of the largest families among the eucotiledonous angiosperms, gathering around 150 genera and 3000 species concentrated in the neotropical region [7]. The species of this family are of great economic importance, being used in food such as potato (Solanum tuberosum), tomato (*Solanum lycopersicum*), eggplant (*Solanum melongena*), and pepper (*Capsicum annum*) [8]. Several of these species have been investigated with great interest by the pharmaceutical industry due to their bioactive metabolites, such as alkaloids and steroids that occur in many genera [9].

The genus *Aureliana* is a small endemic genus in the Solanaceae family, widely distributed in South and Southeastern of Brazil, usually found in mountain forests area, semi-deciduous mesophyllous forests, and reef areas [10]. The species *Aureliana fasciculata* (Vell.) Sendtner var. *fasciculata* is a native Brazilian species rarely studied on a chemical point of view, which is found in Atlantic Forest [11].

Phytochemical studies showed that the steroid derivatives are the major compound metabolites present in *A. fasciculata* leaves [11]. The withasteroids comprises a group of steroidal substances characterized by a moiety ergostane with 28 carbon atoms, where C-22 and C-26 are oxidized to form six-membered lactone [12]. The most abundant type is usually designated as withanolides and these compounds possess an α,β-unsaturated δ-lactone ring in the side chain of the molecule [13]. These steroid derivatives are frequently polyoxygenated, and biogenetic transformations can produce highly modified structures, both in the steroid nucleus and in the side chain [14].

Since the 1960s, about 750 withanolides have been isolated. These substances are found in many genera of the family Solanaceae, such as *Acnistus, Datura, Deprea, Discopodium, Dunalia, Iochroma, Jaborosa, Lycium, Nicandra, Physalis, Solanum, Trechonaetes, Tubocapsicum, Vassobia, Withania*, and *Witheringia* [15]. However, the occurrence of withanolides is not completely restricted to Solanaceous plants and reports of their isolation from marine organisms, and from members of the Taccaceae, Fabaceae (Leguminosae) [16], and Dioscoreaceae [17], Myrtaceae and Lamiaceae [18] families suggest that they are much more widely distributed.

In this work, the antileishmanial property of *Aureliana fasciculata* Vell. Sendtner var. *fasciculata* was first demonstrated and the purification guided by the biological activity pointed two withanolides as the active constituents.

2. Results and Discussion

Following an approach of antileishmanial-guided extraction, the methanolic extract of leaves of *A. fasciculata* was submitted to partition using solvents with crescent polarities (Figure 1). The resulting fractions were evaluated for antipromastigote activity. All fractions showed some degree of promastigote inhibition, but only the dichloromethane fraction (AFfPDi) had IC_{50} below than 10 μM (1.85 μM), the threshold considered in this study. Thus, this fraction was successively chromatographed, originating two purified withanolides, Aurelianolide A and Aurelianolide B.

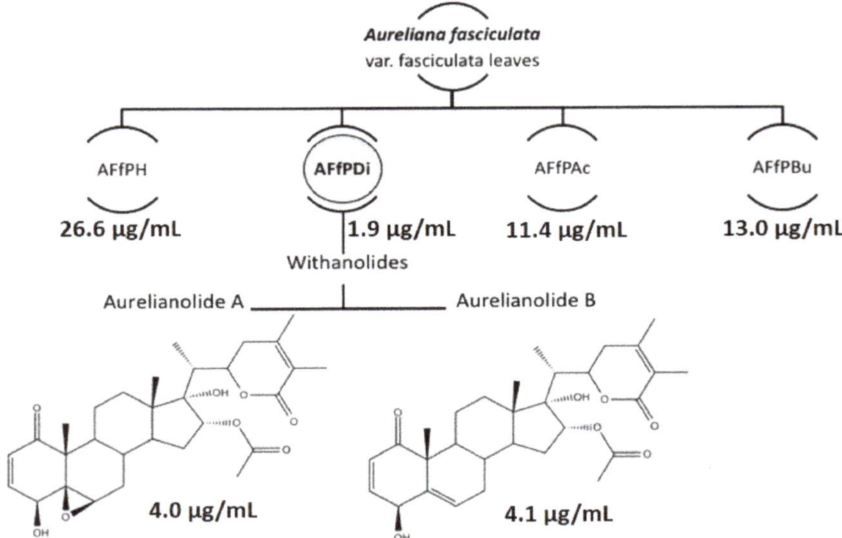

Figure 1. Antileishmanial-guided extraction of leaves of *Aureliana fasciculata*. AFfPH, hexane partition; AFfPDi, dichloromethane partition; AFfPAc, ethyl acetate partition; AFfPBu, butanol partition. The numbers refer to antipromastigote ICs$_{50}$.

These compounds were first isolated from the same species and described by Almeida-Lafetá in 2010 [11] (Figure 1).

The Aurelianolide A (MW: 528,64) was obtained as a white amorphous solid and its molecular formula was deduced as $C_{32}H_{38}O_8$. The mass spectra of HRMS showed an ion with m/z 551.2682 resulting from the formation of adducts of the Aurelianolide A with sodium ions. The Aurelianolide B (MW: 512,64) was obtained as white crystals having its molecular formula deduced as $C_{32}H_{39}O_7$. The mass spectra of HRMS showed an ion with m/z 535.2682 resulting from the formation of adducts of one substance with sodium ions. The NMR data were compared to the literature data [11]. Up to this moment, no biological studies were performed with these steroid-derivatives metabolites.

The purified compounds conserved the antipromastigote activity, although being slightly less active separately, with IC_{50} of 4.0 μg/ml (7.6 μM) and 4.1 μg/ml (7.9 μM), for Aurelianolides A and B, respectively (Figure 1, Table 1).

Following these results, Aurelianolide A and Aurelianolide B were evaluated for antimastigote activity, the clinically relevant form. Both compounds were active but, differently of the antipromastigote action, they had distinct potencies. Aurelianolide A was more potent, with an IC_{50} 1.2 μg/ml (2.3 μM), while Aurelianolide B showed an IC_{50} 3.3 μg/ml (6.43 μM).

Withanolides have already shown to possess various biological activities such as anti-inflammatory [19,20], antitumoral [20], trypanosomicidal [21], antileishmanial [22], immunomodulatory [23], and antibacterial [24]. Such substances also exhibit insecticidal activity and phytotoxicity [25]. Some withanolides with highlighted leishmanicidal activity were isolated from *Withania coagulans* [26], *Physalis minima* [27], and *Dunalia brachyachantha* [28].

Table 1. Antileishmanial activity, cytotoxicity, and selectivity index for fractions and withanolides from *Aureliana fasciculata*.

	L. amazonensis		J774 Macrophages (CC_{50}) *	Selectivity Index (SI)
	Promastigotes (IC_{50}) *	Intracellular Amastigotes (IC_{50}) *		
AFfPH	26.6 ± 0.1	N.D.	N.D.	N.D.
AFfDi	1.9 ± 0.7	N.D.	N.D.	N.D.
AFfAc	11.4 ± 0.1	N.D.	N.D.	N.D.
AFfBu	13.0 ± 0.1	N.D.	N.D.	N.D.
Aurelianolide A	4.0 ± 0.1 (7.6 ± 0.1)	1.2 ± 0.1 (2.3 ± 0.1)	6.7 ± 0.2 (12.7 ± 0.2)	5.6
Aurelianolide B	4.1 ± 0.3 (7.9 ± 0.7)	3.3 ± 0.1 (6.4 ± 0.1)	6.7 ± 0.1 (13.1 ± 0.1)	2.0
Pentamidine	2.8 ± 0.1 (4.8 ± 0.1)	1.1 ± 0.1 (1.9 ± 0.1)	5.0 ± 0.7 (8.5 ± 1.2)	4.5

SI = CC_{50}/IC_{50} in amastigotes; * μg/ml (μM).

To evaluate the selectivity, the cytotoxic profile against J774 cells was evaluated with resazurin. Both Aurelianolides A and B showed the same cytotoxic activity to J774 macrophages, with CC_{50} 6.7 μg/ml (12.7 μM and 13.1 μM, respectively) (Table 1). Note that, in this case, the presence of the epoxide did not influence in the activity, as well as in promastigotes. The selectivity index (SI) was also calculated, revealing an SI for Aurelianolide A of 5.6 and for Aurelianolide B of 2.0. The calculated SI for Aurelanolide A was higher than that found for the reference drug, pentamidine (4.5, Table 1).

The higher activity of withanolides in intracellular amastigotes, mainly the Aurelianolide A, is suggestive that the host cell could be playing a role in clearing the parasites. Nitric oxide (NO) is an important tool for killing intracellular parasites. The outcome of treatment with withanolides was also examined by measuring the NO concentration on the culture supernatant. Figure 2 shows that the NO level increased significantly when infected macrophages were treated with twice the IC_{50} values of antipromastigote activity for both withanolides. The NO level was low among the infected macrophages treated with a quarter the IC_{50} value of Aurelianolide A, but there were no significant differences.

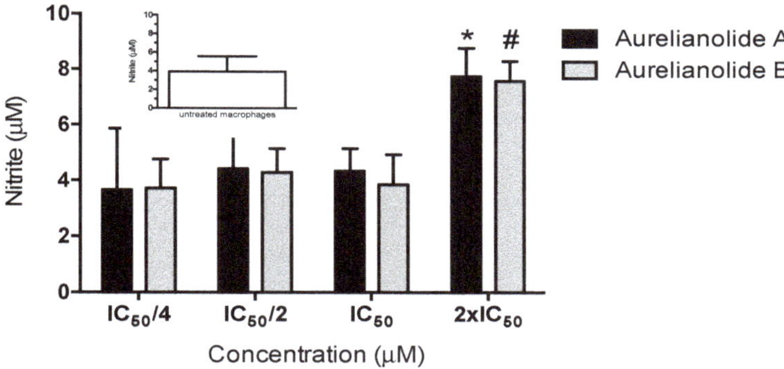

Figure 2. Effect of Aurelianolides on Nitric Oxide production by macrophages infected with *L. amazonensis*. Concentrations are related with a quarter, half, twice and the IC_{50} values of antipromastigote activity.

NO, produced by the nitric oxide synthase (iNOS) enzyme, is a product of macrophages activated by cytokines and is one of the most important molecules responsible for the killing of *Leishmania* parasites [29,30]. It was observed that IFN-γ has been shown to synergize with

TNF in murine systems, leading NO production by iNOS, resulting in eradication of intracellular parasites. In the infected host organisms, functions of NO described to date include antiviral, antimicrobial, immunostimulatory (proinflammatory), immunosuppressive (anti-inflammatory), cytotoxic (tissue-damaging), and cytoprotective (tissue-preserving) effects. The antimicrobial activity of NO was originally thought to result from mutation of DNA, inhibition of DNA repair and synthesis, inhibition of protein synthesis and alteration of proteins by S-nitrosylation, ADP-ribosylation, or tyrosine nitration [31]. Previous findings show a higher NO expression in monocytes from human CL patients comparing to expression in monocytes from healthy patients. It was observed a huge correlation between NO production and lesion size of CL patients. Further, NO alone is not sufficient to control infection and may contribute to the tissue damage observed in human CL [32].

Three other withanolides isolated from leaves of Solanaceae family were found to be responsible for inhibiting NO production by activated macrophages [33]. In this work, NO production in infected macrophages treated with a quarter, half, and the IC_{50} value of withanolides did not show significant differences comparing the NO levels of non-treated infected macrophages. The NO production by infected cytokines-activated macrophages and its consequences in killing parasites should be further investigated.

The theoretical analysis of the physicochemical parameters, Lipinski's rule of five (Ro5) and ADMET (absorption, distribution, metabolism, excretion, and toxicity) properties of withanolides and miltefosine were performed using the Predicting Small-Molecule Pharmacokinetic and Toxicity Properties (PkCSm tool). In short, the rule of five predicts absorption or drug permeability parameters which are values multiple of 5 or even 5 [34]. The comparison with miltefosine was made since is the only oral drug available for leishmaniasis treatment [35]. In the first analysis, we observed that both withanolides violated for a few tens only one parameter; the molecular weight that should be below 500. Also, a little violation was observed for miltefosine in the logP parameter with a value above 5, while Aurelianolide A and Aurelianolide B followed the rule proposal. Taken together with the water solubility values found, these results suggest a profile of water-soluble drugs for withanolides (Table 2). Mckerrow and Lipinski (2017) [35] highlighted that they never intend the "rule of 5" as a mainstay fixed rule for a new drug, but a parameter to be carefully evaluated. Miltefosine represented a breakthrough, as the first orally active compound for Leishmaniasis in clinics and since then screening efforts continue the search for an oral improved drug when it comes parasitic diseases drug discovery in general [36–38].

Table 2. Physicochemical parameters and Lipinski's rule of five of withanolides and miltefosine using pkCMS tool*.

Parameters	Aurelianolide A	Aurelianolide B	Miltefosine
MW	528.642	512.643	407.576
LogP	3.037	3.825	5.6755
#ACCEPTORS	8	7	4
#DONORS	2	2	0
Water solubility (log mg/l)	−4.924	−5.329	−5.673

* MW, molecular weight; LogP, logarithm of the compound partition coefficient between *n*-octanol and water; #, NHB (number of hydrogen bonds)

The in silico ADMET analysis (Table 3) showed a good probability of permeability on Caco2 cells, with values above of the adopted threshold of 0.9 for Aurelianolide A and B. High human intestinal absorption probability was observed for Aurelianolide A (91%) and Aurelianolide B (90.53%), near to values found for miltefosine (94.987%). The withanolides showed low distribution volume (VDss less than 0.56 l/kg), suggesting low output from blood to the tissues. Furthermore, both withanolides are unlikely to penetrate the central nervous system (CNS) (logBBB < 0 and logPS < −2), which helps to reduce side effects and toxicity. The prediction suggests withanolides should be metabolized by cytochrome P3A4 (CYP3A4), and there is no indication of inhibition of the main cytochrome P450 (CYP P450) oxidases.

Table 3. In silico ADME (absorption, distribution, metabolism, excretion) properties of Aurelianolides and miltefosine using pkCMS tool *.

Parameters	Aurelianolide A	Aurelianolide B	Miltefosine
Absorption			
Caco2 permeability (log Papp in 10^{-6} cm/s)	1.31	1.474	1.153
Intestinal absorption (human, %)	91.459	90.543	94.987
Skin Permeability (log Kp)	−3.101	−3.638	−2.702
Distribution			
VDss (human, l/kg)	0.04	0.121	0.96
Fraction unbound (human)	0.223	0.171	0.238
BBB permeability (log BB)	−0.801	−0.602	−0.345
CNS permeability (log PS)	−3.278	−3.033	−3.172
Metabolism			
CYP2D6 substrate	No	No	No
CYP3A4 substrate	Yes	Yes	Yes
CYP1A2 inhibitor	No	No	No
CYP2C19 inhibitor	No	No	No
CYP2C9 inhibitior	No	No	No
CYP2D6 inhibitior	No	No	No
CYP3A4 inhibitior	No	No	No
Excretion			
Total Clearance (log ml/min/kg)	0.275	0.36	1.156

* VDss, steady-state volume of distribution; BBB, blood-brain barrier; CNS, central nervous system.

Toxicity predictions (Table 4) pointed out that Aurelianolides and miltefosine are not likely to be mutagenic or cause skin sensitization. Predictions suggest that equally to miltefosine, Aurelianolides are not expected to inhibit human ether-à-go-go related genes (hERGI), but probably inhibit hERGII. These predictions indicate the need for evaluation of cardiac markers in the biochemical analysis when we proceed to tests in animal model. Aurelianolides were not predicted to be hepatotoxic unlikely miltefosine. Miltefosine is mainly used for visceral leishmaniasis and in some countries to treat CL, but gastrointestinal side effects, teratogenicity alert for young females and hepato- and nephrotoxicity, require patient monitoring; in addition, it was not considered a good candidate for topical treatment [39,40].

Table 4. In silico Toxicity of Aurelianolides and miltefosine using the pkCMS tool.

Parameters	Aurelianolide A	Aurelianolide B	Miltefosine
AMES toxicity	No	No	No
Max. tolerated dose (human, log mg/kg/day)	−1.053	−0.858	1.079
hERG I inhibitor	No	No	No
hERG II inhibitor	Yes	Yes	Yes
Oral Rat Acute Toxicity (LD50) (mol/kg)	2.518	2.284	2.211
Oral Rat Chronic Toxicity (LOAEL) (log mg/kg_bw/day)	1.786	1.692	1.34
Hepatotoxicity	No	No	Yes
Skin Sensitisation	No	No	No
T. Pyriformis toxicity pIGC50 (log µg/l)	0.299	0.340	1.054
Minnow toxicity LC50 (log mM)	0.503	0.109	−2.403

3. Material and Methods

3.1. Botanical Material

The species *Aureliana fasciculata* (Vell.) Sendtner var. *fasciculata* was collected in the city of Simão Pereira, MG, Brazil and was identified by botanist Dr. Rita de Cassia Almeida-Lafetá. A voucher specimen was deposited in RFA Herbarium (UFRJ, Rio de Janeiro, Brazil) under number 40829.

3.2. Extraction and Isolation

Leaves were weighed and dried in an oven with circulating air at 40 °C. The dried plant organ was reduced to small fragments in a knife mill (Tecnal 048, Piracicaba, Brazil). After drying, the fragmented leaves were extracted by static maceration at room temperature with methanol. The extract was concentrated by using a rotary evaporator (Buchi, Flawil, Switzerland) into the dry extract. The crude methanol extract (50g) from leaves of *Aureliana fasciculata* var. *fasciculata* was suspended in MeOH/H_2O (3:7) and subjected to the liquid-liquid partitions in sequences with solvents such as hexane, dichloromethane, ethyl acetate, and butanol. The dichloromethane fraction (2.0 g) was chromatographed on a column of XAD using methanol as eluent. The fractions 2–3 were submitted to chromatography on a column of silica gel, using as solvent systems, binary mixtures of dichloromethane, and methanol in increasing polarity gradient, as also pure hexane, dichloromethane, and methanol. The fractions collected 8–11 were reunited and chromatographed on a preparative plate to obtain the two withanolides: Aurelianolide A (3 mg) and Aurelianolide B (11 mg).

3.3. ESI-MS Analysis

Mass spectra were obtained from the High-Resolution device in MicroTOFII Bruker electrospray ionization (Bruker, Bremen, Germany). The samples were diluted with spectroscopic grade MeOH (Tedia, Fairfield, OH, USA) concentration of 500 µg/ml being injected in the device flow 5 min/l. The analysis was performed in the positive mode. Change in spectral window of m/z 50 to 2000. The obtained data were compared to literature data [11].

Aurelianolide A: white amorphous solid. ESI-MS (positive): m/z 551.2625 [M+Na].

Aurelianolide B: white crystals. ESI-MS (positive): m/z 535.2682 [M+Na].

3.4. NMR Analysis

Nuclear magnetic resonance spectra of hydrogen and carbon (^{13}C and ^{1}H NMR) were obtained on Varian device VNMRS-Gemini 500 spectrometer (NMR Associates, Fitchburg, MA, USA) operating at a frequency of 400MHz/100MHz using CD_3OD as the solvent. Special techniques and bi-dimensional such as COSY, HMBC, and HSQC were also performed. The chemical shift values (δ) in dimensionless units, were referred to an internal standard (TMS), is represented in parts per million (ppm) of the applied frequency for each experiment and coupling constants (J) were measured in Hz. The obtained data were compared to literature data [11].

Aurelianolide A: RMN ^{1}H (400MHz, CD_3OD): (δ, ppm): 6,17 (H2, d, J = 9,92 Hz, 1H), 7,06 (H3, dd, J = 9,98 Hz and 6,28 Hz, 1H), 3,64 (H4, d, J = 6,28 Hz, 1H), 3,15 (H6, s, 1H), 2,07 (H7a, m, 1H), 1,35 (H7b, m, 1H), 1,50 (H8, m, 1H), 1,21 (H9, m, 1H), 1,92 (H11a, m, 1H), 1,47 (H11b, m, 1H), 1,71 (H12a, m, 1H), 1,68 (H12b, m, 1H), 1,75 (H14, m, 1H), 1,89 (H15a, m, 1H), 1,48 (H15b, m, 1H), 5,13 (H16, dd = 8,56 Hz and 2,24 Hz, 1H), 0,85 (H18, s, 3H), 1,37 (H19, s, 3H), 2,25 (H20, dq, J = 6,72 Hz and 4,00 Hz, 3H), 1,04 (H21, d, J = 6,96 Hz, 1H), 4,36 (H22, dt, J = 11,08 and 4,0 Hz, 1H), 2,50 (H23a, m, 1H), 2,12 (H23b, m, 1H), 1,82 (H27, s, 3H), 1,97 (H28, s, 3H), 1,99 (OCH_3, s, 3H). RMN ^{13}C (100MHz, CD_3OD): (δ, ppm): 204,17 (C1), 133,17 (C2), 145,32 (C3), 71,14 (C4), 64,75 (C5), 61,16 (C6), 32,46 (C7), 31,08 (C8), 43,60 (C9), 51,73 (C10), 21,78 (C11), 33,24 (C12), 49,67 (C13), 49,25 (C14), 33,78 (C15), 80,13 (C16), 84,46 (C17), 15,34 (C18), 16,94 (C19), 45,22 (C20), 9,59 (C21), 79,94 (C22), 34,59 (C23), 153,19 (C24), 122,15 (C25), 169,20 (C26), 12,39 (C27), 20,50 (C28), 171,61 (COAc), 21,11 (OCH_3).

Aurelianolide B: RMN ^{1}H (400MHz, CD_3OD): (δ,ppm): 5,86 (H2, d, J = 10,48 Hz, 1H), 6,85 (H3, dd, J = 10,00 Hz and 4,60 Hz, 1H), 4,55 (H4, d, J = 4,60 Hz, 1H), 5,89 (H6, d, J = 4,2 Hz, 1H), 2,07 (H7a, m, 1H), 1,46 (H7b, m, 1H), 1,66 (H8, m, 1H), 1,16 (H9, m, 1H), 2,15 (H11a, m, 1H), 1,55 (H11b, m, 1H), 1,93 (H12a, m, 1H), 1,69 (H12b, m, 1H), 1,86 (H14, m, 1H), 1,90 (H15a, m, 1H), 1,50 (H15b, m, 1H), 5,15 (H16, dd, J = 9,0 Hz and 2,56 Hz, 1H), 0,93 (H18, s, 3H), 1,43 (H19, s, 3H), 2,27 (H20, dq, J = 6,96 Hz and 3,96 Hz, 1H), 1,08 (H21, d, J = 6,96 Hz, 1H), 4,36 (H22, dt, J = 13,92 Hz and 3,96 Hz, 1H), 2,51 (H23a. m, 1H), 2,12 (H23b, m, 1H), 1,83 (H27, s, 3H), 1,97 (H28, s, 3H), 1,99 (OCH_3, s, 3H). RMN ^{13}C (100MHz, CD_3OD): (δ, ppm): 205,84 (C1), 131,19 (C2), 146,18 (C3), 69,86 (C4), 139,72 (C5), 129,28 (C6), 32,11 (C7), 33,83 (C8), 43,65 (C9), 49,74 (C10), 23,68 (C11), 33,59 (C12), 50,48 (C13), 49,72 (C14), 33,78 (C15), 80,22 (C16), 84,61 (C17), 15,69 (C18), 23,01 (C19), 44,20 (C20), 9,61 (C21), 80,04 (C22), 34,59 (C23), 153,25 (C24), 122,12 (C25), 169,22 (C26), 12,40 (C27), 20,52 (C28), 171,70 (COAc), 21,15 (OCH_3).

3.5. Parasites

Leishmania amazonensis promastigotes (MHOM/BR/77/LTB/0016) were maintained at 26°C in Schneider's medium (Sigma-Aldrich, St. Louis, MO, USA) supplemented with 10% bovine serum (FBS), 100 mg/ml streptomycin and 100 U/ml penicillin. Subcultures were performed twice a week until the seventh passage. Subsequently, old cultures were discarded and fresh parasites were obtained from BALB/c mice lesions.

3.5.1. Antipromastigote Activity

To evaluate the antileishmanial activity, promastigotes of *L. amazonensis* were maintained in cell culture flasks at 26 °C in Schneider's medium (Sigma-Aldrich, St. Louis, MO, USA), supplemented as described above. Experiments were performed in 96-well plates for 72 h at 26 °C with an initial inoculum of 1.0×10^6 cells/ml and varying concentrations of plant extracts or withanolides. Pentamidine was used as a control, varying from 0.39 to 25μM. After 72h of incubation, parasites viability was assessed adding resazurin (50 μM) for additional 3h. After this time, fluorescence was quantified (excitation λ = 560 nm; emission λ = 590 nm) and the data obtained from three experiments were expressed as the mean ± standard error of the mean (Mean ± S.E.M.). The half maximal inhibitory concentration (IC_{50}) was determined by logarithmic non-linear regression analysis using GraphPrism software (Version 5, GraphPad, San Diego, CA, USA).

3.5.2. Antiamastigote Activity

Resident macrophages were harvested from the peritoneum of BALB/c mice in ice-cold RPMI supplemented with 1% glutamine and 1% pyruvate. The cells were plated at 2.0×10^6/ml (0.4 ml/well) on circular 13mm glass diameter coverslips in 24 well plates and kept in a 5% CO_2 atmosphere at 37°C, for 1h. Nonadherent cells were removed by washing with pre-warmed complete medium. Macrophages were infected with promastigotes of *L. amazonensis* on stationary phase at a 3:1 parasite/macrophage ratio. After 3h of incubation, the monolayers were washed three times with pre-warmed complete medium to remove free parasites. The withanolides were added in duplicates in concentrations based on the antipromastigote IC_{50}, ranging from twice, half and a quarter. The plates were incubated for a further 72 h. Afterward, the coverslips were stained with a Romanowsky stain (Panótico, New Prov, Pinhais, Brazil), according to fabricant instructions. The number of intracellular amastigotes was determined by counting at least 100 macrophages per well. The results were expressed as an infection index (% infected macrophage × number of amastigotes / total number of macrophages) and IC_{50} was determined by logarithmic non-linear regression analysis using GraphPrism software.

3.6. Cytotoxic Study

Mouse macrophages cell line J774 were plated at 2.0×10^6 cells/ml in 96-well plates, in ice-cold RPMI supplemented with 10% FBS, 1% glutamine and 1% pyruvate. The cells were incubated at 37°C under an atmosphere of 5% CO_2 for 1 h. Non-adherent cells were removed by washing with pre-warmed complete medium. Tests were performed with concentration ranging from 0 to 100 μM of withanolides. After 72h of incubation, macrophages viability was measured by colorimetric assay using rezasurin (50μM). After 3 h, the fluorescence was quantified (excitation λ = 560 nm; emission λ = 590 nm) and CC_{50} value (concentration that reduces in 50% the cells viability) was determined by logarithmic non-linear regression analysis using GraphPrism software. Selectivity index (SI) was expressed by the ratio between CC_{50} value over the host cells and the IC_{50} obtained over intracellular amastigotes.

3.7. Nitric Oxide Production

Supernatant from antiamastigote assay was collected after 72h and the nitric oxide (NO) concentration was indirectly measured using Griess reagent, as described by Green et al. [41]. Griess reagent is 0.1% *N*-(1-Naphthyl) ethylenediamine (under orthophosphoric acid conditions—5%) and 1% sulfanamide solution. The reaction was realized by the addition of 50μl of Griess reagent and 50μl of supernatants obtained from the antiamastigote assay. After 10 min of incubation at room temperature, the absorbance at 540 nm was measured and the nitrite concentration was determined from a sodium nitrite ($NaNO_2$) solution standard curve.

3.8. In Silico ADMET properties

We performed some theoretical analysis of the drug-likeness of withanolides. The pharmacokinetic profile of a compound defines its absorption, distribution, metabolism, excretion, and toxicity (ADMET). The ADMET properties of withanolides were evaluated using the admetSAR tool [42] and the Lipinski's rule of the compounds was also calculated.

4. Conclusions

Aureliana fasciculata (Vell.) Sendtner var. *fasciculata* is a Brazilian species of Atlantic Forest that has been no longer investigated on the chemical and biological activity point of view. The phytochemical study of *A. fasciculata* var. *fasciculata* (Vell.) Sendtner var. *fasciculata* resulted in the isolation of two potential leishmanicidal withanolides: Aurelianolide **A** and Aurelianolide **B**, which show the trend of the species subfamily Solanoideae to withanolides in the family Solanaceae.

When outlining a strategy to screen the antileishmanial activity of plant extracts or other compounds, the first step involves choosing an approach: target-driven or phenotypic assays. Both assays have advantages and disadvantages. The target-driven assay allows to detect compounds with a mechanism of action previously chosen but limited to only one target. Phenotypic assay allows to explore all the molecular targets in whole and live parasites. However, when an active compound is found, the discovery of the mechanism of action is challenging. Here we decided to use a phenotypic assay with promastigotes and intracellular amastigotes to maximize the probability of finding an active compound. The extraction guided by the antileishmanial activity revealed Aureanolides A and B as the active compounds present in *A. fasciculata*.

These compounds showed direct activity on the parasite, as shown by antipromastigote assay, but the lower IC_{50} for intracellular amastigotes and the enhancement in NO production in the highest concentration suggest also an additional mechanism involving the host cell. The in-silico predictions pointed to a high probability for good bioavailability by oral route and low toxicity. The scaffold of aurelianolides and the findings of its in-silico pharmacokinetics properties are the major guides for optimization possibilities that could support their future preclinical and clinical applications to leishmaniasis. The low VDss might suggest plasma protein binding (PPB) and this phenomenon influences the absorption, biodistribution, metabolism, and excretion of drugs. In silico results of our study also corroborate with the one performed by Singh et al. [43] in which withanolide A from *Withania somnifera* (L.), with neuropharmacological activity, presented high PPB, passive permeability and a fast and wide distribution kinetics in vitro. Also, Dubay et al. [44] demonstrated that withanolides and withanosides of *W. somnifera* have a strong binding to serum albumin. In addition, further structure-based drug design with withanolides is required. Initially, to improve the distribution volume and make aurelianolides more available to the tissues affected by the high parasitic load in patients with leishmaniasis, changes in the structure to increase liposolubility, without resulting in the loss of its biological effect, should be performed. Along with that, combined techniques to pursue their molecular target like scaffold-based virtual ligand screening and Quantitative structure–activity relationship (QSAR) are also complementary attempts that could help us to find a ligand and exploit pharmacokinetic and pharmacodynamics properties of aurelianolides to achieve an improved lead compound.

Author Contributions: E.C.T.S and M.R.F conceived and designed the experiments; S.C.L., J.S.P., R.C.A.L., and A.M.M. performed the experiments; E.C.T.S., M.A.C.K., and M.R.F analyzed the data; all the group staff wrote the paper.

Funding: This study was financed by the Coordenação de Aperfeiçoamento de Pessoal de Nível Superior–Brazil (CAPES) (Finance Code 001), Conselho Nacional de Desenvolvimento Científico e Tecnológico–Brazil (CNPq) and Fundação Carlos Chagas Filho de Amparo à Pesquisa do Estado do Rio de Janeiro–Brazil (FAPERJ).

Acknowledgments: The authors would like to thank Farmanguinhos and Instituto Oswaldo Cruz for technical support.

Conflicts of Interest: The authors declare no conflict of interest.

References

1. Alvar, J.; Vélez, I.D.; Bern, C.; Herrero, M.; Desjeux, P.; Cano, J.; Jannin, J.; den Boer, M.; WHO Leishmaniasis Control Team. Leishmaniasis worldwide and global estimates of its incidence. *PLoS ONE* **2012**, *7*, e35671. [CrossRef] [PubMed]
2. Sundar, S.; Singh, A. Chemotherapeutics of visceral leishmaniasis: Present and future developments. *Parasitology* **2017**, *7*, 1–7. [CrossRef] [PubMed]
3. de Vries, H.J.C.; Reedjik, S.H.; Schallig, H.D.F.H. Cutaneous Leishmaniasis: Recent developments in diagnosis and management. *Am. J. Clin. Dermatol.* **2015**, *16*, 99–109. [CrossRef] [PubMed]
4. Berman, J.N. Chemotherapy of leishmaniasis: Recent advances in the treatment of visceral disease. *Curr. Opin. Infect Dis.* **1998**, *11*, 707–710. [CrossRef] [PubMed]
5. Gontijo, B.; Carvalho, M.L. American cutaneous leishmaniasis. *Ver. Soc. Bras. Med. Trop.* **2003**, *36*, 71–80. [CrossRef]
6. Jhingran, A.; Chawla, B.; Saxena, S.; Barret, M.P.; Madhubala, R. Paronomycin: Uptake and resistance in *Leishmania donovani*. *Mol. Biochem. Parasitol.* **2009**, *164*, 111–117. [CrossRef] [PubMed]
7. Souza, V.C.; Lorenzi, H. *Botânica Sistemática: Guia ilustrado para identificação das famílias de Angiospermas da flora brasileira, baseado em APG II.*; Instituto Plantarum de Estudos da Flora: Nova Odessa, Brazil, 2005.
8. Elabbar, F.A.; Nawill, M.A.B.; Ashraf, T.M.E. Extraction, separation and identification of compounds from leaves of *Solanum elaeagnifolium* Cav. (Solanaceae). *Int. Curr. Pharm. J.* **2014**, *3*, 234–239. [CrossRef]
9. Hawkes, J.G. *The economic importance of the family Solanaceae, In Solanaceae IV. Advances in Botany and Utilization*; Royal Botanic Gardens: London, UK, 1999; pp. 1–8.
10. Hunziker, A.T.; Barbosa, G. Estudios sobre Solanaceae XXX: Revisión de Aureliana. *Darwiniana* **1991**, *30*, 95–112.
11. Almeida-Lafetá, R.; Ferreira, M.J.P.; Emerenciano, V.P.; Kaplan, M.A.C. Withanolides from *Aureliana fasciculata var. fasciculata*. *Helv. Chim. Acta* **2010**, *93*, 2478–2487.
12. Dhar, N.; Razdan, S.; Rana, S.; Bhat, W.W.; Vishwakarma, R.; Lattoo, S.K.A. Decade of molecular understanding of withanolide biosynthesis and in vitro studies in *Withania somnifera* (L.) Dunal: Prospects and perspectives for pathway engineering. *Front. Plant Sci.* **2015**, *6*, 1031. [CrossRef] [PubMed]
13. Vaishnavi, K.; Saxena, N.; Shah, N.; Singh, R.; Manjunath, K.; Uthayakumar, M.; Kanaujia, S.P.; Kaul, S.C.; Sekar, K.; Wadhwa, R. Differential activities of the two closely related withanolides, withaferin A and withanone: Bioinformatics and experimental evidences. *PLoS ONE* **2012**, *7*, e0044419. [CrossRef] [PubMed]
14. Olmstead, R.G.; Bohs, L.; Migid, H.A.; Santiago-Valentin, E.; Garcia, V.F.; Collier, S.M. A molecular phylogeny of the Solanaceae. *Taxon* **2008**, *54*, 1159–1181.

15. Zhang, H.; Samadi, A.K.; Cohen, M.S.; Timmermann, B.N. Antiproliferative withanolides from the Solanaceae: A structure–activity study. *Pure Appl. Chem.* **2012**, *84*, 1353–1367. [CrossRef] [PubMed]
16. Glotter, E. Withanolides and related ergostane-type steroids. *Nat. Prod. Rep.* **1991**, *8*, 415–440. [CrossRef] [PubMed]
17. Kim, K.H.; Choi, S.U.; Choi, S.Z.; Son, M.W.; Lee, K.R. Withanolides from the rhizomes of *Dioscorea japonica* and their cytotoxicity. *J. Agric. Food Chem.* **2011**, *59*, 6980–6984. [CrossRef] [PubMed]
18. Chao, C.H.; Chou, K.J.; Wen, Z.H.; Wang, G.H.; Wu, Y.C.; Dai, C.F.; Sheu, J.H. Paraminabeolides A.−F, cytotoxic and anti-inflammatory marine withanolides from the soft coral *Paraminabea acronocephala*. *J. Nat. Prod.* **2011**, *74*, 1132–1141. [CrossRef] [PubMed]
19. Kaileh, M.; Vanden, B.W.; Heyerick, A.; Horion, J.; Piette, J.; Libert, C.; De Keukeleire, D.; Essawi, T.; Haegeman, G. Withaferin A strongly elicits I kappa B kinase beta hyperphosphorylation concomitant with potent inhibition of its kinase activity. *J. Biol. Chem.* **2007**, *282*, 4253–4264. [CrossRef] [PubMed]
20. Mulabagal, V.; Subbaraju, G.V.; Rao, C.V.; Sivaramakrishna, C.; DeWitt, D.L.; Holmes, D.; Sung, B.; Aggarwal, B.B.; Tsay, H.S.; Nair, M.G. Withanolide sulfoxide from aswagandha roots inhibits nuclear transcription factor-kappa-B, cyclooxygenase and tumor cell proliferation. *Phytother. Res.* **2009**, *23*, 987–992. [CrossRef] [PubMed]
21. Nagafuji, S.; Okabe, H.; Akahane, H.; Abe, F. Trypanocidal constituents in plants. *Bio Pharm. Bull.* **2004**, *27*, 193–197. [CrossRef]
22. Chandrasekaran, S.; Dayakar, A.; Veronica, J.; Sundar, S.; Maurya, R. An in vitro study of apoptotic like death in *Leishmania donovani* promastigotes by withanolides. *Parasitol. Int.* **2013**, *62*, 253–261. [CrossRef] [PubMed]
23. Budhiraja, R.D.; Krishan, P.; Sudhir, S. Biological activity of withanolides. *J. Sci. Ind. Res.* **2000**, *59*, 904–911.
24. Nicolás, F.G.; Reyes, G.; Audisio, M.C.; Uriburu, M.L.; González, S.; Barboza, G.E.; Nicotra, V.E. Withanolides with antibacterial activity from Nicandra john-tyleriana. *J. Nat. Prod.* **2015**, *78*, 250–257. [CrossRef] [PubMed]
25. Misico, R.I.; Viviana, E.; Nicotra, J.C.; Oberti, G.B.; Gil, R.R.; Burton, G. Withanolides and related steroids. *Prog. Chem. Org. Nat. Prod.* **2011**, *94*, 127–229. [PubMed]
26. Kuroyanagi, M.I.; Murata, M.; Nakane, T.; Shirota, O.; Sekita, S.; Fuchino, H.; Shinwari, Z.K. Leishmanicidal activity withanolides from a Pakistani medicinal plant, *Withania coagulans*. *Chem. Pharm. Bull.* **2012**, *60*, 892–897. [CrossRef] [PubMed]
27. Choudary, M.I.; Yousaf, S.; Ahmed, S.; Samreen, Y.K.; Rahman, A. Antileishmanial physalins from *Physalis minima*. *Chem. Biodivers.* **2005**, *2*, 1164–1173. [CrossRef] [PubMed]
28. Bravo, B.J.A.; Sauvain, M.; Gimenez, T.A.; Balanza, E.; Serani, L.; Laprévote, O. Trypanocidal withanolides and withanolide Glycosides from *Dunalia brachyacantha*. *J. Nat. Prod.* **2001**, *64*, 720–725. [CrossRef]
29. Murray, H.W.; Nathan, C.F. Macrophage microbicidal mechanisms in vivo: Reactive nitrogen versus oxygen intermediates in the killing of intracellular visceral *Leishmania donovani*. *J. Exp. Med.* **1999**, *189*, 741–746. [CrossRef] [PubMed]
30. Bogdan, C.; Rollinghoff, M.; Diefenbach, A. Reactive oxygen and reactive nitrogen intermediates in innate and specific immunity. *Curr. Opin. Immunol.* **2000**, *12*, 64–76. [CrossRef]
31. Carneiro, P.P.; Conceição, J.; Macedo, M.; Magalhães, V.; Carvalho, E.M.; Bacellar, O. The role of nitric oxide and reactive oxygen species in the killing of *Leishmania braziliensis* by monocytes from patients with Cutaneous Leishmaniasis. *PLoS ONE* **2016**, *11*, e0148084. [CrossRef] [PubMed]
32. Yang, B.Y.; Guo, R.; Li, T.; Liu, Y.; Wang, C.F.; Shu, Z.P.; Wang, Z.B.; Zhang, J.; Xia, Y.G. Five withanolides from the leaves of *Datura metel* L. and their inhibitory effects on Nitric Oxide production. *Molecules* **2014**, *19*, 4548–4559. [CrossRef] [PubMed]
33. Lipinski, C.A.; Lombardo, F.; Dominy, B.W.; Feeney, P.J. Experimental and computational approaches to estimate solubility and permeability in drug discovery and development settings. *Adv. Drug Deliv. Rev.* **1997**, *23*, 3e25. [CrossRef]
34. Sundar, S.; Olliaro, P.L. Miltefosine in the treatment of leishmaniasis: Clinical evidence for informed clinical risk management. *Ther. Clin. Risk Manag.* **2007**, *3*, 733–740. [PubMed]
35. McKerrow, J.H.; Lipinski, C.A. The rule of five should not impede anti-parasitic drug development. *Int. J. Parasitol. Drugs Drug Resist.* **2017**, *7*, 248–249. [CrossRef] [PubMed]
36. Bhattacharya, S.K.; Sinha, P.K.; Sundar, S.; Thakur, C.P.; Jha, T.K.; Pandey, K.; Das, V.R. Phase 4 trial of miltefosine for the treatment of Indian visceral leishmaniasis. *J. Infect. Dis.* **2007**, *196*, 591–598. [CrossRef] [PubMed]

37. Field, M.C.; Horn, D.; Fairlamb, A.H.; Ferguson, M.A.; Gray, D.W.; Read, K.D.; De Rycker, M.; Torrie, L.S.; Wyatt, P.G.; Wyllie, S.; et al. Anti-trypanossomtid drug discovery: An ongoing challenge and a continuing need. *Nat. Rev. Microbiol.* **2017**, *15*, 217–231. [CrossRef] [PubMed]
38. Soto, J.; Rea, J.; Balderrama, M.; Toledo, J.; Soto, P.; Valda, L.; Berman, J.D. Efficacy of miltefonsine for Bolivian cutaneous leishmaniasis. *Am. J. Trop. Med. Hyg.* **2008**, *70*, 210–211. [CrossRef]
39. Uranw, S.; Ostyn, B.; Dorlo, T.P.; Hasker, E.; Dujardin, B.; Dujardin, J.C.; Rijal, S.; Boelaert, M. Adherence to miltefosine treatment for visceral leishmaniasis under routine conditions in Nepal. *Trop. Med. Int. Health* **2013**, *18*, 179–187. [CrossRef] [PubMed]
40. Van Bocxlaer, K.; Yardley, V.; Murdan, S.; Croft, S.L. Opical formulations of miltefosine for cutaneous leishmaniasis in a BALB/c mouse model. *J. Pharm. Pharmacol.* **2016**, *68*, 862–872. [CrossRef] [PubMed]
41. Green, L.C.; Wagner, D.A.; Glogowski, J.; Skipper, P.L.; Wishnok, J.S.; Tannebaum, S.R. Analysis of nitrate, nitrite, and [15N] nitrate in biological fluids. *Anal. Biochem.* **1982**, *126*, 131–138. [CrossRef]
42. Cheng, F.; Li, W.; Zhou, Y.; Shen, J.; Wu, Z.; Liu, G.; Lee, PW. AdmetSAR: A comprehensive source and free tool for assessment of chemical ADMET properties. *J. Chem. Inf. Model.* **2012**, *52*, 3099–3105. [CrossRef] [PubMed]
43. Singh, S.K.; Valicherla, G.R.; Joshi, P.; Shahi, S.; Syed, A.A.; Gupta, A.P.; Hossain, Z.; Italiya, K.; Makadia, V.; Singh, S.K.; Wahajuddin, M.; Gayen, J.R. Determination of permeability, plasma protein binding, blood partitioning, pharmacokinetics and tissue distribution of Withanolide A in rats: A neuroprotective steroidal lactone. *Drug Dev Res.* **2018**, *79*, 339–351. [CrossRef] [PubMed]
44. Dubey, S.; Kallubai, M.; Sarkar, A.; Subramanyam, R. Elucidating the active interaction mechanism of phytochemicals withanolide and withanoside derivatives with human serum albumin. *PLoS ONE* **2018**, *13*, e0200053. [CrossRef] [PubMed]

Sample Availability: Samples of the isolated compounds are not available from the authors.

Article

Sterol Composition of Clinically Relevant Mucorales and Changes Resulting from Posaconazole Treatment

Christoph Müller [1], Thomas Neugebauer [2], Patrizia Zill [1], Cornelia Lass-Flörl [2], Franz Bracher [1] and Ulrike Binder [2,*]

1 Department of Pharmacy-Center for Drug Research, Ludwig-Maximilians University of Munich, Butenandtstr. 5-13, 81377 Munich, Germany; christoph.mueller@cup.uni-muenchen.de (C.M.); patrizia.zill@campus.lmu.de (P.Z.); franz.bracher@cup.uni-muenchen.de (F.B.)

2 Department of Hygiene, Microbiology and Public Health, Division of Hygiene and Medical Microbiology, Medical University Innsbruck, Schöpfstr. 41, 6020 Innsbruck, Austria; tneugebauer@gmx.at (T.N.); cornelia.lass-floerl@i-med.ac.at (C.L.-F.)

* Correspondence: ulrike.binder@i-med.ac.at; Tel.: +43-512-9003-70748

Academic Editors: Wenxu Zhou and De-An Guo
Received: 3 May 2018; Accepted: 17 May 2018; Published: 19 May 2018

Abstract: Mucorales are fungi with increasing importance in the clinics. Infections take a rapidly progressive course resulting in high mortality rates. The ergosterol biosynthesis pathway and sterol composition are of interest, since they are targeted by currently applied antifungal drugs. Nevertheless, Mucorales often exhibit resistance to these drugs, resulting in therapeutic failure. Here, sterol patterns of six clinically relevant Mucorales (*Lichtheimia corymbifera*, *Lichtheimia ramosa*, *Mucor circinelloides*, *Rhizomucor pusillus*, *Rhizopus arrhizus*, and *Rhizopus microsporus*) were analysed in a targeted metabolomics fashion after derivatization by gas chromatography-mass spectrometry. Additionally, the effect of posaconazole (POS) treatment on the sterol pattern of *R. arrhizus* was evaluated. Overall, fifteen different sterols were detected with species dependent variations in the total and relative sterol amount. Sterol analysis from *R. arrhizus* hyphae confronted with sublethal concentrations of posaconazole revealed the accumulation of 14-methylergosta-8,24-diene-3,6-diol, which is a toxic sterol that was previously only detected in yeasts. Sterol content and composition were further compared to the well-characterized pathogenic mold *Aspergillus fumigatus*. This work contributes to a better understanding of the ergosterol biosynthesis pathway of Mucorales, which is essential to improve antifungal efficacy, the identification of targets for novel drug design, and to investigate the combinatorial effects of drugs targeting this pathway.

Keywords: Mucorales; *Rhizopus arrhizus*; sterol pattern; antifungal effectivity; gas chromatography-mass spectrometry (GC-MS); posaconazole

1. Introduction

The order Mucorales represent the most prominent order of zygospore-forming fungi, which was formerly placed in the phylum Zygomycota and was referred to as Zygomycetes. Over the years, this phylum has undergone constant taxonomic rearrangements. Now, the Mucorales, or mucormycetes, are placed in the phylum Glomeromycota and subphylum Mucormycotina [1–3]. Mucorales are mainly saprophytes commonly found in soil and decomposing material. Members of this group play various important roles for human life—e.g., *Mucor* spp., are of biotechnological importance due to their high growth rates and their great potential in the production of secondary metabolites [4]. Other Mucorales, such as *Rhizopus* spp., have been used for food fermentation in Asia for centuries [5]. Contrary to these positive effects of mucoralean fungi on human life, they also cause a wide range of diseases to plants, animals, and human beings. In the last ten years, the number of mucormycosis cases (also known as zygomycosis) [3,6] has increased significantly in the clinics. Mucorales that were isolated in high abundance from patient material belong to the following genera: *Rhizopus*, *Lichtheimia* (formerly *Absidia*), *Mucor*, *Rhizomucor*, and *Cunninghamella*. Worldwide, *Rhizopus arrhizus* is the most common species that was isolated from clinical specimen. The clinical presentation of mucormycosis varies, with mostly rhinocerebral, pulmonary and gastrointestinal manifestations. Primarily, immunocompromised patients are at risk to develop mucormycosis, but some forms, such as cutaneous mucormycosis, have also been seen in otherwise healthy persons [1,7]. The aggressive course of the disease, delayed diagnosis, and poor treatment options result in unacceptably high mortality rates (40–70%), even with antifungal therapy. Often, mucormycosis is seen as so called breakthrough infection when voriconazole is applied as prophylactic treatment regime [6]. Antifungal drugs currently licensed for treatment of mucormycosis are the polyene antifungal amphotericin B (AMB), preferably the liposomal formulation as first line treatment, and the triazole posaconazole as salvage treatment. In 2015, another azole, isavuconazole (ISA), was approved by the Food and Drug Administration (FDA) as a stand-alone treatment for mucormycosis [1,8–13]. AMB forms 1:1 adducts with fungal ergosterol and induces an accumulation of reactive oxygen species in the cytoplasm [14,15]. Azoles conduct their antifungal activity by inhibiting the enzyme sterol C14-demethylase (named ERG11 in yeasts, and CYP51 in *Aspergillus* spp., Figure 1, enzyme **A**) [16,17]. POS is also very active against *Aspergillus* isolates and used in the therapy of invasive aspergillosis. In contrast, another second-generation triazole, voriconazole (VRC), was shown to have no in vitro nor in vivo activity against Mucormycetes [13,15,18–20]. The limited arsenal of highly active anti-mucoralean drugs, plus the occurrence of AMB and azole resistant strains highlight the importance of illuminating the mode of action of antifungal drugs on, and resistance mechanisms in this group of fungi [13,15]. As the target of the currently used antifungal drugs against Mucorales is ergosterol or its biosynthesis, respectively, it is of great importance to understand the biosynthetic pathway and decipher the differences to other human pathogenic fungi.

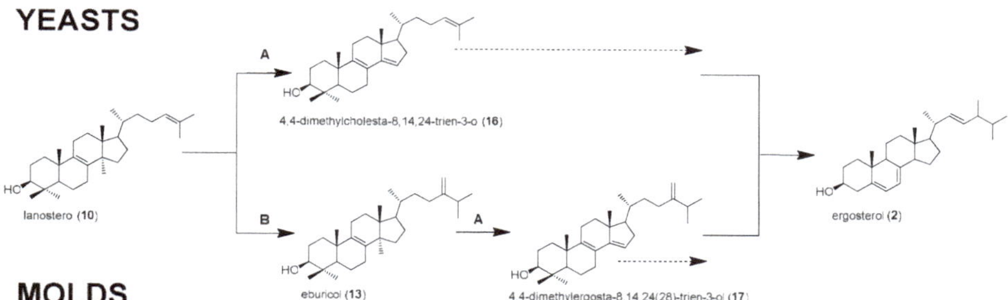

Figure 1. Preferred ergosterol biosynthesis pathways in yeasts (*S. cerevisiae*) and molds (*A. fumigatus*) starting from lanosterol. Enzymes: (**A**) sterol C14-demethylase, (**B**) sterol C24-methyltransferase.

For two major human pathogenic fungi, the yeast *Candida albicans*, and the mold *Aspergillus fumigatus*, differences in the ergosterol biosynthesis pathways have already been elucidated. In *Candida* spp., and this is similar to baker´s yeast *S. cerevisiae* (Figure 1, upper panel), lanosterol is the preferred substrate for the sterol C14-demethylase (Figure 1, enzyme **A**), which converts lanosterol into 4,4-dimethylcholesta-8,14,24-trien-3β-ol [21–25]. In molds, e.g., *Aspergillus fumigatus*, the preferred ergosterol biosynthesis route from lanosterol starts with a methylation at C-24 by sterol C24-methyltransferase (enzyme **B**) in order to give eburicol, and is then followed by a demethylation at C-14 by sterol C14-demethylase to give 4,4-dimethylergosta-8,14,24(28)-trien-3β-ol [24–26] (Figure 1, lower panel).

In the presence of azoles, the accumulation of physiological substrates for sterol C14-demethylase and a depletion of ergosterol are observed in both yeasts and molds [16,25,27–30]. The accumulation of C14-methylated sterols (e.g., lanosterol (**10**) and eburicol (**13**)) and lower levels of ergosterol as the predominant native sterol in fungi or changes in the relative sterol composition results in alterations of the plasma membrane that impact its function and the activity of membrane-bound enzymes [31–33].

So far, very little is known about the ergosterol biosynthesis pathway, the sterol content, and the composition of clinically relevant Mucorales, and most studies were carried out with one species or one strain only, which makes the comparison in-between the group of Mucorales difficult [26,34–39]. Therefore, a full understanding of the ergosterol biosynthesis pathway and the sterol pattern of this group of pathogenic fungi is essential to determine the efficacy of antifungal therapy, to identify targets for novel drug design, and to investigate the combinatorial effects of drugs targeting the ergosterol biosynthesis pathway. Here, the sterol pattern and sterol content of six clinically relevant Mucorales (*Rhizopus (R.) arrhizus, Rhizopus microsporus Lichtheimia (L.) corymbifera, Lichtheimia ramosa, Mucor (M.) circinelloides,* and *Rhizomucor (Rh.) pusillus*) were analysed in a targeted metabolomics fashion by gas chromatography-mass spectrometry (GC-MS). In addition, sterols in hyphae of *R. arrhizus* that was confronted with sub-lethal concentrations of posaconazole were analysed and compared to untreated hyphae. *R. arrhizus* was chosen for this test because it represents the most abundant Mucorales species from patient material worldwide. Furthermore, the obtained data were compared to the well-studied ergosterol biosynthetic pathway of *Aspergillus fumigatus* [16,24,25,27,30,40–46].

2. Results and Discussion

2.1. Relative and Absolute Amounts of Sterol Intermediates in Mucorales

In total, 14 different sterols were detected (Table 1) in the six investigated *Mucormycetes* (without azole treatment), which are discussed in the order of their retention time below (Figure 2, blue chromatogram).

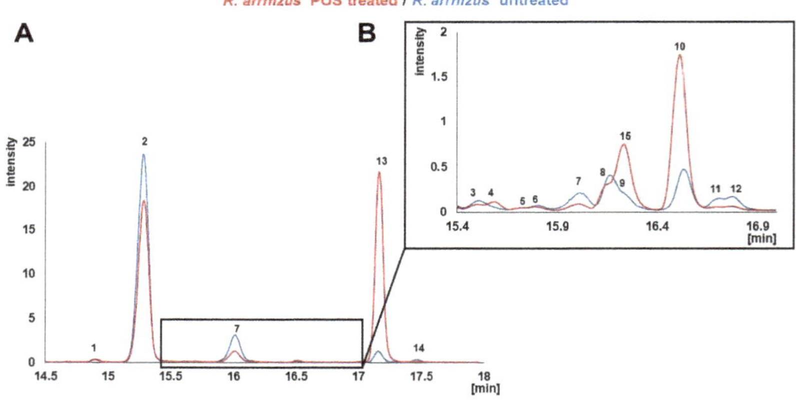

Figure 2. Representative selected ion chromatograms of untreated *R. arrhizus* samples (blue) and hyphae confronted with posaconazole (red). Hyphae were confronted with sublethal concentrations (0.5 µg/mL) for 4 h. Numbers in the diagram represent the sterol intermediates as given in Table 1. X-axis presents retention time. Selected ions for chromatogram (**A**) m/z 363 + 366 − 365 + 407 + 408 and selected ions for chromatogram (**B**) m/z 343 + 377 + 379 + 393 + 472 + 486.

Table 1. Sterol content and sterol composition of six different mucormycete species and *A. fumigatus*. Cultures were grown in $RPMI_{1640}$ medium over night at 37 °C before sterol extraction. The results are presented as the average of two independent experiments, comprising 6 technical replicates in total. Sterol composition is given as the relative amount of the respective sterol (%, in bold letters) of all the sterols detected. Sterol content is expressed as µg sterol intermediate/mg biomass (dry weight). Standard deviation is given in brackets. * n.d. = not detected, ** i.t. = in traces (< 0.04 µg/mg); major sterols are indicated in red.

	Compound		Relative [%] and Absolute [µg/mg] Amount of Sterols in Each Species						
No.	IUPAC Name	Common Name	*L. corymbifera*	*L. ramosa*	*M. circinelloides*	*Rh. pusillus*	*R. arrhizus*	*R. microsporus*	*A. fumigatus*
1	ergosta-5,8,22-trien-3β-ol	Lichesterol	**0.2** 0.02 (± 0.00)	**0.4** 0.07 ± 0.02)	**0.3** 0.07 (± 0.00)	**0.2** 0.05 (± 0.00)	**0.8** 0.07 (± 0.01	**0.9** 0.13 (± 0.02	**1.2** 0.12 (± 0.03)
2	ergosta-5,7,22-trien-3β-ol	Ergosterol	**57.5** 6.45 (± 0.66)	**69.2** 13.69 (± 3.76)	**65.8** 12.93 (± 0.37)	**38.0** 8.13 (± 0.33)	**76.3** 7.07 (± 0.04)	**70.3** 9.91 (± 1.85)	**95.0** 9.29 (± 2.28)
3	ergosta-7,22-dien-3β-ol		**3.5** 0.39 (± 0.04)	**i.t. ****	**n.d. ***	**1.9** 0.41 (± 0.02)	**0.2** 0.02 (± 0.01)	**0.2** 0.03 (± 0.02)	**n.d.**
4	ergosta-5,7,22,24(28)-tetraen-3β-ol		**0.4** 0.04 (± 0.00)	**0.6** 0.11 (± 0.00)	**0.5** 0.10 (± 0.01)	**0.3** 0.06 (± 0.00)	**1.0** 0.09± 0.02)	**1.2** 0.17 (± 0.03)	**0.7** 0.06 (± 0.01)
5	ergosta-7,22,24(28)-trien-3β-ol		**0.2** 0.02 (± 0.01)	**0.2** 0.04 (± 0.03)	**0.2** 0.04 (± 0.00)	**0.2** 0.03 (± 0.02)	**0.2** 0.03 ± 0.02)	**0.2** 0.03 (± 0.02)	**0.2** 0.02 (± 0.00)
6	ergosta-5,7,24(28)-trien-3β-ol		**n.d. ***	**7.4** 1.50 (± 1.94)	**10.4** 2.05 (± 0.37)	**n.d. ****	**2.6** 0.24 ± 0.22)	**3.3** 0.46 (± 0.39)	**0.6** 0.06 (± 0.01)
7	ergosta-5,7-dien-3β-ol		**28.5** 3.19 (± 0.36)	**12.3** 2.48 (± 0.11)	**10.6** 2.07 (± 0.26)	**37.2** 9.96 (± 0.07)	**10.6** 0.98 (± 0.19)	**12.9** 1.82 (± 0.43)	**0.3** 0.03 (± 0.00)
8	ergosta-7,24(28)-dien-3β-ol	Episterol	**2.4** 0.27 (± 0.04)	**1.9** 0.38 (± 0.13)	**1.8** 0.35 (± 0.01)	**1.4** 0.29 (± 0.01)	**2.1** 0.20 (± 0.13)	**2.4** 0.34 (± 0.19)	**0.8** 0.08 (± 0.01)
9	ergost-7-en-3β-ol		**3.7** 0.41 (± 0.02)	**0.1** 0.02 (± 0.03)	**i.t. ****	**3.0** 0.64 (± 0.02)	**0.3** 0.03 (± 0.02)	**0.5** 0.07 (± 0.03)	**n.d. ***
10	4,4,14-trimethylcholesta-8,24-dien-3β-ol	Lanosterol	**0.3** 0.03 (± 0.00)	**1.1** 0.21 (± 0.24)	**1.3** 0.26 (± 0.04)	**0.7** 0.07 (± 0.03)	**0.7** 0.07 (± 0.03)	**1.0** 0.14 (± 0.05)	**0.2** 0.02 (± 0.00)
11	4,4-dimethylcholesta-8,24-dien-3β-ol	T-MAS	**0.2** 0.04 (± 0.00)	**0.6** 0.12 (± 0.13)	**0.7** 0.15 (± 0.01)	**0.6** 0.13 (± 0.01)	**0.2** 0.02 (± 0.00)	**0.2** 0.03 (± 0.00)	**0.4** 0.04 (± 0.00)
12	4-methylergost-8-en-3β-ol		**0.4** 0.05 (± 0.00)	**0.1** 0.01 (± 0.02)	**i.t. ****	**0.9** 0.18 (± 0.00)	**0.2** 0.02 (± 0.02)	**0.2** 0.02 (± 0.02)	**n.d. ***
13	4,4,14-trimethylergosta-8,24(28)-dien-3β-ol	Eburicol	**2.0** 0.22 (± 0.01)	**5.3** 1.07 (± 1.33)	**7.2** 1.41 (± 1.16)	**15.4** 3.29 (± 0.18)	**3.7** 0.34 (± 0.12)	**5.5** 0.77 (± 0.15)	**n.d. ***
14	4,4-dimethylergosta-8,24(28)-dien-3β-ol		**0.7** 0.08 (± 0.01)	**0.9** 0.17 (± 0.19)	**1.1** 0.21 (± 0.01)	**0.7** 0.15 (± 0.00)	**0.3** 0.03 (± 0.01)	**0.3** 0.05 (± 0.01)	**0.6** 0.06 (± 0.00)
	total sterol content		**100** 11.22 (± 1.16)	**100** 20.17 (± 7.83)	**100** 19.64 (± 0.45)	**100** 21.39 (± 0.61)	**100** 9.26 (± 0.66)	**100** 14.09 (± 2.96)	**100** 9.78 (± 2.36)

Ergosta-5,8,22-trien-3β-ol (lichesterol, **1**) was detected in all of the six species tested. Among the Mucormycetes, highest relative amounts of this intermediate were detected in *Rhizopus* spp., although these amounts were still lower than in *A. fumigatus*, which was included for comparison.

As expected, ergosta-5,7,22-trien-3β-ol (ergosterol, **2**) was identified as the dominating sterol in all the species studied. This is in agreement with a study that was carried out with *R. arrhizus* [39], in which ergosterol was also identified as the sterol exhibiting highest abundance. This finding is supported by a study that was investigating the phylogenetic distribution of fungal sterols [39]. One contrary result was shown by Weete et al. [37], explaining that ergosta-7,22-dien-3β-ol (**3**) is the major sterol (56.0 to 59.9%) and ergosterol (**2**) is only the second most abundant sterol (21.1–28.4%) in *R. arrhizus*. Importantly, the relative content of ergosterol was highly variable between the species, ranging from 38.0% in *Rh. pusillus* to 76.3% in *R. arrhizus*. In *A. fumigatus*, the relative amount of ergosterol reached 95.0% of all the sterols in our experimental setup. This amount is slightly higher than what was shown by a comprehensive study of Alcazar-Fuoli et al. [24], where *A. fumigatus* isolates reached relative ergosterol amounts of 75.8 to 88.4%. Differences in experimental setup, like growth media and the age of the cultures used for sterol extraction, might explain these differences.

Only minor amounts of ergosta-7,22-dien-3β-ol (**3**) were detected in most species. Only *L. corymbifera* and *Rh. pusillus* showed little higher accumulation of **3**, representing 3.5% and 1.9%, respectively, whereas this intermediate was neither detected in *M. circinelloides* nor in *A. fumigatus*.

The relative amount of ergosta-5,7,22,24(28)-tetraen-3β-ol (**4**) and ergosta-7,22,24(28)-trien-3β-ol (**5**) was very low in all of the samples (0.2–1.2%). This is in agreement with the value of 1.2% of **4** reported for *M. rouxii* [36].

Sterol **6**, ergosta-5,7,24(28)-trien-3β-ol, was found in five out of the seven species tested (*A. fumigatus* included), it exhibited the third highest abundancy among all the sterols in *L. ramosa* (7.4%) and *M. circinelloides* (10.4%). Interestingly, sterol **6** was missing in *L. corymbifera* and *Rh. pusillus*.

In all Mucorales strains, significant amounts of ergosta-5,7-dien-3β-ol (**7**), the saturated side chain analogue of ergosterol (**2**) were found (10.6–37.2%), making it the second most abundant sterol after ergosterol in the Mucorales, whereas only a small amount (0.3%) of this sterol was detected in *A. fumigatus*. Interestingly, in *Rh. pusillus* the relative amount of ergosterol (**2**) and ergosta-5,7-dien-3β-ol (**7**) were nearly equal (38.0% and 37.2%), a characteristic that seems unique for this species. McCorkindale et al. [34] also detected larger amounts of **7** in Mucorales (mean 45%; n = 8) and also Weete et al. [39] identified **7** in *Rhizomucor pusillus* (formerly named *Mucor pusillus*). This indicates a putative role of ergosta-5,7-dien-3β-ol (**7**) as a marker sterol for the identification and classification of Mucorales species. This sterol was not found in studies evaluating sterols of *A. fumigatus* [24,40], while we detected minor amounts (0.3%) in our cultures. From previous studies, we learned that ergosta-5,7-dien-3β-ol (**7**) is only found in very little amounts in molds [25,41,42] and yeasts [25,29,43,44].

Episterol, ergosta-7,24(28)-dien-3β-ol (**8**), was evident in all of the samples, ranging from 1.4% (*Rh. pusillus*) to 2.4% in (*L. corymbifera* and *R. microsporus*). This sterol was also detected by Safe [36] in another mucoralean fungus, *M. rouxii*, in which the amount of free and bounded **8** was shown to be strongly dependent on the growth conditions (1.1–25.8%).

All of the Mucorales strains produced ergost-7-en-3β-ol (**9**) with maximum levels in *L. corymbifera* (3.7%) and *Rh. pusillus* (3.0%). In contrast to the results of Weete et al. [37] (mean 13.5%; n = 6), we detected clearly lower levels (0.3%) of **9** in *R. arrhizus*.

Low percentages ($\leq$ 1.3%) of lanosterol (4,4,14-trimethylcholesta-8,24-dien-3β-ol, **10**), T-MAS (4,4-dimethylcholesta-8,24-dien-3β-ol, **11**), and 4-methylergost-8-en-3β-ol (**12**) were ubiquitously found in all of the samples.

Most interestingly, a high amount of eburicol (4,4,14-trimethylergosta-8,24(28)-dien-3β-ol, **13**) was detected to a degree that is usually only observed under azole treatment in fungi [25,29,41,43]. Levels reached up to 15.4% in *Rh. pusillus*, all other species also showed considerably high eburicol levels, ranging from 2.0% (*L. corymbifera*) to 7.2% (*M. circinelloides*). In our experimental setup, we did not detect eburicol in *A. fumigatus*, which is in contrast to the sterol profile of *A. fumigatus* in Alcazar-Fuoli et al. [24], where an eburicol content of approx. 1.9% was reported. Different growth conditions and different growth media might explain this discrepancy.

The last detected sterol, 4,4-dimethylergosta-8,24(28)-dien-3β-ol (**14**), was present in all of the samples, ranging from 0.3% (*R. arrhizus*) to 1.1% (*M. circinelloides*).

Sterol **1** and the sterols **4**, **6**, **10–14** were not explicitly mentioned in previous reports on the sterol composition of Mucormycetes [26,34–39], but the percentages that we obtained in our study were in the range of other fungi [22,24,40–42].

Additional to the relative amount of sterol intermediates, we also determined the actual amount of each sterol, which was expressed as μg/mg biomass, (dry weight, Table 1). The total sterol content was proportionately 1–2% of the total mycelial biomass, which is in agreement with what has been detected for other *Mucor* species before [36]. Lowest amount of total sterols was found in *R. arrhizus* (9.26 μg/mg) and highest in *Rh. pusillus* (21.39 μg/mg). The values of each intermediate determined correlate with the relative amounts that were previously described. In *A. fumigatus*, a total sterol content of 9.78 μg/mg was determined, of which 9.25 μg/mg was ergosterol, an amount that is comparable to previously reported data [24].

2.2. Sterol Composition and Sterol Content of Posaconazole Treated R. arrhizus

To investigate the effect of the sterol C14-demethylase inhibitor POS on the sterol composition and content of Mucorales, *R. arrhizus* hyphae confronted with sublethal concentrations (0.5 μg/mL) of POS were compared to the sterol pattern of untreated hyphae Figures 2 and 3, Table 2. For comparison and also to validate our results that were obtained with Mucorales, we included *A. fumigatus*, in which the effect of azole treatment on sterol pattern has been extensively studied [24,25,40,41,45].

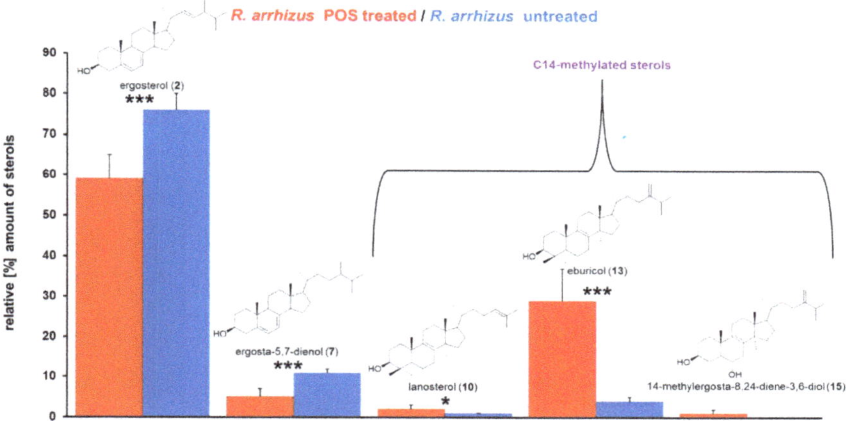

Figure 3. Relative amounts [%] of the most prominent sterols in posaconazole treated (red) and untreated (blue) *R. arrhizus* hyphae. Cultures, pre-grown for 16 h, were confronted with 0.5 μg/mL posaconazole for 4 h before sterol extraction. Sterol pattern was compared to untreated controls, which were incubated under identical conditions. Error bars represent standard deviation out of two independent experiments, comprising six technical replicates. (* $p < 0.05$; *** $p < 0.001$: student's *t*-test). For detailed information on all sterols extracted, see Table 2.

Table 2. Sterol content and sterol composition of *R. arrhizus* and *A. fumigatus* confronted with sublethal concentrations of posaconazole (POS) when compared to the untreated controls. Cultures were grown in $RPMI_{1640}$ medium over night at 37 °C before being transferred to fresh media containing 0.5 µg/mL POS, or no POS. Cultures were incubated for additional 4 h before sterol extraction. The results are presented as the average of two independent experiments, comprising six technical replicates in total. Sterol composition is given as the relative amount of the respective sterol (%, in bold letters) of all sterols detected. Sterol content is expressed as µg sterol intermediate/mg biomass (dry weight). Standard deviation is given in brackets.* n.d. = not detected, ** i.t. = in traces (< 0.04 µg/mg), major sterols are indicated in red.

Compound			Relative [%] and Absolute [µg/mg] Amount of Sterols			
			R. arrhizus		*A. fumigatus*	
No.	IUPAC Name	Common Name	POS Treated	Untreated	POS Treated	Untreated
1	ergosta-5,8,22-trien-3β-ol	Lichesterol	**0.7** 0.09 (± 0.01)	**0.8** 0.07(± 0.01)	**1.2** 0.10 (± 0.00)	**1.2** 0.12 (± 0.03)
2	ergosta-5,7,22-trien-3β-ol	Ergosterol	58.5 7.52 (± 0.20)	76.3 7.07(SD ± 0.04)	84.1 7.41 (± 1.05)	95.0 9.29 (± 2.28)
3	ergosta-7,22-dien-3β-ol		**0.1** 0.01 (± 0.01)	**0.2** 0.02 (± 0.01)	**0.1** 0.01 (± 0.01)	n.d. *
4	ergosta-5,7,22,24(28)-tetraen-3β-ol		**0.7** 0.09 (± 0.02)	**1.0** 0.09 (± 0.02)	**0.8** 0.07 (± 0.01)	**0.7** 0.06 (± 0.01)
5	ergosta-7,22,24(28)-trien-3β-ol		**0.1** 0.01 (± 0.00)	**0.2** 0.03 (± 0.02)	**0.2** 0.01 (± 0.01)	**0.2** 0.02 (± 0.00)
6	ergosta-5,7,24(28)-trien-3β-ol		**0.9** 0.12 (± 0.12)	**2.6** 0.24 (± 0.22)	**0.6** 0.05 (± 0.04)	**0.6** 0.06 (± 0.01)
7	ergosta-5,7-dien-3β-ol		**5.1** 0.66 (± 0.33)	**10.6** 0.98 (± 0.19)	**1.9** 0.17 (± 0.10)	**0.3** 0.03 (± 0.00)
8	ergosta-7,24(28)-dien-3β-ol	Episterol	**0.8** 0.10 (± 0.00)	**2.1** 0.20 (± 0.13)	**1.0** 0.09 (± 0.02)	**0.8** 0.08 (± 0.01)
9	ergost-7-en-3β-ol		**0.1** 0.01(± 0.01)	**0.3** 0.03 (± 0.02)	**i.t. ****	n.d. *
10	4,4,14-trimethylcholesta-8,24-dien-3β-ol	Lanosterol	2.5 0.33 (± 0.26)	0.7 0.07 (± 0.03)	1.3 0.12 (± 0.12)	0.2 0.02 (± 0.00)

Table 2. *Cont.*

Compound			Relative [%] and Absolute [μg/mg] Amount of Sterols			
			R. arrhizus		*A. fumigatus*	
No.	**IUPAC Name**	**Common Name**	**POS Treated**	**Untreated**	**POS Treated**	**Untreated**
11	4,4-dimethylcholesta-8,24-dien-3β-ol	T-MAS	**i.t. ****	**0.2** 0.02 (± 0.00)	**0.4** 0.04 (± 0.00)	**0.4** 0.04 (± 0.00)
12	4-methylergost-8-en-3β-ol		**0.1** 0.01(± 0.01)	**0.2** 0.02 (± 0.02)	n.d. *	n.d. *
13	4,4,14-trimethylergosta-8,24(28)-dien-3β-ol	Eburicol	**29.1** 3.74 (± 1.58)	**3.7** 0.34 (± 0.12)	**6.9** 0.61 (± 0.80)	n.d.*
14	4,4-dimethylergosta-8,24(28)-dien-3β-ol		**i.t. ****	**0.3** 0.03 (± 0.01)	**0.7** 0.06 (± 0.02)	**0.6** 0.06 (SD ± 0.00)
15	14-methylergosta-8,24(28)-diene-3β,6α-diol		**0.7** 0.09 (± 0.11)	**n.d. ***	**n.d. ***	n.d.*
	total sterol content		**100** 12.86 (± 1.30)	**100** 9.26 (± 0.66)	**100** 8.81 (± 0.08)	**100** 9.78 (± 2.36)

As expected, the relative amount of ergosterol (**2**) was significantly reduced by 17.8% in *R. arrhizus* due to POS treatment. On the other hand, eburicol (**13**), which is one of the substrates of the azoles' target enzyme C14-demethylase, increased by 25.4% (Figure 3), which underlines the enzyme inhibiting properties of POS also in *R. arrhizus* and it further indicates that **13** is the favored substrate for sterol C14-demethylase in Mucorales (Figure 4).

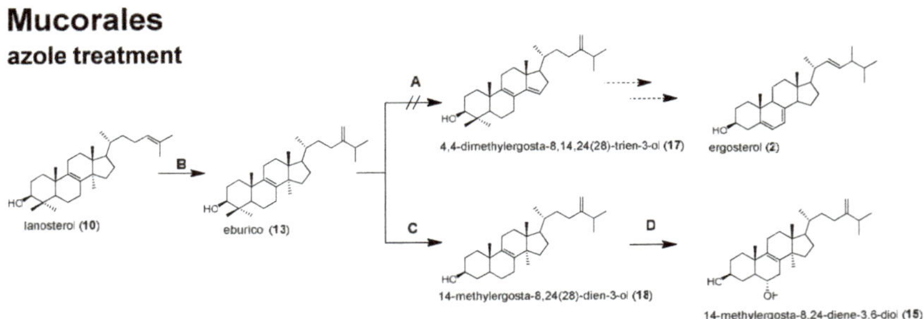

Figure 4. Putative alternative sterol biosynthesis pathway from lanosterol in Mucorales in the presence of posaconazole. Enzymes: (**A**) sterol C14-demethylase, (**B**) sterol C24-methyltransferase, (**C**) sterol C4-demethylase complex, (**D**) sterol C5-desturase.

Surprisingly, the actual amount of ergosterol (**2**) relative to the biomass significantly increased in POS treated *R. arrhizus* hyphae (from 7.07 μg/mg to 7.52 μg/mg). Furthermore, a significant increase of the total sterol content (from 9.26 μg/mg to 12.86 μg/mg) was evident, which is most likely due to the accumulation of new sterols and the increased amount of ergosterol precursors, such as eburicol, which is increased eight-fold when compared to the untreated samples. This finding reflects what was shown by Weete and Wise [26] in propiconazole treated *M. rouxii*. However, this is contrary to our observations in *A. fumigatus*, where inhibition of sterol C14-demethylase resulted in ergosterol reduction (from 9.29 μg/mg to 7.41 μg/mg), and in this course resulted in a decrease of the total sterol content from 9.78 μg/mg to 8.81 μg/mg, although this decrease was not significant. Our results that were obtained for *A. fumigatus* correlate with the results of Alcazar-Fuoli et al. [27], showing that in strains with defective sterol C14-demethylase, caused by mutations in one of the two *cyp51* genes, statistically lower amounts of total ergosterol were detected than in the wildtype strain.

Another sterol, ergosta-5,7-dien-3β-ol (**7**), was clearly reduced from 10.6% to 5.1% in *R. arrhizus*. In *A. fumigatus*, the opposite effect was observed for **7**, resulting in a slight increase from 0.3% to 1.9%. Ergosta-5,7-dien-3β-ol (**7**) is a substrate for sterol C22-desaturase [25], the final enzyme in ergosterol biosynthesis, which converts **7** into ergosterol (**2**). The decrease of non-C14-methylated sterol levels under azole treatment can be explained by an up-regulation of ergosterol biosynthesis enzymes, which is aimed at converting all of the intermediates into ergosterol (e.g., **7**), to avoid the impairment of membrane function. In both, *R. arrhizus* and *A. fumigatus*, a significant accumulation (relative and total amount) of lanosterol (**10**, Figures 1–4) was observed. Lanosterol (**10**) is a further physiological substrate for the inhibited enzyme sterol C14-demethylase (Figures 1 and 4), which explains the accumulation due to enzyme inhibition, even though eburciol (**13**) was shown to be the favored enzyme substrate in molds (Mucorales and *A. fumigatus*) [24–27].

Most interestingly, the non-physiological 14-methylergosta-8,24(28)-diene-3β,6α-diol (**15**) was found (0.7%) in POS that was treated *R. arrhizus*. Accumulation of this toxic intermediate [28,29,46] under the inhibition of sterol C14-demethylase was only detected in yeasts so far, but not in *A. fumigatus* or another filamentous fungus (Figure 4). Further studies are needed to verify if the accumulation of 14-methylergosta-8,24(28)-diene-3β,6α-diol (**15**) is unique for *R. arrhizus* or is ubiquitously found in clinically relevant Mucorales upon azole treatment.

The relative amounts of sterols **1**, **3–6**, **8**, **9**, **11**, **12**, **14** were not, or only to a minor extent, affected by POS treatment Table 2.

In conclusion, the azole activity in Mucorales results in significant alterations of the sterol composition, but it does not subsequently lead to a reduction of the total sterol content. The high amount of eburicol, even in untreated hyphae of the Mucorales, let us hypothesize that these fungi could be less affected by the accumulation of non-physiological intermediates due to azole treatment, which is reflected in their lower sensitivity to azole drugs when compared to other molds. In the genome of *R. arrhizus* two paralogous genes encoding for sterol C14-demethylase have been found [20]. It remains to be elucidated if the expression of both the genes is effected to a similar extent in the presence of azoles. So far, our results in *R. arrhizus* let us hypothesize that sterol C14-demethylase is only partly inhibited, as *R. arrhizus* is still able to synthesize ergosterol. This has been shown for other fungi as well, and it reflects that molds inherit a wide repertoire of adaption mechanisms to overcome such drawbacks. From the sterol pattern, we learn that in Mucorales the entry into the post-lanosterol pathway of ergosterol biosynthesis is similar to other molds, whereas under azole treatment, differences to molds, e.g., *A. fumigatus* were detected, resulting in the accumulation of a non-physiological sterol (14-methylergosta-8,24(28)-diene-3β,6α-diol, **15**) which has so far only been found in azole treated yeasts (Figure 4). Because the Mucorales exhibit high variability in their susceptibility to azoles, it is of special interest to decipher the effect of azoles on sterol composition and the content of other clinically relevant mucoralean fungi.

3. Material and Methods

3.1. Fungal Strains Used in This Study

The following fungal strains, all being obtained from the strain collection of the Division of Hygiene and Medical Microbiology, Medical University Innsbruck (Innsbruck, Austria), were used for sterol analysis: *Aspergillus fumigatus (ATCC46645)*, *Lichtheimia corymbifera* (CBS 109940), *Lichtheimia ramosa* (CBS 101.55), *Mucor circinelloides* (CBS 394.68), *Rhizomucor pusillus* (CBS 219.31), *Rhizopus arrhizus* (CBS 126971), and *Rhizopus microsporus* (CBS 102277).

3.2. Fungal Growth and Culture Conditions

All of the strains were cultivated on supplemented minimal agar (SUP) at 37 °C until sporulation (five days). Spores were obtained by harvesting them with sterile spore suspension buffer (0.9% NaCl, 0.01% Tween80). To obtain mycelia for sterol extraction, all of the strains were cultivated in $RPMI_{1640}$ cell culture medium (Sigmaaldrich, Vienna, Austria) at 37 °C with shaking overnight. Cultures of *R. arrhizus* and *R. microsporus* with optimal growth (fine mycelia, no formation of hyphal pellets) were harvested and transferred to a new shake flask containing $RPMI_{1640}$ plus 0.5 μg/mL posaconazole (POS). Cultures were further incubated in the presence of posaconazole for 4 h. Untreated controls were transferred into new medium without posaconazole. Then, the cultures were harvested by filtration, washed, and freeze dried to determine the fungal biomass on dry weight basis.

3.3. Sterol Extraction

Six mg of dry fungal biomass were used for sterol extraction as described by Müller et al. [25].

3.4. Gas Chromatography-Mass Spectrometry (GC-MS) Analysis of Sterol TMS Ethers

Sterol pattern was determined by GC-MS, according to Müller et al. [25,43]. The quantification, managed with an external calibration with ergosterol, consists of six levels with concentrations up to 20 μg/mg. The base peak of each sterol TMS ether were taken as a quantifier ion for calculating the peak areas for IS cholestane m/z 217, **1** ergosta-5,8,22-trien-3β-ol (lichesterol) m/z 363, **2** ergosta-5,7,22-trien-3β-ol (ergosterol) m/z 363, **3** ergosta-7,22-dien-3β-ol m/z 343, **4** ergosta-5,8,22,24(28)-tetraen-3β-ol m/z 466, **5** ergosta-7,22,24(28)-trien-3β-ol m/z 343, **6** ergosta-5,7,24(28)-trien-3β-ol m/z 363, **7** ergosta-5,7-dien-3β-ol *m/z* 365, **8** ergosta-7,24(28)-dien-3β-ol (episterol) *m/z* 343, **9** ergost-7-en-3β-ol *m/z* 472, **10** 4,4,14-trimethylcholesta-8,24-dien-3β-ol (lanosterol) *m/z* 393, **11** 4,4-dimethylcholesta-8,24-dien-3β-ol (T-MAS) *m/z* 379, **12** 4-methylergost-8-en-3β-ol *m/z* 486, **13** 4,4,14-trimethylergosta-8,24(28)-dien-3β-ol (eburicol) *m/z* 407, **14** 4,4-dimethylergosta-8,24(28)-dien-3β-ol *m/z* 408, and **15** 14-methylergosta-8,24(28)-diene-3β,6α-diol *m/z* 377. Sterols 4,4-dimethylcholesta-8,14,24-trien-3β-ol (FF-MAS) 4,4-dimethylergosta-8,14,24(28)-trien-3β-ol, and 14-methylergosta-8,24(28)-dien-3β-ol were not detected in this study. For detailed information about these sterols, see literature [25]. Representative selected ion chromatograms of sterol fractions from *R. arrhizus* are given in Figure 2.

The amount of each sterol was expressed as μg per mg dry weight. The results represent the mean (μg/mg ± S.D.) of two independent biological replicates, including six technical parallels. Experiments involving POS treatment were carried out in triplicate. The sterol composition is expressed as relative amount of total sterols (%).

Author Contributions: Conceptualization, C.M. and U.B.; Methodology, C.M., T.N., P.Z. and U.B.; Formal Analysis, C.M., T.N., P.Z. and U.B.; Investigation, C.M., T.N., P.Z. and U.B; Resources, F.B. and C.L.-F.; Data Curation, C.M. and U.B.; Writing-Original Draft Preparation, C.M. and U.B.; Writing-Review & Editing, C.M., F.B., C.L.-F., U.B.; Visualization, C.M. and U.B.; Supervision, F.B., U.B. and C.L.-F.; Funding Acquisition, C.L.-F.

Funding: This work was financially supported by the "Christian Doppler Forschungsgesellschaft" (CD-Labor Invasive Pilzinfektionen) to C.L.-F.

Acknowledgments: The authors would like to thank Denise Grässle and Verena Naschberger for excellent technical assistance. C.L.-F. has received grant support from the Austrian Science Fund (FWF), MFF Tirol, Astellas Pharma, Gilead Sciences, Pfizer, Schering Plough, and Merck Sharp & Dohme. She has been an advisor/consultant to Gilead Sciences, Merck Sharp & Dohme, Pfizer, and Schering Plough. She has received travel/accommodation expenses from Gilead Sciences, Merck Sharp & Dohme, Pfizer, Astellas, and Schering Plough and has been paid for talks on behalf of Gilead Sciences, Merck Sharp & Dohme, Pfizer, Astellas, and Schering Plough. U.B. has received an independent research grant from Gilead Sciences. All other authors have no conflicts of interest to declare.

Conflicts of Interest: The authors declare no conflict of interest.

References

1. Spellberg, B.; Edwards, J.; Ibrahim, A. Novel perspectives on mucormycosis: Pathophysiology, presentation and management. *Clin. Microbiol. Rev.* **2005**, *18*, 556–569. [CrossRef] [PubMed]
2. Hibbett, D.S.; Binder, M.; Bischoff, J.F.; Blackwell, M.; Cannon, P.F.; Eriksson, O.E.; Huhndorf, S.; James, T.; Kirk, P.M.; Lücking, R.; et al. A higher-level phylogenetic classification of the Fungi. *Mycol. Res.* **2007**, *111*, 509–547. [CrossRef] [PubMed]
3. Kwon-Chung, K.J. Taxonomy of fungi causing mucormycosis and entomophthoramycosis (zygomycosis) and nomenclature of the disease: Molecular mycologic perspectives. *Clin. Infect. Dis.* **2012**, *54*, S8–S15. [CrossRef] [PubMed]
4. Morin-Sardin, S.; Nodet, P.; Coton, E.; Jany, J.-L. *Mucor*: A Janus-faced fungal genus with human health impact and industrial applications. *Fungal Biol. Rev.* **2017**, *31*, 12–32. [CrossRef]
5. Dolatabadi, S.; Scherlach, K.; Figge, M.; Hertweck, C.; Dijksterhuis, J.; Menken, S.B.; de Hoog, G.S. Food preparation with mucoralean fungi: A potential biosafety issue. *Fungal Biol.* **2016**, *120*, 393–401. [CrossRef] [PubMed]
6. Lewis, R.E.; Kontoyiannis, D.P. Epidemiology and treatment of mucormycosis. *Future Microbiol.* **2013**, *8*, 1163–1175. [CrossRef] [PubMed]
7. Austin, C.L.; Finley, P.J.; Mikkelson, D.R.; Tibbs, B. Mucormycosis: A rare fungal infection in tornado victims. *J. Burn Care Res.* **2014**, *35*, 164–171. [CrossRef] [PubMed]
8. Walsh, T.J.; Hiemenz, J.W.; Seibel, N.L.; Perfect, J.R.; Horwith, G.; Lee, L.; Silber, J.L.; DiNubile, M.J.; Reboli, A.; Bow, E.; et al. Ampothericin B lipid complex for invasive fungal infections: Analysis of safety and efficacy in 556 cases. *Clin. Infect. Dis.* **1998**, *26*, 1383–1396. [CrossRef] [PubMed]
9. Peixoto, D.; Gagne, L.S.; Hammond, S.P.; Gilmore, E.T.; Joyce, A.C.; Soiffer, R.J.; Marty, F.M. Isavuconazole treatment of a patient with disseminated mucormycosis. *J. Clin. Microbiol.* **2014**, *52*, 1016–1019. [CrossRef] [PubMed]

10. Cornely, O.A.; Arikan-Akdagli, S.; Dannaoui, E.; Groll, A.H.; Lagrou, K.; Chakrabarti, A.; Lanternier, F.; Pagano, L.; Skiada, A.; Akova, M.; et al. ESCMID and ECMM joint clinical guidelines for the diagnosis and management of mucormycosis 2013. *Clin. Microbiol. Infect.* **2014**, *20*, 5–26. [CrossRef] [PubMed]
11. Maurer, E.; Binder, U.; Sparber, M.; Lackner, M.; Caramalho, R.; Lass-Flörl, C. Susceptibility profiles of amphotericin B and posaconazole against clinical relevant Mucorales species under hypoxic conditions. *Antimicrob. Agents Chemother.* **2015**, *59*, 1344–1346. [CrossRef] [PubMed]
12. Marty, F.M.; Ostrosky-Zeichner, L.; Cornely, O.A.; Mullane, K.M.; Perfect, J.R.; Thompson, G.R.; Alangaden, G.J.; Brown, J.M.; Fredricks, D.N.; Heinz, W.J.; et al. Isavuconazole treatment for mucormycosis: A single-arm open-label trial and case-control analysis. *Lancet Infect. Dis.* **2016**, *16*, 828–837. [CrossRef]
13. McCreary, E.K.; Schulz, L.T.; Lepak, A.J. Isavuconazole: Has it saved us? A pharmacotherapy review and update on clinical experience. *Curr. Treat. Options Infect. Dis.* **2017**, *9*, 356–370. [CrossRef]
14. Gray, K.C.; Palacios, D.S.; Dailey, I.; Endo, M.M.; Uno, B.E.; Wilcock, B.C.; Burke, M.D. Amphotericin primarily kills yeast by simply binding ergosterol. *Proc. Natl. Acad. Sci. USA* **2012**, *109*, 2234–2239. [CrossRef] [PubMed]
15. Mesa-Arango, C.; Trevijano-Contador, N.; Román, E.; Sánchez-Fresneda, R.; Casas, C.; Herrero, E.; Argüelles, J.C.; Pla, J.; Cuenca-Estrella, M.; Zaragoza, O. The production of reactive oxygen species is a universal action mechanism of amphotericin B against pathogenic yeasts and contributes to the fungicidal effect of this drugs. *Antimicrob. Agents Chemother.* **2014**, *58*, 6627–6638. [CrossRef] [PubMed]
16. Munayyer, H.K.; Mann, P.A.; Chau, A.S.; Yarosh-Tomaine, T.; Greene, J.R.; Hare, R.S.; Heimark, L.; Palermo, R.E.; Loenberg, D.; McNicholas, P.M. Posaconazole is a potent inhibitor of sterol 14α-demethylation in yeasts and molds. *Antimicrob. Agents Chemother.* **2004**, *48*, 3690–3696. [CrossRef] [PubMed]
17. Miceli, M.H.; Kauffman, C.A. Isavconazole: A new broad-spectrum triazole antifungal agent. *Clin. Infect. Dis.* **2015**, *61*, 1558–1565. [CrossRef] [PubMed]
18. Alastruey-Izquierdo, A.; Castelli, M.V.; Cuesta, I.; Monzon, A.; Cuenca-Estrella, M.; Rodriguez-Tudela, J.L. Activity of posaconazole and other antifungal agents against Mucorales strains identified by sequencing of internal transcribed spacers. *Antimicrob. Agents Chemother.* **2009**, *53*, 1686–1689. [CrossRef] [PubMed]
19. Pagano, L.; Cornely, O.A.; Busca, A.; Caira, M.; Cesaro, S.; Gasbarrino, C.; Girmenia, C.; Heinz, W.J.; Herbrecht, R.; Lass-Flörl, C.; et al. Combined antifungal approach for the treatment of invasive mucormycosis in patients with hematologic diseases: A report from the SEIFEM and FUNGISCOPE registries. *Haematologica* **2013**, *98*, e127–e129. [CrossRef] [PubMed]
20. Caramalho, R.; Tyndall, J.D.A.; Monk, B.C.; Larentis, T.; Lass-Flörl, C.; Lackner, M. Intrinsic short-tailed azole resistance in mucormycetes s due to an evolutionary conserved aminoacid substitution of the lanosterol 14α-demethylase. *Sci. Rep.* **2017**, *7*, 15898. [CrossRef] [PubMed]
21. Fryberg, M.; Oehlschlager, A.C.; Unrau, A.M. Biosynthesis of ergosterol in yeast. Evidence for multiple pathways. *J. Am. Chem. Soc.* **1973**, *95*, 5747–5757. [CrossRef] [PubMed]
22. Osumi, T.; Taketani, S.; Katsuki, H.; Kuhara, T.; Matsumoto, I. Ergosterol biosynthesis in yeast: Pathways in the late stages and their variation under various conditions. *J. Biochem.* **1978**, *83*, 681–691. [CrossRef] [PubMed]
23. Parks, L.P. Physiological implications of sterol biosynthesis in yeast. *Annu. Rev. Microbiol.* **1995**, *49*, 95–116. [CrossRef] [PubMed]
24. Alcazar-Fuoli, L.; Mellado, E.; Garcia-Effron, G.; Lopez, J.F.; Grimalt, O.J.; Cuenca-Estrella, J.M.; Rodriguez-Tudela, J.L. Ergosterol biosynthesis pathway in *Aspergillus fumigatus*. *Steroids* **2008**, *73*, 339–347. [CrossRef] [PubMed]
25. Müller, C.; Binder, U.; Bracher, F.; Giera, M. Antifungal drug testing: Combining minimal inhibitory concentration testing with target identification by the use of gas chromatography—Mass spectrometry. *Nat. Protoc.* **2017**, *12*, 947–963. [CrossRef] [PubMed]
26. Weete, J.D.; Wise, M.L. Effects of triazoles on fungi. *Exp. Mycol.* **1987**, *11*, 214–222. [CrossRef]
27. Alcazar-Fuoli, L.; Mellado, E. Ergosterol biosynthesis in *Aspergillus fumigatus*: Its relevance as an antifungal target and role in antifungal drug resistance. *Front. Microbiol.* **2013**, *3*, 1–5. [CrossRef] [PubMed]
28. Martel, C.M.; Parker, J.E.; Bader, O.; Weig, M.; Warrilow, A.G.S.; Kelly, D.E.; Kelly, S.L. A clinical isolate of *Candida albicans* with mutations in ERG11 (encoding sterol 14α-demethylase) and ERG5 (encoding C22-desturase) is cross resistant to azoles and amphotericin B. *Antimicrob. Agents Chemother.* **2010**, *54*, 3578–3583. [CrossRef] [PubMed]

29. Keller, P.; Müller, C.; Engelhardt, I.; Hiller, E.; Lemuth, K.; Eickhoff, H.; Wiesmüller, K.H.; Burger-Kentischer, A.; Bracher, F.; Rupp, S. An antifungal benzimidazole derivative inhibits ergosterol biosynthesis and reveals novel sterols. *Antimicrob. Agents Chemother.* **2015**, *59*, 6296–6307. [CrossRef] [PubMed]
30. Colley, T.; Alanio, A.; Kelly, S.L.; Sehra, G.; Kizawa, Y.; Warilow, A.G.S.; Parker, J.E.; Kelly, D.E.; Kimura, G.; Anderson-Dring, L.; et al. In vitro and in vivo antifungal profile of a novel and long acting inhaled azole PC945, on *Aspergillus fumigatus* infection. *Antimicrob. Agents Chemother.* **2017**, *61*, 1–14. [CrossRef] [PubMed]
31. Georgopapadakou, N.H. Antifungals: Mechanism of action and resistance, established and novel drugs. *Curr. Opin. Microbiol.* **1998**, *1*, 547–557. [CrossRef]
32. Ghannoum, M.A.; Rice, L.B. Antifungal agents: Mode of action, mechanisms of resistance, and correlation of these mechanisms with bacterial resistance. *Clin. Microbiol. Rev.* **1999**, *12*, 501–517. [PubMed]
33. Lupetti, A.; Danesi, R.; Campa, M.; del Tacca, M.; Kelly, S. Molecular basis of resistance to azole antifungals. *Trends Mol. Med.* **2002**, *8*, 76–81. [CrossRef]
34. McCorkindale, N.J.; Hutchinson, S.A.; Pursey, B.A.; Scott, W.T.; Wheeler, R. A comparison of the types of sterol found in species of the Saprolegniales and Leptomitales with those found in some other Phycomycetes. *Phytochemistry* **1969**, *8*, 861–867. [CrossRef]
35. Gordon, P.A.; Stewart, P.R.; Clark-Walker, G.D. Fatty acids and sterol composition of *Mucor genevensis* in relation to dimorphism and anaerobic growth. *J. Bacteriol.* **1971**, *107*, 114–120. [PubMed]
36. Safe, S. The effect of environment on the free and hydrosoluble sterols of *Mucor rouxii*. *Biochim. Biophys. Acta* **1973**, *326*, 471–475. [CrossRef]
37. Weete, J.D.; Lawler, G.C.; Laseter, J.L. Total lipid and sterol components of *Rhizopus arrhizus*: Identification and metabolism. *Arch. Biochem. Biophys.* **1973**, *155*, 411–419. [CrossRef]
38. Weete, J.D.; Gandhi, S.R. Sterols of the phylum Zygomycota: Pylogenetic implications. *Lipids* **1997**, *32*, 1309–1316. [CrossRef] [PubMed]
39. Weete, J.D.; Abril, M.; Blackwell, M. Phylogenetic distribution of fungal sterols. *PLoS ONE* **2010**, *5*, e10899. [CrossRef] [PubMed]
40. Dannaoui, E.; Persat, F.; Borel, E.; Piens, M.-A.; Picot, S. Sterol composition of itraconazole-resistant and itraconazole-susceptible isolates of *Aspergillus fumigatus*. *Can. J. Microbiol.* **2001**, *47*, 706–710. [CrossRef] [PubMed]
41. Gsaller, F.; Hortschansky, P.; Furukawa, T.; Carr, P.D.; Rash, B.; Capilla, J.; Müller, C.; Bracher, F.; Bowyer, P.; Haas, H.; et al. Sterol biosynthesis and azole tolerance is governed by the opposing actions of SrbA and the CCAAT binding complex. *PloS Pathog.* **2016**, *12*, e1005775.
42. Misslinger, M.; Gsaller, F.; Hortschansky, P.; Müller, C.; Bracher, F.; Bromley, M.J.; Haas, H. The cytochrome b_5 CybE is regulated by iron availability and is crucial for azole resistance in *A. fumigatus*. *Metallomics* **2017**, *9*, 1655–1665. [CrossRef] [PubMed]
43. Müller, C.; Staudacher, V.; Krauss, J.; Giera, M.; Bracher, F. A convenient cellular assay for the identification of the molecular target of ergosterol biosynthesis inhibitors and quantification of their effects on total ergosterol biosynthesis. *Steroids* **2013**, *78*, 483–493. [CrossRef] [PubMed]
44. Müller, C.; Binder, U.; Maurer, E.; Grimm, C.; Giera, M.; Bracher, F. Fungal sterol C22-desaturase is not an antimycotic target as shown by selective inhibitors and testing on clinical isolates. *Steroids* **2015**, *101*, 1–6. [CrossRef] [PubMed]
45. Nemec, T.; Jernejec, K.; Cimerman, A. Sterols and fatty acids of different *Aspergillus* species. *FEMS Microbiol. Lett.* **1997**, *149*, 201–205. [CrossRef]
46. Kelly, S.L.; Lamb, D.C.; Corran, A.J.; Baldwin, B.C.; Kelly, D.E. Mode of action and resistance to azole antifungals associated with the formation of 14α-methylergosta-8,24(28)-dien-3β,6α-diol. *Biochem. Biohys. Res. Commun.* **1995**, *207*, 910–915. [CrossRef] [PubMed]

Sample Availability: Samples of the compounds are not available from the authors.

Communication

Developing an Enzyme-Assisted Derivatization Method for Analysis of C_{27} Bile Alcohols and Acids by Electrospray Ionization-Mass Spectrometry

Jonas Abdel-Khalik, Peter J. Crick, Eylan Yutuc, Yuqin Wang and William J. Griffiths *

Swansea University Medical School, ILS1 Building, Swansea University, Singleton Park, Swansea SA2 8PP, Wales, UK; jonas.abdelkhalik@gmail.com (J.A.-K.); peter.crick@gmail.com (P.J.C.); eylan.yutuc@swansea.ac.uk (E.Y.); y.wang@swansea.ac.uk (Y.W.)

* Correspondence: w.j.griffiths@swansea.ac.uk; Tel.: +44-1792-295562

Received: 21 December 2018; Accepted: 30 January 2019; Published: 7 February 2019

Abstract: Enzyme-assisted derivatization for sterol analysis (EADSA) is a technology designed to enhance sensitivity and specificity for sterol analysis using electrospray ionization–mass spectrometry. To date it has only been exploited on sterols with a 3β-hydroxy-5-ene or 3β-hydroxy-5α-hydrogen structure, using bacterial cholesterol oxidase enzyme to convert the 3β-hydroxy group to a 3-oxo group for subsequent derivatization with the positively charged Girard hydrazine reagents, or on substrates with a native oxo group. Here we describe an extension of the technology by substituting 3α-hydroxysteroid dehydrogenase (3α-HSD) for cholesterol oxidase, making the method applicable to sterols with a 3α-hydroxy-5β-hydrogen structure. The 3α-HSD enzyme works efficiently on bile alcohols and bile acids with this stereochemistry. However, as found by others, derivatization of the resultant 3-oxo group with a hydrazine reagent does not go to completion in the absence of a conjugating double bond in the sterol structure. Nevertheless, Girard P derivatives of bile alcohols and C_{27} acids give an intense molecular ion ($[M]^+$) upon electrospray ionization and informative fragmentation spectra. The method shows promise for analysis of bile alcohols and 3α-hydroxy-5β-C_{27}-acids, enhancing the range of sterols that can be analyzed at high sensitivity in sterolomic studies.

Keywords: bile alcohol; cholestanoic acid; oxysterol; sterolomics; enzyme-assisted derivatization; electrospray ionization-mass spectrometry; Girard reagent

---

1. Introduction

Sterols represent one of the major classes of lipids found in living systems [1]. In mammals, cholesterol represents the archetypal sterol. It is metabolized through a myriad of intermediates to C_{21-18} steroids and to C_{24} bile acids [2–9]. For decades there was little interest in these intermediates, however, in recent years the situation has changed with the realization that intermediates in bile acid biosynthesis are ligands to nuclear receptors, including the liver X receptors (LXRs, NR1H3, NR1H2) [10–13], farnesoid X receptor (FXR, NR1H4) [14], pregnane X receptor (PXR, also known as xenobiotic sensing nuclear receptor, SXR, NR1I2) [15,16], RAR-related orphan receptor γt (RORγt, NR1F3) [17], and estrogen receptors (ERs, NR3A1, NR3A2) [18]. They are also related to G protein-coupled receptors (e.g., Epstein-Barr virus induced gene 2 (EBI2, GPR183) [19,20] and smoothened (SMO, FZD11) [21,22]), and are involved in the regulation of cholesterol biosynthesis by binding to INSIG1 (insulin induced gene 1) [23]. Cholesterol metabolites have traditionally been analyzed by gas-chromatography-mass spectrometry (GC-MS) [3,4,6,24,25], however, liquid chromatography (LC)-MS, is currently taking a dominant role in their analysis [5,26].

Analysis of cholesterol metabolites is valuable for the diagnosis of rare inborn errors of metabolism, and defects in bile acid and steroid synthesizing enzymes are efficiently characterized by LC-MS or GC-MS analysis of plasma or urine [6,27–29]. By performing these analyses, unexpected metabolites are identified, which are normally only minor components of the sterolome, but which are abundant in the disease state [7–9]. When these unexpected metabolites are considered, it becomes evident that the complexity of the sterolome is enormous. Sterolomics is one of the subdivisions of lipidomics, however, in most lipidomic studies, sterols are underrepresented; this is because, other than cholesterol and its esters, they are not abundant, and their ionization characteristics in positive-ion electrospray ionization (ESI)-MS (the dominant ionization method in lipidomics) are poor [30,31]. To improve the sensitivity for sterol analysis, many groups adopt a derivatization strategy where sterols are chemically modified to improve their ionization characteristics [32–39].

One derivatization strategy that has lately become popular is enzyme-assisted derivatization for sterol analysis (EADSA, Scheme 1) [8,9,12,13,17,40–47]. EADSA technology was designed to add specificity and sensitivity to sterol analysis [40,44–46]. This is achieved by specifically targeting the 3β-hydroxy-5-ene or 3β-hydroxy-5α-hydrogen function in sterols and converting the 3β-hydroxy to a 3-oxo group with bacterial cholesterol oxidase from *Brevibacterium* or *Streptomyces* sp. [40,41]. Once introduced, the 3-oxo group is derivatized with the positively charged Girard hydrazine reagent, introducing a charge-tag to the target analyte and improving sensitivity in ESI-MS. A limitation of the existing protocol is that it is not applicable to sterols with a 3α-hydroxy group. Here we describe how this limitation is overcome for the analysis of bile alcohols and C_{27} bile acids with this stereochemistry. The methodology is potentially applicable for C_{24} bile acids but requires further optimization to achieve similar sensitivity as for C_{27} alcohols and acids.

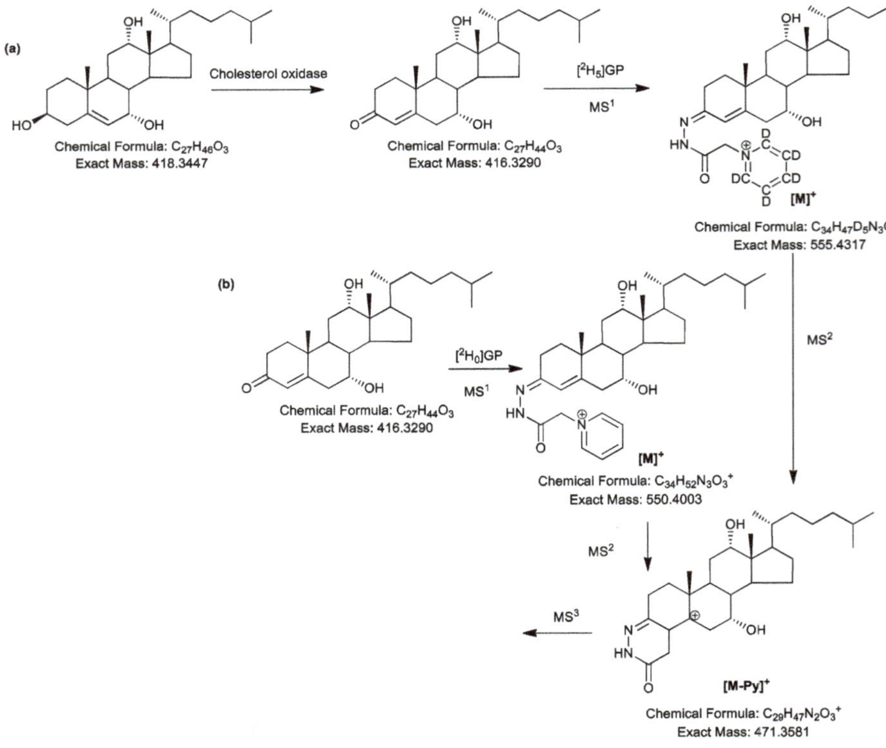

(a)

Cholesterol oxidase

Chemical Formula: $C_{27}H_{46}O_3$
Exact Mass: 418.3447

Chemical Formula: $C_{27}H_{44}O_3$
Exact Mass: 416.3290

[$^2H_5$]GP
MS^1

[M]$^+$
Chemical Formula: $C_{34}H_{47}D_5N_3O_3^+$
Exact Mass: 555.4317

MS^2

(b)

Chemical Formula: $C_{27}H_{44}O_3$
Exact Mass: 416.3290

[$^2H_0$]GP
MS^1

[M]$^+$
Chemical Formula: $C_{34}H_{52}N_3O_3^+$
Exact Mass: 550.4003

MS^2

MS^3

[M-Py]$^+$
Chemical Formula: $C_{29}H_{47}N_2O_3^+$
Exact Mass: 471.3581

Scheme 1. *Cont.*

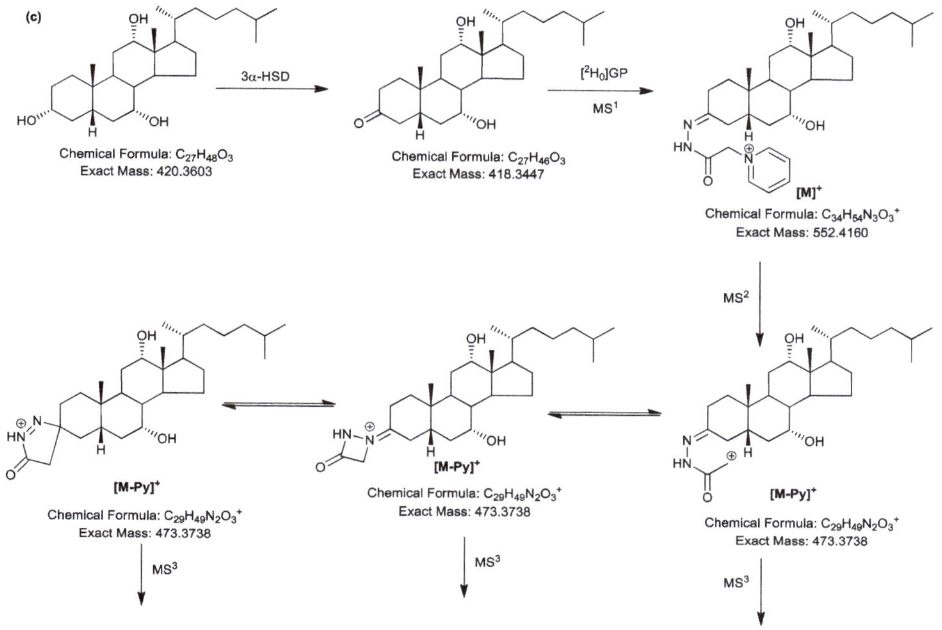

Scheme 1. Enzyme-assisted derivatization for sterol analysis (EASDA): (**a**) EADSA of the 3β-hydroxy-5-ene function using cholesterol oxidase and [$^2H_5$] Girard P reagent ([$^2H_5$]GP); (**b**) derivatization of the 3-oxo-4-ene function with [$^2H_0$]GP; (**c**) EADSA of the 3α-hydroxy-5β-hydrogen function with 3α-hydroxysteroid dehydrogenase (3α-HSD) and [$^2H_0$]GP. Derivatives with the [$^2H_5$]GP and [$^2H_0$]GP reagents can be combined and analyzed in a single LC-MS run. Other isotope-coded GP reagents have been synthesized to allow triplex analysis if required [45].

2. Results

2.1. Oxidation Efficiency of 3α-Hydroxysteroid Dehydrogenase (3α-HSD)

The efficiency of oxidation of 3α-HSD towards 3α-hydroxy-5β-substrates was evaluated using cholic acid (3α,7α,12α-trihydroxy-5β-cholan-24-oic acid, BA-3α,7α,12α-triol) and 3α,7α,12α-trihydroxy-5β-cholestan-26-oic acid (CA-3α,7α,12α-triol) as representative analytes, because unlike neutral sterols, C_{24} and C_{27} bile acids are readily ionized by ESI (negative-ion mode) and can be detected in LC-MS analysis in both their 3α-hydroxy or 3-oxo forms. To determine oxidization efficiency, the amounts of unoxidized (3α-hydroxy) and oxidized (3-oxo) acids were determined after different incubation periods and incorporated into Equation (1).

$$\text{Oxidation efficiency (\%)} = \text{amount oxidized}/(\text{amount oxidized} + \text{amount unoxidized}) \times 100\% \quad (1)$$

Using similar reaction conditions to those employed in the current study (see Section 4.2.1), Une et al. have shown that an incubation time of 20 h leads to about 95% conversion of the 3α-hydroxy group to the 3-one in cholic acid and in most bile alcohols [48]. We found that reducing the incubation time to 14 h gave a similarly high yield (>95%) of product for CA-3α,7α,12α-triol and cholic acid. Either incubation time, 14 h or 20 h, gave satisfactory results. In the current study, similarly high efficiencies of oxidation (>95%) were found for both the glycine and taurine conjugates of cholic acid and for unconjugated chenodeoxycholic (3α,7α-dihydroxy-5β-cholan-24-oic), deoxycholic (3α,12α-dihydroxy-5β-cholan-24-oic) and lithocholic (3α-hydroxy-5β-cholan-24-oic) acids.

2.2. Girard P (GP)-Derivatisation of 3-Oxo Groups

Previous studies have shown that hydrazone formation is very efficient towards 3-oxo-4-ene substrates and to other α,β-unsaturated ketones [40,41,49,50]. In earlier EADSA studies using cholesterol oxidase in phosphate buffer to convert 3β-hydroxy-5-ene sterols to their 3-oxo-4-ene equivalents, GP derivatization was achieved by simply adding methanol to the incubation solution to give a 70% methanol solution and then adding acetic acid and GP hydrazine reagent [40,41,45,46,50–52]. However, the buffer required for 3α-HSD oxidation of the 3α-hydroxy group to the 3-one is 100 mM pyrophosphate buffer, pH 8.9, and upon methanol addition, necessary for subsequent hydrazone formation, a precipitate is formed. This can be avoided by limiting the methanol content to 5%, however, under these conditions hydrazone formation is reversed back to the hydrazine and free carbonyl. For this reason, following incubation with 3α-HSD, samples were desalted on an Oasis HLB reversed-phase column and eluted in methanol, a solvent suitable for subsequent GP hydrazone formation. The GP derivatization efficiency was assessed by LC-MS in the negative-ion mode by comparing the amount of underivatized oxidized acid present before and after the GP derivatization step and incorporating the data into Equation (2).

$$\text{Derivatization efficiency (\%)} = 100\% - [(\text{amount underivatized acid after derivatization}/\text{amount underivatized acid before derivatization}) \times 100\%] \quad (2)$$

Unlike 3-oxo-4-ene sterols, which are derivatized with 100% efficiency in acidic methanol [40], the derivatization efficiency for the 3-oxo-5β-hydrogen compounds formed from their 3α-hydroxy-5β-hydrogen substrates (cholic, chenodeoxycholic, deoxycholic acids and CA-3α,7α,12α-triol), was only 45–60%. Taurine- and glycine-conjugated cholic acid gave a similar degree of derivatization efficiency after 3α-HSD oxidation. Despite the moderate yield of derivatization products, the high sensitivity provided by GP derivatization of unconjugated substrates (see Section 2.3.) negates this imperfection.

2.3. LC-MS Analysis of Oxidised/Derivatised 3α-Hydroxy-5β-Hydrogen Substrates

In this preliminary study we have not optimized the chromatographic or MS conditions for the GP-derivatized target compounds but rather used existing LC-MS conditions used previously for GP-derivatized sterols [44–46,52]. The logic behind this is that by using isotope-labelled GP reagent, the ultimate aim will be to analyze sterols oxidized with cholesterol oxidase or 3α-HSD in a single LC-MS run. Neither have we performed detailed investigations of limits of quantification or linearity of dynamic range in the current study. However, we find that the sensitivity obtained here for 3α-HSD oxidized/GP-derivatized C_{27} sterols with a 3α-hydroxy-5β-stereochemistry is of the same order of magnitude to that obtained for GP derivatives generated after cholesterol oxidase treatment of 3β-hydroxy-5-ene substrates. For the C_{27} substrates an on-column limit of detection (LOD, signal/noise, 5:1) of 250 fg was achieved. The on-column LOD for the C_{24} acid, cholic acid, was 250 pg. More work is required to explain this discrepancy in sensitivity and the even poorer sensitivity with glycine- and taurine-conjugated acids. Optimization of the ion-source conditions for different groups of analytes, or at least compromise in the settings chosen, is likely to be necessary.

2.4. MS^n Fragmentation

A major driver for the current study was the poor fragmentation properties of unconjugated C_{24} and C_{27} bile acids under conditions of ESI–tandem MS (MS/MS) at low collision energy (<100 eV) [5,53–56] (see also MassBank of North America http://mona.fiehnlab.ucdavis.edu/). This has led to many studies in which the precursor ion at unit mass resolution is also used as the "product ion" for generation of LC–multiple reaction monitoring (MRM) chromatograms. Once derivatized with the GP reagent, both bile acids and bile alcohols fragment under low-energy conditions with the loss of the pyridine group, resulting in $[M\text{-}Py]^+$ ions (see Scheme 1). These ions can be fragmented further in ion-trap instruments to give multistage fragmentation (MS^3, [M]$^+$→[M-Py]$^+$→) spectra rich in fragment ions. The advantage of MS^3 is that it provides an extra dimension of separation compared to MS^2, where spectra are a composite of fragment ions derived from desired and undesired coselected precursor ions.

2.4.1. Triols and Tetrols

Shown in Figure 1 are representative reconstructed-ion chromatograms (RICs) and MS^3 ([M]$^+$→[M-Py]$^+$→) spectra of oxidized/GP-derivatized C_{27} bile alcohols 5β-cholestane-3α,7α,12α-triol (C-3α,7α,12α-triol) and 5β-cholestane-3α,7α,12α,26-tetrol (C-3α,7α,12α,26-tetrol)), the C_{27} acid (CA-3α,7α,12α-triol), and the C_{24} trihydroxy bile acid (cholic acid).

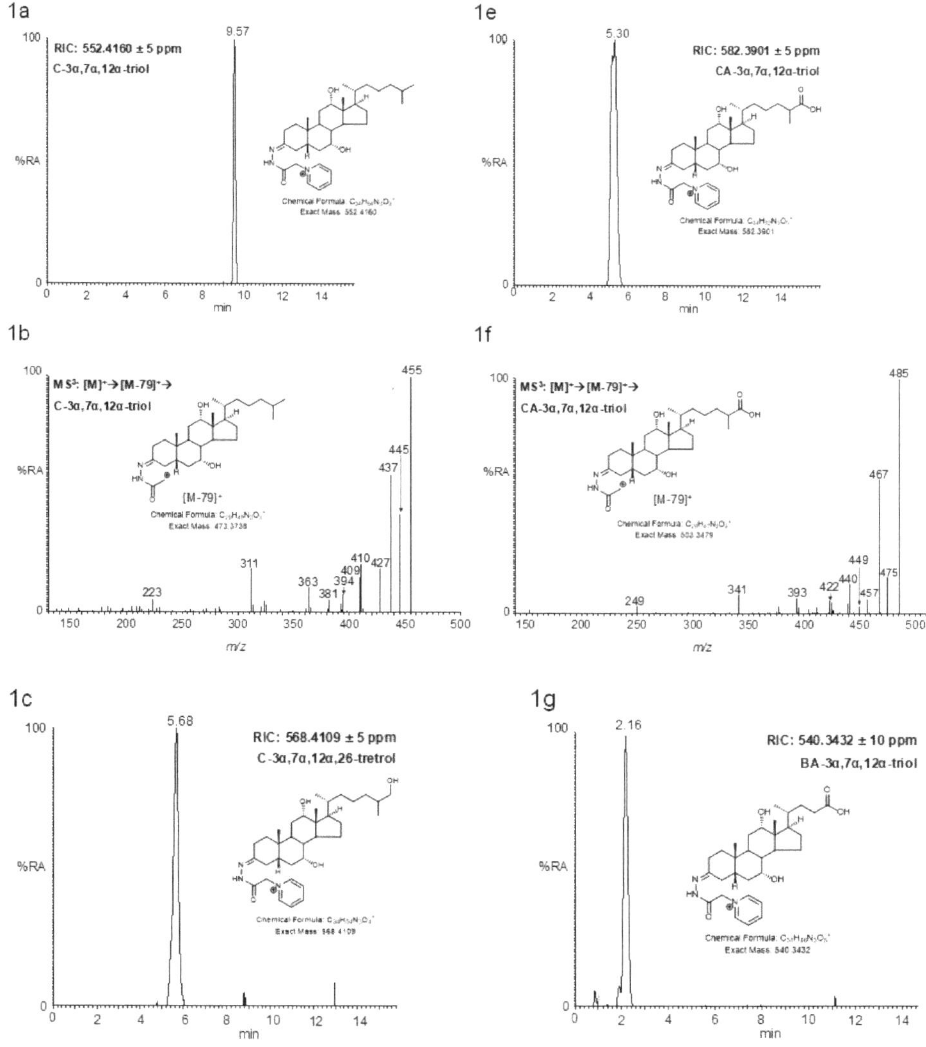

Figure 1. *Cont.*

1d
100
MS³: [M]⁺→[M-79]⁺→
C-3α,7α,12α,26-tetrol
%RA
[M-79]⁺
Chemical Formula: C₃₄H₄₅N₂O₄⁺
Exact Mass: 489.3667
453
471
426
461
443
327
379
408
397
209
249
0
150
200
250
300
350
400
450
500
m/z
1h
100
MS³: [M]⁺→[M-79]⁺→
BA-3α,7α,12α-triol
%RA
[M-79]⁺
Exact Mass: 461.3010
443
433
425
415
398
380
351
362
368
407
299
299
249
0
150
200
250
300
350
400
450
500
m/z

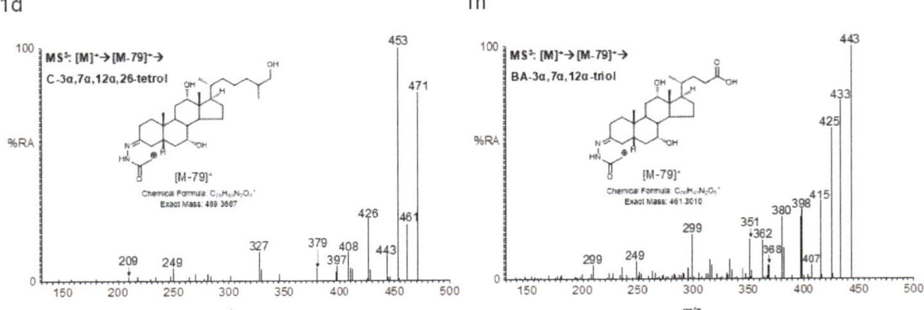

Figure 1. Reconstructed-ion chromatograms (RICs) and multistage fragmentation (MS^3) ($[M]^+ \rightarrow [M\text{-}Py]^+ \rightarrow$) spectra of oxidized/GP-derivatized 3α-hydroxy-5β-bile alcohols and acids: (**a**,**b**) C-3α,7α,12α-triol; (**c**,**d**) C-3α,7α,12α,26-tetrol; (**e**,**f**) CA-3α,7α,12α-triol; (**g**,**h**) cholic acid. The RICs were generated from mass spectra recorded in the Orbitrap mass analyzer at a resolution of 120,000 (FWHM definition at m/z 400), with an m/z window of ± 5 ppm. MS^3 spectra were generated in the linear ion-trap and recorded at the ion-trap detector of the Orbitrap-Elite mass spectrometer. Mass accuracy for fragment ion measurements made with the linear ion-trap is typically ± 0.3 Da. Postulated compositions of fragment ions are listed in Table 1. Note that the data for cholic acid was generated on an earlier version of instrument (i.e., Orbitrap-LTQ) at lower resolution and with reduced mass accuracy.

The MS^3 spectra show considerable similarity, with many fragment ions in the spectrum of the C_{26} acid and tetrol being shifted by m/z 30 and m/z 16, respectively from the corresponding triol. This is explained by the introduction of a carboxylic acid group (+ O_2 − H_2) or hydroxy (+ O) group to the terminal carbon (C-26) of the sterol side-chain (Scheme 2). Postulated structures of fragment ions for C-3α,7α,12α-triol are shown in Scheme 3 and for C-3α,7α,12α,26-tetrol and CA-3α,7α,12α-triol in Supplemental Schemes S1 and S2, and are listed in Table 1.

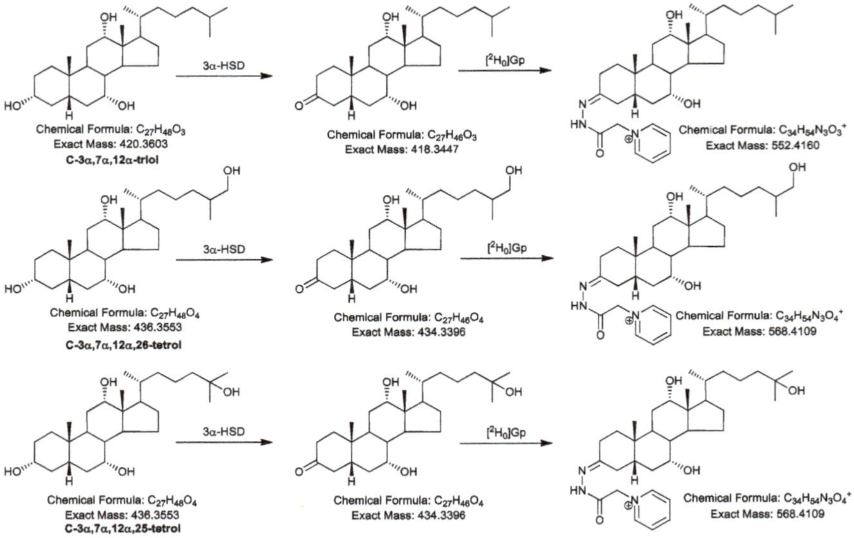

3α-HSD
[²H₀]Gp
Chemical Formula: C₂₇H₄₈O₃
Exact Mass: 420.3603
C-3α,7α,12α-triol
Chemical Formula: C₂₇H₄₆O₃
Exact Mass: 418.3447
Chemical Formula: C₃₄H₅₄N₃O₃⁺
Exact Mass: 552.4160
Chemical Formula: C₂₇H₄₈O₄
Exact Mass: 436.3553
C-3α,7α,12α,26-tetrol
Chemical Formula: C₂₇H₄₆O₄
Exact Mass: 434.3396
Chemical Formula: C₃₄H₅₄N₃O₄⁺
Exact Mass: 568.4109
Chemical Formula: C₂₇H₄₈O₄
Exact Mass: 436.3553
C-3α,7α,12α,25-tetrol
Chemical Formula: C₂₇H₄₆O₄
Exact Mass: 434.3396
Chemical Formula: C₃₄H₅₄N₃O₄⁺
Exact Mass: 568.4109

Scheme 2. *Cont.*

+2O

3α-HSD

[^{2}H$_0$]Gp

Chemical Formula: $C_{27}H_{48}O_5$
Exact Mass: 452.3502
C-3α,7α,12α,x,y-pentol

Chemical Formula: $C_{27}H_{46}O_5$
Exact Mass: 450.3345

Chemical Formula: $C_{34}H_{54}N_3O_5^+$
Exact Mass: 584.4058

Chemical Formula: $C_{27}H_{46}O_5$
Exact Mass: 450.3345
CA-3α,7α,12α-triol

Chemical Formula: $C_{27}H_{44}O_5$
Exact Mass: 448.3189

Chemical Formula: $C_{34}H_{52}N_3O_5^+$
Exact Mass: 582.3901

Chemical Formula: $C_{24}H_{40}O_5$
Exact Mass: 408.2876
BA-3α,7α,12α-triol

Chemical Formula: $C_{24}H_{38}O_5$
Exact Mass: 406.2719

Chemical Formula: $C_{31}H_{46}N_3O_5^+$
Exact Mass: 540.3432

Scheme 2. Structures of bile alcohols and acids, and their products of 3α-HSD oxidation and GP-derivatization.

Table 1. Composition of fragment ions generated from oxidised/GP-derivatised C_{27} bile alcohols and C_{27} and C_{24} acids.

Ion	Composition	C-triol m/z	[^{2}H$_7$]C-triol m/z	C-tetrol m/z	C-pentol m/z	CA-triol m/z	BA-triol m/z
[M]$^+$	[M]$^+$	552.4160	559.4599	568.4109	584.4058	582.3901	540.3432
[M-79]$^+$	[M-Py]$^+$	473.3738	480.4177	489.3687	505.3636	503.3479	461.3010
[M-79-18]$^+$	[M-Py-H$_2$O]$^+$	455.3632	462.4071	471.3581	487.3530	485.3373	443.2904
[M-79-28]$^+$	[M-Py-CO]$^+$	445.3789	452.4228	461.3738	477.3687	475.3530	433.3061
[M-79-36]$^+$	[M-Py-(H$_2$O)$_2$]$^+$	437.3526	444.3965	453.3475	469.3424	467.3267	425.2798
[M-79-28-18]$^+$	[M-Py-CO-H$_2$O]$^+$	427.3683	434.4122	443.3632	459.3581	457.3424	415.2955
[M-79-36-18]$^+$	[M-Py-(H$_2$O)$_3$]$^+$	419.3421	426.3860	435.3370	451.3319	449.3162	407.2693
[M-79-28-18-15]$^+$	[M-Py-CO-H$_2$O-NH]$^+$	412.3574	419.4013	428.3523	444.3472	442.3315	400.2846
[M-79-28-18-17]$^+$	[M-Py-CO-H$_2$O-NH$_3$]$^+$	410.3417	417.3856	426.3366	442.3315	440.3158	398.2689
[M-79-28-36]$^+$	[M-Py-CO-(H$_2$O)$_2$]$^+$	409.3577	416.4016	425.3526	441.3475	439.3318	397.2849
[M-79-36-36]$^+$	[M-Py-(H$_2$O)$_4$]$^+$			417.3264	433.3213	431.3056	389.2587
[M-79-28-15-31]$^+$	[M-Py-CO-NH-CH$_2$NH$_3$]$^+$	399.3258	406.3697	415.3207	431.3156	429.2999	387.2530
[M-79-36-36-2]$^+$	[M-Py-(H$_2$O)$_4$-H$_2$]$^+$			415.3108	431.3057	429.2900	387.2431
[M-79-28-36-15]$^+$	[M-Py-CO-(H$_2$O)$_2$-NH]$^+$	394.3468	401.3907	410.3417	426.3366	424.3209	382.2740
[M-79-28-36-17]$^+$	[M-Py-CO-(H$_2$O)$_2$-NH$_3$]$^+$	392.3312	399.3751	408.3261	424.3210	422.3053	380.2584
[M-79-28-36-18]$^+$	[M-Py-CO-(H$_2$O)$_3$]$^+$			407.3421	423.3370	421.3213	379.2744
[M-79-28-18-15-31]$^+$	[M-Py-CO-H$_2$O-NH-CH$_2$NH$_3$]$^+$	381.3152	388.3591	397.3101	413.3050	411.2893	369.2424
[M-79-36-36-18-2]$^+$	[M-Py-(H$_2$O)$_5$-H$_2$]$^+$				413.2951	411.2794	369.2325
[M-79-28-36-15-18+2]$^+$	[M-Py-CO-(H$_2$O)$_3$-NH+H$_2$]$^+$	378.3519	385.3958	394.3468	410.3417	408.3260	366.2791
[M-79-28-36-15-18]$^+$	[M-Py-CO-(H$_2$O)$_3$-NH]$^+$			392.3312	408.3261	406.3104	364.2635
[M-79-28-36-17-18]$^+$	[M-Py-CO-(H$_2$O)$_3$-NH$_3$]$^+$			390.3155	406.3104	404.2947	362.2478
[M-79-28-36-15-31]$^+$	[M-Py-CO-(H$_2$O)$_2$-NH-CH$_2$NH$_3$]$^+$	363.3046	370.3485	379.2995	395.2944	393.2787	351.2318
[M-79-28-36-15-36+2]$^+$	[M-Py-CO-(H$_2$O)$_4$-NH+H$_2$]$^+$				392.3312	390.3155	348.2686
[M-79-28-36-15-31-18]$^+$	[M-Py-CO-(H$_2$O)$_3$-NH-CH$_2$NH$_3$]$^+$			361.2890	377.2839	375.2682	333.2213
M-79-28-36-15-31-36]$^+$	[M-Py-CO-(H$_2$O)$_4$-NH-CH$_2$NH$_3$]$^+$				359.2733	357.2576	315.2107
[A$_3$+H-36]$^+$	[A$_3$+H-(H$_2$O)$_2$]$^+$	313.2890	320.3329	329.2839	345.2788	343.2631	301.2162
[A$_3$-H-36]$^+$	[A$_3$-H-(H$_2$O)$_2$]$^+$	311.2733	318.3172	327.2682	343.2631	341.2474	299.2005
[A$_3$-H-36-18]$^+$	[A$_3$-H-(H$_2$O)$_3$]$^+$			309.2577	325.2526	323.2369	281.1900
[A$_3$-H-26]$^+$		285.2577	292.3016	301.2526	317.2475	315.2318	273.1849
[A$_3$-H-36-36]$^+$	[A$_3$-H-(H$_2$O)$_4$]$^+$				307.2420	305.2263	263.1794
[ABCD-Δx5-H]$^+$		249.1638	249.1638	249.1638	249.1638	249.1638	249.1638
[ABC-Δx5-H]$^+$		209.1325	209.1325	209.1325	209.1325	209.1325	209.1325

Note: m/z values are calculated for the chemical compositions. ABCD corresponds to the intact ring structure including C-19 and C-18. ABC corresponds to the intact ring structure including C-19. Δx5 corresponds to 5 double bonds. The fragmenation-route A_3 is depicted in the inset to Scheme 3. Colour Code: Red, loss of H_2O; green, loss of CO; blue, loss of NH or NH_3; purple, loss of CH_2NH_3.

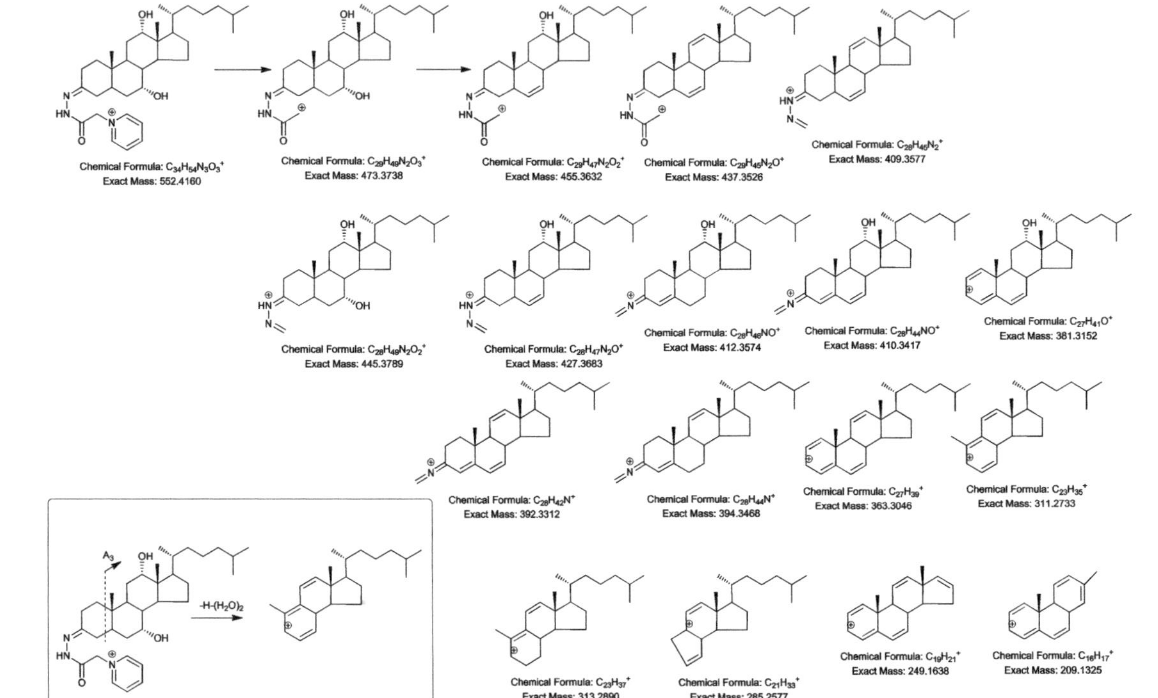

Scheme 3. MS^2 and MS^3 fragmentation of oxidized/GP derivatized bile alcohols as illustrated by C-3α,7α,12α-triol. For simplicity the fragmented Girard derivatizing group is shown in its linear isomeric form. The inset shows fragmentation route A_3 leading to the $[A_3\text{-H-}(H_2O)_2]^+$ fragment ion. Cyclic isomers are depicted in Scheme 1. See Supplemental Schemes S1–S4 for fragmentation schemes of other trihydroxy- and tetrahydroxy-bile alcohols and trihydroxy-bile acids. Table 1 correlates m/z with fragment ion composition.

When a hydroxy group is positioned at C-25 rather than C-26, the lability of the hydroxy group leads to a more intense [M-Py-18]$^+$ than [M-Py]$^+$ ion in the MS^2 ([M]$^+$$\rightarrow$) spectrum of 5β-cholestane-3α,7α,12α,25-tetrol (C-3α,7α,12α,25-tetrol, Figure 2b) than in its epimer C-3α,7α,12α,26-tetrol (Supplemental Figure S1b). See Table 1 to correlate m/z with fragment ion composition. The MS^3 ([M]$^+$$\rightarrow$[M-Py-18]$^+$$\rightarrow$) spectrum of C-3α,7α,12α,25-tetrol (Figure 2d) is almost identical to the MS^3 ([M]$^+$$\rightarrow$[M-Py]$^+$$\rightarrow$) of C-3α,7α,12α-triol (Figure 1b) but with an offset of m/z -2 (+ O – H_2O, see also Supplemental Scheme S3).

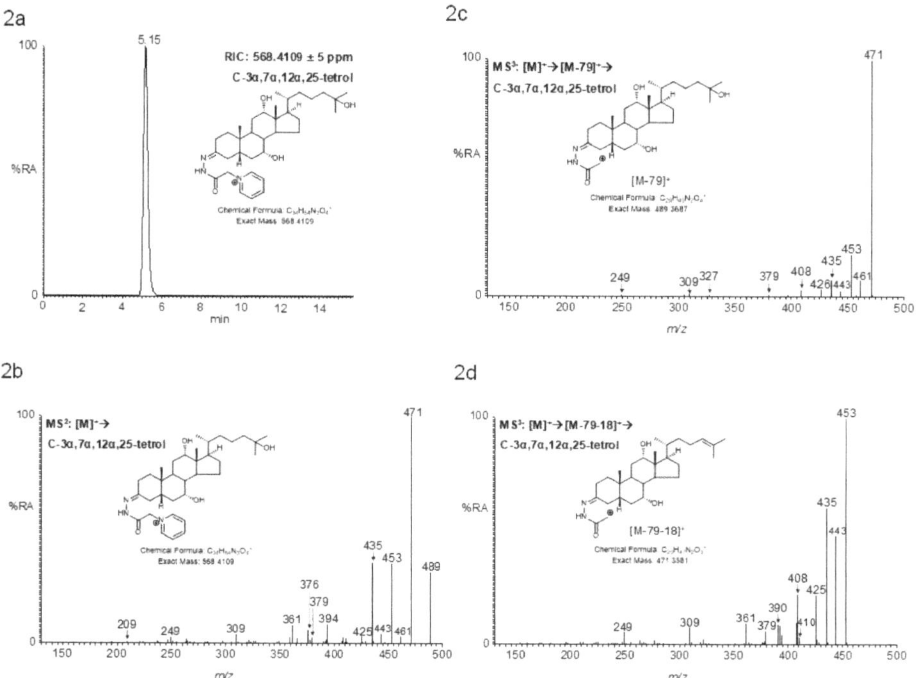

Figure 2. (**a**) RIC and (**b**) MS^2 ([M]$^+$$\rightarrow$), (**c**) MS^3 ([M]$^+$$\rightarrow$[M-Py]$^+$$\rightarrow$), (**d**) ([M]$^+$$\rightarrow$[M-Py-18]$^+$$\rightarrow$) spectra of oxidized/GP-derivatized C-3α,7α,12α,25-tetrol. For comparison, MS^2 ([M]$^+$$\rightarrow$) spectra of C-3α,7α,12α-triol and C-3α,7α,12α,26-tetrol and the MS^3 ([M]$^+$$\rightarrow$[M-Py-18]$^+$$\rightarrow$) spectrum of C-3α,7α,12α,26-tetrol are shown in Supplemental Figure S1a–c respectively. Mass spectra recorded at the peak of the RIC for these and other sterols analyzed are shown in Supplemental Figure S2. Data was generated on the Orbitrap-Elite mass spectrometer as in Figure 1. See Table 1 to correlate m/z with fragment ion composition.

Unsurprisingly, the MS^3 ([M$^+$$\rightarrow$[M-Py]$^+$$\rightarrow$) spectrum of the C_{24} acid, cholic acid (Figure 1h), shows the same pattern of fragment ions as the C_{27} acid CA-3α,7α,12α-triol (Figure 1f) but offset by m/z -42 (-C_3H_6), corresponding to the mass difference between equivalent C_{27} and C_{24} acids (cf. Supplemental Schemes S2 and S4).

A key structurally distinct fragment ion for all the 3α,7α,12α-triols is the [A_3-H-(H_2O)$_2$]$^+$ ion (or [A_3-H-(H_2O)$_3$]$^+$ when an additional hydroxy group is at C-25, Table 1), a triply unsaturated carbonium ion consisting of B-, C- and D-rings plus the C_{17} side-chain, where charge is delocalized across the three double bonds in the ring system (Scheme 3, inset). An equivalent fragment ion is not observed in cholesterol oxidase-oxidized/GP-derivatized 3β,5α,6β-triols (Figure 3).

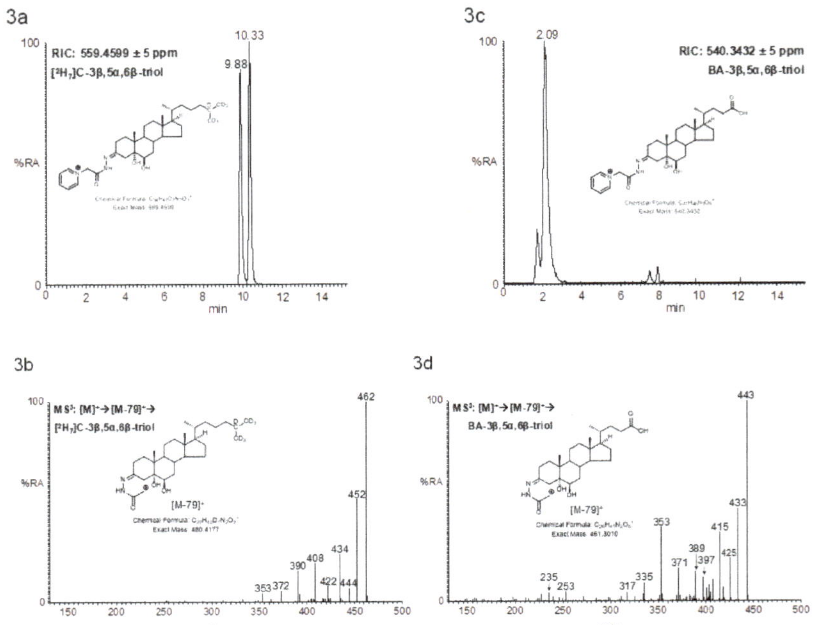

Figure 3. RICs and MS^3 ([M]$^+$→[M-Py]$^+$→) spectra of cholesterol oxidase-oxidized/GP-derivatized 3β,5α,6β-triols: (**a**,**b**) [25,26,26,26,27,27,27-$^2H_7$]cholestane-3β,5α,6β-triol ([$^2H_7$]C-3β,5α,6β-triol); (**c**,**d**) 3β,5α,6β-trihydroxycholanoic acid (BA-3β,5α,6β-triol). See reference [8] for a description of fragmentation pathways. GP-derivatized sterols can give syn and anti conformers resulting in twin chromatographic peaks which may or may not be resolved. Data was generated on the Orbitrap-Elite mass spectrometer as in Figure 1.

2.4.2. Pentols

The human autosomal recessive disease, cerebrotendinous xanthomatosis (CTX), results from a deficiency in cytochrome P450 27A1 (CYP27A1) [57], a key enzyme in the conversion of cholesterol to bile acids [28,29]. As a consequence of this deficiency, polyhydroxy-bile alcohols are produced [58,59], providing an alternative route for bile acid biosynthesis and cholesterol removal [60].

As with C-3α,7α,12α,25-tetrol, the presence of a labile 25-hydroxy group in the epimers 5β-cholestane-3α,7α,12α,24R,25-pentol (C-3α,7α,12α,24R,25-pentol) and 5β-cholestane-3α,7α,12α,24S,25-pentol (C-3α,7α,12α,24S,25-pentol) results in abundant [M-Py-18]$^+$ ions in the MS^2 ([M]$^+$→) spectra, and the MS^3 ([M]$^+$→[M-Py-18]$^+$→) spectra resembles the MS^3 ([M]$^+$→[M-Py]$^+$→) spectra of cholestanetetrols but is offset by m/z -2 (+ O – H_2O) (Figure 4). Although the MS^n spectra are very similar, the two epimers are readily separated on the LC column. Here in the MS^2 ([M]$^+$→) and MS^3 ([M]$^+$→[M-Py]$^+$) spectra, in addition to the [A_3-H-$(H_2O)_2$]$^+$ fragment ion, a [A_3-H-$(H_2O)_3$]$^+$ fragment ion is also prominent (See Table 1 to correlate m/z with fragment ion composition).

Movement of the hydroxy group from C-24 to C-26 as in 5β-cholestane-3α,7α,12α,25,26-pentol (C-3α,7α,12α,25,26-pentol) results in a small delay in retention time and subtle changes to the MS^2 ([M]$^+$→) and MS^3 ([M]$^+$→[M-Py]$^+$→ and [M]$^+$→[M-Py-18]$^+$→) spectra, for example, reduced abundance of fragment ions having lost four water molecules compared the equivalent having lost three water molecules (i.e., [M-79-36-36]$^+$/[M-79-36-18]$^+$) (Figure 5b–d, cf. Figure 4b–d,f–h). The bile alcohol 5β-cholestane-3α,7α,12α,26,27-pentol (C-3α,7α,12α,26,27-pentol) elutes between the 24R- and 25S-epimers of C-3α,7α,12α,24,25-pentol, but gives very different MS^n spectra to the other cholestanepentols on account of the absence of a labile C-25-hydroxy group (Figure 5f–h, Supplemental

Scheme S6). This is reflected in the ratio of fragment ions [M-79-36-18]$^+$/[M-79-36]$^+$ which is greatly reduced compared to pentols with a 25-hydroxy group. The comparative stability of the primary hydroxy groups at the termini, C-26 and C-27, is further reflected by an absence of fragment ions having lost four water molecules (e.g., [M-79-36-36]$^+$).

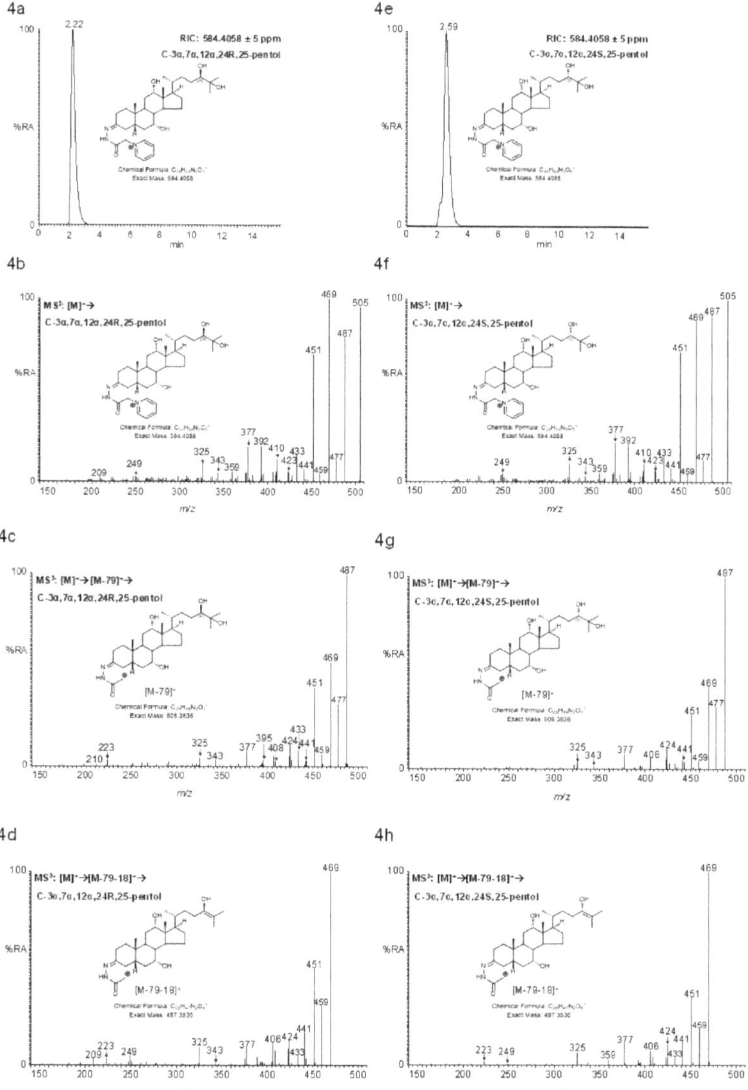

Figure 4. LC–(MS)MS^n analysis of oxidized/GP-derivatized C-3α,7α,12α,24R,25-pentol and C-3α,7α,12α,24S,25-pentol: (**a**) RIC, (**b**) MS^2 ([M]$^+$→), (**c**) MS^3 ([M]$^+$→[M-Py]$^+$→), and (**d**) ([M]$^+$→[M-Py-18]$^+$→) from the analysis of C-3α,7α,12α,24R,25-pentol. (**e**) RIC, (**f**) MS^2 ([M]$^+$→), (**g**) MS^3 ([M]$^+$→[M-Py]$^+$→), and (**h**) ([M]$^+$→[M-Py-18]$^+$→) from the analysis of C-3α,7α,12α,24S,25-pentol. Data was generated on the Orbitrap-Elite mass spectrometer as in Figure 1. Fragment ions are described in Table 1 and postulated structures are shown in Scheme 4.

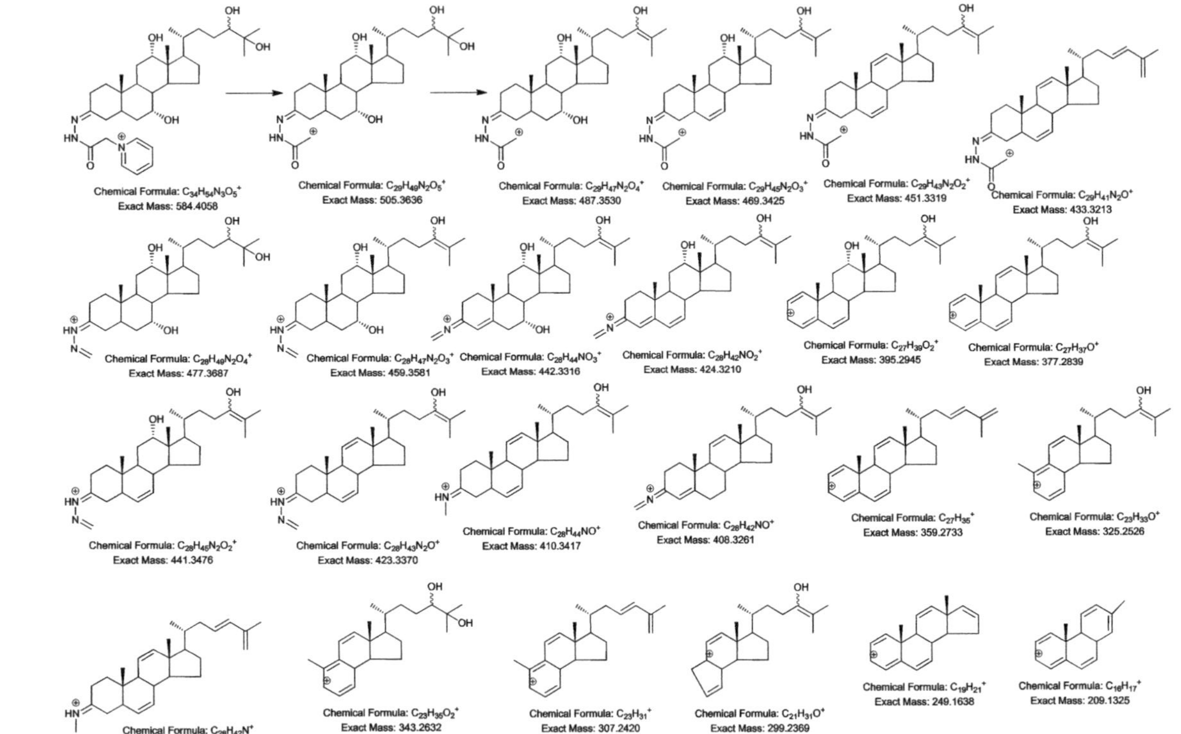

Scheme 4. MS^2 and MS^3 fragmentation of oxidized/GP-derivatized pentahydroxy-bile alcohols as illustrated by C-3α,7α,12α,24,25-pentol. Fragment ions with a 24-hydroxy-24-ene structure are likely to rearrange to the 24-ketone. For simplicity the fragmented Girard derivatizing group is shown in its linear isomeric form. Cyclic isomers are similar to those depicted in Scheme 1. See Supplemental Schemes S5–S7 for fragmentation schemes of other pentahydroxy-bile alcohols.

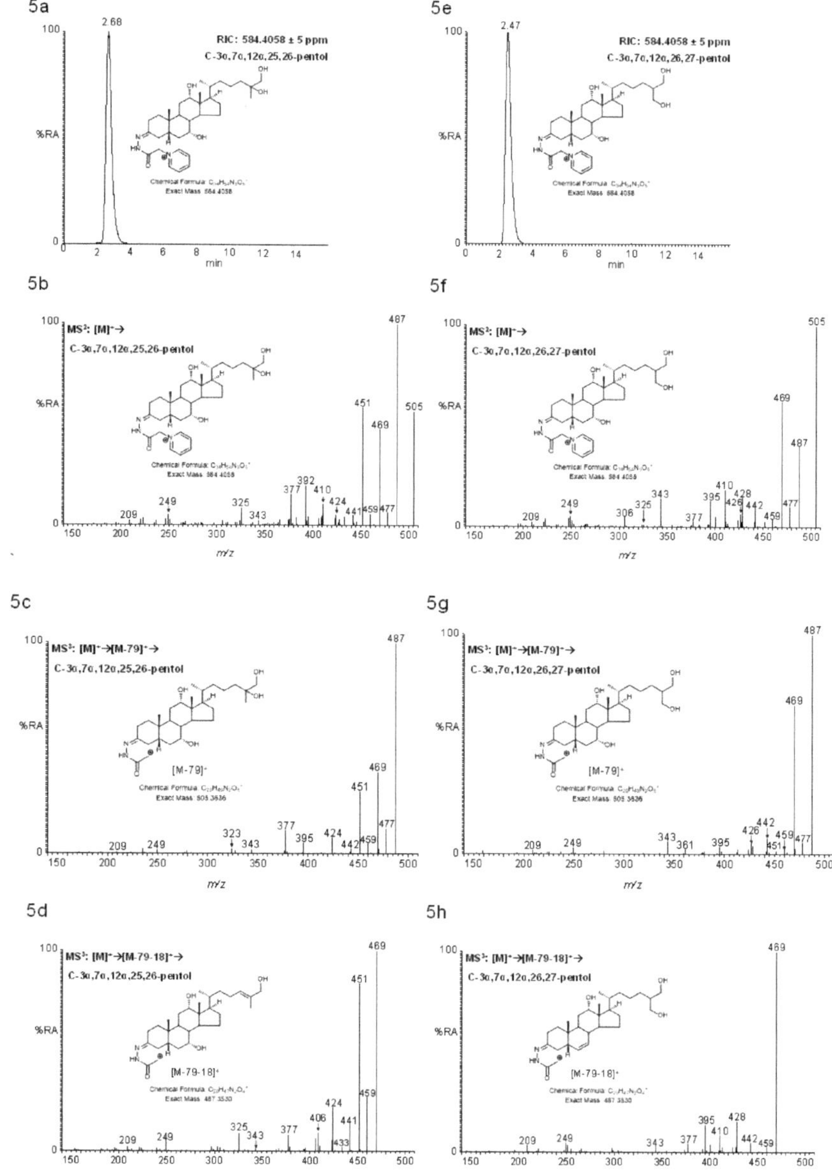

Figure 5. LC–(MS)MS^n analysis of oxidized/GP-derivatized C-3α,7α,12α,25,26-pentol and C-3α,7α,12α,26,27-pentol: (**a**) RIC, (**b**) MS^2 ([M]$^+$→), (**c**) MS^3 ([M]$^+$→[M-Py]$^+$→), and (**d**) ([M]$^+$→[M-Py-18]$^+$→) from the analysis of C-3α,7α,12α,25,26-pentol. (**e**) RIC, (**f**) MS^2 ([M]$^+$→), (**g**) MS^3 ([M]$^+$→[M-Py]$^+$→), and (**h**) ([M]$^+$→[M-Py-18]$^+$→) from the analysis of C-3α,7α,12α,26,27-pentol. Data was generated on the Orbitrap-Elite mass spectrometer as in Figure 1. Fragment ions are described in Table 1 and postulated structures are shown in Supplemental Schemes S5 and S6.

Although it co-elutes with C-3α,7α,12α,24S,25-pentol, 5β-cholestane-3α,7α,12α,23,25-pentol (C-3α,7α,12α,23,25-pentol) gives unique MS^n spectra (Figure 6). Each of the MS^2 ($[M]^+\rightarrow$), MS^3 ($[M]^+\rightarrow[M\text{-}Py]^+\rightarrow$) and MS^3 ($[M]^+\rightarrow[M\text{-}Py\text{-}18]^+\rightarrow$) spectra show an unusual pattern of fragment ions at m/z 431.3, 413.3, and 395.3. It is not immediately obvious why this triad of fragment ions is so distinct for this molecule. Neither are the structures or chemical compositions of all these fragments easy to reconcile with the MS^3 ($[M]^+\rightarrow[M\text{-}Py\text{-}18]^+\rightarrow$) spectrum.

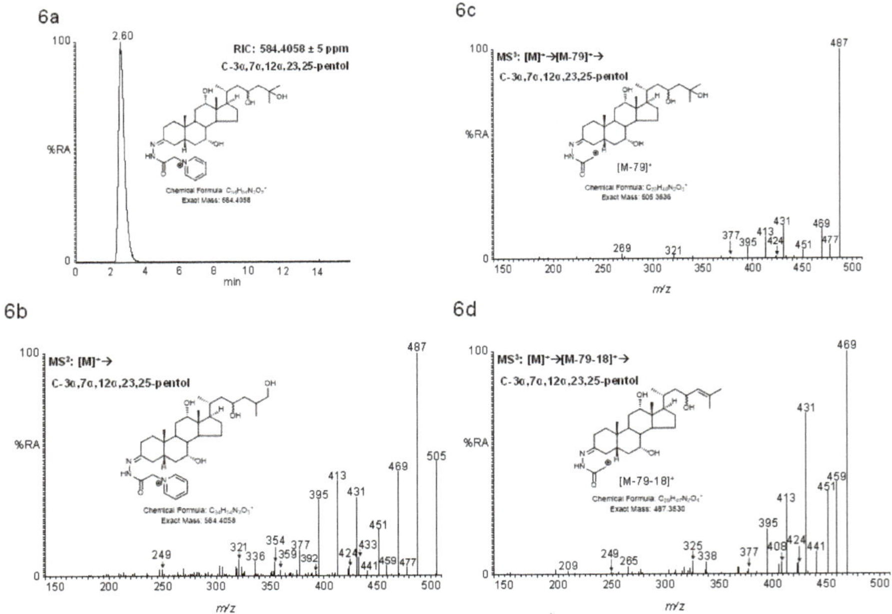

Figure 6. LC–(MS)MS^n analysis of oxidized/GP-derivatized C-3α,7α,12α,23,25-pentol. (**a**) RIC, (**b**) MS^2 ($[M]^+\rightarrow$), (**c**) MS^3 ($[M]^+\rightarrow[M\text{-}Py]^+\rightarrow$), and (**d**) ($[M]^+\rightarrow[M\text{-}Py\text{-}18]^+\rightarrow$). Data was generated on the Orbitrap-Elite mass spectrometer as in Figure 1. Fragment ions are described in Table 1 and postulated structures are shown in Supplemental Schemes S7.

3. Discussion

In this communication we describe preliminary studies to develop an enzyme-assisted derivatization for C_{27} bile alcohols and acids with a 3α-hydroxy-5β-hydrogen stereochemistry. The method still requires further optimization, particularly with respect to the GP-derivatization step which only gave a 45%–60% yield. Despite this, the considerable sensitivity of GP-derivatives makes the moderate yield tolerable. Although the LC-MS sensitivity for detection of C_{24} acids was not as good as for C_{27} acids, the rich MS^3 fragment ion spectra provide a significant advantage over conventional MS/MS spectra of unconjugated acids where few fragment ions are observed. The on-column detection limit of 250 fg for C_{27} analytes translates to a limit of detection of about 0.2 ng/mL if 100 μL of biological fluid is worked up and 1% injected on-column, as in our usual procedure with EADSA [52]. For comparison, Johnson et al. could measure CA-3α,7α,12α-triol, after derivatization to the dimethylaminoethyl ester, at a concentration of about 60 ng/mL in as little as 5 μL of plasma, with 20% injected-on column [38], while DeBarber et al. determined the limit of quantification of 7α,12α-dihydroxy-5β-cholestan-3-one, the 3-oxo form of C-3α,7α,12α-triol, to be 20 ng/mL from 4 μL of plasma after derivatization to the oxime with (*O*-(3-trimethylammoniumpropyl) hydroxylamine) bromide [39]. We have not yet rigorously tested the repeatability of the EADSA methodology in biological samples. This will become relevant with the availability of isotope-labelled standards, which can be synthesized by methods described by Johnson et al. and by Shoda et al. [61]. Isotope-labelled internal standards will similarly facilitate the progression of the method to a quantitative format. We did not attempt to optimize LC-MS conditions for the GP-derivatives analyzed in this study; instead we used previously optimized conditions for derivatized oxysterols. The logic behind this was to allow the expansion of our sterol profiling method to include bile acids and alcohols derivatized with [$^2H_0$]GP after 3α-HSD treatment and oxysterols, and cholestenoic and cholenoic acids derivatized with [$^2H_5$]GP after cholesterol oxidase treatment, or vice versa, in a single LC-MS run. At present there are challenges with this strategy, as efficient ionization of glycine- and taurine-conjugated bile acids requires different ion-source conditions from the unconjugated GP-derivatives.

4. Materials and Methods

4.1. Materials

CA-3α,7α,12α-triol (LMST04030001) and [$^2H_7$]C-3β,5α,6β-triol were from Avanti Polar Lipids (Alabaster, AL, USA). Bile alcohols, C-3α,7α,12α-triol (LMST04030035), C-3α,7α,12α,25-tetrol (LMST04030037) and C-3α,7α,12α,26-tetrol, (LMST04030159 or LMST04030160), C-3α,7α,12α,24R,25-pentol (LMST04030177), C-3α,7α,12α,24S,25-pentol (LMST04030039), C-3α,7α,12α,25,26-pentol (LMST04030016), 3α,7α,12α,26,27-pentol (LMST04030041) and C-3α,7α,12α,23,25-pentol (LMST01010240 or LMST01010241) were kind gifts from Professor Jan Sjövall, Karolinska Institutet, Stockholm. BA-3β,5α,6β-triol (LMST04010339) was a kind gift from Professor Douglas Covey, Washington University. Other C_{24} bile acids were from Sigma-Aldrich (Dorset, UK) or Fluka Chemie (Buchs, Switzerland). 3α-Hydroxysteroid dehydrogenase (3α-HSD) from *Pseudomonas testosteroni* was from Sigma-Aldrich (Dorset, UK). β-Nicotinamide adenine dinucleotide hydrate and sodium pyrophosphate decahydrate were from Sigma-Aldrich. [$^2H_0$]GP ([1-(carboxymethyl)pyridinium chloride hydrazide]) reagent was from TCI Europe (Oxford, UK). [$^2H_5$]GP reagent was synthesized as the bromide salt as described in Crick et al. [45]. Solid phase extraction (SPE) columns, certified Sep-Pak C_{18}, 200 mg (3 cm^3), and 60 mg Oasis HLB (3 cm^3), were from Waters Inc. (Elstree, UK). Solvents were obtained from Fisher Scientific (Loughborough, UK). Acetic acid and formic acid were of AnalaR NORMAPUR grade (BDH, VWR, Lutterworth, UK).

4.2. Methods

4.2.1. Oxidation and Derivatization

Oxidation of analytes by 3α-HSD was essentially as described by Une et al. [48]. β-NAD^+ hydrate (19.8 mg) was dissolved in 100 mM pyrophosphate buffer pH 8.9 (1 mL). Analyte (40–400 ng) dissolved in ethanol (10 μL) was added to the buffered solution giving a final concentration of 1% ethanol, followed by addition of 3α-HSD (0.06 units). After incubation at room temperature for 20 h, methanol (40 μL) was added (giving an organic content of 5%). To separate oxidized analyte from buffer, the solution was loaded onto a HLB column (60 mg, previously washed with methanol, 6 mL, and conditioned with 5% methanol, 6 mL) followed by a rinse with 5% methanol (0.5 mL). The column was then further washed with 5% methanol (6 mL). Analytes were eluted with methanol (2 mL). For samples to be analyzed by ESI in the negative-ion mode, to monitor oxidation efficiency, the methanol eluate was diluted with water to give a 60% methanol solution and was analyzed by LC-MS on the Orbitrap-Elite high resolution mass spectrometer (Thermo Fisher Scientific, Waltham, MA, USA) at 120,000 resolution (full width at half maximum height at m/z 400). To derivatize samples with GP reagent, glacial acetic acid (150 μL) was added followed by GP reagent (150 mg chloride salt, 190 mg bromide salt) and the mixture was left at room temperature overnight. The next day, water (1 mL) was added immediately prior to a second SPE step. This second SPE step was performed with recycling on an Oasis HLB column (60 mg) to remove excess derivatization reagent and was carried out as described in Abdel-Khalik et al. [52].

4.2.2. LC-MS(MS^n) Analysis

LC-MS(MS^n) was performed in the positive-ion mode as described in Abdel-Khalik et al. utilizing the Orbitrap-Elite hybrid MS preceded by a Dionex Ultimate 3000 LC system (Dionex, now Thermo Fisher Scientific) [52]. For analysis of underivatized acids in the negative-ion mode, other than for polarity reversal and a change of column from a Hypersil Gold C_{18} to a Kinetex core-shell technology XB-C_{18} column (2.6 μm, 2.1 mm × 50 mm, Phenomenex, Macclesfield, UK), the method was as for positive-ion mode LC-ESI-MS(MS^n) as described in Abdel-Khalik et al. [52].

5. Patents

The derivatization method described in this manuscript is patented by Swansea University (US9851368B2) and licensed by Swansea Innovations to Avanti Polar Lipids and to Cayman Chemical Company.

Supplementary Materials: The following are available online. Supplemental Figure S1. MS^2 ($[M]^+\rightarrow$) spectra of (a) C-3α,7α,12α-triol and (b) C-3α,7α,12α,26-tetrol, and (c) MS^3 ($[M]^+\rightarrow[M\text{-}Py\text{-}18]^+\rightarrow$) spectrum of C-3α,7α,12α,26-tetrol. Data were generated on the Orbitrap-Elite mass spectrometer as in Figure 1. Supplemental Figure S2. Mass spectra of oxidized/GP-derivatized bile alcohols and acids recorded at the peak of their chromatographic elution. Except for C-3α,7α,12α,26-tetrol, C-3α,7α,12α,25,26-pentol, C-3α,7α,12α,26,27-pentol and C-3α,7α,12α,23,25-pentol which were injected (35 μL) on-column at 0.2 pg/μL all other analytes were inject on-column at 2 pg/μL. (a) C-3α,7α,12α-triol, (b) C-3α,7α,12α,26-tetrol, (c) CA-3α,7α,12α-triol, (d) C-3α,7α,12α,25-tetrol, (e) [$^2H_7$]C-3β,5α,6β-triol, (f) BA-3β,5α,6β-triol, (g) C-3α,7α,12α,24R,25-pentol, (h) C-3α,7α,12α,24S,25-pentol, (i) C-3α,7α,12α,25,26-pentol, (j) C-3α,7α,12α,26,27-pentol and (k) C-3α,7α,12α,23,25-pentol. Data were generated on the Orbitrap-Elite mass spectrometer as in Figure 1. Scheme S1: MS^2 and MS^3 fragmentation of oxidized/GP derivatized C-3α,7α,12α,26-tetrol. For simplicity the fragmented Girard derivatizing group is shown in its linear isomeric form. Cyclic forms are depicted in Scheme 1. Scheme S2: MS^2 and MS^3 fragmentation of oxidized/GP derivatized CA-3α,7α,12α-triol. For simplicity the fragmented Girard derivatizing group is shown in its linear isomeric form. Cyclic forms are depicted in Scheme 1. Scheme S3: MS^2 and MS^3 fragmentation of oxidized/GP derivatized C-3α,7α,12α,25-tetrol. For simplicity the fragmented Girard derivatizing group is shown in its linear isomeric form. Cyclic forms are depicted in Scheme 1. Scheme S4: MS^2 and MS^3 fragmentation of oxidized/GP derivatized BA-3α,7α,12α-triol. For simplicity the fragmented Girard derivatizing group is shown in its linear isomeric form. Cyclic forms are depicted in Scheme 1. Scheme S5: MS^2 and MS^3 fragmentation of oxidized/GP derivatized C-3α,7α,12α,25,26-pentol. For simplicity the fragmented Girard derivatizing group is shown in its linear isomeric form. Cyclic forms are depicted in Scheme 1. Scheme S6: MS^2 and MS^3 fragmentation of oxidized/GP derivatized C-3α,7α,12α,26,27-pentol. For simplicity the fragmented Girard derivatizing group is shown in its linear isomeric form. Cyclic forms are depicted in Scheme 1. Scheme S7: MS^2 and MS^3 fragmentation of oxidized/GP derivatized C-3α,7α,12α,23,25-pentol. For simplicity the fragmented Girard derivatizing group is shown in its linear isomeric form. Cyclic forms are depicted in Scheme 1.

Author Contributions: All authors contributed to conceptualization of the study, generation and analysis of data, writing reviewing and editing the manuscript.

Funding: This research was funded by the UK Biotechnology and Biological Sciences Research Council (BBSRC, grant numbers BB/I001735/1 to WJG, BB/L001942/1 to YW). JA-K was supported by a PhD studentship from Imperial College Healthcare Charities.

Acknowledgments: Professors Jan Sjövall, Karolinska Institutet, Stockholm and Douglas Covey, Washington University, St Louis are thanked for kind gifts of bile alcohols and/or acids. Members of the European Network for Oxysterol Research (ENOR, https://www.oxysterols.net/) are thanked for informative discussions.

Conflicts of Interest: W.J.G., P.J.C. and Y.W. are listed as inventors on the patent "Kit and method for quantitative detection of steroids" US9851368B2. The funders had no role in the design of the study; in the collection, analyses, or interpretation of data; in the writing of the manuscript; or in the decision to publish the results.

References

1. Fahy, E.; Subramaniam, S.; Brown, H.A.; Glass, C.K.; Merrill, A.H., Jr.; Murphy, R.C.; Raetz, C.R.; Russell, D.W.; Seyama, Y.; Shaw, W.; et al. A comprehensive classification system for lipids. *J. Lipid Res.* **2005**, *46*, 839–861. [CrossRef] [PubMed]
2. Schroepfer, G.J., Jr. Oxysterols: Modulators of cholesterol metabolism and other processes. *Physiol. Rev.* **2000**, *80*, 361–554. [CrossRef] [PubMed]
3. Sjovall, J. Fifty years with bile acids and steroids in health and disease. *Lipids* **2004**, *39*, 703–722. [CrossRef] [PubMed]
4. Bjorkhem, I. Rediscovery of cerebrosterol. *Lipids* **2007**, *42*, 5–14. [CrossRef] [PubMed]
5. Griffiths, W.J.; Sjovall, J. Bile acids: Analysis in biological fluids and tissues. *J. Lipid Res.* **2010**, *51*, 23–41. [CrossRef] [PubMed]
6. Shackleton, C.H. Role of a disordered steroid metabolome in the elucidation of sterol and steroid biosynthesis. *Lipids* **2012**, *47*, 1–12. [CrossRef] [PubMed]
7. Wang, Y.; Griffiths, W.J. Unravelling new pathways of sterol metabolism: Lessons learned from in-born errors and cancer. *Curr. Opin. Clin. Nutr. Metab. Care* **2018**, *21*, 90–96. [CrossRef]
8. Griffiths, W.J.; Gilmore, I.; Yutuc, E.; Abdel-Khalik, J.; Crick, P.J.; Hearn, T.; Dickson, A.; Bigger, B.W.; Wu, T.H.; Goenka, A.; et al. Identification of unusual oxysterols and bile acids with 7-oxo or 3beta,5alpha,6beta-trihydroxy functions in human plasma by charge-tagging mass spectrometry with multistage fragmentation. *J. Lipid Res.* **2018**, *59*, 1058–1070. [CrossRef]
9. Griffiths, W.J.; Crick, P.J.; Meljon, A.; Theofilopoulos, S.; Abdel-Khalik, J.; Yutuc, E.; Parker, J.E.; Kelly, D.E.; Kelly, S.L.; Arenas, E.; et al. Additional pathways of sterol metabolism: Evidence from analysis of Cyp27a1-/- mouse brain and plasma. *Biochim. Biophys. Acta Mol. Cell Biol. Lipids* **2019**, *1864*, 191–211. [CrossRef]
10. Lehmann, J.M.; Kliewer, S.A.; Moore, L.B.; Smith-Oliver, T.A.; Oliver, B.B.; Su, J.L.; Sundseth, S.S.; Winegar, D.A.; Blanchard, D.E.; Spencer, T.A.; et al. Activation of the nuclear receptor LXR by oxysterols defines a new hormone response pathway. *J. Biol. Chem.* **1997**, *272*, 3137–3140. [CrossRef]
11. Song, C.; Liao, S. Cholestenoic acid is a naturally occurring ligand for liver X receptor alpha. *Endocrinology* **2000**, *141*, 4180–4184. [CrossRef] [PubMed]
12. Ogundare, M.; Theofilopoulos, S.; Lockhart, A.; Hall, L.J.; Arenas, E.; Sjovall, J.; Brenton, A.G.; Wang, Y.; Griffiths, W.J. Cerebrospinal fluid steroidomics: Are bioactive bile acids present in brain? *J. Biol. Chem.* **2010**, *285*, 4666–4679. [CrossRef] [PubMed]
13. Theofilopoulos, S.; Griffiths, W.J.; Crick, P.J.; Yang, S.; Meljon, A.; Ogundare, M.; Kitambi, S.S.; Lockhart, A.; Tuschl, K.; Clayton, P.T.; et al. Cholestenoic acids regulate motor neuron survival via liver X receptors. *J. Clin. Investig.* **2014**, *124*, 4829–4842. [CrossRef] [PubMed]
14. Nishimaki-Mogami, T.; Une, M.; Fujino, T.; Sato, Y.; Tamehiro, N.; Kawahara, Y.; Shudo, K.; Inoue, K. Identification of intermediates in the bile acid synthetic pathway as ligands for the farnesoid X receptor. *J. Lipid Res.* **2004**, *45*, 1538–1545. [CrossRef] [PubMed]
15. Goodwin, B.; Gauthier, K.C.; Umetani, M.; Watson, M.A.; Lochansky, M.I.; Collins, J.L.; Leitersdorf, E.; Mangelsdorf, D.J.; Kliewer, S.A.; Repa, J.J. Identification of bile acid precursors as endogenous ligands for the nuclear xenobiotic pregnane X receptor. *Proc. Natl. Acad. Sci. USA* **2003**, *100*, 223–228. [CrossRef]
16. Dussault, I.; Yoo, H.D.; Lin, M.; Wang, E.; Fan, M.; Batta, A.K.; Salen, G.; Erickson, S.K.; Forman, B.M. Identification of an endogenous ligand that activates pregnane X receptor-mediated sterol clearance. *Proc. Natl. Acad. Sci. USA* **2003**, *100*, 833–838. [CrossRef]
17. Soroosh, P.; Wu, J.; Xue, X.; Song, J.; Sutton, S.W.; Sablad, M.; Yu, J.; Nelen, M.I.; Liu, X.; Castro, G.; et al. Oxysterols are agonist ligands of RORgammat and drive Th17 cell differentiation. *Proc. Natl. Acad. Sci. USA* **2014**, *111*, 12163–12168. [CrossRef]
18. DuSell, C.D.; Umetani, M.; Shaul, P.W.; Mangelsdorf, D.J.; McDonnell, D.P. 27-hydroxycholesterol is an endogenous selective estrogen receptor modulator. *Mol. Endocrinol.* **2008**, *22*, 65–77. [CrossRef]
19. Liu, C.; Yang, X.V.; Wu, J.; Kuei, C.; Mani, N.S.; Zhang, L.; Yu, J.; Sutton, S.W.; Qin, N.; Banie, H.; et al. Oxysterols direct B-cell migration through EBI2. *Nature* **2011**, *475*, 519–523. [CrossRef]
20. Hannedouche, S.; Zhang, J.; Yi, T.; Shen, W.; Nguyen, D.; Pereira, J.P.; Guerini, D.; Baumgarten, B.U.; Roggo, S.; Wen, B.; et al. Oxysterols direct immune cell migration via EBI2. *Nature* **2011**, *475*, 524–527. [CrossRef]

21. Myers, B.R.; Sever, N.; Chong, Y.C.; Kim, J.; Belani, J.D.; Rychnovsky, S.; Bazan, J.F.; Beachy, P.A. Hedgehog pathway modulation by multiple lipid binding sites on the smoothened effector of signal response. *Dev. Cell* **2013**, *26*, 346–357. [CrossRef] [PubMed]
22. Raleigh, D.R.; Sever, N.; Choksi, P.K.; Sigg, M.A.; Hines, K.M.; Thompson, B.M.; Elnatan, D.; Jaishankar, P.; Bisignano, P.; Garcia-Gonzalo, F.R.; et al. Cilia-Associated Oxysterols Activate Smoothened. *Mol. Cell* **2018**, *72*, 316–327 e5. [CrossRef] [PubMed]
23. Radhakrishnan, A.; Ikeda, Y.; Kwon, H.J.; Brown, M.S.; Goldstein, J.L. Sterol-regulated transport of SREBPs from endoplasmic reticulum to Golgi: Oxysterols block transport by binding to Insig. *Proc. Natl. Acad. Sci. USA* **2007**, *104*, 6511–6518. [CrossRef] [PubMed]
24. Bjorkhem, I. Five decades with oxysterols. *Biochimie* **2013**, *95*, 448–454. [CrossRef] [PubMed]
25. Dzeletovic, S.; Breuer, O.; Lund, E.; Diczfalusy, U. Determination of cholesterol oxidation products in human plasma by isotope dilution-mass spectrometry. *Anal. Biochem.* **1995**, *225*, 73–80. [CrossRef] [PubMed]
26. Griffiths, W.J.; Sjovall, J. Analytical strategies for characterization of bile acid and oxysterol metabolomes. *Biochem. Biophys. Res. Commun.* **2010**, *396*, 80–84. [CrossRef] [PubMed]
27. Heubi, J.E.; Setchell, K.D.; Bove, K.E. Inborn errors of bile acid metabolism. *Semin. Liver Dis.* **2007**, *27*, 282–294. [CrossRef]
28. Clayton, P.T. Disorders of bile acid synthesis. *J. Inherit. Metab. Dis.* **2011**, *34*, 593–604. [CrossRef]
29. Vaz, F.M.; Ferdinandusse, S. Bile acid analysis in human disorders of bile acid biosynthesis. *Mol. Asp. Med.* **2017**, *56*, 10–24. [CrossRef]
30. Quehenberger, O.; Armando, A.M.; Brown, A.H.; Milne, S.B.; Myers, D.S.; Merrill, A.H.; Bandyopadhyay, S.; Jones, K.N.; Kelly, S.; Shaner, R.L.; et al. Lipidomics reveals a remarkable diversity of lipids in human plasma. *J. Lipid Res.* **2010**, *51*, 3299–3305. [CrossRef]
31. Bowden, J.A.; Heckert, A.; Ulmer, C.Z.; Jones, C.M.; Koelmel, J.P.; Abdullah, L.; Ahonen, L.; Alnouti, Y.; Armando, A.M.; Asara, J.M.; et al. Harmonizing lipidomics: NIST interlaboratory comparison exercise for lipidomics using SRM 1950-Metabolites in Frozen Human Plasma. *J. Lipid Res.* **2017**, *58*, 2275–2288. [CrossRef] [PubMed]
32. Liu, S.; Sjovall, J.; Griffiths, W.J. Analysis of oxosteroids by nano-electrospray mass spectrometry of their oximes. *Rapid. Commun. Mass. Spectrom.* **2000**, *14*, 390–400. [CrossRef]
33. Johnson, D.W.; ten Brink, H.J.; Jakobs, C. A rapid screening procedure for cholesterol and dehydrocholesterol by electrospray ionization tandem mass spectrometry. *J. Lipid Res.* **2001**, *42*, 1699–1705. [PubMed]
34. Jiang, X.; Ory, D.S.; Han, X. Characterization of oxysterols by electrospray ionization tandem mass spectrometry after one-step derivatization with dimethylglycine. *Rapid. Commun. Mass. Spectrom.* **2007**, *21*, 141–152. [CrossRef]
35. Honda, A.; Yamashita, K.; Hara, T.; Ikegami, T.; Miyazaki, T.; Shirai, M.; Xu, G.; Numazawa, M.; Matsuzaki, Y. Highly sensitive quantification of key regulatory oxysterols in biological samples by LC-ESI-MS/MS. *J. Lipid Res.* **2009**, *50*, 350–357. [CrossRef] [PubMed]
36. Sidhu, R.; Jiang, H.; Farhat, N.Y.; Carrillo-Carrasco, N.; Woolery, M.; Ottinger, E.; Porter, F.D.; Schaffer, J.E.; Ory, D.S.; Jiang, X. A validated LC-MS/MS assay for quantification of 24(S)-hydroxycholesterol in plasma and cerebrospinal fluid. *J. Lipid Res.* **2015**, *56*, 1222–1233. [CrossRef]
37. Faqehi, A.M.M.; Cobice, D.F.; Naredo, G.; Mak, T.C.S.; Upreti, R.; Gibb, F.W.; Beckett, G.J.; Walker, B.R.; Homer, N.Z.M.; Andrew, R. Derivatization of estrogens enhances specificity and sensitivity of analysis of human plasma and serum by liquid chromatography tandem mass spectrometry. *Talanta* **2016**, *151*, 148–156. [CrossRef]
38. Johnson, D.W.; ten Brink, H.J.; Schuit, R.C.; Jakobs, C. Rapid and quantitative analysis of unconjugated C(27) bile acids in plasma and blood samples by tandem mass spectrometry. *J. Lipid Res.* **2001**, *42*, 9–16.
39. DeBarber, A.E.; Luo, J.; Giugliani, R.; Souza, C.F.; Chiang, J.P.; Merkens, L.S.; Pappu, A.S.; Steiner, R.D. A useful multi-analyte blood test for cerebrotendinous xanthomatosis. *Clin. Biochem.* **2014**, *47*, 860–863. [CrossRef]
40. Griffiths, W.J.; Wang, Y.; Alvelius, G.; Liu, S.; Bodin, K.; Sjovall, J. Analysis of oxysterols by electrospray tandem mass spectrometry. *J. Am. Soc. Mass. Spectrom.* **2006**, *17*, 341–362. [CrossRef]
41. Karu, K.; Hornshaw, M.; Woffendin, G.; Bodin, K.; Hamberg, M.; Alvelius, G.; Sjovall, J.; Turton, J.; Wang, Y.; Griffiths, W.J. Liquid chromatography-mass spectrometry utilizing multi-stage fragmentation for the identification of oxysterols. *J. Lipid Res.* **2007**, *48*, 976–987. [CrossRef]

42. Ali, F.; Zakkar, M.; Karu, K.; Lidington, E.A.; Hamdulay, S.S.; Boyle, J.J.; Zloh, M.; Bauer, A.; Haskard, D.O.; Evans, P.C.; et al. Induction of the cytoprotective enzyme heme oxygenase-1 by statins is enhanced in vascular endothelium exposed to laminar shear stress and impaired by disturbed flow. *J. Biol. Chem.* **2009**, *284*, 18882–18892. [CrossRef] [PubMed]
43. Roberg-Larsen, H.; Strand, M.F.; Grimsmo, A.; Olsen, P.A.; Dembinski, J.L.; Rise, F.; Lundanes, E.; Greibrokk, T.; Krauss, S.; Wilson, S.R. High sensitivity measurements of active oxysterols with automated filtration/filter backflush-solid phase extraction-liquid chromatography-mass spectrometry. *J. Chromatogr. A* **2012**, *1255*, 291–297. [CrossRef] [PubMed]
44. Griffiths, W.J.; Crick, P.J.; Wang, Y.; Ogundare, M.; Tuschl, K.; Morris, A.A.; Bigger, B.W.; Clayton, P.T.; Wang, Y. Analytical strategies for characterization of oxysterol lipidomes: Liver X receptor ligands in plasma. *Free Radic Biol. Med.* **2013**, *59*, 69–84. [CrossRef] [PubMed]
45. Crick, P.J.; William Bentley, T.; Abdel-Khalik, J.; Matthews, I.; Clayton, P.T.; Morris, A.A.; Bigger, B.W.; Zerbinati, C.; Tritapepe, L.; Iuliano, L.; et al. Quantitative charge-tags for sterol and oxysterol analysis. *Clin. Chem.* **2015**, *61*, 400–411. [CrossRef] [PubMed]
46. Crick, P.J.; Bentley, T.W.; Wang, Y.; Griffiths, W.J. Revised sample preparation for the analysis of oxysterols by enzyme-assisted derivatisation for sterol analysis (EADSA). *Anal. Bioanal. Chem.* **2015**, *407*, 5235–5239. [CrossRef]
47. Soncini, M.; Corna, G.; Moresco, M.; Coltella, N.; Restuccia, U.; Maggioni, D.; Raccosta, L.; Lin, C.Y.; Invernizzi, F.; Crocchiolo, R.; et al. 24-Hydroxycholesterol participates in pancreatic neuroendocrine tumor development. *Proc. Natl. Acad. Sci. USA* **2016**, *113*, E6219–E6227. [CrossRef]
48. Une, M.; Harada, J.; Mikami, T.; Hoshita, T. High-performance liquid chromatographic separation of ultraviolet-absorbing bile alcohol derivatives. *J. Chromatogr. B Biomed. Appl.* **1996**, *682*, 157–161. [CrossRef]
49. Wheeler, O.H. The Girard Reagents. *Chem. Educ.* **1968**, *45*, 435. [CrossRef]
50. Griffiths, W.J.; Liu, S.; Alvelius, G.; Sjovall, J. Derivatisation for the characterisation of neutral oxosteroids by electrospray and matrix-assisted laser desorption/ionisation tandem mass spectrometry: The Girard P derivative. *Rapid. Commun. Mass. Spectrom.* **2003**, *17*, 924–935. [CrossRef]
51. Theofilopoulos, S.; Wang, Y.; Kitambi, S.S.; Sacchetti, P.; Sousa, K.M.; Bodin, K.; Kirk, J.; Salto, C.; Gustafsson, M.; Toledo, E.M.; et al. Brain endogenous liver X receptor ligands selectively promote midbrain neurogenesis. *Nat. Chem. Biol.* **2013**, *9*, 126–133. [CrossRef] [PubMed]
52. Abdel-Khalik, J.; Yutuc, E.; Crick, P.J.; Gustafsson, J.A.; Warner, M.; Roman, G.; Talbot, K.; Gray, E.; Griffiths, W.J.; Turner, M.R.; et al. Defective cholesterol metabolism in amyotrophic lateral sclerosis. *J. Lipid Res.* **2017**, *58*, 267–278. [CrossRef] [PubMed]
53. Griffiths, W.J. Tandem mass spectrometry in the study of fatty acids, bile acids, and steroids. *Mass. Spectrom. Rev.* **2003**, *22*, 81–152. [CrossRef]
54. Vilarinho, S.; Sari, S.; Mazzacuva, F.; Bilguvar, K.; Esendagli-Yilmaz, G.; Jain, D.; Akyol, G.; Dalgic, B.; Gunel, M.; Clayton, P.T.; et al. ACOX2 deficiency: A disorder of bile acid synthesis with transaminase elevation, liver fibrosis, ataxia, and cognitive impairment. *Proc. Natl. Acad. Sci. USA* **2016**, *113*, 11289–11293. [CrossRef] [PubMed]
55. Jiang, X.; Sidhu, R.; Mydock-McGrane, L.; Hsu, F.F.; Covey, D.F.; Scherrer, D.E.; Earley, B.; Gale, S.E.; Farhat, N.Y.; Porter, F.D.; et al. Development of a bile acid-based newborn screen for Niemann-Pick disease type C. *Sci. Transl. Med.* **2016**, *8*, 337ra63. [CrossRef] [PubMed]
56. Fang, N.; Yu, S.; Adams, S.H.; Ronis, M.J.; Badger, T.M. Profiling of urinary bile acids in piglets by a combination of enzymatic deconjugation and targeted LC-MRM-MS. *J. Lipid Res.* **2016**, *57*, 1917–1933. [CrossRef] [PubMed]
57. Bjorkhem, I. Cerebrotendinous xanthomatosis. *Curr. Opin. Lipidol.* **2013**, *24*, 283–287. [CrossRef] [PubMed]
58. Honda, A.; Salen, G.; Matsuzaki, Y.; Batta, A.K.; Xu, G.; Leitersdorf, E.; Tint, G.S.; Erickson, S.K.; Tanaka, N.; Shefer, S. Side chain hydroxylations in bile acid biosynthesis catalyzed by CYP3A are markedly up-regulated in Cyp27-/- mice but not in cerebrotendinous xanthomatosis. *J. Biol. Chem.* **2001**, *276*, 34579–34585. [CrossRef]
59. Honda, A.; Salen, G.; Matsuzaki, Y.; Batta, A.K.; Xu, G.; Leitersdorf, E.; Tint, G.S.; Erickson, S.K.; Tanaka, N.; Shefer, S. Differences in hepatic levels of intermediates in bile acid biosynthesis between Cyp27(-/-) mice and CTX. *J. Lipid Res.* **2001**, *42*, 291–300.

60. Duane, W.C.; Pooler, P.A.; Hamilton, J.N. Bile acid synthesis in man. In vivo activity of the 25-hydroxylation pathway. *J. Clin. Investig.* **1988**, *82*, 82–85. [CrossRef]
61. Shoda, J.; Axelson, M.; Sjovall, J. Synthesis of potential C27-intermediates in bile acid biosynthesis and their deuterium-labeled analogs. *Steroids* **1993**, *58*, 119–125. [CrossRef]

Sample Availability: Samples of the compounds C-3α,7α,12α-triol, C-3α,7α,12α,25-tetrol, C-3α,7α,12α,26-tetrol, C-3α,7α,12α,24R,25-pentol, C-3α,7α,12α,24S,25-pentol, C-3α,7α,12α,25,26-pentol, C-3α,7α,12α,26,27-pentol and C-3α,7α,12α,23,25-pentol are available from the authors.

MDPI

Review

Microbial Sterolomics as a Chemical Biology Tool

Brad A. Haubrich

Department of Chemistry, University of Nevada, Reno, Reno, NV 89557, USA; bhaubrich@unr.edu;
Tel.: +1-775-784-4857

Received: 5 October 2018; Accepted: 23 October 2018; Published: 25 October 2018

Abstract: Metabolomics has become a powerful tool in chemical biology. Profiling the human sterolome has resulted in the discovery of noncanonical sterols, including oxysterols and meiosis-activating sterols. They are important to immune responses and development, and have been reviewed extensively. The triterpenoid metabolite fusidic acid has developed clinical relevance, and many steroidal metabolites from microbial sources possess varying bioactivities. Beyond the prospect of pharmacognostical agents, the profiling of minor metabolites can provide insight into an organism's biosynthesis and phylogeny, as well as inform drug discovery about infectious diseases. This review aims to highlight recent discoveries from detailed sterolomic profiling in microorganisms and their phylogenic and pharmacological implications.

Keywords: algal sterols; ergosterol biosynthesis; infectious disease; lipidomics; oxyphytosterol; pharmacognosy; phytosterol; sterolomics

---

1. Introduction

Sterols, like cholesterol **1**, ergosterol **2**, and sitosterol **3**, as well as secondary metabolites, are amphipathic lipids that contain a 1,2-cyclopentanoperhydrophenanthrene ring nucleus (Figure 1). Sterols are ubiquitous molecules found in all eukaryotic life, serving a multitude of crucial biological functions [1]. Some prokaryotes synthesize sterols as well, and some prokaryotes contain enzymes with incomplete Δ^5 sterol biosynthesis [1–5]. While sterol biosynthesis may predate eukaryotes [6], it is often hypothesized that aside from the protomitochondrial lineage, most bacteria have gained these genes via lateral gene transfer [3,4]. The end product of Δ^5 sterols such as cholesterol **1** and ergosterol **2** (Figure 1a) contribute to cell membrane fluidity in their bulk insert role in mammals and fungi, respectively [1,7]. Steroidal secondary metabolites of the steroid hormone and bile acid classes serve well-known important roles in inflammation, sex characteristics, and lipid absorption [7].

Minor components within the human sterol metabolome, which serve unusual but essential functions, have also been identified. For instance, the meiosis-activating sterols (MASs) 4,4-dimethylcholesta-8(9),14(15),24-trienol (follicular fluid meiosis-activating sterol; FF-MAS) **4** and 4,4-dimethylcholesta-8(9),24-dienol (testicular meiosis-activating sterol; T-MAS) **5** are biosynthetic intermediates in the cholesterol pathway that signal meiosis in mammalian oocytes and spermatozoa [7]. Various minor metabolites occurring both upstream and downstream of cholesterol have been demonstrated as ligands for nuclear hormone receptors and play critical roles in development and immunology, including 25-hydroxycholesterol **6** (Figure 1b) [8–11]. Recent advances in methodologies in lipidomics have expedited discoveries with regard to these necessary minor human sterols and steroids, as well as provided new diagnostic screens for patients with dysregulated sterol biosynthesis, as in Niemann-Pick and Smith-Lemli-Opitz syndrome. Contemporary discoveries in human sterolomics [11–14], as well as plant sterolomics [15], have been reviewed extensively elsewhere.

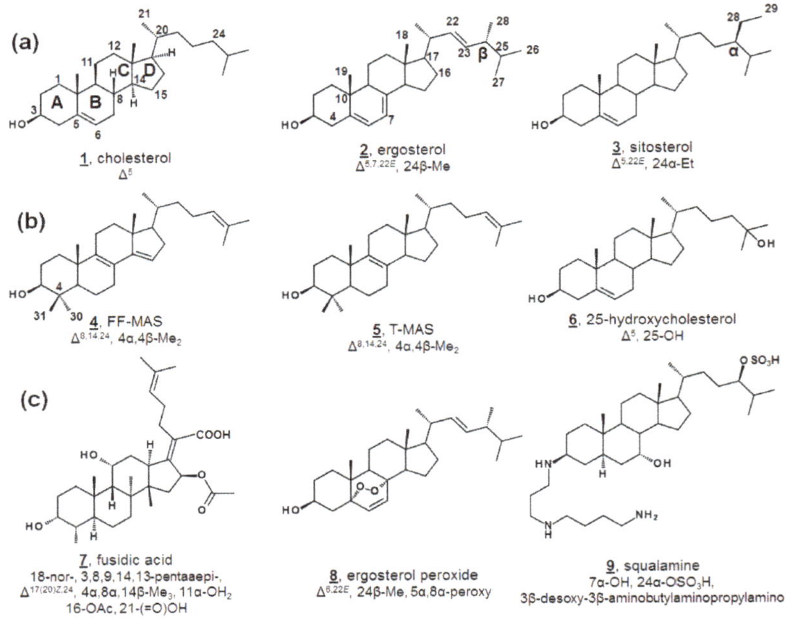

Figure 1. Structure and numbering systems of sterols and steroids. (**a**) Δ^5 end product inserts from mammals, fungi, and vascular plants, respectively, cholesterol 1, ergosterol 2, and sitosterol 3. (**b**) Examples of steroidal metabolites important in human biology for F-MAS 4, TT-MAS 5, 25-hydroxycholesterol 6. (**c**) Examples of steroidal metabolites from nonhuman sources with bioactivity, fusidic acid 7, ergosterol peroxide 8, and squalamine 9. The numbering system shown here, and used in this manuscript, is the conventional system [1]. Designations of α and β within the sterol nucleus signify below and above the plane. Unrelated to nucleus α and β, substituents on C24 are also designated α and β to reflect the C24 stereochemistries of sitosterol and ergosterol, respectively, as drawn above. Carbon numbering is provided on 1–4, and stereochemistries at C8, C9, C14, and C16 on structure 1 are hereafter implied on structures, unless otherwise annotated as in fusidic acid. Molecular features for each structure are provided relative to 5α-cholestanol for clarity. For a complete list of systematic names of compounds, see Table A1.

Metabolites can also be used to classify organisms and explore evolutionary relationships. Sterol distribution has long been used for chemotaxonomic purposes in plants [16], fungi [17–19], and other microorganisms [20–22]. There is also potential for sterols to serve as biomarkers, and sterol composition can play a role in feedstocks.

Small molecule ligands for ergosterol biosynthetic enzymes in fungi have long been clinically and agriculturally relevant [23–25]. Marketed antimycotics include molecules in such classes as allylamines, which target squalene epoxidase (SqE); azoles, which target sterol C14-demethylase (14-SDM = CYP51, =Erg11p in fungi); and morpholines, which target both sterol C14-reductase (14-SR, =Erg3p in fungi) and sterol C8(7)-isomerase (8(7)-SI, =Erg2p in fungi) (Figure 2) [24]. There is further interest in the design and discovery of inhibitors of other sterol enzymes, particularly sterol C24-methyltransferase (24-SMT, =Erg6p in fungi), which is absent from humans' cholesterol biosynthesis [1,25,26]. Understanding sterol biosynthesis in non-fungal microbes may provide new insights for treating infections by eukaryotic pathogens.

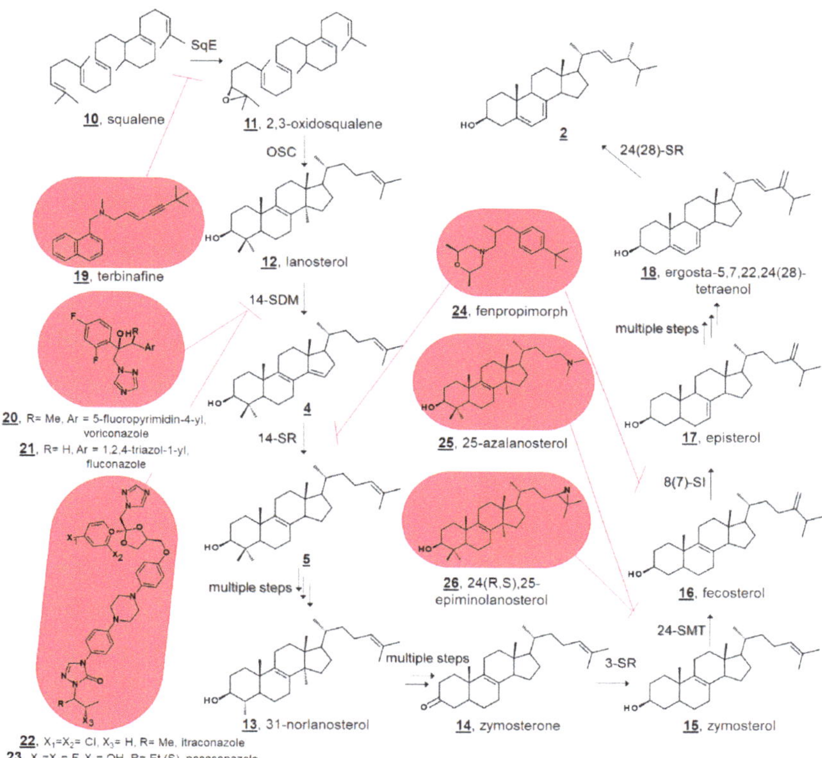

Figure 2. Truncated hypothetical pathway of fungal ergosterol 2 biosynthesis from squalene 10. Inhibitor targets of squalene epoxidase (SqE) by allylamines, e.g., terbinafine 19, sterol C14-demethylase (14-SDM = CYP51) by azoles, e.g., voriconazole 20, fluconazole 21, itraconazole 22, and posaconazole 23, sterol C14-reductase (14-SR) and sterol C8(7)-isomerase (8(7)-SI) by morpholines, e.g., fenpropimorph 24, and sterol C24-methyltransferase (24-SMT) by 25-azalanosterol 25 or 24(R,S),25-epiminolanosterol 26 are highlighted at the biosynthetic steps they block. 3-SR; sterol C3 reductase, 24-SR, sterol C24 reductase.

Novel metabolites isolated from microbial sources are conversely often found to exhibit biological activity. Famously, fusidic acid 7 (Figure 1c), originally isolated from fungal *Fusidium* spp., is a tetracyclic triterpene antibacterial and has been used in the clinic for decades [27–29]. Fusidic acid inhibits growth by restricting protein synthesis via elongation factor G in Gram-positive bacteria, including *Streptococcus* spp., *Clostridium* spp., and penicillin-resistant strains of *Staphylococcus* spp. [28,29]. Structural analogues of fusidic acid, have shown varying antimicrobial, as well as anticholesterolemic and antineoplastic, characteristics [29]. Isolated from a variety of fungi and sponges, as well as vascular plants, ergosterol peroxide 8 possesses broad bioactivity, including anti-tumor, immunomodulatory, inhibitory hemolytic, anti-inflammatory, antioxidant, and antimicrobial properties. Several other endoperoxides of other phytosterols and of cholestenols have been reported to have similar properties, as well [30–34]. Squalamine 9 is a non-microbially derived natural steroidal, which has demonstrated antimicrobial and antiangiogenic properties and has led to interest in synthetic analogues for structure-activity improvement [35].

This short review aims to highlight new findings in microbial sterolomics, with respect to phylogeny, ecology, biosynthesis for drug discovery, and discovery of bioactive metabolites.

2. Phylogenic and Ecological Insights

2.1. Algal Phytosterol Biosynthesis

Ergosterol **2**, having long been considered the "fungal sterol", is nevertheless present in every major eukaryotic kingdom [1]. Ergosterol is present in amoebae [21,22,36,37] and trypanosomatids [38–44], and ergosterol is a major sterol of many taxa within green algae [20,45–48]. The unicellular green alga model organism *Chlamydomonas reinhardtii* uses ergosterol and its 24-ethyl analogue, 7-dehydroporiferasterol **35**, as its main Δ^5 sterols [45,46]. Vascular plants, on the other hand, chiefly use campesterol **36** and sitosterol **3** as Δ^5 membrane inserts (Figure 3) [1,49]. Ergosterol and 7-dehydroporiferasterol differ from campesterol and sitosterol by units of unsaturation (double bonds) in the sterol nucleus and side chain, as well as stereochemistry at C24. While all four compounds possess 24*R* stereochemistry, 24-alkylation of ergosterol and 7-dehydroporiferasterol has β-stereochemistry (alkyl groups behind the plane, as drawn), while 24-alkylation of campesterol and sitosterol has α-stereochemistry (above the plane, as drawn) [1,45]. Conversely, the green alga synthesizes sterol from the photosynthetic protosterol. Fungi (nonphotosynthetic lineage) cyclize 2,3-oxidosqualene to lanosterol (Figure 2), while higher plants, and green algae, (photosynthetic lineage) cyclize 2,3-oxidosqualene to the plant protosterol cycloartenol [1,45].

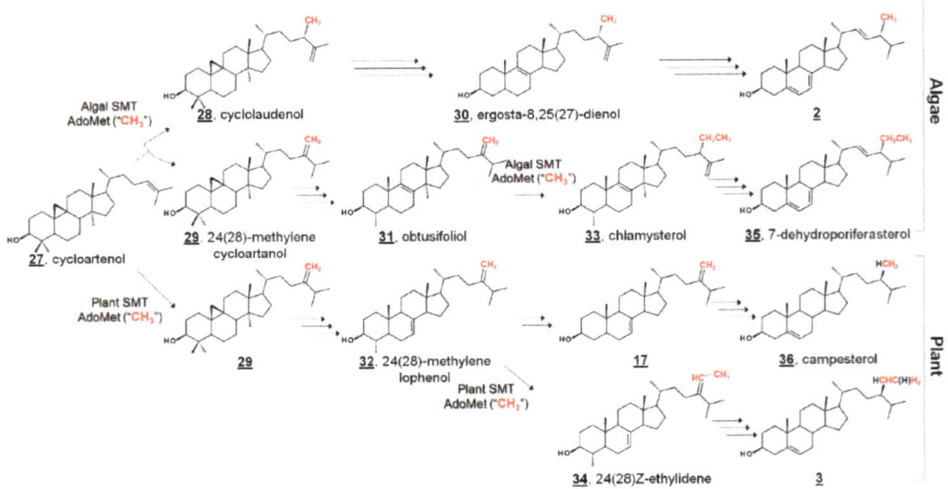

Figure 3. Comparative phytosterol biosynthesis in the photosynthetic lineage from the protosterol cycloartenol **27**. In algae, 24-methyl and 24-ethyl sterols arise from a bifurcation of products of biomethylation by sterol methyltransferase (SMT); In higher plants, they arise from alternate pathways from the intermediate 24(28)-methylene lophenol **30**, which can be methylated again or metabolized to campesterol **36**. Red methyl groups from SMT co-substrate S-adenosyl methionine (AdoMet) are annotated to show hypothetical labeling patterns of Δ^5 sterols as discussed in [45,50]. An additional 15 algal sterols were reported in [45]. Truncated fungal phytosterol biosynthesis from protosterol lanosterol **12** is illustrated in Figure 2.

In *C. reinhardtii*, the biochemical pathway from the "plant" protosterol cycloartenol to the "fungal" Δ^5 end product was investigated by sterolomic experiments of *C. reinhardtii* cultures. Sterol profiling of wild-type, mutant, and inhibitor-treated cultures revealed an additional 21 sterols beyond cycloartenol, ergosterol, and 7-dehydroporiferasterol **33** [45] (Figure 3). *C. reinhardtii* cultures that were not treated with a 24-SMT inhibitor contained only cycloartenol and 24-alkylsterols, indicating that bioalkylation and introduction of C28 by algal 24-SMT occurs upon cycloartenol itself early in the pathway. Further, 24-methylated cycloartenols were 24β-methylcycloart-25(27)-enol (cyclolaudenol) **28** and 24(28)-methylenecycloartanol **29**, signifying a bifurcation of methylated products of algal 24-SMT [45]. The presence of C29 (i.e., a 24-ethyl group) on a 4α,14α-dimethyl sterol **33** led to the identification of obtusifoliol **31** as the substrate for the second biomethylation reaction of the algal sterol side chain, different from the substrate preference in higher plants (Figure 3). Furthermore, the alkylation product in plants has a 24-ethylene substituent, whereas the product in *C. reinhardtii* was found to bear a 24β-ethyl group with desaturation at C25 [45].

This pathway delineates algal biosynthesis of ergosterol disparate from the fungal pathway. In the former $\Delta^{25(27)}$-olefin pathway, *C. reinhardtii* alkylates sterols at C24 in a bifurcated manner to $\Delta^{25(27)}$-olefin and $\Delta^{24(28)}$-olefin products. $\Delta^{24(28)}$-Olefin products are further metabolized and later alkylated at C28 to only 24β-ethyl-$\Delta^{25(27)}$-olefin products. Conversely, fungal bioalkylation of C24 yields only $\Delta^{24(28)}$-olefin products, which are reduced to eventually yield ergosterol. That is, the stereochemistry of C24 in algal ergosterol arises from the methylation steps, whereas the stereochemistry of C24 in fungal ergosterol arises from a successive reduction step [45]. The $\Delta^{25(27)}$-olefin pathway was confirmed by sterol profiling of cultures incubated with isotopically labeled [methyl-$^2H_3$]methionine ([$^2H_3$]Met). These algal cultures incorporated three and five deuterium atoms into ergosterol and 7-dehydroporfierasterol, respectively [45].

The algal pathway was further corroborated by characterization of recombinant *C. reinhardtii* 24-SMT, found to catalyze the methylation of C24 by introduction of C28 and the methylation of C28 with C29. *C. reinhardtii* 24-SMT favored cycloartenol as a substrate, and a bifurcation of products to cyclolaudenol **28** and 24(28)-methylenecycloartanol **29** was found in ratios comparable to in vivo ratios of ergosterol and 7-dehydroporiferasterol [50]. A switch to $\Delta^{25(27)}$-olefin "algal" products of fungal or plant 24-SMT has been noted upon mutagenesis or incubation with electronically modified substrates [49,51]. In addition, obtusifoliol was found to be a substrate for the second methyltransfer of *C. reinhardtii* 24-SMT, 24β-methyl-$\Delta^{25(27)}$-sterols were not substrates, and incubation with [methyl-$^2H_3$]S-adenosyl methionine ($^2H_3$-AdoMet) produced labeled products with three and five deuterium atoms [50].

Green algae from the *Acicularia* spp. and *Acetabularia* spp. are macroscopic, yet unicellular. With a long and uninterrupted fossil record, they are often used to provide insight into the evolution of green algae and plants. Δ^5-Bulk sterols of these genera lack Δ^7 desaturations, in contrast to Chlamydomonas. Trimethylsilylated (TMS) sterols extracted from *Acicularia schenckii* and four species of Acetabularia revealed a principal Δ^5 sterol (60–70%) of 24-ethylcholesterol (24α/24R = sitosterol **3**, 24β/24S = clionosterol **37**). Four other minor Δ^5 sterols occurred in all five species: 24-methylcholesterol (24α/24R = campesterol **36**, 24β/24S = 22-dihydrobrassicasterol **38**), 24-ethylcholesta-5,22*E*-dienol (24α/24S = stigmasterol **39**, 24β/24R = poriferasterol **40**), 24-methylcholesta-5,22*E*-dienol (24α/24S = crinosterol **41**, 24β/24R = brassicasterol **42**), and 24-ethylidenecholesterol, which was tentatively assigned by the authors as the $\Delta^{24(28)Z}$ isomer = isofucosterol **44**. Among the TMS-derivatized sterols of *Acetabularia caliculus*, 24-ethylcholest-7-enol **46**/**47** was identified (Figure 4, Table 1) [52]. Prior studies had also identified cholesterol and 24-methylenecholesterol **45** in cultures of *Acetabularia mediterranea*, suggested by the authors to potentially be a result of differences in algal cultivation. *Acetabularia caliculus* also contained 24-ethylcholesterol in the sterol ester fraction, while the other Acetabularia species and *Acicularia schenckii* did not contain sterols in the ester fraction. These nearly identical sets of sterols from the five species, with a large separation in their geographical origin, illustrate a lack of divergence in sterol composition. It was thus hypothesized that these sterols represent an ancient biochemical trait within the photosynthetic lineage [52].

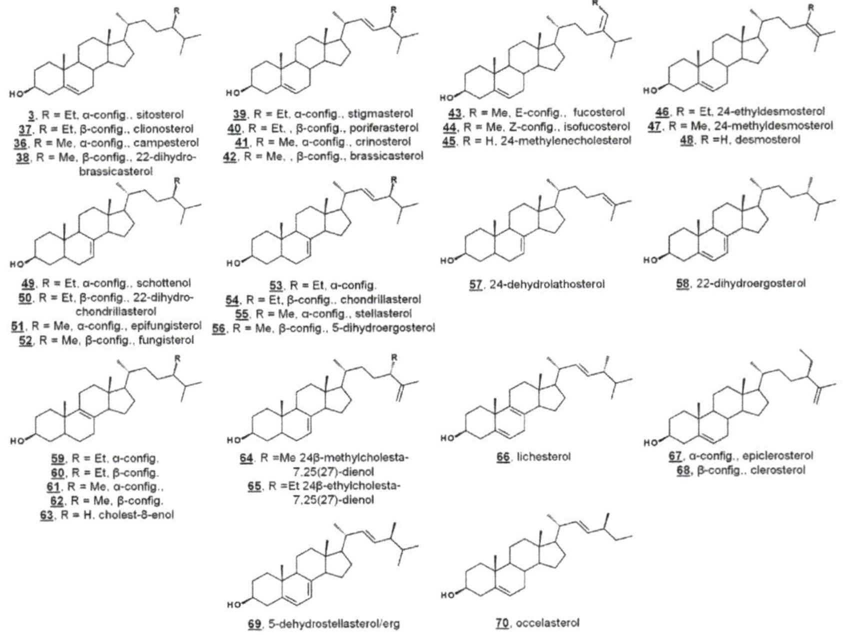

Figure 4. Molecular structures of algal sterols.

Table 1. Recently reported sterol profiles from algae across classes.

Algal Organism	Major Sterols [1] (>40%)	Semi-Major Sterols [1] (>20%)	Minor Sterols [1] (<20%)	Reference [2]
		Ulvophyceae		
Acetabularia caliculus	(3/37)		(36/38), (39/40), (41/42), 44, (48/49)	[52]
Acicularia schenckii	(3/37)		(36/38), (39/40), (41/42), 44	[52]
		Trebouxiophyceae		
Chlorella vulgaris	2	52	56, 58, 64, 62, 66	[53]
Chlorella luteoviridis	40	38	37, 42, 52	[53]
		Eustigmatophyceae		
Nannochloropsis limnetica	1		44, (3/37), 45, (67/68)	[53]
		Bacillariophyceae (diatoms)		
Stephanodiscus hantzschii	45		48, 17, 57, 46	[53]
Gomphonema parvulum	41		(69/2), (59/60), (36,38)	[53]
Cyclotella meneghiniana	45		38, 43, 48	[53]
		Raphidophyceae		
Chloromorum toxicum	40		37, 1, 42, 38, 48	[54]
Chattonella marina	3		1, 63	[54]
Heterosigma akashiwo	37		38	[54]
		Dictyochophyceae		
Verrucophora farcimen	70			[54]
		Chlorophyceae (see also Figure 3)		
Scenedesmus obliquus	54	52	50, 62, 60	[53]
Monoraphidium mintutum	50	52	65, 54, 62, 60, 64	[53]
		Cryptophyceae		
Cryptomonas sp.	39, 41			[53]
Rhodomonas sp.	41			[55]

[1] Major, semi-major, and minor components of algal sterols as a percentage of total sterol. Numbers refer to structures in Figure 4 and earlier. Parenthetical pairs are provided for epimers, for which C24 stereochemistry was not reported. [2] Reference.

A study investigating the sterolome via free sterols and TMS derivatives from various classes of microalgae showed two species of the green algae Chlorella, *C. vulgaris* and *C. luteoviridis*, possessing different sterol profiles [53]. *C. vulgaris* contained chiefly ergosterol and fungisterol **52**. In the past, *C. vulgaris* has been reported to also contain 7-dehydroporiferasterol. The reported minor components included 5-dihydroergosterol **56**, 22-dihydroergosterol **58**, 24β-methylcholesta-7,25(27)-dienol **64**, 24β-methylcholest-8(9)-enol **62**, and lichesterol **66**. Conversely, the profile of *Chlorella luteoviridis* was dominated by poriferasterol **40** and 22-dihydrobrassicasterol **38**, with minor composition by clionasterol **35**, brassicasterol, and fungisterol [53]. The predominant sterol from *Nanochloropsis limnetica* was cholesterol, while its minor components were isofucosterol **44**, 24-ethylcholesterol (**3**/**37**), 24-methylenecholesterol **45**, and clerosterol **68** [53]. This report included the sterol profiling of several species of diatoms. The diatom *Stephanodiscus hantzschii*, whose sterols had not been studied prior to this report, had a composition of mostly 24-methylenecholesterol, with minor components of desmosterol **48**, 24-methylenelathosterol (Δ^7, rather than Δ^5, termed episterol above) **17**, and traces of two other sterols. Sterols from diatoms *Cyclotella meneghiniana* and *Gomphonema parvulum* were analyzed, with principal sterols of 24-methylenecholesterol and epibrassicasterol (called crinosterol, above; for list of trivial and systematic names, see Table A1) **41**, respectively. *C. meneghiniana* also contained desmosterol **48**, 24-methylenelathosterol, 24-dehydrolathosterol **57**, and 24-ethyldesmosterol, and *G. parvulum* contained 5-dehydrostellasterol/ergosterol **69**/**2**, 24α/β-ethylcholest-8(9)-enol **58**/**59**, and campesterol/22-dihydrobrassicasterol **36**/**38** (C24 alkyl group was presumably α-oriented) [53]. A brief list of recently reported algal sterols by taxonomic class is presented in Figure 4 and Table 1; for more comprehensive and historical lists, see Refs. [47,48].

2.2. Trophic and Limnological Sterols

In the cross-class algal study [53], the researchers presented these profiles, along with their quantification, as references to the algal sterolome. As prey, Δ^7 and $\Delta^{7,22}$ sterols are often nutritionally inadequate to invertebrate consumers [53,56,57]. Many invertebrates are auxotrophic for sterols and rely on diet to fulfill their sterol needs for cell membrane and hormonal requirements. Several of these specimens contain alternate enzymes, which dealkylate side chains of phytosterols, yet they lack the enzymes to desaturate C5–C6 or reduce C7–C8 (Figure 5) [57,58]. It has been proposed that these quantitated algal sterolome references can be used for studies involving the nutritional content of aquatic microorganisms for aquatic invertebrates [53]. Another study monitored the sterol profiles of an algal diet and the amphipod consumer *Gammarus roeselii*. Prey alga *N. limnetica*, rich in cholesterol, and alga *S. obliquus*, lacking cholesterol but rich in Δ^7 sterols (See Table 1), were fed to *G. roeselii*. The sterol profile of *S. obliquus*-fed *G. roeselii* decreased in cholesterol, and increased in the Δ^7 metabolite lathosterol **69**, detectable when the diet was 50% *S. obliquus* (Figure 5) [56].

Isotopically labeled sterolomic experiments have been used to explore trophic modifications by the Northern Bay scallop *Argopecten irradians irradians*. Dietary alga Rhodomonas was supplemented with sterols enriched with ^{13}C at the C22 position. The ^{13}C-label was noted on new sterol metabolites, including those newly desaturated with Δ^7 and those bearing an introduced 4α-methyl group. The mollusk's ability to synthesize cholesterol from food was noted to correlate to Δ^5 double bonds in the dietary sterols. They were more likely to dealkylate side chains possessing 24-ethyl groups. The only 24-methyl sterols dealkylated by *A. irradians* contained a $\Delta^{24(28)}$ olefin (i.e., 24-methylene, rather than 24-methyl) [55].

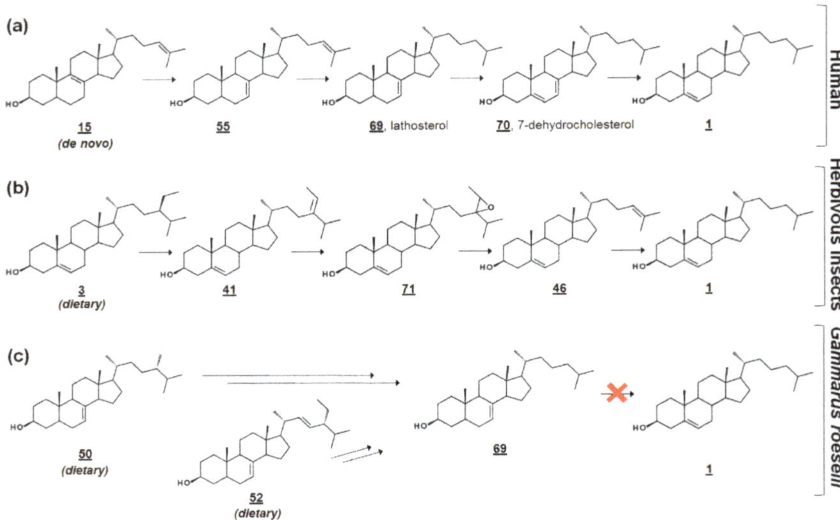

Figure 5. Comparative cholesterol biosynthesis between humans and arthropods. (**a**) Late-stage cholesterol biosynthesis in humans from *de novo* zymosterol **15**. (**b**) Proposed synthesis of cholesterol in herbivorous insects via dealkylation of dietary plant sterols (sitosterol) [58]. (**c**) Amphipod *Gammarus roeselii* can dealkylate the side chain of Δ^7 algal sterols, such as fungisterol and chondrillasterol, but cannot produce cholesterol [56].

A recent study investigated the lipid content of 37 strains within 10 classes of phytoplankton. Four classes, Cryptophyceae, Chlorophyceae, Treouciophyceae, and the diatoms are additionally represented in Table 1; this study additionally included dinoflagellates, euglenoids and the conjugatophyceae. Of the 37 strains, 29 sterols were detected, with notable variability of profile as a function of taxonomic class. The authors suggested $\Delta^{5,22}$ sterols as a potential biomarker for Chlorophyceae *Sphaerocystis* sp. and ergosterol as a potential biomarker for *Chlamydomonas* in habitats lacking other aquatic ergosterol-synthesizing microorganisms [59].

While sterol metabolites of toxic blooms are likely non-toxic to fish populations, these metabolites may have a stronger influence on marine invertebrates. Toxic bloom-causing algae *Chloromorum toxicum*, *Chattonella marina*, *Heterosigma akashiwo*, and *Verrucophora farcimen* [54] have sterol profiles given in Table 1. *Verrucophora* sp. were found to produce the rare 27-nor sterol occelasterol **68** (Figure 4) [54]. It has been proposed that isofucosterol **44** is a potential biomarker for the green-tide forming multicellular alga *Ulva prolifera*, and that dinosterol **74** and 24*Z*-propylidienecholesterol **75** are potential biomarkers for bloom-forming dinoflagellates [60] (Figure 6). Toxic bloom-causing dinoflagellate *Cochlodinium polykrikoides* had a sterol profile including prevalent sterols of dinosterol **74** (40%), dihydrodinosterol **76** (32%), and the rare 4α-methyl sterol amphisterol **77** (23%). Small amounts of 4-methylergost-24(28)-enol **78** (5%) were detected [61]. Two isolates of the bioluminescent dinoflagellate *Pyrodinium bahamense* had a sterol profile of largely cholesterol (74–75%), but also components of dinosterol **74** (13–14%) and 4α-methylgorgosterol **79** (11–13%), analyzed as their TMS derivatives. 4α-Methylgorgosterol is uncommon in dinoflagellates and has potential as a biomarker (Figure 6) [62].

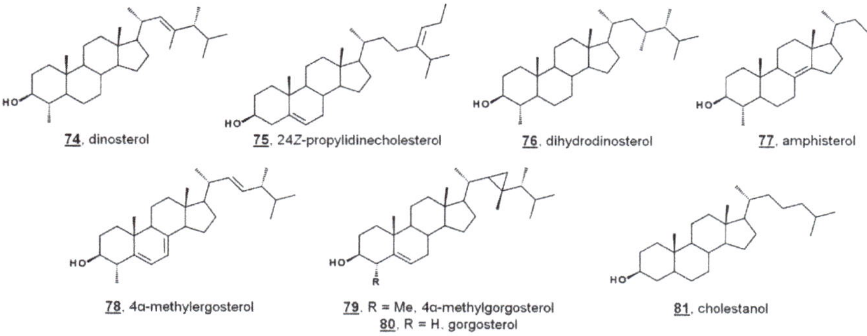

Figure 6. Sterol structures from various dinoflagellates.

Lipidomic study of the coral *Dendrophyllia cornigera* revealed a geographical correlation to diet. *D. cornigera* analyzed from the Cantabrian Sea in the Northeast Atlantic reflected a productive environment, and the coral contained a high diversity of phytosterols. *D. cornigera* sampled from the Menorca Channel in the Mediterranean had a lower sterol content per dry weight and had less phytosterols. The Mediterranean coral had a higher relative abundance of occelasterol **70**, brassicasterol **42**, and cholestanol **81**, or cholesterol and ergosterol, depending on the sample. The difference in the geographic profiles was attributed to a diet high in phytoplankton and herbivorous grazers in the Cantabrian coral, and a diet primarily consisting of dinoflagellates in the Mediterranean coral [63]. Specimens of the coral *Agaricia* spp. taken from shallow waters and deep waters were found to have markedly different sterol profiles from one another. From shallow Caribbean waters, *Agaricia* contained mostly cholesterol and 24-methylenecholesterol, with lower abundances of other phytosterols. Samples from deep waters contained mostly cholesterol and 24-ethylcholesterol. No gorgosterol was detected in either set. The Caribbean coral *Montastraea cavernosa* contained mostly 24-methylcholesterol, followed by cholesterol and gorgosterol, and variation in subsurface depth did not cause a significant change in sterol content. It was concluded that *Agaricia* spp. relies primarily on heterotrophy, even at greater depths [64].

3. Sterolome-Informed Antimicrobial Targets

3.1. *Trypanosoma brucei*

Trypanosomatids are flagellated protozoa, all of which are parasitic. Some examples from this clade are *Crithidia fasciculata*, solely parasitic to insect hosts, *Phytomonas serpens*, soley phytopathogenic, and a number of human pathogens, including *Trypanosoma cruzi*, *Leishmania* spp., and *Trypanosoma brucei*, which are the etiological agents of the following human diseases: leishmaniasis, Chagas' disease, and human African trypanosomiasis (also known as African sleeping sickness), respectively. Most of the species, *C. fasciculata* [38], *P. serpens* [40], *T. cruzi* [38,44], and *Leishmania* spp. [39] synthesize ergosterol and other 24β-methyl/24(28)-methylene-sterols (ergostenols) *de novo* as their Δ^5 end products. In light of this de novo biosynthesis, there has been interest in using ergosterol biosynthesis inhibitors (EBIs) to treat Chagas' disease and leishmaniasis, and some molecules have even progressed to the clinic [25,44]. *Trypanosoma brucei*, conversely, synthesizes ergostenols during its life cycle in the insect vector (procyclic form (PCF)), but uses largely cholesterol from the host's blood as its Δ^5 bulk sterol in the human host (bloodstream form (BSF)) [41–43].

In the fly vector, cholesterol comprises a significant portion of the PCF sterol content. The profile contains sterols endogenous to PCF *T. brucei*, including prominent cholesta-5,7,24-trienol **82** and ergosta-5,7,25(27)-trienol **85**. PCF synthesizes trace ergosterol **2**; Ergosta-5,7,24(28)-trienol **85** and ergosta-5,7,24(25)-trienol **84** comprise some of the minor compounds present [41–43] (Figure 7). 24,24-Dimethylcholesta-5,7,25(27)-trienol **86** has also been detected in PCF profiles [42].

Culturing PCF in lipid-depleted media yields a higher composition of endogenous ergostenols and cholesta-5,7,24-trienol **82** relative to cholesterol [42,43]. Treatment of PCF cells with the 24-SMT inhibitor 25-azalanosterol **25** causes an increase in cholestenols in the profile [43].

Sterolomic analysis of PCF revealed a novel biosynthetic network. For instance, T. brucei demethylates protosterol lanosterol **12** at C4 initially (Figure 7), compared to mammalian and fungal pathways demethylating C14 first (cf. Figure 2) [42,43]. Moreover, the side chain methylation patterns of 24-SMT to yield $\Delta^{24(28)}$, $\Delta^{25(27)}$, and $\Delta^{24(25)}$ products, as well as the $\Delta^{25(27)}$ 24,24-dimethyl product **86**, are unique [42,43,65]. Isotopic experiments with ^{13}C-labeled carbon sources leucine, acetate, and glucose were shown to produce variable labeling of Δ^5 endproducts and biosynthetic intermediates. No labelling was noted on cholesterol. Isotopic incorporation was higher with acetate and glucose. The variability of labeling was potentially attributed to the equilibrium of acetyl-CoA pools in the mitochondria and cytosol [42]. Trypanosomal sterols protothecasterol **87** (ergosta-5,7,22*E*,25(27)-tetraenol), cholesta-5,7,24-trienol **82**, and ergosta-5,7,25(27)-trienol **83** have also been noted to incorporate isotope labeled from threonine [66].

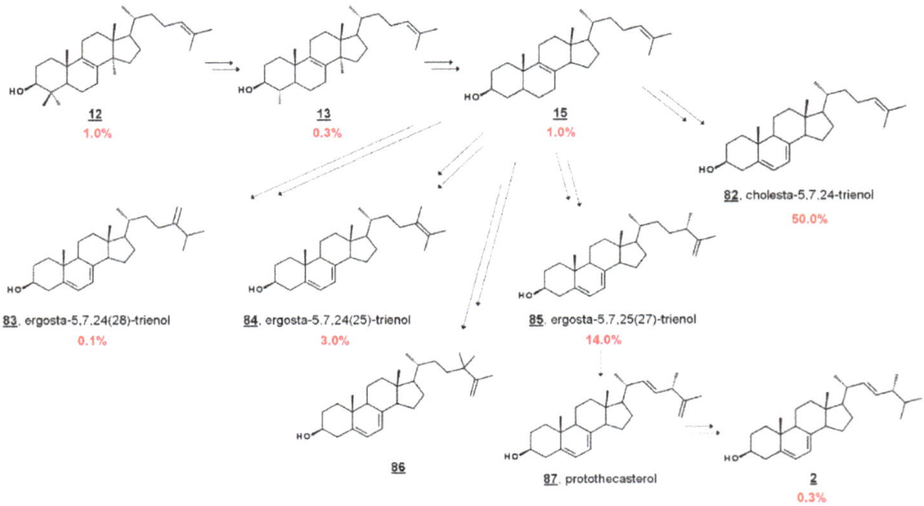

Figure 7. Abbreviated biosynthetic sterol pathway and composition in *T. brucei*. In *T. brucei*, C4 is demethylated before C14, contrary to mammalian and fungal pathways (cf. Figure 2). Values are percentage sterol composition reported by Zhou et al. [43]. Dietary cholesterol 1 accounted for 20.0 %, and other components were 16 (0.1%), 30 (1.0%), 48 (1.0%), 57 (8.0%), and others (0.2%). 24,24-Dimethylcholesta-5,7,25(27)-trienol and 86 and protothecasterol 87 were not detected in this composition, but have been reported in subsequent studies [42,66], respectively.

In BSF *T. brucei*, however, the sterol content is overwhelmingly cholesterol, as well as dietary phytosterols, like sitosterol **3** and campesterol **36**, present in the hosts' blood [41–43]. Single trace ^{13}C-labeled sterol was found in BSF cultures fed [1-^{13}C]glucose [42]. Upon removal of the main sterol component cholesterol, detailed targeted sterolomics of BSF *T. brucei* cells revealed minor components of the sterol profile. Due to the *S*-cis double bond configuration in the B ring of ergosterol and compounds **81**–**87**, UV absorbances of 282 nm can be monitored for the presence of endogenous $\Delta^{5,7}$ sterols, absent in serum. Endogenous cholesta-5,7,24-trienol and ergostenols were found at trace amounts, while they were undetectable when the presence of cholesterol was predominant. The ergosterol requirements for BSF was estimated to be 0.01 fg/cell, compared to the PCF requirement of 6 fg/cell [41]. Consequently, treatment with the EBIs itraconazole **22** and 25-azalanosterol **25** resulted in parasite death and an increased survival rate of infected mice. Correspondingly, the effects of EBIs on cultures were reversed upon supplementation of ergosterol [41].

24-SMT substrate analogues substituted with fluorine at C26, **88** and **89** (Figure 8a), inhibited both PCF cultures and *T. brucei* 24-SMT in vitro. 26-Fluorolanosterol **88** inhibited trypanosome growth with an IC_{50} of about 3 μM, though it was not productively bound in *T. brucei* 24-SMT assays. 26-Fluorolanosterol is a reversible inhibitor of 24-SMT. Conversely, 26-fluorocholesta-5,7,24-trienol **89** is a substrate of 24-SMT, which can be turned over to 24-methylated products or bind irreversibly to the enzyme, with a k_{cat}/k_{inact} of 0.26 min^{-1}/0.24 min^{-1}. Sterol analysis of treated PCF revealed a loss of 24-alkylated sterols as well as a loss of 25(27)-desaturated sterols. Moreover, 26-fluorinated biosynthetic intermediates **90**–**93** downstream from lanosterol (Figure 8b) were detected in 26-fluorolanosterol-treated PCF and human epithelial kidney (HEK) cells. The activity of 26-fluorolanosterol on PCF was attributed to conversion to 26-fluorosterols lacking C4- and C14-methyl groups, capable of irreversibly binding to 24-SMT [67].

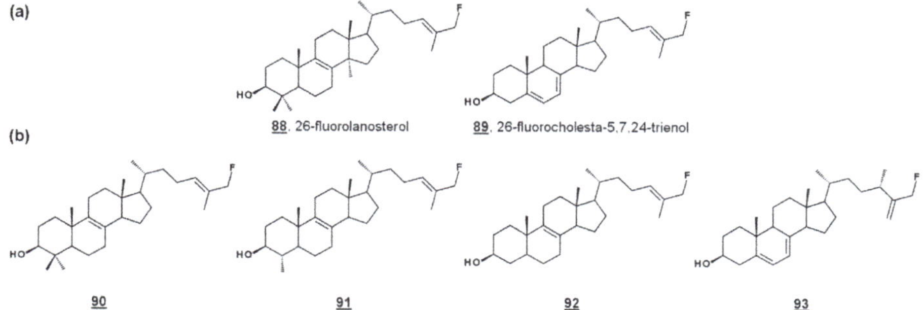

Figure 8. 26-Fluorinated sterol analogues. (**a**) Fluorinated inhibitors of *T. brucei* 24-SMT and growth. (**b**) Metabolites of **88** identified from *T. brucei* and HEK cells [67].

The importance of endogenous synthesis of ergostenols in BSF is accentuated by the effectiveness of other reported EBIs [41,43,67–71].

3.2. *Acanthamoeba* spp.

Ergosterol is a significant Δ^5 bulk sterol in amoebae, as is 7-dehydroporiferasterol. Sterols are synthesized *de novo* in amoebae via a biosynthetic pathway involving the protosterol cycloartenol **25**, as in green algae and higher plants. Amoebae also synthesize 19(10→6)-abeo-sterols containing aromatic B rings called the amebasterols [22,36]. Amebasterol-1 **94**, amebasterol-2 **95**, and amebasterol-4 **98** have been described [22]; trace amebasterols-3 **96**, -5 **99**, and -6 **97** have been identified as of late (Figure 9). These compounds can be selectively monitored at UV absorbances of 270 nm [36].

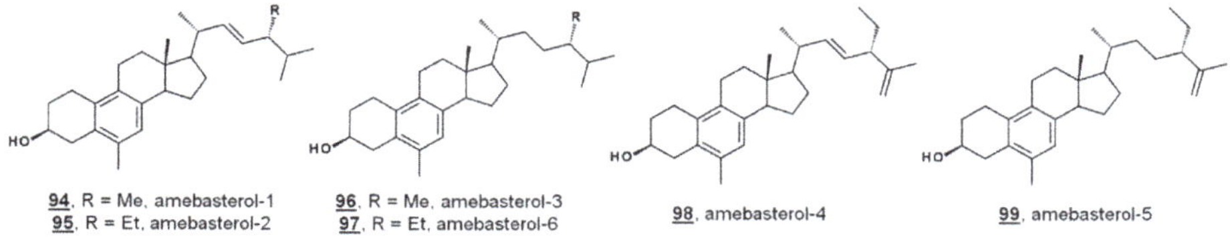

94, R = Me, amebasterol-1
95, R = Et, amebasterol-2
96, R = Me, amebasterol-3
97, R = Et, amebasterol-6
98, amebasterol-4
99, amebasterol-5

Figure 9. Structures of amebasterols.

The sterol profile of was found to be variable as a function of growth and encystment phases. Analysis of the *Acanthamoeba castellanii* sterolome throughout the first week and one month after inoculation revealed a variable composition with changes to cell morphology and viability. At the beginning of the excystment-trophozoite-encystment cycle, in early log phase of growth, an accumulation of protosterol cycloartenol **27** and 24-methylenated cycloartanol **29** was noted. As the cells replicated, trophozoites contained mostly the $\Delta^{5,7}$ products ergosterol and 7-dehydroporiferasterol, whereas, in the stationary growth phase, with a mixture of trophozoites and cysts, sterols shifted to the Δ^5 products brassicasterol and poriferasterol. Supplementation of trophozoite cultures with cholesterol had only a minor stimulation effect on their growth. After one-month incubation, dead cells were mostly comprised of amebasterols, amebasterol-1 **94** and amebasterol-2 **95** (Figure 9). The shift from $\Delta^{5,7}$ products in non-viable encysted cells to the amebasterols was attributed to turnover from stress and a sterol composition associated with altered membrane fluidity affording lysis (Figure 10) [36].

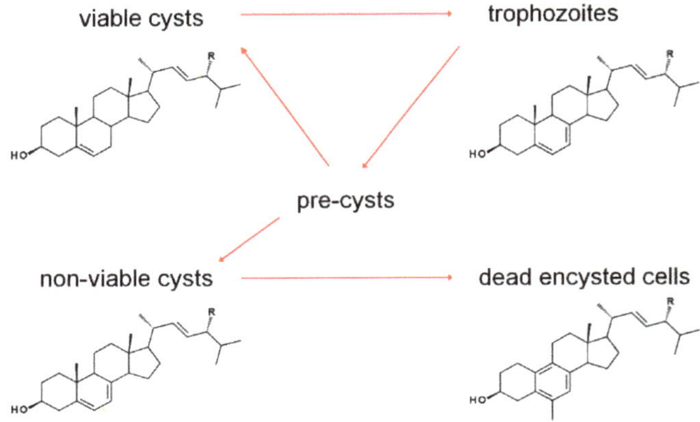

Figure 10. Growth-phase dependence of predominant sterols in *A. castellanii*. R = Me and Et. Adapted from [36].

Beyond the protosterols, ergosterol/poriferasterol pairs, brassicasterol/poriferasterol pairs, and amebasterol-1/amebasterol-2 pairs, this study identified an additional 13 minor sterols in the metabolome of *A. castellanii*. Labeled experiments with $[^2H_3]$Met elucidated labeling patterns of dideuterated ergosterol and pentadeuterated 7-dehydroporfierasterol, consistent with a $\Delta^{24(28)}$ product in its first biomethylation by SMT and a $\Delta^{25(27)}$ product in the second biomethylation (*Vs.* Section 2.1) [36]. Labeling outcomes are supported by in vitro mechanisms with recombinant Acanthamoebic SMTs yielding a single $\Delta^{24(28)}$ product in the first biomethylation (introduction of C28) [72]. While recombinant SMT yielded both $\Delta^{25(27)}$ and $\Delta^{25(27)}$ products for the second biomethylation (introduction of C29) [72], the authors concluded the labeling pattern of sterols from $[^2H_3]$Met-fed cultures, indicating that 24(28)-ethyidene sterols are not incorporated into 7-dehydroporiferasterol under physiological conditions [36].

The noted pairs of cycloartenol and 24(28)-methylenecycloartanol (24-H/24-Me), and pairs of ergosterol/poriferasterol, brassicasterol/poriferasterol, and amebasterol-1/amebasterol-2 (each 24-Me/24-Et) in the various portions of the Acanthamoebic life cycle [36], along with product outcomes being largely determined by biomethylation patterns of *A. castellanii* SMTs [36,72], underscores the crucial nature of SMT function in the pathogen. Subtle alterations in substrate selectivity were noted to have a profound impact on the balance of 24-methyl and 24-ethyl sterols [36]. After treatment with the 24-SMT inhibitor 24(*R*,*S*),25-epiminolatnosterol **26** and the azole 14-SDM inhibitor voriconazole **20** (See Figure 2 for structures), and small increase in amounts of cycloartenol and obtusifoliol were noted [72,73]. Upon treatment with EBIs, trohpozoites were stimulated to encyst, while excystment was insensitive to treatment. The correlation between stage-specific sterol compositions and the physiological effects of EBIs provide insight on opportunities for therapeutics (Figure 10). It is imagined that EBIs targeting the enzyme that reduces the Δ^7 olefin of ergosterol/7-dehydroproferasterol to brassicasterol/poriferasterol could be used to modulate Acanthamoeba growth phases and prevent recurrence of the disease [36].

Azole inhibitors of 14-SDM have been reported to restrict Acanthamoeba growth in the nanomolar to micromolar range [36,37,73–76], and inhibitors of 24-SMT have been reported with nanomolar activity against Acanthamoeba cultures [36,72]. Treatments of 14-SDM- and 24-SMT-inhibitors in combination led to complete eradication of the amoeba parasite at concentrations as low as their respective IC_{50}s [36].

3.3. Fungal Sterol Profiles in Drug-Treated Cultures

EBIs are a staple of antimycotic drug discovery [23–25]. A general hypothetical biosynthetic pathway, as well as popular block points for EBIs, are presented in Figure 2. Sterolomics can be used to confirm the inhibition of ergosterol biosynthesis upon treatment with new small molecules with antifungal properties.

Series of amidoesters substituted with imidazolylmethyl groups were reported to have bioactivity against opportunistic fungal pathogens *Candida albicans*, *Candida tropicalis*, *Cryptococcus neoformans*, and *Aspergillus fumigatus* [77,78]. Some of these compounds, including **100** [77] and **101** [78] (Figure 11a) displayed better antifungal properties than fluconazole **21** (cf. Figure 2). The sterols of *C. albicans* administered with these compounds were analyzed to confirm a mechanism of disrupting ergosterol biosynthesis. Ergosterol normally comprises of the vast majority of the sterol profile in *C. albicans* (>98%), and treatment with **100** [77] or **101** [78] reduced ergosterol in a dose dependent manner. Dose-dependent increases in lanosterol **12** were noted, as well as increases in 14α-methylsterol by-products eburicol **102** and obtusifoliol **31** (Figure 11a). The increase in lanosterol (substrate for *C. albicans* 14-SDM), the increase in 14-methylsterols, and a commensurate decrease in ergosterol itself, suggested 14-SDM as a target for these molecules [77,78].

Many molds, like clinically relevant Mucorales, methylate the side chain of protosterol lanosterol **12**, before demethylating the sterol nucleus, to produce eburicol **102** as a normal intermediate (Figure 11b). Sterols were examined from six pathogenic molds from the order Mucorales, as well as sterols from cultures treated with the azole drug posaconazole **23** (cf. Figure 2). The untreated molds were reported to contain ergosterol, with prominent composition by ergosta-5,7-dienol **58**. An additional 12 sterols from untreated cultures were reported. *Rhizopus arrhizus* contained 76.3% ergosterol and 10.6% ergosta-5,7-dienol within its sterol fraction. Upon administering sub-lethal concentration of 0.5 μg/mL posaconazole **23**, these percentages were reduced to 58.5% and 5.1%, respectively. Correspondingly, lanosterol and eburicol **102** increased with azole, and other 14-methylsterols were noted. Moreover, non-physiological and toxic 14-methylergosta-8,24(28)-dien-3β,6α-diol **103**, which had only been found prior in azole-dosed yeasts, was detected at 0.7% in treated cells (Figure 11b) [79].

Of a set of sesquiterpenes isolated from Chinese liverwort *Tritomaria quinquedentata (Huds.) Buch.*, 5 exhibited activity against strains of *C. albicans*. The most potent of these compounds, *ent*-isoalantolactone **104** suppressed hyphal formation of the yeast and was further investigated for its antifungal mechanism. An increase in lanosterol **12** and zymosterol **15** was noted in *C. albicans* sterol composition when applied with MIC_{80} concentrations of *ent*-isoalantolactone (Figure 11c). The accumulation of zymosterol connotes inhibition of Erg6p (=24-SMT). Subsequent transcriptional analysis of treated *C. albicans* revealed increased expression of the Erg6 gene 9.3-fold and the Erg11 (=14-SDM) gene 2.7-fold, supporting the sterolomic findings [80].

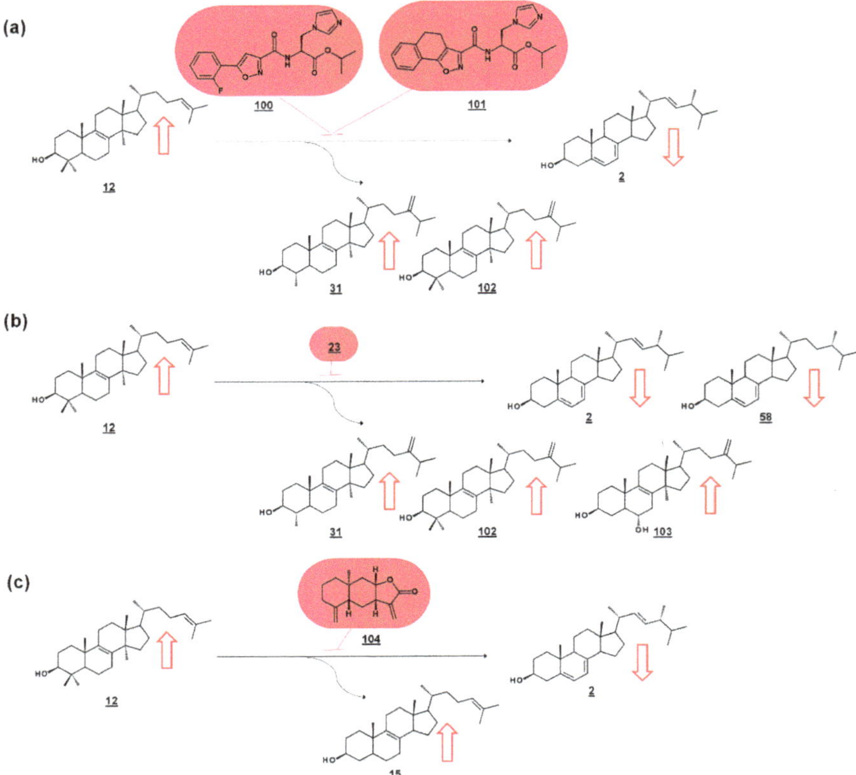

Figure 11. Sterolomic identification of ergosterol biosynthesis inhibitors (EBIs) in fungi. Red arrows signify increase or decrease in sterols within the profile of inhibited cultures relative to non-inhibited cultures. (**a**) Oxazole amidoester-treated cultures of *C. albicans* decrease in ergosterol and increase in lanosterol and by-products obtusifoliol and eburicol, indicating disruption of 14-SDM activity [77,78]. (**b**) Posaconazole-treated cultures of *Rhizopus arrhizus* decrease in ergosterol and ergosta-5,7-dienol and increase in lanosterol, obtusifoliol, and eburicol, and produce toxic 14-methylergosta-8,24(28)-dien-3β,6α-diol [79]. (**c**) *ent*-Isoalantolactone-treated cultures of *C. albicans* decrease in ergosterol and increase in lanosterol and zymosterol, indicating disruption of 24-SMT activity [80].

Bioactive natural product FR171456 **105** (Figure 12) was shown to inhibit ergosterol biosynthesis of *C. albicans*, by a dose-dependent decrease in labeled zymosterol **15** and ergosterol and increase in labeled lanosterol, upon co-incubation with ^{13}C-glucose, ^{13}C-acetate. Similarly, fluconazole **21** –treated cultures also decreased in zymosterol and ergosterol [81]. Likewise, investigative drug VT-1129 **106** (Figure 12) caused an increase in lanosterol, eburicol, obtusifoliol, and its 3-ketone analogue, as well as reduction in ergosterol and fungisterol, in *Cryptococcus* sp. [82].

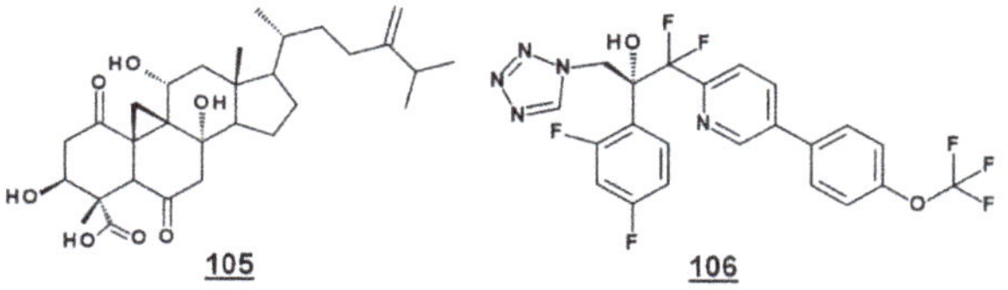

Figure 12. EBIs FR171456 105 and VT-1129 106, confirmed by sterolomic analysis.

4. Bioactive Steroidal Metabolites

Endogenous oxysterols play essential roles in human biology, including signaling, development, and immunology [8–11]. Similarly, several oxysterols isolated from microbial sources have been reported to exhibit therapeutic properties. Many of these compounds from microbes are oxyphytosterols, i.e., unlike human endogenous sterols, they possess alkyl groups at C24 and therefore do not occur in human biology. Bioactivities include those against cancer cell lines, as well as ligands for nuclear receptors, antioxidants, anti-inflammatory agents, and inhibitors of amyloid-β (Aβ) aggregation.

Minor steroidal metabolites often possess bioactivity against other microbes, like fusidic acid, as discussed above. Study of these natural products can further lead to semi-synthetic analogues for structure-activity relationship studies and improvement of antimicrobial agents. For instance, squalamine 9 (Figure 1c), isolated from dogfish shark, is a steroid with polyamine substitution. The cationic polyamine moiety and its polyvalence have been attributed to much of its antimicrobial and anticancer properties [35], and, as a result, a class of synthetic and semi-synthetic analogues, collectively termed cationic steroid antibiotics, have been developed [35,83,84]. For the purposes of this review, only isolated compounds are discussed, though these compounds can inform synthetic and semisynthetic analogues for increased bioactivity. Likewise, steroidal metabolites with a compromised cyclopentanoperhydrophenanthrene nucleus are omitted here.

4.1. Peroxides

Michosterol A **107** (Figure 13) is a newly described polyoxygenated sterol with a C20 hydroperoxyl group and a C25 acetoxyl group, isolated by the ethyl acetate extract of the soft coral *Lobophytum michaelae*. Michosterol A demonstrated moderate cytotoxic effects against A549 cells, with an IC_{50} of 14.9 μg/mL, and was not cytotoxic (IC_{50}s > 20 μg/mL) to DLD-1 and LNCap cell lines. Its anti-inflammatory activity was examined by assaying against superoxide formation in human neutrophils and against elastase release. Michosterol A had IC_{50}s of 7.1 μM and 4.5 μM for superoxide anion generation and elastase release, respectively. A second hydroperoxyl polyoxygenated sterol (C15 hydroperoxyl, and $\Delta^{17(20)}$), named michosterol B **108** (Figure 13) was discovered in this extract. Michosterol B did not display cytotoxicity against the cell lines tested, but inhibited superoxide anion generation 14.7% and elastase release 31.8% each at 10 μM michosterol B [85].

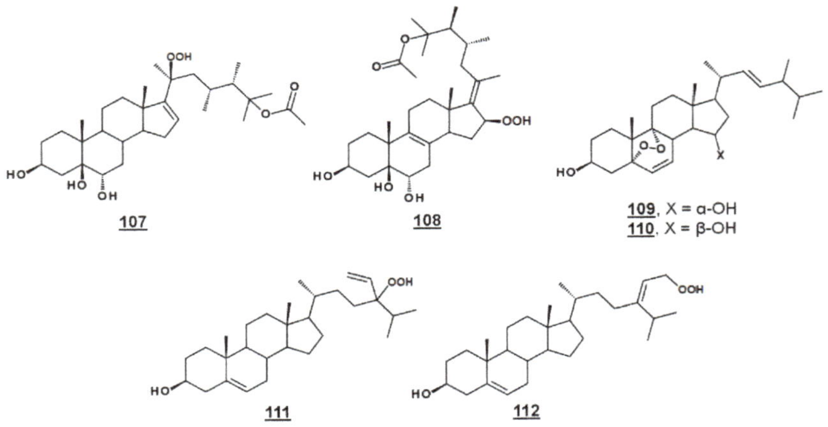

Figure 13. Steryl peroxides discussed in text.

Nigerasterol A **109** and nigerasterol B **110** (Figure 13) are C15 epimers of 3,15-diols containing a 5,α,9α-peroxide obtained from *Aspergillus niger* MA-132, an endophytic fungus isolated from the mangrove plant *Avicennia marina*. Nigerasterol A and nigerasterol B inhibited cell growth in cancer cell lines HL-60 (IC_{50}s 0.3 μM and 1.50 μM, respectively) and A549 (IC_{50}s 1.82 μM and 5.41 μM, respectively) [32].

24-Vinyl-24-hydroperoxycholesterol **111** (Figure 13) has been isolated from *Xestponsgia* sp. [33,86]. It had an IC_{50} in an NF-κB-luciferase assay of 31.3 μg/mL [33] and restricted growth of various human cell lines, including A549 (IC_{50} 29.0 μM) and WI-38 (IC_{50} 43.4 μM) [86]. From Xestospongia, the 29-hydroperoxyl derivative **112** (Figure 13) of isofucosterol has also been reported, with broad activity against such targets as NF-κB-luciferase (IC_{50} 12.6 μg/mL), 3-hydroxy-3-methylglutaryl CoA reductase (HMGR)-green fluorescent protein (IC_{50} 3.8 μg/mL) and protein tyrosine phosphatase 1B (IC_{50} 5.8 μg/mL) [33].

4.2. Acetates

A third michosterol, michosterol C **113** (Figure 14), isolated from the soft coral *Lobophytum michaelae* (*Vs.* 4.1. peroxides) lacked a peroxyl moiety, but contained a 6α-acetoxyl group. Michosterol C was not cytotoxic on cell lines tested, but inhibited superoxide anion generation 17.8% at 10 μM and had an IC_{50} for elastase release of 0.9 μM [85].

Anicequol **114** (Figure 14), also known as NGA0187, is a polyhydroxylated ergost-6-one first described in 2002. Originally isolated from the fungi *Penicillium aurantiogriseum Dierckx* TP-F0213 [87] and *Acremonium* sp. TF-0356 [88], Anicequol inhibited anchorage-dependent growth of human colon cancer DLD-1 cells with an IC_{50} of 1.2 μM [87]. Anicequol was found to induce anoikis, or apoptosis by loss of cell adhesion to the extracellular matrix. Induction of anoikis by anicequol, as well as 25-hydroxycholesterol, was additionally found to involve p38 mitogen-activated protein kinase (p38MAPK) and Rho-associated, coiled-coil containing kinase (ROCK), suggesting new therapeutic strategies against cancer [89]. Anicequol has neurotrophic activity and induced significant neurite outgrowth at 30 μg/mL in PC12 cells [88]. Aniceuquol has also been isolated from *Aspergillus terreus* (No. GX7-3B) [90] and *Penicillum chrysogeum* QEN-24S [91], and supplementary activities against α-acetylcholinesterase (AchE) with an IC_{50} of 1.89 μM [90] and against other fungi, with a zone of inhibition (ZOI) of cultures of the pathogen *Alternaria brassicae* of 6 mm compared to 16 mm by amphotericin B [91].

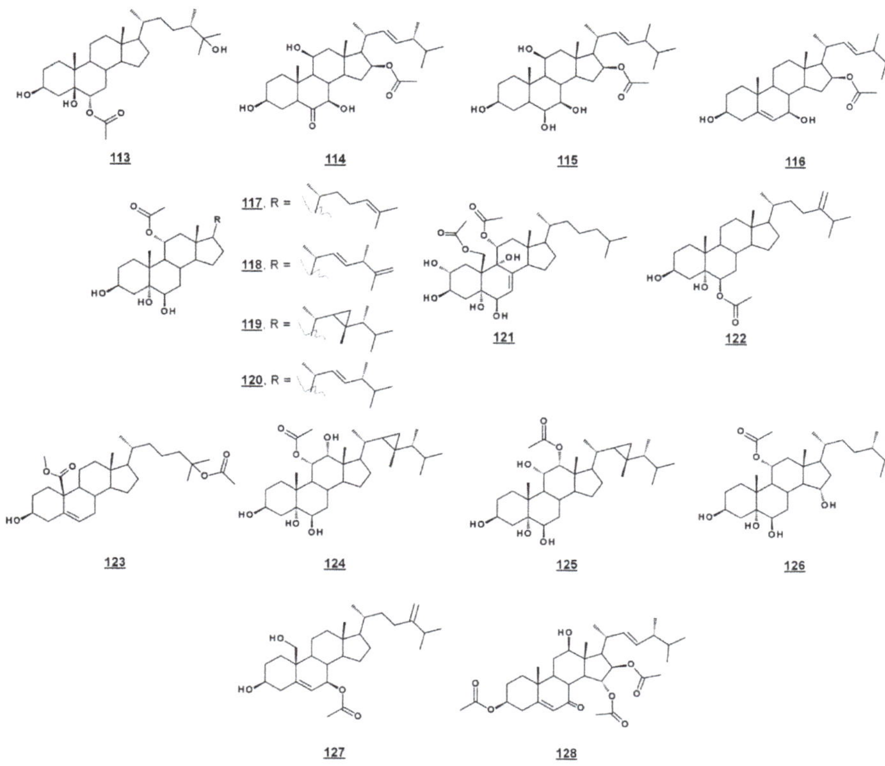

Figure 14. Steryl acetates discussed in text.

Penicisteroid A **115** (Figure 14) is an analogue of anicequol bearing a 7α-hydroxyl rather than a 7-oxo-group. Extracted from *Penicillium chrysogenum* QEN-24S, an endophytic fungus isolated from a red alga of the genus Laurencia, penicisteroid A exhibited both antimycotic and cytotoxic effects. Against the pathogenic fungi *Aspergillus niger* and *Alternaria brassicae*, penicisteroid A gave ZOIs (20 μg) of 18 mm and 9 mm, respectively, compared to 24 mm and 16 mm for control amphotericin B. Penicisteroid A also inhibited HeLa, SW1990, and NCI-H460 cancer cell lines with IC_{50}s of 15 μg/mL, 31 μg/mL, and 40 μg/mL, showing selectivity variable from the anicequol parent compound [91]. Penicisteroid C **116** (Figure 14) also has a C16 acetate, but is less oxygenated than penicisteroid A. It was isolated from a co-cultivation of bacteria *Streptomyces piomogenus* AS63D and fungus *Aspergillus niger* using solid-state fermentation on rice medium. Penicisteroid C displayed selective antimicrobial activity against tested organisms. ZOIs for penicisteroid C were 7 mm, 9 mm, and 10 mm for bacterial cultures *Staphylococcus aureus*, *Bacillus cereus*, and *Bacillus subtilis*, respectively, and were 8 mm and 12 mm for fungal cultures *Candida albicans* and *Saccharomyces cerevisiae*, respectively [92].

A study of the oxysterols from the marine sponge *Haliclona crassiloba* (Figure 14) identified two steryl acetates with antibacterial properties. Newly identified halicrasterol D **120** had minimum inhibition constants (MICs) against tested Gram-positive bacteria ranging from 4 μg/mL against *Enterococcus faecalis* to 128 μg/mL against *S. aureus*. The known diacetate compound **121**, additionally isolated from *H. crassiloba*, had MICs ranging from 8 μg/mL against *S. aureus* to 32 μg/mL *E. faecalis*, in the bacteria tested [93]. A newly identified phytosterol acetate **122** from the soft coral *Sinularia conferta* exhibited low micromolar IC_{50}s against cell lines PANC-1 (1.78 μM), A549 (IC_{50} 3.64 μM), and HeLa (19.34 μM) [94]. From Xestospongia, 25-acetoxyl sterol with an oxidized C19 (carboxylate substitution on C10), **123**, was identified and exhibited an IC_{50} against AMP activated protein kinase of 8.5 μg/mL [33]. Acetates **117–119** and **124–126** (Figure 14) isolated from the coral *Sacrophyton* sp. inhibited Gram-positive and Gram-negative bacteria, with ZOIs ranging from 7.0–14.5 mm for *Escherichia coli* and from 7.5–12.0 mm for *Bacillus megaterium*. They also displayed antifungal properties, inhibiting *Septoria tritci* growth 4.5–10.5 mm [95]. Acetate **127** from the coral *Nephthea erecta* stimulated cytC release and inhibited Akt and mTOR phosphorylation in small cell lung cancer cells, as well as inhibiting tumor growth in the mouse xenograft model [96]. Halymeniaol **128**, an triacetoxyl steroid from the rhodophyte *Halymenia floresii*, was recently reported to have antiplasmodial activity with an IC_{50} of 3.0 μM [97].

4.3. Cyclopropanes

From the marine sponge *Xestospongia testudinaria*, oxyphytosterols **129** and **130** (Figure 15), with a side chain cyclized at C26–C27 were recently reported to posess anti-adhesion properties against bacteria *Pseudoalteromonas* spp. and *Polaribacter* sp. New compounds **129** and **133**, as well as known compounds xestokerol A **130**, 7α-hydroxypetrosterol **132**, and aragusterol B **143** (Figure 15), had antifouling EC_{50}s ranging from 10 to 171 μM. New compound **133** and petrosterol **135** had an EC_{50} > 200 μM [98]. Some of these compounds, other known analogues, and seven new analogues have also been extracted from the marine sponge *Petrosia* (*Strongylophora*) sp. Compounds **130**, **131**, **134–141**, and **143–147** (Figure 15) displayed micromolar inhibition across various human cancer cell lines tested, with the ketal **139** showing weaker activity [99]. Representatives from this class of steroids from *Xestospongia* spp. tested against human cancer cell line K562 yielded IC_{50}s for aragusterol J **149** of IC_{50} 34.31 μM and for aragusterol A **146** of 24.19 μM. Compounds **141–143** and **148** had IC_{50}s >10 μM [100].

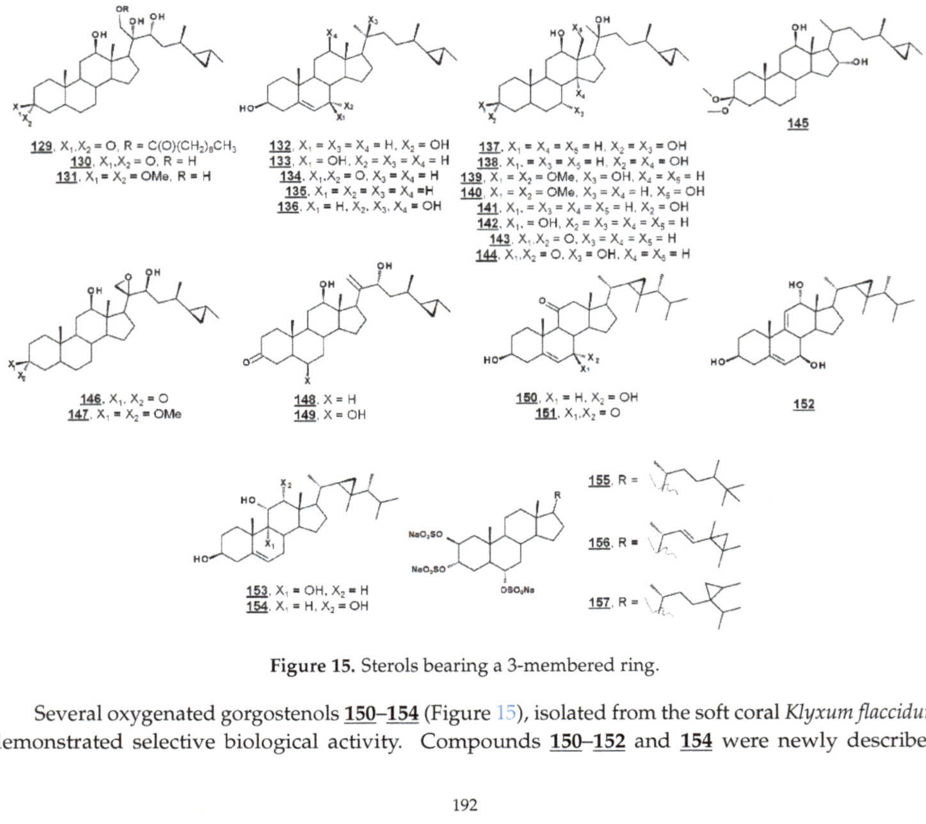

129. X_1,X_2 = O, R = $C(O)(CH_2)_8CH_3$
130. X_1,X_2 = O, R = H
131. $X_1 = X_2$ = OMe, R = H

132. $X_1 = X_3 = X_4$ = H, X_2 = OH
133. X_1 = OH, $X_2 = X_3 = X_4$ = H
134. X_1,X_2 = O, $X_3 = X_4$ = H
135. $X_1 = X_2 = X_3 = X_4$ = H
136. X_1 = H, X_2, X_3, X_4 = OH

137. $X_1 = X_4 = X_5$ = H, $X_2 = X_3$ = OH
138. $X_1 = X_3 = X_5$ = H, $X_2 = X_4$ = OH
139. $X_1 = X_2$ = OMe, X_3 = OH, $X_4 = X_5$ = H
140. $X_1 = X_2$ = OMe, $X_3 = X_4$ = H, X_5 = OH
141. $X_1 = X_3 = X_4 = X_5$ = H, X_2 = OH
142. X_1 = OH, $X_2 = X_3 = X_4 = X_5$ = H
143. X_1,X_2 = O, $X_3 = X_4 = X_5$ = H
144. X_1,X_2 = O, X_3 = OH, $X_4 = X_5$ = H

145

146. X_1, X_2 = O
147. $X_1 = X_2$ = OMe

148. X = H
149. X = OH

150. X_1 = H, X_2 = OH
151. X_1,X_2 = O

152

153. X_1 = OH, X_2 = H
154. X_1 = H, X_2 = OH

155. R =
156. R =
157. R =

Figure 15. Sterols bearing a 3-membered ring.

Several oxygenated gorgostenols **150–154** (Figure 15), isolated from the soft coral *Klyxum flaccidum* demonstrated selective biological activity. Compounds **150–152** and **154** were newly described and named klyflaccisteroids C-E [101] and klyflaccisteroid H [102], respectively. These compounds demonstrated variable inhibition across human cancer cell lines, as well as inhibition of superoxide anion generation and elastase release [101,102]. New analogues of the known trisulfate compound halistanol sulfate **155** bearing cyclopropyl rings on their side chains have been isolated from the marine sponge *Halichondria* sp. Halistanol sulfates I **156** and J **157** had IC_{50} values for sirtuins 1–3 of 45.9, 18.9, and 32.6 μM and 67.9, 21.1, and 37.5 μM, respectively, compared to IC_{50}s of the parent structure **105** of 49.1, 19.2, and 21.8 μM [103].

4.4. Other Bioactive Steroids

Several sponge sterols, such as solomonsterol A **158** and B **159**, theonellasterol **160**, conicasterol **161** (Figure 16), and their analogues, can serve as ligands for human nuclear receptors; many of these compounds and their activities have been reviewed [104]. Ganoderic acid A **163** (Figure 16) and related compounds isolated from the higher fungus *Ganoderma* sp. possess broad therapeutic properties, including those of anti-tumor and anti-inflammation [105]. Additional recently reported bioactivities of sterols from microorganisms and algae are presented in Table 2 and Figure 16.

Table 2. Recently reported biological activities from microbial steroids.

No.[1]	Microbial Source	Biological Target [2]	Biological Activity	Reference
		Fungi		
163	*Nigrospora sphaerica*	Cryptococcus neoformans	IC_{50} 14.81 μg/mL	[106]
164	*Gymnoascus reessii*	NCI-H187	IC_{50} 16.3 μg/mL	[34]
		Plasmodium falciparum	IC_{50} 3.3 μg/mL	[34]
165	*Gymnoascus reessii*	NCI-H187	IC_{50} 47.9 μg/mL	[34]
		Plasmodium falciparum	IC_{50} 4.5 μg/mL	[34]
166	*Gymnoascus reessii*	NCI-H187	IC_{50} 1.9 μg/mL	[34]
		Plasmodium falciparum	IC_{50} 3.4 μg/mL	[34]
167	*Gymnoascus reessii*	NCI-H187	IC_{50} 12.5 μg/mL	[34]
		Plasmodium falciparum	IC_{50} 3.4 μg/mL	[34]
168	*Aspergillus* sp.	*Balanus amphitrite* biofouling	EC_{50} 18.40 μg/mL	[107]
169	*Nodulisporium* sp.	$A\beta_{42}$ aggregation	IC_{50} 10.1 μM	[108]
170	*Nodulisporium* sp.	$A\beta_{42}$ aggregation	54.6% relative inhibitory activity at 100 μM	[108]
171	*Nodulisporium* sp.	$A\beta_{42}$ aggregation	IC_{50} 1.2 μM	[108]
172	*Nodulisporium* sp.	$A\beta_{42}$ aggregation	IC_{50} 43.5 μM	[108]
173	*Dichotomomyces cejpii*	$A\beta_{42}$ aggregation	pretreatment with 10 μM reduced production of Aβ peptides to 3.8-fold increase with 100 μM Aftin-5 compared to 9.4-fold increase with only Aftin-5 and no inhibitor	[109]
		Coral		
174	*Sinularia nanolobata*	HL-60	IC_{50} 33.53 μM	[110]
		HepG2	IC_{50} 64.35 μM	[110]
175	*Sinularia microspiculata*	HL-60	IC_{50} 82.80 μM	[111]
		SK-Mel2	IC_{50} 72.32 μM	[111]
176	*Sinularia leptoclados*	HL-60	IC_{50} 13.45 μM	[112]
		SW480	IC_{50} 14.42 μM	[112]
		LNCaP	IC_{50} 17.13 μM	[112]
		MCF-7	IC_{50} 17.29 μM	[112]
177	*Sinularia leptoclados*	HL-60	IC_{50} 20.53 μM	[112]
		SW480	IC_{50} 26.61 μM	[112]
		KB	IC_{50} 32.86 μM	[112]
178	*Sinularia conferta*	A549	IC_{50} 78.73 μM	[94]
		HeLa	IC_{50} 30.5 μM	[94]
		PANC-1	IC_{50} 9.35 μM	[94]

Table 2. *Cont.*

179	*Sinularia conferta*	A549	IC_{50} 27.12 μM	[94]
		HeLa	IC_{50} 24.64 μM	[94]
		PANC-1	IC_{50} 20.51 μM	[94]
180	*Sinularia brassica*	PANC-1	IC_{50} 15.24 μM	[113]
		A549	IC_{50} 39.36 μM	[113]
181	*Sinularia brassica*	PANC-1	IC_{50} 22.47 μM	[113]
		A549	IC_{50} 41.20 μM	[113]
182	*Sinularia brassica*	A549	IC_{50} 47.31 μM	[113]
183	*Sinularia brassica*	PANC-1	IC_{50} 15.39 μM	[113]
		A549	IC_{50} 47.46 μM	[113]
184	*Sinularia brassica*	PANC-1	IC_{50} 38.12 μM	[113]
		A549	IC_{50} 23.73 μM	[113]
185	*Sinularia brassica*	A549	IC_{50} 92.53 μM	[113]
	Sacrcophyton sp.	*E. coli*	0.05 mg ZOI [3] 10.0 mm	[95]
		S. tritici	0.05 mg ZOI 7.5 mm	[95]
187	*Sinularia microspiculata*	HL-60	IC_{50} 89.02 μM	[111]
188	*Sinularia* sp.	HL-60	IC_{50} 1.79 μM	[114]
189	*Sinularia* sp.	HL-60	IC_{50} 4.03 μM	[114]
190	*Sinularia* sp.	HL-60	IC_{50} 0.69 μM	[114]
191	*Sinularia brassica*	P388D1	IC_{50} 37.2 μM	[115]
		MOLT-4	IC_{50} 37.8 μM	[115]
192	*Sinularia brassica*	P388D1	IC_{50} 9.7 μM	[115]
		MOLT-4	IC_{50} 6.0 μM	[115]
193	*Sinularia brassica*	P388D1	IC_{50} 5.7 μM	[115]
		MOLT-4	IC_{50} 5.3 μM	[115]
194	*Sinularia brassica*	P388D1	IC_{50} 24.4 μM	[115]
		MOLT-4	IC_{50} 31.2 μM	[115]
186	*Sacrcophyton* sp.	*E. coli*	0.05 mg ZOI 5.0 mm	[95]
		S. tritici	0.05 mg ZOI 7.0 mm	[95]
195	*Sacrcophyton* sp.	*E. coli*	0.05 mg ZOI 7.5 mm	[95]
		S. tritici	0.05 mg ZOI 10.5 mm	[95]
196	*Sacrcophyton* sp.	*E. coli*	0.05 mg ZOI 4.5 mm	[95]
		S. tritici	0.05 mg ZOI 6.5 mm	[95]
197	*Sacrcophyton* sp.	*E. coli*	0.05 mg ZOI 6.0 mm	[95]
		S. tritici	0.05 mg ZOI 4.5 mm	[95]
198	*Sacrcophyton* sp.	*E. coli*	0.05 mg ZOI 6.0 mm	[95]
		S. tritici	0.05 mg ZOI 6.0 mm	[95]
199	*Sacrcophyton* sp.	*E. coli*	0.05 mg ZOI 6.0 mm	[95]
		S. tritici	0.05 mg ZOI 9.0 mm	[95]
200	*Klyxum flaccidum*	A549	ED_{50} 7.7 μg/mL	[101]
201	*Klyxum flaccidum*	K562	IC_{50} 12.7 μM	[102]
		elastase release	IC_{50} 4.40 μM	[102]
202	*Klyxum flaccidum*	P388	IC_{50} 31.8 μM	[116]
		elastase release	IC_{50} 5.84 μM	[116]
203	*Subergorgia suberosa*	Influenza virus strain A/WSN/33 (H1N1)	IC_{50} 37.73 μM	[117]
204	*Subergorgia suberosa*	Influenza virus strain A/WSN/33 (H1N1)	IC_{50} 50.95 μM	[117]
		Sponges		
205	*Petrosia* sp.	MOLT-3	IC_{50} 36.57 μM	[99]
		A549	IC_{50} 54.26 μM	[99]
206	*Xestospongia testudinaria*	MCF-7	IC_{50} 55.8 μM	[86]
		A549	IC_{50} 63.1 μM	[86]
207	*Xestospongia testudinaria*	PTP1B [4]	IC_{50} 4.27 μM	[118]
208	*Xestospongia* sp.	K562	IC_{50} 18.32 μM	[100]
209	Xestospongia sp.	K562	25.73% inhibition at 10 μM	[100]
210	Xestospongia sp.	K562	41.32% inhibition at 10 μM	[100]
158	*Theonella swinhoei*	arthritis	30% reduction in clinical arthritis scores in mice treated with 10 mg/kg	[119]

Table 2. *Cont.*

211	*Theonella swinhoei*	U937 PC-9	IC_{50} 8.8 μM IC_{50} 7.7 μM	[120] [120]
212	*Theonella swinhoei*	U937 PC-9	IC_{50} 2.0 μM IC_{50} 9.7 μM	[120] [120]
213	*Theonella swinhoei*	U937 PC-9	IC_{50} 3.2 μM IC_{50} 1.6 μM	[120] [120]
214	*Callyspongia* aff. *implexa*	*Chlamydia trachomatis*	IC_{50} 2.3 μM	[121]
		Brown Algae		
44	*Sargassum linearifolium*	*Plasmodium falciparum*	IC_{50} 7.48 μg/mL	[122]
215	*Sargassum muticum*	obesity	decreased lipid accumulation and dose-dependent suppression of PPARγ [5]	[123]
	Sargassum fusiform	depression	32.67/53.60 and 32.06/50.83 percentage decrease in immobility duration for forced swimmin and tail suspension test in the mouse model at 10 mg/kg/30 mg/kg	[124]
216	*Dictyopteris undulata* Holmes	PTP1B [4]	IC_{50} 7.92 μM	[125]
217	*Dictyopteris undulata* Holmes	PTP1B [4]	IC_{50} 7.78 μM	[125]
218	*Dictyopteris undulata* Holmes	PTP1B [4]	IC_{50} 3.03 μM	[125]
219	*Dictyopteris undulata* Holmes	PTP1B [4]	IC_{50} 3.72 μM	[125]
220	*Dictyopteris undulata* Holmes	PTP1B [4]	IC_{50} 15.01μM	[125]
221	*Dictyopteris undulata* Holmes	PTP1B [4]	IC_{50} 35.01 μM	[126]
222	*Dictyopteris undulata* Holmes	PTP1B [4] HL-60	IC_{50} 1.88 μM IC_{50} 2.08 μM	[126] [126]
223	*Dictyopteris undulata* Holmes	HL-60	IC_{50} 2.45 μM	[126]
224	*Dictyopteris undulata* Holmes	PTP1B [4] HL-60	IC_{50} 38.15 μM IC_{50} 2.70 μM	[126] [126]
225	*Dictyopteris undulata* Holmes	PTP1B [4] HL-60	IC_{50} 48.21 μM IC_{50} 1.02 μM	[126] [126]
226	*Dictyopteris undulata* Holmes	PTP1B [4] HL-60	IC_{50} 3.47 μM IC_{50} 1.26 μM	[126] [126]
227	*Dictyopteris undulata* Holmes	PTP1B [4] HL-60	IC_{50} 16.03 μM IC_{50} 1.17 μM	[126] [126]

[1] Compound number. Structures are given in Figure 16. [2] Cancer cell lines include A549, lung adenocarcinoma; HeLa, cervical adenocarcinoma; HepG2, hepatocellular carcinoma; HL-60, promyelocytic leukemia; KB, epidermoid carcinoma; K562, bone marrow myelogenous leukemia; LNCaP, prostate cancer; MCF-7, breast adenocarcinoma; MOLT-4, lymphoblastic leukemia; NCI-H187, lung carcinoma; PANC-1, pancreatic epithelioid carcinoma; PC-9, lung adenocarcinoma; P388, murine leukemia; P388D1, lymphoma; SK-Mel2, melanoma; SW480, colorectal adenocarcinoma; U937, histiocytic lymphoma. [3] ZOI, zone of inhibition. [4] PTP1B, protein tyrosine phosphatase 1B. [5] PPARγ, peroxisome proliferator-activated receptor γ.

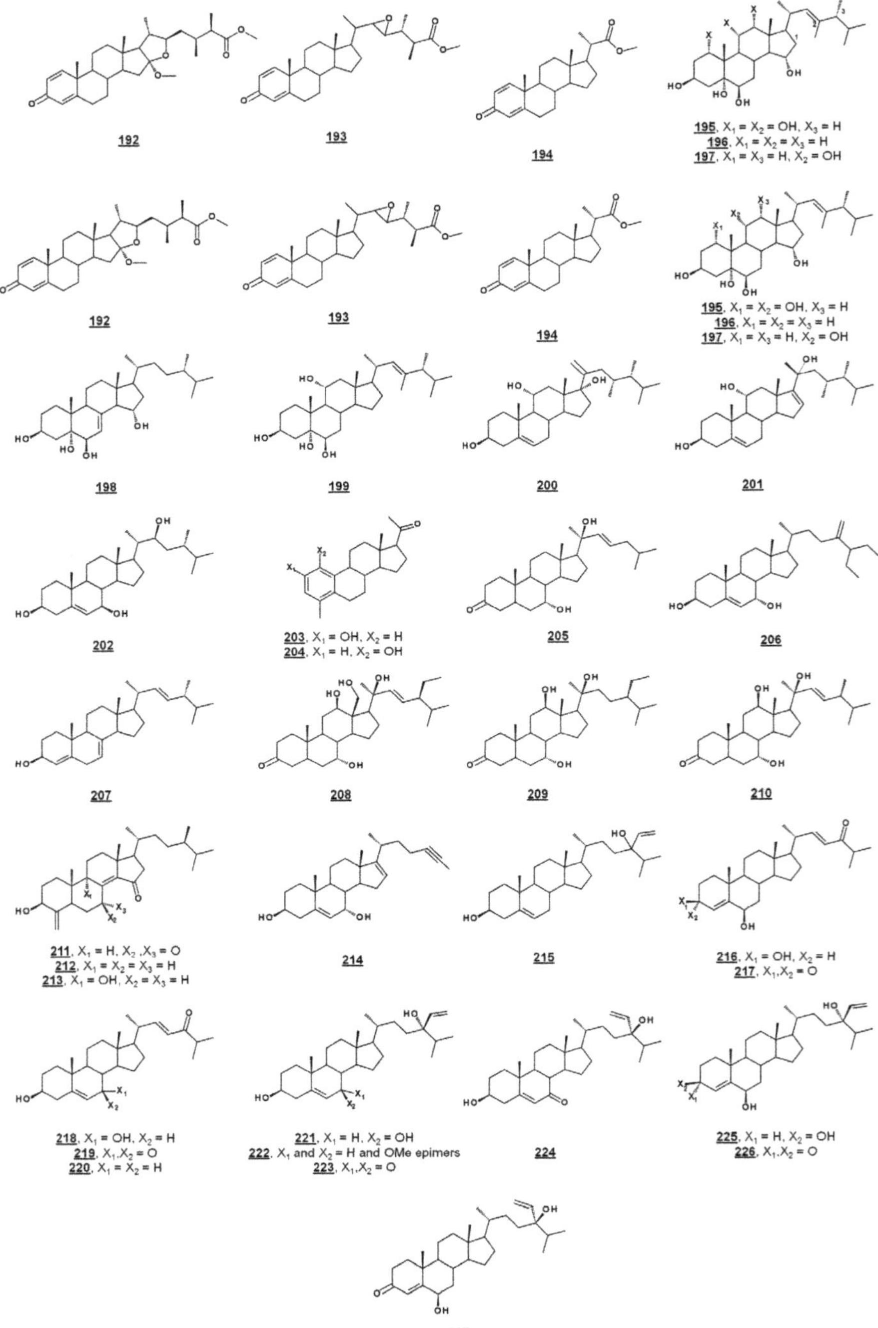

Figure 16. Bioactive sterols and steroids. Activities are given in Table 2.

5. Conclusions

Metabolomics of sterols in microorganisms have provided insight into the biology of microorganisms. Steroidal chemotaxonomy can be used to elucidate phylogenic relationships and steroidal biomarkers can be used to monitor microbial growth and biomass production. Sterolomics additionally plays an influential role in drug discovery, through validation of drug targets, by confirmation of small molecule mechanisms, and by biological testing of microbial metabolites. The sterolome of microbiota can inform chemical biology, evolutionary traits, ecology, and pharmacology.

Author Contributions: B.A.H. wrote the paper.

Funding: This research received no external funding.

Conflicts of Interest: The author declares no conflict of interest.

Abbreviations

$[^2H_3]$AdoMet	methyl-trideuterated *S*-adenosylmethionine
$[^2H_3]$Met	methyl-trideuterated methionine
8(7)-SI	sterol C8(7)-isomerase
14-SDM	sterol C14-demethylase
14-SR	sterol C14-sterol reductase
24-SMT	sterol C24-methyltransferase
Aβ	amyloid-β
AchE	α-acetylcholinesterase
BSF	bloodstream form
EBIs	ergosterol biosynthesis inhibitors
FF-MAS	follicular fluid meiosis-activating sterol
HEK	human epithelial kidney
HMGR	3-hydroxy-3-methylglutaryl CoA reductase
MASs	meiosis-activating sterols
T-MAS	testicular meiosis-activating sterol
OSC	oxidosqualene cyclase
p38MAPK	p38 mitogen-activated protein kinase
PCF	procyclic form
PPARγ	peroxisome proliferator-activated receptor γ
PTP1B	protein tyrosine phosphatase 1B
ROCK	Rho-associated coiled-coil containing kinase
SqE	squalene epoxidase; TMS, trimethylsilyl(ated)
ZOI	zone of inhibition

Appendix

Table A1 gives the systematic names of all steroids discussed in the text.

Table A1. Trivial and systematic names of sterols depicted in figures and discussed in text.

No.[1]	Trivial Name, If Applicable (Secondary Trivial Name, If Applicable)	Systematic Name [2] (Systematic Name Relative to 5α-Cholestane)	PubChem CID
<u>1</u>	cholesterol	cholest-5-en-3β-ol	5997
<u>2</u>	ergosterol	ergosta-5,7,22*E*-trien-3β-ol (24β-methylcholesta-5,7,22*E*-trien-3β-ol)	444679
<u>3</u>	sitosterol	stigmast-5-en-3β-ol (24α-methylcholest-5-en-3β-ol)	222284
<u>4</u>	FF-MAS	4α,4β-dimethylcholesta-8,14,24-trien-3β-ol	443212

Table A1. *Cont.*

5	T-MAS	4α,4β-dimethylcholesta-8,24-dien-3β-ol	50990081
6	25-hydroxycholesterol	cholest-5-en-3β,25-diol	65094
8	ergosterol peroxide	5α,8α-epidioxyergosta-6,22*E*-dien-3β-ol (24β-methyl-5α,8α-epidioxycholesta-6,22*E*-dien-3β-ol)	5351516
9	squalamine	3β-[3-(4-aminobutyl)amino]propyl-7α-hydroxycholestan -24β-hydrosulfate	72495
12	lanosterol	lanosta-8,24-dien-3β-ol (4α,4β,14α-trimethylcholesta-8,24-dien-3β-ol)	246983
13	31-norlanosterol	4α,14α-dimethylcholesta-8,24-dien-3β-ol	15101557
14	zymosterone	cholesta-8,24-dien-3-one	22298942
15	zymosterol	cholesta-8,24-dien-3β-ol	92746
16	fecosterol	ergosta-8,24(28)-dien-3β-ol (24-methylideneholest-8-en-3β-ol)	440371
17	episterol	ergosta-7,24(28)-dien-3β-ol (24-methylideneholest-7-en-3β-ol)	5283662
18		ergosta-5,7,22*E*,24(28)-tetraen-3β-ol (24-methylideneholesta-5,7,22*E*-trien-3β-ol)	11090531
25	25-azalanosterol	25-azalanost-8(9)-en-3β-ol 4α,4β,14α-trimethyl-25-azacholest-8-en-3β-ol	66746490
26	24(R,S),25-epiminolanosterol	24(*R*,*S*),25-epiminolanost8(9)-en-**3**β-ol (4α,4β,14α-trimethyl-24,25-azanetriylcholest-8-en-3β-ol)	163740
27	cycloartenol	cycloart-24(25)-en-**3**β-ol (4α,4β,14α-trimethyl-9β,19-cyclocholest-24-en-3β-ol)	92110
28	cyclolaudenol	24β-methylcycloart-25(27)-en-**3**β-ol (4α,4β,14α,24β-tetramethyl-9β,19-cyclocholest-25(27)-en-3β-ol)	101729
29	24-methylenecycloartanol	24-methylidenecycloartan-**3**β-ol (4α,4β,14α-trimethyl-24-methylidene-9β,19-cyclocholestan-3β-ol)	94204
30		ergosta-8,25(27)-dien-**3**β-ol (24β-methylcholest-8,25(27)-dien-3β-ol)	102515129
31	obtusifoliol	4α,14α-dimethylergosta-8,24(28)-dien-3β-ol (4α,14α-dimethyl-24-methylideneholest-8-en-3β-ol)	65252
32	24(28)-methylenelophenol	4α-methylergosta-7,24(28)-dien-3β-ol (4α-methyl-24-methylideneholest-7-en-3β-ol)	5283640
33	chlamysterol	4α,14α-dimethylporiferasta-8,25(27)-dien-3β-ol (24β-ethyl-4α,14α-dimethyl-cholesta-8,25(27)-dien-3β-ol)	90657605
34	24(28)*Z*-ethylidene lophenol (citrostadienol)	4α-methylstigmasta-7,24(28)*Z*-dien-3β-ol (4α-methyl-24*Z*-ethylideneholest-7-en-3β-ol)	9548595
35	7-dehydroporiferasterol	poriferasta-5,7,22*E*-trien-**3**β-ol (24β-ethylcholesta-5,7,22*E*-trien-3β-ol)	20843308
36	campesterol	campest-5-en-**3**β-ol (24α-methylcholest-5-en-3β-ol)	173183
37	clionosterol (22-dihydroporiferasterol)	poriferast-5-en-**3**β-ol (24β-ethylcholest-5-en-3β-ol)	457801
38	22-dihydrobrassicasterol	ergost-5-en-**3**β-ol (24β-methylcholesta-5-en-3β-ol)	312822
39	stigmasterol	stigmasta-5,22*E*-dien-**3**β-ol (24α-ethylcholesta-5,22*E*-dien-3β-ol)	5280794
40	poriferasterol	poriferasta-5,22*E*-dien-**3**β-ol (24β-ethylcholesta-5,22*E*-dien-3β-ol)	5281330
41	crinosterol (epibrassiasterol)	campesta-5,22*E*-dien-**3**β-ol (24α-methylcholesta-5,22*E*-dien-3β-ol)	5283660
42	brassicasterol	ergosta-5,22*E*-**3**β-ol (24β-methylcholesta-5,22*E*-dien-3β-ol)	5281327
43	fucosterol	stigmasta-5,24(28)*E*-dien-**3**β-ol (24*E*-ethylideneholest-7-en-3β-ol)	5281328
44	isofucosterol	stigmasta-5,24(28)*Z*-dien-**3**β-ol (24*Z*-ethylideneholest-5-en-3β-ol)	5281326

Table A1. *Cont.*

45	24(28)-methylenecholesterol (chalinasterol)	ergosta-5,24(28)-dien-3β-ol (24-methylideneholest-5-en-3β-ol)	92113
46	24-ethyldesmosterol	stigmasta-5,24(25)-dien-**3**β-ol (24-ethylcholesta-5,24(25)-dien-**3**β-ol)	22848721
47	24-methyldesmosterol	ergosta-5,24(25)-dien-**3**β-ol (24-methylcholesta-5,24(25)-dien-**3**β-ol)	193567
48	desmosterol	cholesta-5,24-dien-3β-ol	439577
49	schottenol	stigmast-7-en-**3**β-ol (24α-ethylcholest-7-en-3β-ol)	441837
50	22-dihydrochondrillasterol	poriferast-7-en-**3**β-ol (24β-ethylcholest-7-en-3β-ol)	5283639
51	epifungisterol (22-dihydrostellasterol)	campest-7-en-**3**β-ol (24α-methylcholest-7-en-3β-ol)	90889779
52	fungisterol	ergost-7-en-**3**β-ol (24β-methylcholest-7-en-3β-ol)	5283646
53		stigmasta-7,22*E*-dien-**3**β-ol (24α-ethylcholesta-7,22*E*-dien-3β-ol)	125122456
54	chondrillasterol	poriferasta-7,22*E*-dien-**3**β-ol (24β-ethylcholesta-7,22*E*-dien-3β-ol)	5283663
55	stellasterol	campesta-7,22*E*-dien-**3**β-ol (24α-methylcholest-7,22*E*-dien-3β-ol)	5283669
56	5-dihydroergosterol	ergosta-7,22*E*-dien-**3**β-ol (24β-methylcholest-7,22*E*-dien-3β-ol)	13889661
57	24-dehydrolathosterol	cholesta-7,24-dien-**3**β-ol	5459827
58	22-dihydroergosterol	ergosta-5,7-dien-**3**β-ol (24β-methylcholest-5,7-dien-3β-ol)	5326970
59		stigmast-8-en-**3**β-ol (24α-ethylcholest-8-en-3β-ol)	23424905
60		poriferast-8-en-**3**β-ol (24β-ethylcholest-8-en-3β-ol)	101826503
61		campest-8-en-**3**β-ol (24α-methylcholest-8-en-3β-ol)	-
62		ergost-8-en-**3**β-ol (24β-methylcholest-8-en-3β-ol)	60077053
63	cholest-8-enol (24-dihydrozymosterol)	cholest-8-en-3β-ol	101770
64		ergosta-7,25(27)-dien-3β-ol (24β-methylcholesta-7,25(27)-dien-3β-ol)	60077052
65		poriferasta-7,25(27)-dien-3β-ol (24β-ethylcholesta-7,25(27)-dien-3β-ol)	5283655
66	lichesterol	ergosta-5,8,22*E*-trien-3β-ol (24β-methylcholesta-5,8,22*E*-trien-3β-ol)	5281329
67	clerosterol	poriferasta-5,25(27)-dien-3β-ol (24β-ethylcholesta-5,25(27)-dien-3β-ol)	5283638
68	epiclerosterol	stigmasta-5,25(27)-dien-3β-ol (24α-ethylcholesta-5,25(27)-dien-3β-ol)	185472
69	5-dehydrostellasterol (epiergosterol)	campesta-5,7,22*E*-trien-3β-ol (24α-methylcholesta-5,7,22*E*-trien-3β-ol)	124427258
70	occelasterol	27-norcholesta-5,22*E*-dien-3β-ol	15481847
71	lathosterol	cholest-7-en-3β-ol	65728
72	7-dehydrocholesterol (provitamin D3)	cholesta-5,7-dien-3β-ol	439423
73	fucosteryl epoxide	24,28-epoxyergost-5-en-3β-ol 24-(3-methyloxiran-2-yl)-cholest-5-en-3β-ol	3082427
74	dinosterol	4α,23-dimethylergost-22*E*-en-3β-ol (4α,23,24β-trimethylcholest-22*E*-en-3β-ol)	44263330
75		24*Z*-propylidenecholesterol	6443745

Table A1. *Cont.*

76	dihydrodinosterol	(23*R*)-4α,23-dimethylergostan-3β-ol ((23*R*)-4α,23,24β-trimethylcholestan-3β-ol)	133309
77	amphisterol	4α-methylergosta-8(14),24(28)-dien-3β-ol (4α-methyl-24-methylidenechoest-8(14)-en-3β-ol)	60077061
78	4α-methylergosterol	4α-methylergosta-5,7,22E-trien-3β-ol (4α,24β-dimethylcholesta-5,7,22E-trien-3β-ol)	-
79	4α-methylgorgosterol	4α-methylgorgost-5-en-3β-ol ((22*R*,23*R*)-4α,23,24β-trimethyl-22,23-methanocholest-5-en-3β-ol)	-
80	gorgosterol	gorgost-5-en-3β-ol ((22*R*,23*R*)-23,24β-dimethyl-22,23-methanocholest-5-en-3β-ol)	52931413
81	cholestanol	(5α) cholestan-3β-ol	6710664
82	cholesta-5,7,24-trienol 7-dehydrodesmosterol	cholesta-5,7,24-trien-3β-ol	440558
83		ergosta-5,7,24(28)-trien-3β-ol (24β-methylcholesta-5,7,24(28)-trien-3β-ol)	10894570
84		ergosta-5,7,24(25)-trien-3β-ol (24β-methylcholesta-5,7,24(25)-trien-3β-ol)	58104987
85		ergosta-5,7,25(27)-trien-3β-ol (24β-methylcholesta-5,7,25(27)-trien-3β-ol)	101600336
86		24,24-dimethylcholesta-5,7,25(27)-trien-3β-ol	-
87	protothecasterol	ergosta-5,7,22*E*,25(27)-tetraen-3β-ol (24β-methylcholesta-5,7,22*E*,25(27)-tetraen-3β-ol)	101600338
88	26-fluorolanosterol	26-fluorolanosta-8,24-dien-3β-ol (26-fluoro-4α,4β,14α-trimethylcholesta-8,24-dien-3β-ol)	-
89		26-fluorocholesta-5,7,24-trien-3β-ol	-
90		26-fluoro-4α,4β-dimethylcholesta-8,24-dien-3β-ol	-
91		26-fluoro-4α-methylcholesta-8,24-dien-3β-ol	-
92		26-fluorocholesta-8,24-dien-3β-ol	-
93		26-fluoroergosta-8,25(27)-dien-3β-ol 26-fluoro-24β-methylcholesta-8,25(27)-dien-3β-ol	-
94	amebasterol-1	19(10→6)-abeo-ergosta-5,7,9,22*E*-tetraen-3β-ol (24β-methyl-19(10→6)-abeo-cholesta-5,7,9,22*E*-tetraen-3β-ol)	11596359
95	amebasterol-2	19(10→6)-abeo-poriferasta-5,7,9,22*E*-tetraen-3β-ol (24β-ethyl-19(10→6)-abeo-cholesta-5,7,9,22*E*-tetraen-3β-ol)	-
96	amebasterol-3	19(10→6)-abeo-ergosta-5,7,9-trien-3β-ol (24β-methyl-19(10→6)-abeo-ergosta-5,7,9-trien-3β-ol)	-
97	amebasterol-4	19(10→6)-abeo-poriferasta-5,7,9,22*E*,25-pentaen-3β-ol (24β-ethyl-19(10→6)-abeo-cholesta-5,7,9,22*E*,25-pentaen-3β-ol)	-
98	amebasterol-5	19(10→6)-abeo-poriferasta-5,7,9,25(27)-tetraen-3β-ol (24β-ethyl-19(10→6)-abeo-cholesta-5,7,9,25(27)-tetraen-3β-ol)	-
99	amebasterol-6	19(10→6)-abeo-poriferasta-5,7,9-trien-3β-ol (24β-ethyl-19(10→6)-abeo-poriferasta-5,7,9-trien-3β-ol)	-
102	eburicol	24-methylidenelanost-8-en-3β-ol 4α,4β,14α-trimethyl-24-methylidenecholest-8-en-3β-ol	9803310
103		14-methylergosta-8,24(28)-dien-3β,6α-diol 14,24β-dimethyl-24-methylidenecholest-8-en–3β,6α-diol	148910
105	FR171456	24-methylidene-3β,8α,11α-trihydroxy-1,6-dioxocycloartan-30-oic acid (4α,14α-dimethyl-24-methylidene-9β,19-cyclo-3β,8α,11α-trihydroxy -1,6-dioxocholestan-4β-carboxylic acid)	-
107	michosterol A	(20*S*,23*R*)-23-methyl-20-hydroperoxy-25-acetoxyergost -16-en-3β,5β,6α-triol ((20*S*,23*R*)-23,24β-dimethyl-20-hydroperoxy-25 -acetoxycholest-16-en-3β,5β,6α-triol)	-
108	michosterol B	(17*E*,23*R*)-23-methyl-16-hydroperoxy-25-acetoxyergost- 17-en-3β,5β,6α-triol ((17*E*,23*R*)-23,24β-di-methyl-16-hydroperoxy- 25-acetoxycholest-17-en-3β,5β,6α-triol)	-

Table A1. *Cont.*

109	nigerasterol A	5α,9α-epidioxyergosta-6,8(14),22*E*-trien-3β,15α-diol (24β-methyl-5α,9α-epidioxycholesta-6,8(14), 22*E*-trien-3β,15α-diol)	-
110	nigerasterol B	5α,9α-epidioxyergosta-6,8(14),22*E*-trien-3β,15β-diol (24β-methyl-5α,9α-epidioxycholesta -6,8(14),22*E*-trien-3β,15β-diol)	-
111		24-ethenyl-24-hydroperoxycholest-5-en-3β-ol	10411225
112	29-hydroperoxyisofucosterol	(24*Z*)-29-hydroperoxystigmasta-5,24(28)-dien-3β-ol 24*Z*-(2-hydroperoxyethylidnene)cholest-5-en-3β-ol	46224335
113	michosterol C	6α-acetoxyergostan-3β,5β,25-triol (24β-methyl-6α-acetoxycholestan-3β,5β,25-triol)	-
114	anicequol (NGA0187)	16β-acetoxy-3β,7β,11β-trihydroxyergost-22*E*-en-6-one (24β-methyl-16β-acetoxy- 3β,7β,11β-trihydroxycholest-22*E*-en-6-one)	10413810
115	penicisteroid A	24-methyl-16β-acetoxycholest-22*E*-en-3β,6β,7β,11β-tetrol	-
116	penicisteroid C	24-methyl-16β-acetoxycholesta-5,22*E*-dien-3β,6β-diol	-
117		11α-acetoxycholest-24-en-3β,5α,6β-triol	-
118		11α-acetoxyergosta-22*E*,25-dien-3β,5α,6β-triol 24β-methyl-11α-acetoxycholesta-22*E*,25-dien-3β,5α,6β-triol	-
119		11α-acetoxygorgostan-3β,5α,6β-triol ((22*R*,23*R*)-23,24β-dimethyl-22,23-methano -11α-acetoxycholestan-3β,5α,6β-triol)	54769262
120	halicrasterol D	11α-acetoxyergost-22*E*-en-3β,5α,6β-triol 24β-methyl-11α-acetoxycholest-22*E*-en-3β,5α,6β-triol	-
121		11α,19-diacetoxycholest-7-en-2α,3β,5α,6β,9α-pentol	-
122		6β-acetoxyergost-24(28)-en-3β,5α-diol 24-methylidene-6β-acetoxycholestan -3β,5α-diol	101687891
123		methyl 25-acetoxy-3β-hydroxycholest-5-en-19-carboxylate	-
124		11α-acetoxygorgostan-3β,5α,6β,12α-tetrol ((22*R*,23*R*)-23,24β-dimethyl-22,23-methano- 11α-acetoxycholest-5-en-3β,5α,6β,12α-tetrol)	56962930
125		12α-acetoxygorgostan-3β,5α,6β,11α-tetrol ((22*R*,23*R*)-23,24β-dimethyl-22,23-methano- 12α-acetoxycholestan-3β,5α,6β,11α-tetrol)	-
126		11α-acetoxygorgostan-3β,5α,6β,15α-tetrol ((22*R*,23*R*)-23,24β-dimethyl-22,23-methano- 12α-acetoxycholestan-3β,5α,6β,15α-tetrol)	-
127		7β-acetoxyergosta-5,24(28)-dien-3β,19-diol 24-methylidene-7β-acetoxycholest-5-en-3β,19-diol	477494
128	halymeniaol	3β,15α,16β-triacetoxy-12β-hydroxycholest-5-en-7-one	-
129	21-*O*-octadecanoyl-xestokerol A	(20*S*,21*R*)-21-octadecanoyl-11β,20,22-trihydroxypetrostan-3-one ((20*S*,21*R*,25*R*,26*R*)-24α,26-dimethyl-26, 27-cyclo-21-octadecanoyl-11β,20,22-trihydroxycholestan-3-one)	71747680
130	xestokerol A	(20*S*,21*R*)-11β,20,21,22-tetrahydroxypetrostan-3-one ((20*S*,21*R*,25*R*,26*R*)-24α,26-dimethyl-26, 27-cyclo-11β,20,21,22-tetrahydroxycholestan-3-one)	44584465
131	xestokerol A dimethyl ketal	(20*S*,21*R*)-3,3-dimethoxypetrostan-11β,20,21,22-tetrol ((20*S*,21*R*,25*R*,26*R*)-24α,26-dimethyl-26, 27-cyclo-3,3-dimethoxycholestan–11β,20,21,22-tetrol)	-
132	7α-hydroxypetrosterol	petrost-5-en-3β,7α-diol ((25*R*,26*R*)-24α,26-dimethyl-26, 27-cyclocholest-5-en-3β,7α-diol)	101209535
133	7β-hydroxypetrosterol	petrost-5-en-3β,7β-diol ((25*R*,26*R*)-24α,26-dimethyl-26,27-cyclocholest-5-en-3β,7β-diol)	71747681
134	7-ketopetrosterol	3β-hydroxypetrost-5-en-7-one ((25*R*,26*R*)-24α,26-dimethyl-26, 27-cyclo-3β-hydroxycholest-5-en-7-one)	101209534
135	petrosterol	petrost-5-en-3β-ol ((25*R*,26*R*)-24α,26-dimethyl-26,27-cyclocholest-5-en-3β-ol)	194249

Table A1. *Cont.*

136	11β-hydroxypetrosterol	petrost-5-en-3β,11βα-diol ((25*R*,26*R*)-24α,26-dimethyl-26,27-cyclocholest-5-en-3β,11β-diol)	-
137		(20*S*)-petrostan-3α,7α,12β,20-tetrol ((20*S*,25*R*,26*R*)-24α,26-dimethyl-26, 27-cyclocholestan-3α,7α,12β,20-tetrol)	-
138		(20*S*)-petrostan-3α,12β,14α,20-tetrol ((20*S*,25*R*,26*R*)-24α,26-dimethyl-26, 27-cyclocholestan-3α,12β,14α,20-tetrol)	-
139		(20*S*)-3,3-dimethoxypetrostan-7α,12β,20-triol ((20*S*,25*R*,26*R*)-24α,26-dimethyl-26,27-cyclo-3,3-dimethoxycholestan-7α,12β,20-triol)	-
140		(20*S*)-3,3-dimethoxypetrostan-7α,12β,19,20-tetrol ((20*S*,25*R*,26*R*)-24α,26-dimethyl-26,27-cyclo -3,3-dimethoxycholestan-7α,12β,19,20-tetrol)	-
141		(20*S*)-petrostan-3α,12β,20-triol ((20*S*,25*R*,26*R*)-24α,26-dimethyl-26, 27-cyclocholestan-3α,12β,20-triol)	-
142		(20*S*)-petrostan-3β,12β,20-triol ((20*S*,25*R*,26*R*)-24α,26-dimethyl-26,27-cyclocholestan-3β,12β,20-triol)	-
143	aragusterol B	(20*S*)-12β,20-dihydroxypetrostan-3-one ((20*S*,25*R*,26*R*)-24α,26-dimethyl-26,27-cyclo-12β,20-dihydroxycholestan-3-one)	44566420
144		(20*S*)-7α,12β,20-trihydroxypetrostan-3-one ((20*S*,25*R*,26*R*)-24α,26-dimethyl-26,27-cyclo-7α,12β,20-trihydroxycholestan-3-one)	-
145		3,3-dimethoxypetrostan-12β,16α-diol ((25*R*,26*R*)-24α,26-dimethyl-26,27-cyclo-3,3-dimethoxycholestan-12β,16α-diol)	-
146	aragusterol A	(20*R*,22*S*)-20,21-epoxy-12β,22-dihydroxypetrostan-3-one ((20*R*,22*S*,25*R*,26*R*)-24α,26-dimethyl-20,21-epoxy-26, 27-cyclo-12β,22-dihydroxycholestan-3-one)	9933873
147		(20*R*,22*S*)-20,21-epoxy-3,3-dimethoxypetrostan-12β,22-diol ((20*R*,22S,25*R*,26*R*)-24α,26-dimethyl-20,21-epoxy-26, 27-cyclo-3,3-dimethoxycholestan-12β,22-diol)	10696885
148		(22*R*)-12β,22-dihydroxypetrost-20(21)-en-3-one ((22*R*,25*R*,26*R*)-24α,26-dimethyl-26,27-cyclo-12β,22-dihydroxycholest-20(21)-en–3-one)	-
149	aragusterol J	(22*R*)-7β,12β,22-trihydroxypetrost-20(21)-en-3-one ((22*R*,25*R*,26*R*)-24α,26-dimethyl-26,27-cyclo-7β,12β,22-trihydroxycholest-20(21)-en–3-one)	-
150	klyflaccisteroid C	3β,7α-dihydroxygorgost-5-en-11-one ((22*R*,23*R*)-23,24β-dimethyl-22,23-methano-3β,7α-dihyroxycholest-5-en-11-one)	-
151	klyflaccisteroid D	3β-hydroxygorgost-5-en-7,11-dione ((22*R*,23*R*)-23,24β-dimethyl-22,23-methano-3β, 7α-dihyroxycholest-5-en-7,11-dione)	-
152	klyflaccisteroid E	gorgosta-5,9(11)-dien-3β,7β,12α-triol ((22*R*,23*R*)-23,24β-dimethyl-22,23-methanocholesta-5, 9(11)-dien-3β,7β,12α-triol)	-
153		gorgost-5-en-3β,9α,11α-triol ((22*R*,23*R*)-23,24β-dimethyl-22,23-methanocholest-5,-dien-3β,9α,11α-triol)	10742556
154	klyfaccisteroid H	gorgost-5-en-3β,11α,12α-triol ((22*R*,23*R*)-23,24β-dimethyl-22,23-methanocholesta-5,9(11)-dien-3β,11α,12α-triol)	-
155	halistanol sulfate	24,25-dimethylcholestane-2β,3α,6α-trisulfate	73361
156	halistanol sulfate I	24-methyl-24,25-methanocholestane-2β,3α,6α-trisulfate	-
157	halistanol sulfate J	24,24-(methylethano)cholestane-2β,3α,6α-trisulfate	-
158	solomonsterol A	trisodium cholane-2β,3α,24-trisulfate (trisodium 25,26,27-trinorcholestane-2β,3α,24-trisulfate)	50925451

Table A1. *Cont.*

159	solomonsterol B	trisodium 24-norcholane-2β,3α,23-trisulfate (trisodium 24,25,26,27-tetranorcholestane-2β,3α,24-trisulfate)	53318073
160	theonellasterol	4-methylidineporiferast-8(14)-en-3β-ol (24β-ethyl-4-methylidinecholest-8(14)-en-3β-ol)	52931395
161	conicasterol	4-methylidinecampest-8(14)-en-3β-ol (24α-methyl-4-methylidinecholest-8(14)-en-3β-ol)	21670674
162	ganoderic acid A	7β,15α-dihydroxy-3,11,23-trioxolanost-8-en-26-oic acid 4α,4β,14α-trimethyl-7β,15α-dihydroxy-3,11,23-trioxocholest-8-en-26-oic acid)	471002
163		ergosta-7,9(11),22*E*-trien-3β-ol 24β-methylcholesta-7,9(11),22*E*-trien-3β-ol	12308954
164		ergosta-4,7,22*E*-trien-3-one 24β-methylcholesta-4,7,22*E*-trien-3-one	11003773
165		ergosta-4,6,8(14),22*E*-tetraen-3-one 24β-methylcholesta-4,6,8(14),22*E*-tetraen-3-one	6441416
166		14α-hydroxyergosta-4,7,9(11),22*E*-tetraen-3,6-dione 24β-methyl-14α-hydroxycholesta-4,7,9(11),22*E*-tetraen-3,6-dione	10251684
167		9α,14α-dihydroxyergosta-4,7,22*E*-trien-3,6-dione 24β-methyl-9α,14α-dihydroxycholesta-4,7,22*E*-trien-3,6-dione	-
168		ergosta-4,6,8(14),22*E*,24(28)-pentaen-3-one 24-methylidenecholesta-4,6,8(14),22*E*-tetraen-3-one	-
169	nodulisporiviridin E	18-nor-1α,3β-dihydroxy-4,5,6-[2,3,4]furanoandrosta-5,8,11,13(14)-tetraen-7,17-dione 18,20,21,22,23,24,25,26,27-nonanor-1α,3β-dihydroxy-4,5,6-[2,3,4]furanoandrosta-5,8,11,13(14)-tetraen-7,17-dione	122179368
170	nodulisporiviridin F	3β,11β-dihydroxy-4,5,6-[2,3,4]furanoandrosta-5,8-dien-7,17-dione 20,21,22,23,24,25,26,27-octanor-3β,11β-dihydroxy-4,5,6-[2,3,4]furanocholesta-5,8-dien-7,17-dione	122179369
171	nodulisporiviridin G	11β-hydroxy-4,5,6-[2,3,4]furanoandrosta-5,8-dien-3,7,17-trione 20,21,22,23,24,25,26,27-octanor-11β-hydroxy-4,5,6-[2,3,4]furanocholesta-5,8-dien-3,7,17-trione	122179370
172	nodulisporiviridin H	3β,12β-dihydroxy-4,5,6-[2,3,4]furanoandrosta-5,8-dien-7,17-dione 20,21,22,23,24,25,26,27-octanor-3β,12β-dihydroxy-4,5,6-[2,3,4]furanocholesta-5,8-dien-7,17-dione	122179371
173	16-*O*-desmethylasporyergosterol-β-D-mannoside	β-D-mannosyloxyergosta-6,8(14),17(20)*E*,22*E*-tetraen-3β-ol 24β-methyl-β-D-mannosyloxycholesta-6,8(14),17(20)*E*,22*E*-tetraen-3β-ol	-
174		(24*S*)-24,28-epoxyergost-5-en-3β,4α-diol (24*S*)-24-oxyranylcholest-5-en-3β,4α-diol	44575614
175		ergosta-5,24(28)-dien-3β,7α-diol (24-methylidenecholest-5-en-3β,7α-diol)	10949727
176		ergosta-5,24(28)-dien-3β,7β-diol (24-methylidenecholest-5-en-3β,7β-diol)	11373355
177		ergost-5-en-3β,7β-diol (24β-methylcholest-5-en-3β,7β-diol)	11475561
178		ergost-24(28)-en-3β,5α,6β-triol (24-methylidenecholestan-3β,5α,6β-triol)	21775108
179		ergostan-3β,5α,6β-triol (24β-methylcholestan-3β,5α,6β-triol)	44558918
180		3β,5α,6β,11α-tetrahydroxyergostan-1-one (24β-methyl-3β,5α,6β,11α-tetrahydroxycholestan-1-one)	-
181		ergostan-1α,3β,5α,6β,11α-pentol (24β-methylcholestan-1α,3β,5α,6β,11α-pentol)	-
182	sarcoaldesterol B	ergostan-3β,5α,6β,11α-tetrol (24β-methylcholestan-3β,5α,6β,11α-tetrol)	10718409
183		ergostan-1β,3β,5α,6β-tetrol (24β-methylcholestan-1β,3β,5α,6β-tetrol)	-

Table A1. *Cont.*

184	pregnedioside A	4α-O-β-D-arabinopyranosyloxypregn-20-en-3β-ol22,23,24,25,26, 27-hexanor-4α-O-β-D-arabinopyranosyloxycholest-20-en-3β-ol	21673267
185		gorgostan-1α,3β,5α,6β,11α-pentol ((22R,23R)-23,24β-dimethyl-22,23- methanocholestan-1α,3β,5α,6β,11α-pentol)	23426029
186	sarcoaldesterol A	gorgostan-3β,5α,6β,11α-tetrol ((22R,23R)-23,24β-dimethyl-22,23-methanocholestan- 3β,5α,6β,11α-tetrol)	10790775
187		(20R,23R)-23-methylergost-16-en-3β,20-diol (20R,23R)-23,24β-methylcholest-16-en-3β,20-diol	-
188	ximaosteroid E	(16S)-16,22-epoxycholesta-1,22E-dien-3-one	-
189	ximaosteroid F	(20R,22R)-20,22-dihydroxycholesta-1,4-dien-3-one	-
190		(20S)-20-hydroxycholest-1-en-3,16-dione	53997071
191	sinubrasone A	methyl (22R)-22-O-β-D-xylopyranosyloxy- 3-oxoergosta-1,4-diene-26-carboxylate methyl (22R)-24β-methyl-22-O-β-D-xylopyranosyloxy- 3-oxocholesta-1,4-diene-26-carboxylate	-
192	sinubrasone B	methyl (16S,22R)-16-methoxy-16,22-epoxy- 3-oxoergosta-1,4-diene-26-carboxylate methyl (16S,22R)-24β-methyl-16-methoxy-16,22-epoxy -3-oxocholesta-1,4-diene-26-carboxylate	-
193	sinubrasone C	methyl (22R,23R)-22,23-epoxy- 3-oxoergosta-1,4-diene-26-carboxylate methyl (22R,23R)-24β-methyl-22,23-epoxy- 3-oxocholesta-1,4-diene-26-carboxylate	-
194	sinubrasone D	methyl (20S)-20-methyl-3-oxopregna-1, 4-diene-21-carboxylate methyl 23,24,25,26,27-pentanor-3-oxocholesta- 1,4-diene-21-carboxylate	15929041
195		ergostan-1α,3β,5α,6β,11α,15α-hexol (24β-methylcholestan-1α,3β,5α,6β,11α,15α-hexol)	-
196		ergostan-3β,5α,6β,15α-tetrol (24β-methylcholestan-3β,5α,6β,15α-tetrol)	-
197		ergostan-3β,5α,6β,11α,15α-pentol (24β-methylcholestan-3β,5α,6β,11α,15α-pentol)	-
198		ergost-7-en-3β,5α,6β,15α-tetrol (24β-methylcholest-7-en-3β,5α,6β,15α-tetrol)	-
199		23-methylergost-22E-en-3β,5α,6β,11α-tetrol (23,24β-dimethylcholest-22E-en-3β,5α,6β,11α-tetrol)	-
200	klyflaccisteroid A	(17S,23R)-23-methylergosta-5,20(21)-dien-3β,17α-diol (17S,23R)-23,24β-dimethylcholesta-5,20(21)-dien-3β,17α-diol	-
201	klyfaccisteroid J	(20R,23R)-23-methylergosta-5,16-dien-3β,11α,20-triol (20R,23R)-23,24β-dimethylcholesta-5,16-dien-3β,11α,20-triol	-
202	klyflaccisteroid M	(22S)-ergost-5-en-3β,7β,22-triol ((22S)-24β-methylcholest-5-en-3β,7β,22-triol)	-
203	subergorgol U	19(10→4)-abeo-2-hydroxypregna-2,4,1(10)-trien-20-one (22,23,24,25,26,27-hexnor-19(10→4)-abeo-2-hydroxycholesta- 2,4,1(10)-trien-20-one)	132918691
204		19(10→4)-abeo-1-hydroxypregna-2,4,1(10)-trien-20-one (22,23,24,25,26,27-hexnor-19(10→4)- abeo-2-hydroxycholesta-2,4,1(10)-trien-20-one)	54484024
205		(20S)-7α,12β,20-trihydroxycholest-22E-en-3-one	-
206	langcosterol A	26,27-dimethylergosta-5,24(28)-dien-3β,7α-diol (26,27-dimethyl-24-methylidenecholest-5-en-3β,7α-diol)	23426186
207		ergosta-4,7,22E,25-tetraen-3-one (24β-methylcholesta-4,7,22E,25-tetraen-3-one)	132280531
208		7α,12β,18-trihydoxystigmast-22E-en-3-one (24α-methyl-7α,12β,18-trihydoxycholest-22E-en-3-one)	-
209		(20S)-24-ethyl-7α,12β,20-trihydroxycholestan-3-one	-
210		(20S)-24-methyl-7α,12β,20-trihydroxycholest-22E-en-3-one	-

Table A1. *Cont.*

211	7,15-dioxoconicasterol	4-methylidene-3β-hydroxycampest-8(14)-en-7,15-dione (24α-methyl-4-methylidene-3β-hydroxycholest-8(14)-en-7,15-dione)	-
212	15-oxoconicasterol	4-methylidene-3β-hydroxycampest-8(14)-en-15-one (24α-methyl-4-methylidene-3β-hydroxycholest-8(14)-en-15-one)	-
213		4-methylidene-3β,9α-dihydroxycampest-8(14)-en-15-one (24α-methyl-4-methylidene-3β, 9α-dihydroxycholest-8(14)-en-15-one)	-
214	gelliusterol E	24-methylchola-5,16-dien-23-yn-3β,7α-diol (26,27-dinorcholesta-5,16-dien-23-yn-3β,7α-diol)	-
215	saringosterol	24-ethenylcholest-5-en-3β,24-diol	14161394
216	dictyosterol A	3β,6β-dihydroxycholesta-4,22*E*-dien-24-one	-
217	dictyosterol B	6β-hydroxycholesta-4,22*E*-dien-3,24-dione	-
218	dictyosterol C	3β,7α-dihydroxycholesta-5,22*E*-dien-24-one	-
219		3β-hydroxycholesta-5,22*E*-dien-7,24-dione	-
220		3β-hydroxycholesta-5,22*E*-dien-24-one	-
221	dictyopterisin C	(24*R*)-stigmasta-4,28(29)-dien-3β,7β,24-triol (24*R*)-24-ethenylcholest-4-en-3β,7β,24-triol	-
222		(24*R*)-7-methoxystigmasta-4,28(29)-dien-3β,24-diol (24*R*)-7-methoxy-24-ethenylcholest-4-en-3β,24-diol	-
223	dictyopterisin F	(24*R*)-3β,24-dihydroxystigmasta-4,28(29)-dien-7-one (24*R*)-24-ethenyl-3β,24-dihydroxycholest-4-en-7-one	-
224	dictyopterisin G	(24*S*)-3β,24-dihydroxyporiferasta-4,28(29)-dien-7-one (24*S*)-24-ethenyl-3β,24-dihydroxycholest-4-en-7-one	-
225	dictyopterisin H	(24*R*)-stigmasta-4,28(29)-dien-3β,6β,24-triol (24*R*)-24-ethenylcholest-4-en-3β,6β,24-triol	-
226	dictyopterisin I	(24R)-6β,24-dihydroxystigmasta-4,28(29)-dien-3-one (24R)-24-ethenyl-6β,24-dihydroxycholest-4-en-3-one	-
227	dictyopterisin J	(24*S*)-6β,24-dihydroxyporiferasta-4,28(29)-dien-3-one (24*S*)-24-ethenyl-6β,24-dihydroxycholest-4-en-3-one	-

[1] Compound number. [2] Systematic names use carbon numbering and side chain α/β designations of the Nes system presented in Figure 1 and Ref. [1].

References

1. Nes, W.D. Biosynthesis of cholesterol and other sterols. *Chem. Rev.* **2011**, *111*, 6423–6451. [CrossRef] [PubMed]
2. Volkman, J.K. Sterols and other triterpenoids: Source specificity and evolution of biosynthetic pathways. *Org. Geochem.* **2005**, *36*, 139–159. [CrossRef]
3. Cheng, L.L.; Wang, G.Z.; Zhang, H.Y. Sterol biosynthesis and prokaryotes-to-eukaryotes evolution. *Biochem. Biophys. Res. Commun.* **2007**, *363*, 885–888. [CrossRef] [PubMed]
4. Desmond, E.; Gribaldo, S. Phylogenomics of sterol synthesis: Insights into the origin, evolution, and diversity of a key eukaryotic feature. *Genome Biol. Evol.* **2009**, *1*, 364–381. [CrossRef] [PubMed]
5. Wei, J.H.; Yin, X.; Welander, P.V. Sterol synthesis in diverse bacteria. *Front. Microbiol.* **2016**, *7*, 990. [CrossRef] [PubMed]
6. Blackstone, N.W. An evolutionary framework for understanding the origin of eukaryotes. *Biology* **2016**, *5*, 18. [CrossRef] [PubMed]
7. Tabas, I. Cholesterol in health and disease. *J. Clin. Investig.* **2002**, *110*, 583–590. [CrossRef] [PubMed]
8. Santori, F.R.; Huang, P.; van de Pavert, S.A.; Douglass, E.F., Jr.; Leaver, D.J.; Haubrich, B.A.; Keber, R.; Lorbek, G.; Konijn, T.; Rosales, B.N.; et al. Identification of natural RORgamma ligands that regulate the development of lymphoid cells. *Cell. Metab.* **2015**, *21*, 286–298. [CrossRef] [PubMed]
9. Poirot, M.; Silvente-Poirot, S. The tumor-suppressor cholesterol metabolite, dendrogenin A, is a new class of LXR modulator activating lethal autophagy in cancers. *Biochem. Pharmacol.* **2018**, *153*, 75–81. [CrossRef] [PubMed]

10. Zerbinati, C.; Caponecchia, L.; Puca, R.; Ciacciarelli, M.; Salacone, P.; Sebastianelli, A.; Pastore, A.; Palleschi, G.; Petrozza, V.; Porta, N.; et al. Mass spectrometry profiling of oxysterols in human sperm identifies 25-hydroxycholesterol as a marker of sperm function. *Redox Biol.* **2017**, *11*, 111–117. [CrossRef] [PubMed]
11. Griffiths, W.J.; Abdel-Khalik, J.; Hearn, T.; Yutuc, E.; Morgan, A.H.; Wang, Y. Current trends in oxysterol research. *Biochem. Soc. Trans.* **2016**, *44*, 652–658. [CrossRef] [PubMed]
12. Testa, G.; Rossin, D.; Poli, G.; Biasi, F.; Leonarduzzi, G. Implication of oxysterols in chronic inflammatory human diseases. *Biochimie* **2018**. [CrossRef] [PubMed]
13. Griffiths, W.J.; Wang, Y. Sterolomics: State of the art, developments, limitations and challenges. *Biochim. Biophys. Acta.* **2017**, *1862*, 771–773. [CrossRef] [PubMed]
14. Griffiths, W.J.; Wang, Y. An update on oxysterol biochemistry: New discoveries in lipidomics. *Biochem. Biophys. Res. Commun.* **2018**. [CrossRef] [PubMed]
15. Moreau, R.A.; Nystrom, L.; Whitaker, B.D.; Winkler-Moser, J.K.; Baer, D.J.; Gebauer, S.K.; Hicks, K.B. Phytosterols and their derivatives: Structural diversity, distribution, metabolism, analysis, and health-promoting uses. *Prog. Lipid. Res.* **2018**, *70*, 35–61. [CrossRef] [PubMed]
16. Nes, W.R.; Krevitz, K.; Joseph, J.; Nes, W.D. The phylogenetic distribution of sterols in tracheophytes. *Lipids* **1977**, *12*, 511–527. [CrossRef]
17. Weete, J.D. Structure and function of sterols in fungi. In *Advances in Lipid Research*; Paoletti, R., Kritchevsky, D., Eds.; Elsevier: New York, NY, USA, 1989; Volume 23, pp. 115–167.
18. Weete, J.D.; Gandhi, S.R. Sterols of the phylum zygomycota: Phylogenetic implications. *Lipids* **1997**, *32*, 1309–1316. [CrossRef] [PubMed]
19. Furlong, S.T.; Samia, J.A.; Rose, R.M.; Fishman, J.A. Phytosterols are present in Pneumocystis carinii. *Antimicrob. Agents Chemother.* **1994**, *38*, 2534–2540. [CrossRef] [PubMed]
20. Nes, W.D.; Norton, R.A.; Crumley, F.G.; Madigan, S.J.; Katz, E.R. Sterol phylogenesis and algal evolution. *Proc. Natl. Acad. Sci. USA* **1990**, *87*, 7565–7569. [CrossRef] [PubMed]
21. Halevy, S.; Avivi, L.; Katan, H. Sterols of soil amoebas and Ochromonas danica: Phylogenetic approach. *J. Protozool.* **1966**, *13*, 480–483. [CrossRef] [PubMed]
22. Raederstorff, D.; Rohmer, M. Sterol biosynthesis de nova via cycloartenol by the soil amoeba *Acanthamoeba polyphaga*. *Biochem. J.* **1985**, *231*, 609–615. [CrossRef] [PubMed]
23. Burden, R.S.; Cooke, D.T.; Carter, G.A. Inhibitors of sterol biosynthesis and growth in plants and fungi. *Phytochemistry* **1989**, *28*, 1791–1804. [CrossRef]
24. Odds, F.C.; Brown, A.J.; Gow, N.A. Antifungal agents: Mechanisms of action. *Trends Microbiol.* **2003**, *11*, 272–279. [CrossRef]
25. Leaver, D.J. Synthesis and Biological Activity of Sterol 14alpha-Demethylase and Sterol C24-Methyltransferase Inhibitors. *Molecules* **2018**, *23*, 1753. [CrossRef] [PubMed]
26. Liu, J.; Nes, W.D. Steroidal triterpenes: Design of substrate-based inhibitors of ergosterol and sitosterol synthesis. *Molecules* **2009**, *14*, 4690–4706. [CrossRef] [PubMed]
27. Godtfredsen, W.O.; Jahnsen, S.; Lorck, H.; Roholt, K.; Tybring, L. Fusidic acid: A new antibiotic. *Nature* **1962**, *193*, 987. [CrossRef] [PubMed]
28. Verbist, L. The antimicrobial activity of fusidic acid. *J. Antimicrob. Chemother.* **1990**, *25* (Suppl. B), 1–5. [CrossRef] [PubMed]
29. Zhao, M.; Godecke, T.; Gunn, J.; Duan, J.A.; Che, C.T. Protostane and fusidane triterpenes: A mini-review. *Molecules* **2013**, *18*, 4054–4080. [CrossRef] [PubMed]
30. Bu, M.; Yang, B.B.; Hu, L. Natural bioactive sterol 5α, 8α-endoperoxides as drug lead compounds. *Med. Chem.* **2014**, *4*, 709–716. [CrossRef]
31. Dembitsky, V.M. Bioactive fungal endoperoxides. *Med. Mycol.* **2015**, *1*, 5. [CrossRef]
32. Liu, D.; Li, X.M.; Li, C.S.; Wang, B.G. Nigerasterols A and B, antiproliferative sterols from the mangrove-derived endophytic fungus *Aspergillus niger* MA-132. *Helvetica. Chimica. Acta* **2013**, *96*, 1055–1061. [CrossRef]
33. Zhou, X.; Sun, J.; Ma, W.; Fang, W.; Chen, Z.; Yang, B.; Liu, Y. Bioactivities of six sterols isolated from marine invertebrates. *Pharm. Biol.* **2014**, *52*, 187–190. [CrossRef] [PubMed]
34. Kitchawalit, S.; Kanokmedhakul, K.; Kanokmedhakul, S.; Soytong, K. A new benzyl ester and ergosterol derivatives from the fungus *Gymnoascus reessii*. *Nat. Prod. Res.* **2014**, *28*, 1045–1051. [CrossRef] [PubMed]

35. Brunel, J.M.; Salmi, C.; Loncle, C.; Vidal, N.; Letourneux, Y. Squalamine: A polyvalent drug of the future? *Curr. Cancer Drug Targets* **2005**, *5*, 267–272. [CrossRef] [PubMed]
36. Zhou, W.; Warrilow, A.G.S.; Thomas, C.D.; Ramos, E.; Parker, J.E.; Price, C.L.; Vanderloop, B.H.; Fisher, P.M.; Loftis, M.D.; Kelly, D.E.; et al. Functional importance for developmental regulation of sterol biosynthesis in *Acanthamoeba castellanii. Biochim. Biophys. Acta* **2018**, *1863*, 1164–1178. [CrossRef] [PubMed]
37. Thomson, S.; Rice, C.A.; Zhang, T.; Edrada-Ebel, R.; Henriquez, F.L.; Roberts, C.W. Characterisation of sterol biosynthesis and validation of 14alpha-demethylase as a drug target in *Acanthamoeba. Sci. Rep.* **2017**, *7*, 8247. [CrossRef] [PubMed]
38. Korn, E.D.; Von Brand, T.; Tobie, E.J. The sterols of Trypanosoma cruzi and *Crithidia fasciculata. Comp. Biochem. Physiol.* **1969**, *30*, 601–610. [CrossRef]
39. Goad, L.J.; Holz, G.G., Jr.; Beach, D.H. Sterols of Leishmania species. Implications for biosynthesis. *Mol. Biochem. Parasitol.* **1984**, *10*, 161–170. [CrossRef]
40. Nakamura, C.V.; Waldow, L.; Pelegrinello, S.R.; Ueda-Nakamura, T.; Filho, B.A.; Filho, B.P. Fatty acid and sterol composition of three phytomonas species. *Mem. Inst. Oswaldo Cruz.* **1999**, *94*, 519–525. [CrossRef] [PubMed]
41. Haubrich, B.A.; Singha, U.K.; Miller, M.B.; Nes, C.R.; Anyatonwu, H.; Lecordier, L.; Patkar, P.; Leaver, D.J.; Villalta, F.; Vanhollebeke, B.; et al. Discovery of an ergosterol-signaling factor that regulates Trypanosoma brucei growth. *J. Lipid Res.* **2015**, *56*, 331–341. [CrossRef] [PubMed]
42. Nes, C.R.; Singha, U.K.; Liu, J.; Ganapathy, K.; Villalta, F.; Waterman, M.R.; Lepesheva, G.I.; Chaudhuri, M.; Nes, W.D. Novel sterol metabolic network of Trypanosoma brucei procyclic and bloodstream forms. *Biochem. J.* **2012**, *443*, 267–277. [CrossRef] [PubMed]
43. Zhou, W.; Cross, G.A.; Nes, W.D. Cholesterol import fails to prevent catalyst-based inhibition of ergosterol synthesis and cell proliferation of *Trypanosoma brucei. J. Lipid. Res.* **2007**, *48*, 665–673. [CrossRef] [PubMed]
44. Urbina, J.A. Ergosterol biosynthesis and drug development for Chagas disease. *Mem. Inst. Oswaldo Cruz* **2009**, *104* (Suppl. 1), 311–318. [CrossRef] [PubMed]
45. Miller, M.B.; Haubrich, B.A.; Wang, Q.; Snell, W.J.; Nes, W.D. Evolutionarily conserved Delta(25(27))-olefin ergosterol biosynthesis pathway in the alga *Chlamydomonas reinhardtii. J. Lipid. Res.* **2012**, *53*, 1636–1645. [CrossRef] [PubMed]
46. Gealt, M.A.; Adler, J.H.; Nes, W.R. The sterols and fatty acids from purified flagella of *Chlamydomonas reinhardtii. Lipids* **1981**, *16*, 133–136. [CrossRef]
47. Patterson, G.W. The distribution of sterols in algae. *Lipids* **1971**, *6*, 120–127. [CrossRef]
48. Volkman, J.K. Sterols in microalgae. In *The Physiology of Microalgae*; Borowitzka, M., Beardall, J., Raven, J., Eds.; Springer: Cham, Switzerland, 2016; Volume 6, pp. 485–505.
49. Patkar, P.; Haubrich, B.A.; Qi, M.; Nguyen, T.T.; Thomas, C.D.; Nes, W.D. C-24-methylation of 26-fluorocycloartenols by recombinant sterol C-24-methyltransferase from soybean: Evidence for channel switching and its phylogenetic implications. *Biochem. J.* **2013**, *456*, 253–262. [CrossRef] [PubMed]
50. Haubrich, B.A.; Collins, E.K.; Howard, A.L.; Wang, Q.; Snell, W.J.; Miller, M.B.; Thomas, C.D.; Pleasant, S.K.; Nes, W.D. Characterization, mutagenesis and mechanistic analysis of an ancient algal sterol C24-methyltransferase: Implications for understanding sterol evolution in the green lineage. *Phytochemistry* **2015**, *113*, 64–72. [CrossRef] [PubMed]
51. Nes, W.D.; Marshall, J.A.; Jia, Z.; Jaradat, T.T.; Song, Z.; Jayasimha, P. Active site mapping and substrate channeling in the sterol methyltransferase pathway. *J. Biol. Chem.* **2002**, *277*, 42549–42556. [CrossRef] [PubMed]
52. Dahmen, A.S.; Houle, H.M.; Leblond, J.D. Free sterols of the historically important green algal genera, Acetabularia and Acicularia (Polyphysaceae): A modern interpretation. *Algol. Stud.* **2017**, *2017*, 59–70. [CrossRef]
53. Martin-Creuzburg, D.; Merkel, P. Sterols of freshwater microalgae: Potential implications for zooplankton nutrition. *J. Plankton Res.* **2016**, *38*, 865–877. [CrossRef]
54. Giner, J.L.; Zhao, H.; Tomas, C. Sterols and fatty acids of three harmful algae previously assigned as *Chattonella. Phytochemistry* **2008**, *69*, 2167–2171. [CrossRef] [PubMed]
55. Giner, J.L.; Zhao, H.; Dixon, M.S.; Wikfors, G.H. Bioconversion of (13)C-labeled microalgal phytosterols to cholesterol by the Northern Bay scallop, *Argopecten irradians irradians. Comp. Biochem. Physiol. B Biochem. Mol. Biol.* **2016**, *192*, 1–8. [CrossRef] [PubMed]

56. Gergs, R.; Steinberger, N.; Beck, B.; Basen, T.; Yohannes, E.; Schulz, R.; Martin-Creuzburg, D. Compound-specific delta(13)C analyses reveal sterol metabolic constraints in an aquatic invertebrate. *Rapid. Commun. Mass Spectrom.* **2015**, *29*, 1789–1794. [CrossRef] [PubMed]
57. Martin-Creuzburg, D.; von Elert, E. Ecological significance of sterols in aquatic food webs. In *Lipids in Aquatic Ecosystems*; Kainz, M., Brett, M., Arts, M., Eds.; Springer: New York, NY, USA, 2009.
58. Behmer, S.T.; Nes, W.D. Insect sterol nutrition and physiology: A global overview. *Adv. Insect. Physiol.* **2003**, *31*, 72. [CrossRef]
59. Taipale, S.J.; Hiltunen, M.; Vuorio, K.; Peltomaa, E. Suitability of Phytosterols Alongside Fatty Acids as Chemotaxonomic Biomarkers for Phytoplankton. *Front. Plant. Sci.* **2016**, *7*, 212. [CrossRef] [PubMed]
60. Geng, H.X.; Yu, R.C.; Chen, Z.F.; Peng, Q.C.; Yan, T.; Zhou, M.J. Analysis of sterols in selected bloom-forming algae in China. *Harmful. Algae.* **2017**, *66*, 29–39. [CrossRef] [PubMed]
61. Giner, J.L.; Ceballos, H.; Tang, Y.Z.; Gobler, C.J. Sterols and Fatty Acids of the Harmful Dinoflagellate *Cochlodinium polykrikoides*. *Chem. Biodivers.* **2016**, *13*, 249–252. [CrossRef] [PubMed]
62. Houle, H.M.; Lopez, C.B.; Leblond, J.D. Sterols of the Toxic Marine Dinoflagellate, *Pyrodinium bahamense*. *J. Eukaryot. Microbiol.* **2018**. [CrossRef] [PubMed]
63. Gori, A.; Tolosa, I.; Orejas, C.; Rueda, L.; Viladrich, N.; Grinyo, J.; Flogel, S.; Grover, R.; Ferrier-Pages, C. Biochemical composition of the cold-water coral *Dendrophyllia cornigera* under contrasting productivity regimes: Insights from lipid biomarkers and compound-specific isotopes. *Deep Sea Res. Part 1 Oceanogr. Res. Pap.* **2018**. [CrossRef]
64. Crandell, J.B.; Teece, M.A.; Estes, B.A.; Manfrino, C.; Ciesla, J.H. Nutrient acquisition strategies in mesophotic hard corals using compound specific stable isotope analysis of sterols. *J. Exp. Mar. Bio. Ecol.* **2016**, *474*, 133–141. [CrossRef]
65. Zhou, W.; Lepesheva, G.I.; Waterman, M.R.; Nes, W.D. Mechanistic analysis of a multiple product sterol methyltransferase implicated in ergosterol biosynthesis in *Trypanosoma brucei*. *J. Biol. Chem.* **2006**, *281*, 6290–6296. [CrossRef] [PubMed]
66. Millerioux, Y.; Mazet, M.; Bouyssou, G.; Allmann, S.; Kiema, T.R.; Bertiaux, E.; Fouillen, L.; Thapa, C.; Biran, M.; Plazolles, N.; et al. De novo biosynthesis of sterols and fatty acids in the Trypanosoma brucei procyclic form: Carbon source preferences and metabolic flux redistributions. *PLoS Pathog.* **2018**, *14*, e1007116. [CrossRef] [PubMed]
67. Leaver, D.J.; Patkar, P.; Singha, U.K.; Miller, M.B.; Haubrich, B.A.; Chaudhuri, M.; Nes, W.D. Fluorinated Sterols Are Suicide Inhibitors of Ergosterol Biosynthesis and Growth in *Trypanosoma brucei*. *Chem. Biol.* **2015**, *22*, 1374–1383. [CrossRef] [PubMed]
68. Dauchy, F.A.; Bonhivers, M.; Landrein, N.; Dacheux, D.; Courtois, P.; Lauruol, F.; Daulouede, S.; Vincendeau, P.; Robinson, D.R. *Trypanosoma brucei* CYP51: Essentiality and Targeting Therapy in an Experimental Model. *PLoS Negl. Trop Dis.* **2016**, *10*, e0005125. [CrossRef] [PubMed]
69. Lepesheva, G.I.; Ott, R.D.; Hargrove, T.Y.; Kleshchenko, Y.Y.; Schuster, I.; Nes, W.D.; Hill, G.C.; Villalta, F.; Waterman, M.R. Sterol 14alpha-demethylase as a potential target for antitrypanosomal therapy: Enzyme inhibition and parasite cell growth. *Chem. Biol.* **2007**, *14*, 1283–1293. [CrossRef] [PubMed]
70. Lorente, S.O.; Rodrigues, J.C.; Jimenez Jimenez, C.; Joyce-Menekse, M.; Rodrigues, C.; Croft, S.L.; Yardley, V.; de Luca-Fradley, K.; Ruiz-Perez, L.M.; Urbina, J.; et al. Novel azasterols as potential agents for treatment of leishmaniasis and trypanosomiasis. *Antimicrob. Agents Chemother.* **2004**, *48*, 2937–2950. [CrossRef] [PubMed]
71. Miller, M.B.; Patkar, P.; Singha, U.K.; Chaudhuri, M.; David Nes, W. 24-Methylenecyclopropane steroidal inhibitors: A Trojan horse in ergosterol biosynthesis that prevents growth of Trypanosoma brucei. *Biochim. Biophys. Acta Mol. Cell. Biol. Lipids* **2017**, *1862*, 305–313. [CrossRef] [PubMed]
72. Kidane, M.E.; Vanderloop, B.H.; Zhou, W.; Thomas, C.D.; Ramos, E.; Singha, U.; Chaudhuri, M.; Nes, W.D. Sterol methyltransferase a target for anti-amoeba therapy: Towards transition state analog and suicide substrate drug design. *J. Lipid. Res.* **2017**, *58*, 2310–2323. [CrossRef] [PubMed]
73. Lamb, D.C.; Warrilow, A.G.; Rolley, N.J.; Parker, J.E.; Nes, W.D.; Smith, S.N.; Kelly, D.E.; Kelly, S.L. Azole Antifungal Agents to Treat the Human Pathogens *Acanthamoeba castellanii* and *Acanthamoeba polyphaga* through Inhibition of Sterol 14alpha-Demethylase (CYP51). *Antimicrob. Agents Chemother.* **2015**, *59*, 4707–4713. [CrossRef] [PubMed]

74. Nakaminami, H.; Tanuma, K.; Enomoto, K.; Yoshimura, Y.; Onuki, T.; Nihonyanagi, S.; Hamada, Y.; Noguchi, N. Evaluation of In Vitro Antiamoebic Activity of Antimicrobial Agents Against Clinical Acanthamoeba Isolates. *J. Ocul. Pharmacol. Ther.* **2017**, *33*, 629–634. [CrossRef] [PubMed]
75. Gueudry, J.; Le Goff, L.; Compagnon, P.; Lefevre, S.; Colasse, E.; Aknine, C.; Duval, F.; Francois, A.; Razakandrainibe, R.; Ballet, J.J.; et al. Evaluation of voriconazole anti-Acanthamoeba polyphaga in vitro activity, rat cornea penetration and efficacy against experimental rat *Acanthamoeba keratitis*. *J. Antimicrob. Chemother.* **2018**. [CrossRef] [PubMed]
76. Taravaud, A.; Loiseau, P.M.; Pomel, S. In vitro evaluation of antimicrobial agents on *Acanthamoeba* sp. and evidence of a natural resilience to amphotericin B. *Int. J. Parasitol. Drugs Drug Resist.* **2017**, *7*, 328–336. [CrossRef] [PubMed]
77. Zhao, S.; Zhang, X.; Wei, P.; Su, X.; Zhao, L.; Wu, M.; Hao, C.; Liu, C.; Zhao, D.; Cheng, M. Design, synthesis and evaluation of aromatic heterocyclic derivatives as potent antifungal agents. *Eur. J. Med. Chem.* **2017**, *137*, 96–107. [CrossRef] [PubMed]
78. Zhao, S.; Wei, P.; Wu, M.; Zhang, X.; Zhao, L.; Jiang, X.; Hao, C.; Su, X.; Zhao, D.; Cheng, M. Design, synthesis and evaluation of benzoheterocycle analogues as potent antifungal agents targeting CYP51. *Bioorg. Med. Chem.* **2018**, *26*, 3242–3253. [CrossRef] [PubMed]
79. Muller, C.; Neugebauer, T.; Zill, P.; Lass-Florl, C.; Bracher, F.; Binder, U. Sterol Composition of Clinically Relevant Mucorales and Changes Resulting from Posaconazole Treatment. *Molecules* **2018**, *23*, 1218. [CrossRef] [PubMed]
80. Li, S.; Shi, H.; Chang, W.; Li, Y.; Zhang, M.; Qiao, Y.; Lou, H. Eudesmane sesquiterpenes from Chinese liverwort are substrates of Cdrs and display antifungal activity by targeting Erg6 and Erg11 of *Candida albicans*. *Bioorg. Med. Chem.* **2017**, *25*, 5764–5771. [CrossRef] [PubMed]
81. Helliwell, S.B.; Karkare, S.; Bergdoll, M.; Rahier, A.; Leighton-Davis, J.R.; Fioretto, C.; Aust, T.; Filipuzzi, I.; Frederiksen, M.; Gounarides, J.; et al. FR171456 is a specific inhibitor of mammalian NSDHL and yeast Erg26p. *Nat. Commun.* **2015**, *6*, 8613. [CrossRef] [PubMed]
82. Warrilow, A.G.; Parker, J.E.; Price, C.L.; Nes, W.D.; Garvey, E.P.; Hoekstra, W.J.; Schotzinger, R.J.; Kelly, D.E.; Kelly, S.L. The Investigational Drug VT-1129 Is a Highly Potent Inhibitor of Cryptococcus Species CYP51 but Only Weakly Inhibits the Human Enzyme. *Antimicrob. Agents Chemother.* **2016**, *60*, 4530–4538. [CrossRef] [PubMed]
83. Alhanout, K.; Rolain, J.M.; Brunel, J.M. Squalamine as an example of a new potent antimicrobial agents class: A critical review. *Curr. Med. Chem.* **2010**, *17*, 3909–3917. [CrossRef] [PubMed]
84. Savage, P.B.; Li, C.; Taotafa, U.; Ding, B.; Guan, Q. Antibacterial properties of cationic steroid antibiotics. *FEMS Microbiol. Lett.* **2002**, *217*, 1–7. [CrossRef] [PubMed]
85. Huang, C.Y.; Tseng, W.R.; Ahmed, A.F.; Chiang, P.L.; Tai, C.J.; Hwang, T.L.; Dai, C.F.; Sheu, J.H. Anti-Inflammatory Polyoxygenated Steroids from the Soft Coral *Lobophytum michaelae*. *Mar. Drugs* **2018**, *16*, 93. [CrossRef] [PubMed]
86. Nguyen, H.M.; Ito, T.; Win, N.N.; Vo, H.Q.; Nguyen, H.T.; Morita, H. A new sterol from the Vietnamese marine sponge Xestospongia testudinaria and its biological activities. *Nat. Prod. Res.* **2018**. [CrossRef] [PubMed]
87. Igarashi, Y.; Sekine, A.; Fukazawa, H.; Uehara, Y.; Yamaguchi, K.; Endo, Y.; Okuda, T.; Furumai, T.; Oki, T. Anicequol, a novel inhibitor for anchorage-independent growth of tumor cells from *Penicillium aurantiogriseum* Dierckx TP-F0213. *J. Antibiot.* **2002**, *55*, 371–376. [CrossRef] [PubMed]
88. Nozawa, Y.; Sakai, N.; Matsumoto, K.; Mizoue, K. A novel neuritogenic compound, NGA0187. *J. Antibiot.* **2002**, *55*, 629–634. [CrossRef] [PubMed]
89. Tanaka, A.R.; Noguchi, K.; Fukazawa, H.; Igarashi, Y.; Arai, H.; Uehara, Y. p38MAPK and Rho-dependent kinase are involved in anoikis induced by anicequol or 25-hydroxycholesterol in DLD-1 colon cancer cells. *Biochem. Biophys. Res. Commun.* **2013**, *430*, 1240–1245. [CrossRef] [PubMed]
90. Deng, C.M.; Liu, S.X.; Huang, C.H.; Pang, J.Y.; Lin, Y.C. Secondary metabolites of a mangrove endophytic fungus Aspergillus terreus (No. GX7-3B) from the South China Sea. *Mar. Drugs* **2013**, *11*, 2616–2624. [CrossRef] [PubMed]
91. Gao, S.S.; Li, X.M.; Li, C.S.; Proksch, P.; Wang, B.G. Penicisteroids A and B, antifungal and cytotoxic polyoxygenated steroids from the marine alga-derived endophytic fungus *Penicillium chrysogenum* QEN-24S. *Bioorg. Med. Chem. Lett.* **2011**, *21*, 2894–2897. [CrossRef] [PubMed]

92. Abdel-Razek, A.S.; Hamed, A.; Frese, M.; Sewald, N.; Shaaban, M. Penicisteroid C: New polyoxygenated steroid produced by co-culturing of Streptomyces piomogenus with *Aspergillus niger*. *Steroids* **2018**, *138*, 21–25. [CrossRef] [PubMed]
93. Cheng, Z.B.; Xiao, H.; Fan, C.Q.; Lu, Y.N.; Zhang, G.; Yin, S. Bioactive polyhydroxylated sterols from the marine sponge *Haliclona crassiloba*. *Steroids* **2013**, *78*, 1353–1358. [CrossRef] [PubMed]
94. Ngoc, N.T.; Huong, P.T.; Thanh, N.V.; Chi, N.T.; Dang, N.H.; Cuong, N.X.; Nam, N.H.; Thung, D.C.; Kiem, P.V.; Minh, C.V. Cytotoxic Steroids from the Vietnamese Soft Coral *Sinularia conferta*. *Chem. Pharm. Bull.* **2017**, *65*, 300–305. [CrossRef] [PubMed]
95. Wang, Z.; Tang, H.; Wang, P.; Gong, W.; Xue, M.; Zhang, H.; Liu, T.; Liu, B.; Yi, Y.; Zhang, W. Bioactive polyoxygenated steroids from the South China sea soft coral, *Sarcophyton* sp. *Mar. Drugs* **2013**, *11*, 775–787. [CrossRef] [PubMed]
96. Chung, T.W.; Su, J.H.; Lin, C.C.; Li, Y.R.; Chao, Y.H.; Lin, S.H.; Chan, H.L. 24-Methyl-Cholesta-5,24(28)-Diene-3beta,19-diol-7beta-Monoacetate Inhibits Human Small Cell Lung Cancer Growth In Vitro and In Vivo via Apoptosis Induction. *Mar. Drugs* **2017**, *15*, 210. [CrossRef] [PubMed]
97. Meesala, S.; Gurung, P.; Karmodiya, K.; Subrayan, P.; Watve, M.G. Isolation and structure elucidation of halymeniaol, a new antimalarial sterol derivative from the red alga *Halymenia floresii*. *J. Asian Nat. Prod. Res.* **2018**, *20*, 391–398. [CrossRef] [PubMed]
98. Nguyen, X.C.; Longeon, A.; Pham, V.C.; Urvois, F.; Bressy, C.; Trinh, T.T.; Nguyen, H.N.; Phan, V.K.; Chau, V.M.; Briand, J.F.; et al. Antifouling 26,27-cyclosterols from the Vietnamese marine sponge *Xestospongia testudinaria*. *J. Nat. Prod.* **2013**, *76*, 1313–1318. [CrossRef] [PubMed]
99. Pailee, P.; Mahidol, C.; Ruchirawat, S.; Prachyawarakorn, V. Sterols from Thai Marine Sponge Petrosia (*Strongylophora*) sp. and Their Cytotoxicity. *Mar. Drugs* **2017**, *15*, 54. [CrossRef] [PubMed]
100. Cheng, Z.; Liu, D.; de Voogd, N.J.; Proksch, P.; Lin, W. New sterol derivatives from the marine sponge *Xestospongia* sp. *Helvetica. Chim. Acta* **2016**, *99*, 588–596. [CrossRef]
101. Tsai, C.-R.; Huang, C.Y.; Chen, B.W.; Tsai, Y.Y.; Shih, S.-P.; Hwang, T.-L.; Dai, C.-F.; Wang, S.-Y.; Sheu, J.-H. New bioactive steroids from the soft coral *Klyxum flaccidum*. *RSC Adv.* **2015**, *5*, 12546–12554. [CrossRef]
102. Tseng, W.R.; Huang, C.Y.; Tsai, Y.Y.; Lin, Y.S.; Hwang, T.L.; Su, J.H.; Sung, P.J.; Dai, C.F.; Sheu, J.H. New cytotoxic and anti-inflammatory steroids from the soft coral *Klyxum flaccidum*. *Bioorg. Med. Chem. Lett.* **2016**, *26*, 3253–3257. [CrossRef] [PubMed]
103. Nakamura, F.; Kudo, N.; Tomachi, Y.; Nakata, A.; Takemoto, M.; Ito, A.; Tabei, H.; Arai, D.; de Voogd, N.; Yoshida, M.; et al. Halistanol sulfates I. and J, new SIRT1-3 inhibitory steroid sulfates from a marine sponge of the genus Halichondria. *J. Antibiot.* **2018**, *71*, 273–278. [CrossRef] [PubMed]
104. Fiorucci, S.; Distrutti, E.; Bifulco, G.; D'Auria, M.V.; Zampella, A. Marine sponge steroids as nuclear receptor ligands. *Trends Pharmacol. Sci.* **2012**, *33*, 591–601. [CrossRef] [PubMed]
105. Xia, Q.; Zhang, H.; Sun, X.; Zhao, H.; Wu, L.; Zhu, D.; Yang, G.; Shao, Y.; Zhang, X.; Mao, X.; et al. A comprehensive review of the structure elucidation and biological activity of triterpenoids from *Ganoderma* spp. *Molecules* **2014**, *19*, 17478–17535. [CrossRef] [PubMed]
106. Metwaly, A.M.; Kadry, H.A.; El-Hela, A.A.; Mohammad, A.I.; Ma, G.; Cutler, S.J.; Ross, S.A. Nigrosphaerin A a new isochromene derivative from the endophytic fungus Nigrospora sphaerica. *Phytochem. Lett.* **2014**, *7*, 1–5. [CrossRef] [PubMed]
107. Chen, M.; Wang, K.L.; Liu, M.; She, Z.G.; Wang, C.Y. Bioactive steroid derivatives and butyrolactone derivatives from a gorgonian-derived Aspergillus sp. fungus. *Chem. Biodivers.* **2015**, *12*, 1398–1406. [CrossRef] [PubMed]
108. Zhao, Q.; Chen, G.D.; Feng, X.L.; Yu, Y.; He, R.R.; Li, X.X.; Huang, Y.; Zhou, W.X.; Guo, L.D.; Zheng, Y.Z.; et al. Nodulisporiviridins A-H, Bioactive Viridins from *Nodulisporium* sp. *J. Nat. Prod.* **2015**, *78*, 1221–1230. [CrossRef] [PubMed]
109. Harms, H.; Kehraus, S.; Nesaei-Mosaferan, D.; Hufendieck, P.; Meijer, L.; Konig, G.M. Abeta-42 lowering agents from the marine-derived fungus *Dichotomomyces cejpii*. *Steroids* **2015**, *104*, 182–188. [CrossRef] [PubMed]
110. Ngoc, N.T.; Huong, P.T.; Thanh, N.V.; Cuong, N.X.; Nam, N.H.; Thung do, C.; Kiem, P.V.; Minh, C.V. Steroid Constituents from the Soft Coral *Sinularia nanolobata*. *Chem. Pharm. Bull.* **2016**, *64*, 1417–1419. [CrossRef] [PubMed]

111. Thanh, N.V.; Ngoc, N.T.; Anh Hle, T.; Thung do, C.; Thao do, T.; Cuong, N.X.; Nam, N.H.; Kiem, P.V.; Minh, C.V. Steroid constituents from the soft coral *Sinularia microspiculata*. *J. Asian Nat. Prod. Res.* **2016**, *18*, 938–944. [CrossRef] [PubMed]
112. Ngoc, N.T.; Hanh, T.T.H.; Thanh, N.V.; Thao, D.T.; Cuong, N.X.; Nam, N.H.; Thung, D.C.; Kiem, P.V.; Minh, C.V. Cytotoxic Steroids from the Vietnamese Soft Coral *Sinularia leptoclados*. *Chem. Pharm. Bull.* **2017**, *65*, 593–597. [CrossRef] [PubMed]
113. Tran, H.H.T.; Nguyen Viet, P.; Nguyen Van, T.; Tran, H.T.; Nguyen Xuan, C.; Nguyen Hoai, N.; Do Cong, T.; Phan Van, K.; Chau Van, M. Cytotoxic steroid derivatives from the Vietnamese soft coral *Sinularia brassica*. *J. Asian Nat. Prod. Res.* **2017**, *19*, 1183–1190. [CrossRef] [PubMed]
114. Li, S.W.; Chen, W.T.; Yao, L.G.; Guo, Y.W. Two new cytotoxic steroids from the Chinese soft coral *Sinularia* sp. *Steroids* **2018**, *136*, 17–21. [CrossRef] [PubMed]
115. Huang, C.Y.; Su, J.H.; Liaw, C.C.; Sung, P.J.; Chiang, P.L.; Hwang, T.L.; Dai, C.F.; Sheu, J.H. Bioactive Steroids with Methyl Ester Group in the Side Chain from a Reef Soft Coral *Sinularia brassica* Cultured in a Tank. *Mar. Drugs* **2017**, *15*, 280. [CrossRef] [PubMed]
116. Tsai, Y.Y.; Huang, C.Y.; Tseng, W.R.; Chiang, P.L.; Hwang, T.L.; Su, J.H.; Sung, P.J.; Dai, C.F.; Sheu, J.H. Klyflaccisteroids, K-M bioactive steroidal derivatives from a soft coral *Klyxum flaccidum*. *Bioorg. Med. Chem. Lett.* **2017**, *27*, 1220–1224. [CrossRef] [PubMed]
117. Cheng, W.; Ren, J.; Huang, Q.; Long, H.; Jin, H.; Zhang, L.; Liu, H.; van Ofwegen, L.; Lin, W. Pregnane steroids from a gorgonian coral *Subergorgia suberosa* with anti-flu virus effects. *Steroids* **2016**, *108*, 99–104. [CrossRef] [PubMed]
118. He, W.F.; Xue, D.Q.; Yao, L.G.; Li, J.; Liu, H.L.; Guo, Y.W. A new bioactive steroidal ketone from the South China Sea sponge *Xestospongia testudinaria*. *J. Asian Nat. Prod. Res.* **2016**, *18*, 195–199. [CrossRef] [PubMed]
119. Mencarelli, A.; D'Amore, C.; Renga, B.; Cipriani, S.; Carino, A.; Sepe, V.; Perissutti, E.; D'Auria, M.V.; Zampella, A.; Distrutti, E.; et al. Solomonsterol A, a marine pregnane-X-receptor agonist, attenuates inflammation and immune dysfunction in a mouse model of arthritis. *Mar. Drugs* **2013**, *12*, 36–53. [CrossRef] [PubMed]
120. Yang, F.; Li, Y.Y.; Tang, J.; Sun, F.; Lin, H.W. New 4-methylidene sterols from the marine sponge *Theonella swinhoei*. *Fitoterapia* **2018**, *127*, 279–285. [CrossRef] [PubMed]
121. Abdelmohsen, U.R.; Cheng, C.; Reimer, A.; Kozjak-Pavlovic, V.; Ibrahim, A.K.; Rudel, T.; Hentschel, U.; Edrada-Ebel, R.; Ahmed, S.A. Antichlamydial sterol from the Red Sea sponge Callyspongia aff. implexa. *Planta. Med.* **2015**, *81*, 382–387. [CrossRef] [PubMed]
122. Perumal, P.; Sowmiya, R.; Prasanna Kumar, S.; Ravikumar, S.; Deepak, P.; Balasubramani, G. Isolation, structural elucidation and antiplasmodial activity of fucosterol compound from brown seaweed, *Sargassum linearifolium* against malarial parasite *Plasmodium falciparum*. *Nat. Prod. Res.* **2018**, *32*, 1316–1319. [CrossRef] [PubMed]
123. Lee, J.A.; Cho, Y.R.; Hong, S.S.; Anh, E.K. Anti-obesity activity of saringosterol isolated from *Sargassum muticum* (Yendo) fensholt extract in 3t3-L1 cells. *Phytother. Res.* **2017**, *31*, 1694–1701. [CrossRef] [PubMed]
124. Jin, H.G.; Zhou, M.; Jin, Q.H.; Liu, B.; Guan, L.P. Antidepressant-like effects of saringosterol, a sterol from *Sargassum fusiforme* by performing in vivo behavioral tests. *Med. Chem. Res.* **2017**, *26*, 909–915. [CrossRef]
125. Yang, F.; Zhang, L.W.; Feng, M.T.; Liu, A.H.; Li, J.; Zhao, T.S.; Lai, X.P.; Wang, B.; Guo, Y.W.; Mao, S.C. Dictyoptesterols A–C, Δ^{22}-24-oxo cholestane-type sterols with potent PTP1B inhibitory activity from the brown alga *Dictyopteris undulata* Holmes. *Fitoterapia* **2018**, *130*, 241–246. [CrossRef] [PubMed]
126. Feng, M.T.; Wang, T.; Liu, A.H.; Li, J.; Yao, L.G.; Wang, B.; Guo, Y.W.; Mao, S.C. PTP1B inhibitory and cytotoxic C-24 epimers of Delta(28)-24-hydroxy stigmastane-type steroids from the brown alga *Dictyopteris undulata* Holmes. *Phytochemistry* **2018**, *146*, 25–35. [CrossRef] [PubMed]

Review

Synthesis and Biological Activity of Sterol 14α-Demethylase and Sterol C24-Methyltransferase Inhibitors

David J. Leaver

Department of Biology, Geology and Physical Sciences, Sul Ross State University, Alpine, TX 79832, USA; david.leaver@sulross.edu; Tel.: +1-432-837-8115

Academic Editors: Wenxu Zhou and De-an Guo
Received: 20 June 2018; Accepted: 15 July 2018; Published: 17 July 2018

Abstract: Sterol 14α-demethylase (SDM) is essential for sterol biosynthesis and is the primary molecular target for clinical and agricultural antifungals. SDM has been demonstrated to be a valid drug target for antiprotozoal therapies, and much research has been focused on using SDM inhibitors to treat neglected tropical diseases such as human African trypanosomiasis (HAT), Chagas disease, and leishmaniasis. Sterol C24-methyltransferase (24-SMT) introduces the C24-methyl group of ergosterol and is an enzyme found in pathogenic fungi and protozoa but is absent from animals. This difference in sterol metabolism has the potential to be exploited in the development of selective drugs that specifically target 24-SMT of invasive fungi or protozoa without adversely affecting the human or animal host. The synthesis and biological activity of SDM and 24-SMT inhibitors are reviewed herein.

Keywords: sterol biosynthesis; sterol 14α-demethylase; sterol C24-methyltransferase; mechanism-based inactivators; antifungals; azoles; antiparasitic drugs; human African trypanosomiasis; Chagas disease; synthesis

---

1. Introduction

Sterols such as ergosterol and cholesterol are essential lipid molecules, and they perform numerous cellular roles associated with membrane (bulk) and signal (sparking) functions [1–3]. Cholesterol is biosynthesized in humans, while ergosterol or other 24-alkylated sterols are biosynthesized in opportunistic fungi and parasitic protozoa (Figure 1) [2–4]. This difference in sterol production can be exploited in the development of drugs that are designed to selectively block ergosterol biosynthesis in invasive fungi or protozoan parasites without harming the human host [3–7]. Infectious diseases caused by parasitic protozoa take a heavy toll on human health, are widespread, and are increasing in resistance to current chemotherapies [8]. Leishmaniasis is threatening around 350 million people in more than 98 countries [9], while the World Health Organization (WHO) has estimated that 16–18 million people are infected with Chagas disease [10], and it has been suggested that 70 million people in Africa are at risk of human African trypanosomiasis (HAT; sleeping sickness) [11]. Leishmaniasis is caused by various species of *Leishmania*, while the causative pathogens for Chagas and HAT are *Trypanosoma cruzi* and *Trypanosoma brucei*, respectively. It is well known that these diseases are life-threatening, and without proper treatment they are often fatal. *Leishmania* spp., *T. cruzi*, and *T. brucei* all require ergosterol for growth, and inhibiting the ergosterol biosynthesis pathway in these parasitic protozoa is an ideal approach to treat these infections without harming the human host. It should be noted that fungal infections caused by *Cryptococcus neoformans* have become the leading cause of morbidity and mortality in acquired immune deficiency syndrome (AIDS) patients and other immunocompromised patients, and it is reported that 5–10% of AIDS patients in the United States suffer from these life-threatening infections [12,13].

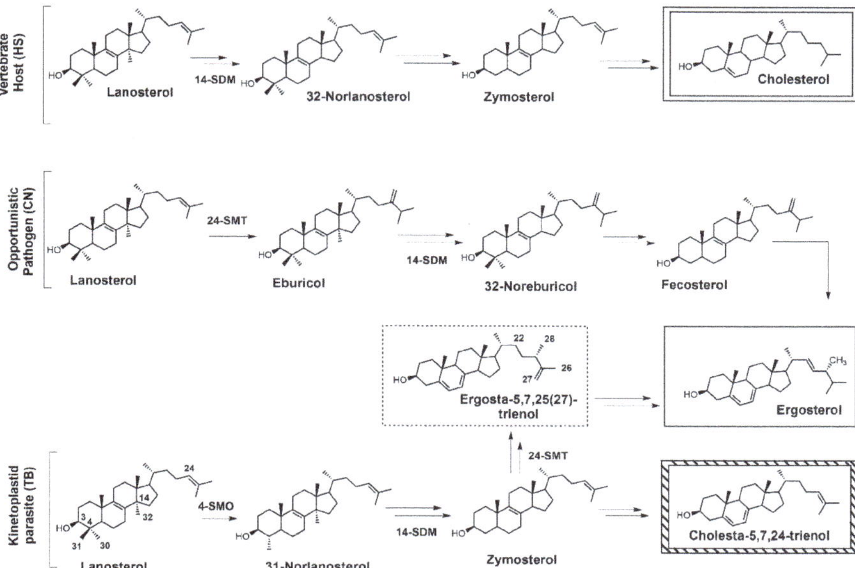

Figure 1. Comparative sterol biosynthesis pathways across kingdoms (adapted from [2]). HS: *Homo sapiens*; CN: *Cryptococcus neoformans*; TB: *Trypanosoma brucei*; 4-SMO: sterol C4-methyl oxidase; 14-SDM: sterol 14α-demethylase; 24-SMT: sterol C24-methyltransferase.

Selective inhibition of fungal SDM is one of the most common ways to treat fungal infections, and the majority of drugs that target fungal sterol 14α-demethylase (SDM) possess an azole side chain [12–17]. For these azole drugs to be efficacious, they need to have greater affinity for fungal SDM versus mammalian SDM [15,16]. Sterol C24-methyltransferase (24-SMT), unlike SDM, is not found in humans but is present in both fungi and protozoa (Figure 1), offering a selective way to inhibit ergosterol biosynthesis [15]. It is important to note that antifungal azoles are heavily used in agriculture [17–19], and azole resistance is becoming a major problem [20]. One mechanism of resistance to these antifungal azoles is the activation of efflux pumps that transport azoles out of fungal cells [21]. There is a huge medical need to develop new fungicides and medicinal drugs that specifically block ergosterol biosynthesis and are not likely to develop resistance.

SDM, also known as $P450_{14DM}$ or CYP51, catalyzes the removal of the C-32 methyl group of lanosterol (**1**) via a repetitive three-step process that uses reduced nicotinamide adenine dinucleotide phosphate (NADPH) and oxygen (Figure 2) to ultimately produce 4,4-dimethyl-5α-cholesta-8,14,24-trien-3β-ol (**4**) [22,23]. SDM transforms the C-32 methyl group of lanosterol into an alcohol (**2**), an aldehyde (**3**), and then formic acid with the insertion of the Δ14–15 double bond (Figure 2) [24]. Each P450 catalytic cycle involves the reduction of heme ferric iron to the ferrous state, the binding of molecular oxygen, and subsequent protonation to form a ferric hydroperoxo intermediate [24]. Further protonation of the distal oxygen atom of the ferric hydroperoxo intermediate causes heterolytic scission of the O–O bond, resulting in the loss of water and the formation of an Fe^{4+} oxo porphyrin cation radical, which is the catalytically active species [24]. Once the substrate is oxygenated, the iron returns to its ferric state ready for another catalytic cycle [24].

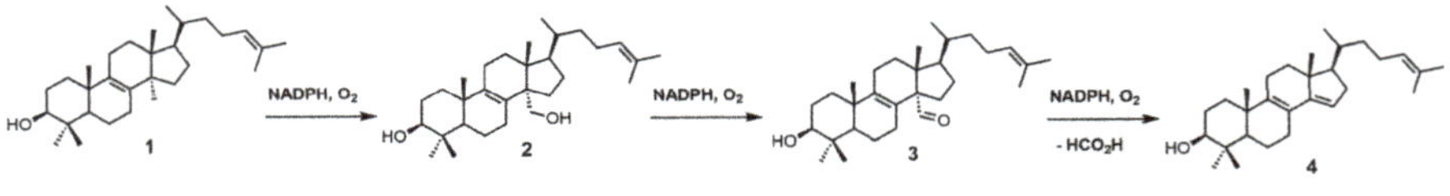

Figure 2. Conversion of lanosterol (**1**) into 4,4-dimethyl-5α-cholesta-8,14,24-trien-3β-ol (**4**) by sterol 14α-demethylase (SDM).

2. Sterol SDM Inhibitors

Sterol-based SDM inhibitors have been reported in the literature [22–30] (Figure 3); however, they are not as commonly reported as azoles [12,17,31–46]. This is likely due to the limited number of functional groups on lanosterol that can be synthetically modified [47], in addition to the difficult and time-consuming syntheses involved with sterol functional-group manipulation. The preferred substrate of SDM is compound **2**, which is the most potent natural inhibitor of SDM [25]. Compound **2** has a half-maximal inhibitory concentration (IC_{50}) value of 7.8 μM against SDM, and it can be made synthetically in nine steps starting from lanosterol [25,48].

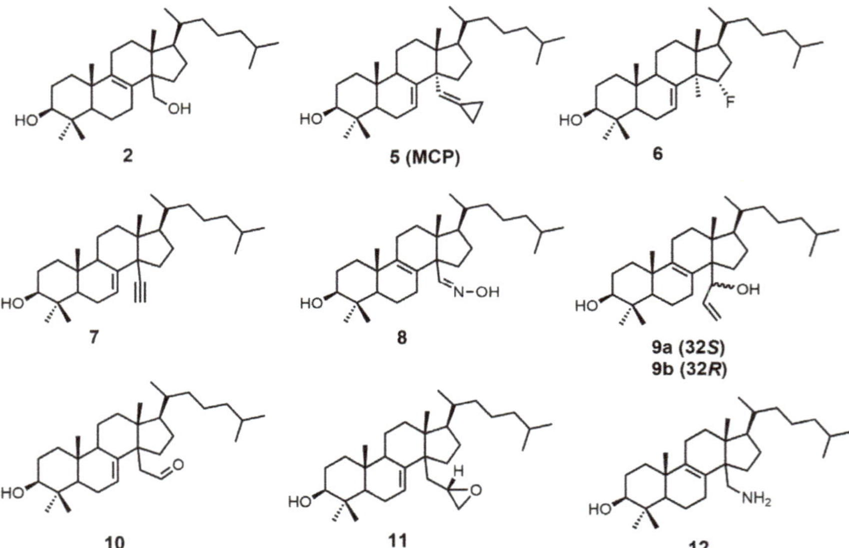

Figure 3. Sterol-based inhibitors of sterol 14α-demethylase (SDM).

14α-Methylenecyclopropyl-Δ7-24,25-dihydrolanosterol (**5** or MCP) was observed to be a competitive inhibitor of F105-containing *T. brucei* CYP51 (*Tb*CYP51)and *Lesihmania infantum* orthologos, while for *T. cruzi,* MCP was presumed to act as a mechanism-based inhibitor (suicide substrate) [24]. The cyclopropyl ring of MCP is presumably opened as MCP binds to *T. cruzi* CYP51 (*Tc*CYP51), forming a covalent bond with the prosthetic heme group [24]. The crystal structure of *Tb*CYP51 covalently bound with MCP has been reported [24]. Despite MCP having the same K_d values of 0.5 μM for both *Tb*CYP51 and *Tc*CYP51, the observed half-maximal effective concentration (EC_{50}) of MCP against *T. brucei* was >50 μM, while *T. cruzi* cell growth was inhibited by 50% at a MCP concentration of 3 μM [49]. MCP inhibits *Tc*CYP51 more than *Tb*CYP51, which is likely due to MCP acting as a suicide substrate for *Tc*CYP51 and as a competitive inhibitor for *Tb*CYP51 [24].

MCP can be synthesized in three steps starting with 3β-acetyloxylanost-7-en-30-ol (**13**) (Figure 4) [24]. Compound **13** can be synthesized directly from lanosterol in 12 steps [29,48]. Oxidation of compound **13** with Fetizon's reagent produced 3β-acetyloxylanost-7-en-30-al (**14**), which was then converted into MCP (**5**) via a Wittig reaction using a cyclopropyltriphenylphosphium ylide, followed by acetyl deprotection with lithium aluminum hydride (LAH) [24].

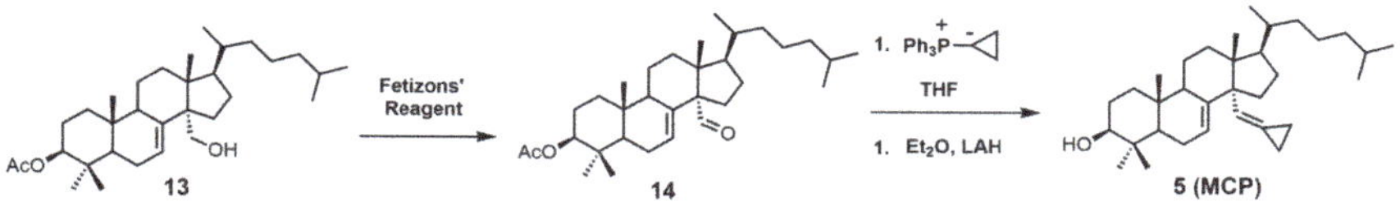

Figure 4. The synthesis of 14α-methylenecyclopropyl-Δ7-24,25-dihydrolanosterol (MCP) (**5**).

15α-Fluorolanost-7-en-3β-ol (**6**) was noted to be a weak competitive inhibitor of SDM with a K_i value of 315 μM [28]. Metabolic studies have indicated that compound **6** is converted to 15α-fluoro-3β-hydroxylanost-7-en-32-al by hepatic microsomal SDM and that the 15α-fluoro substitution blocks further metabolic conversion into other cholesterol biosynthetic intermediates [28]. The starting material used to synthesize compound **6** was 3β-benzoyloxy-lanost-7-en-15α-ol (**15**) (Figure 5) [50,51]. Compound **15** was reacted with diethylaminosulfur trifluoride (DAST) to install the fluorine at C-15, and the benozyl protecting group was removed by LAH [28].

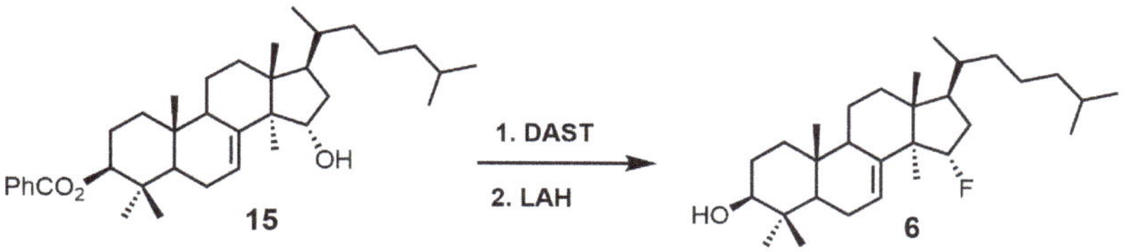

Figure 5. The synthesis of 15α-fluorolanost-7-en-3β-ol (**6**).

4,4-Dimethyl-14α-ethynylcholest-7-en-30-ol (**7**) was observed to have K_d values of 1.2 μM against *Tb*CYP51 and 1.3 μM against *Tc*CYP51 [49]. Compound **7** had a weaker affinity for both *Tb*CYP51 and *Tc*CYP51 in comparison to MCP. Compound **16** (Figure 6) is the starting material required to synthesize compound **7**, and compound **16** can be synthesized in 11 steps starting from lanosterol [23]. Aldehyde **16** was reacted with the ylide of chloromethyltriphenylphosphonium chloride, followed by the addition of *n*-butyllithium to introduce the desired alkyne functionality (Figure 6) [29]. The tetrahydropyran (THP) protecting group was removed by the use of pyridinium *p*-toluenesulfonate (PPTS) in ethanol to yield compound **7** [29].

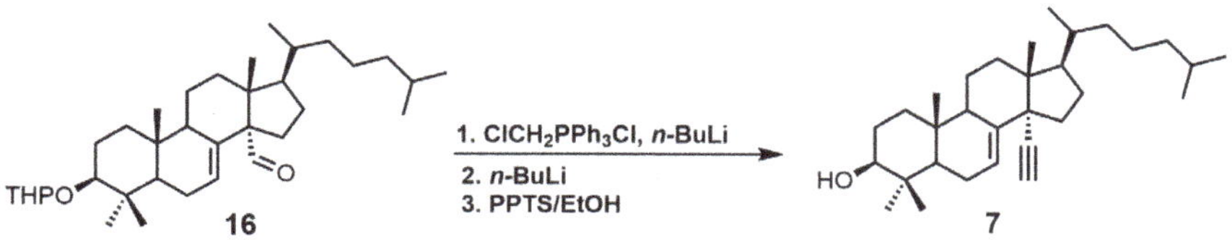

Figure 6. The synthesis of 4,4-dimethyl-14α-ethynylcholest-7-en-30-ol (**7**).

Lanost-8-en-32-alkoxime-3β-ol (**8**) was reported as having an IC_{50} value of 1.1 μM against SDM [25]. Compound **8** was readily prepared from 3β-benzoyloxy-lanost-8-en-32-al (**17**) in two steps (Figure 7) [52]. The aldehyde functional group of compound **17** was converted into an oxime by the use of hydroxylamine hydrochloride in pyridine [52]. The benzoyl protecting group was removed using potassium hydroxide in ethanol to yield compound **8** [52].

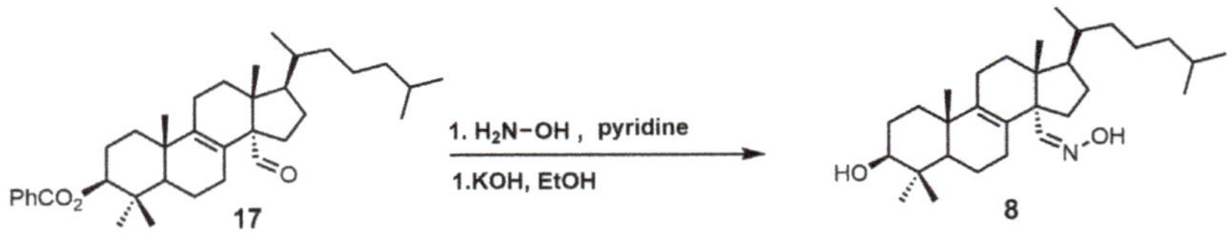

Figure 7. The synthesis of lanost-8-en-32-alkoxime-3β-ol (**8**).

A stereochemical preference of the 32-vinyl alcohols (compounds **9a** and **9b**; Figure 8) was observed against SDM [25]. The 32*S*-isomer (compound **9a**) was observed to have an IC_{50} value of 0.75 μM against SDM in comparison to the 32*R*-isomer (compound **9b**), which has a reported IC_{50} value of 3.20 μM [25]. This result illustrates the importance of having optimized stereochemistry in SDM inhibitors. Compounds **9a** and **9b** were readily prepared from lanost-8-en-32-al-3β-ol (**3**) via a Grignard reaction using vinyl magnesium bromide in THF [52]. The Grignard reaction was compatible without needing a protecting group for the 3β alcohol. The two diastereoisomers **9a** and **9b** were successfully separated by medium pressure liquid chromatography (MPLC) [52].

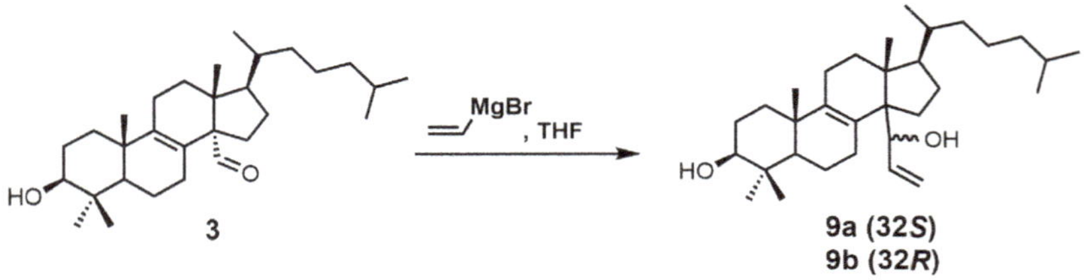

Figure 8. The synthesis of 4,4-dimethyl-14α-(1′hydroxy-2′-vinyl)-5α-cholest-8-en-3β-ols.

Aldehyde **10** and epoxide **11** (Figure 9) were both observed to inhibit total cholesterol and lathosterol biosynthesis by >89% at an inhibitor concentration of 10 μM, and these compounds were believed to inhibit SDM [27]. Compounds **10** and **11** have K_i values of 3 and 0.61 μM, respectively, while the 32*R*-oxiranyllanost-7-en-3β-ol isomer has a K_i value of 2 μM [27]. The synthesis of 32*S*-oxiranyllanost-7-en-3β-ol (**11**) and the 32*R* isomer started with a Wittig reaction between aldehyde **14** and the ylide of (methoxymethyl)triphenylphosphonium chloride to yield compound **18** (Figure 9) [22]. Cleavage of the methyl enol ether of compound **18** was achieved by the use of perchloric acid to yield aldehyde **10** [22]. Corey–Chaykovsky reaction conditions were then used to transform compound **10** into compound **11** [22]. A 6:1 diastereomeric mixture of 32*S*-oxiranyllanost-7-en-3β-ol (**11**) and 32*R*-oxiranyllanost-7-en-3β-ol was obtained, and this mixture was successfully purified by high-performance liquid chromatography (HPLC) [22].

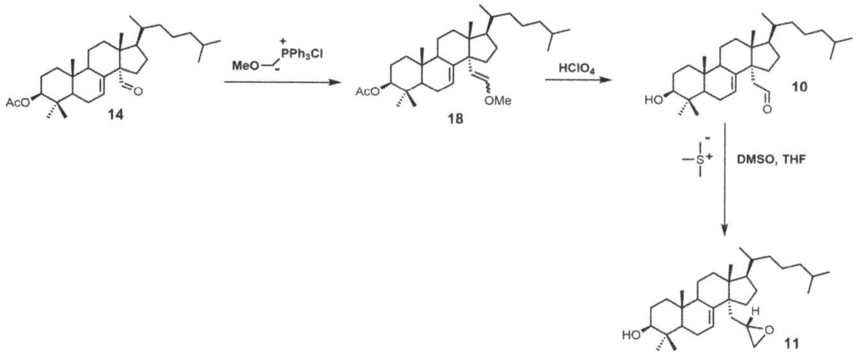

Figure 9. The synthesis of 32*S*-oxiranyllanost-7-en-3β-ol (**11**).

4,4-Dimethyl-14α-aminomethyl-cholest-8-en-3β-ol (**12**) and 4,4-dimethyl-14α-aminomethyl-cholest-7-en-3β-ol (**20**) were observed to be active against *T. cruzi* SDM with apparent K_d values of 5.1 and 1.3 μM, respectively [30]. These two amino derivatives have around a 3-fold stronger inhibitory effect on *Candida albicans* SDM in comparison to *T. cruzi* SDM and produce IC_{50} values of around 4 μM against *C. albicans* growth [30]. 4,4-Dimethyl-14α-aminomethyl-cholest-8-en-3β-ol (**12**) can be synthesized starting with compound **16** (Figure 10) [30]. The aldehyde functional group of compound **16** was converted into an oxime with hydroxylamine hydrochloride, which in turn was transformed into nitrile **19** with acetic anhydride and pyridine [30]. Nitrile **19** was then reduced to a primary amine with lithium aluminum hydride and aluminum trichloride to yield compound **20**, which was easily isomerized into compound **12** with acidic methanol [30].

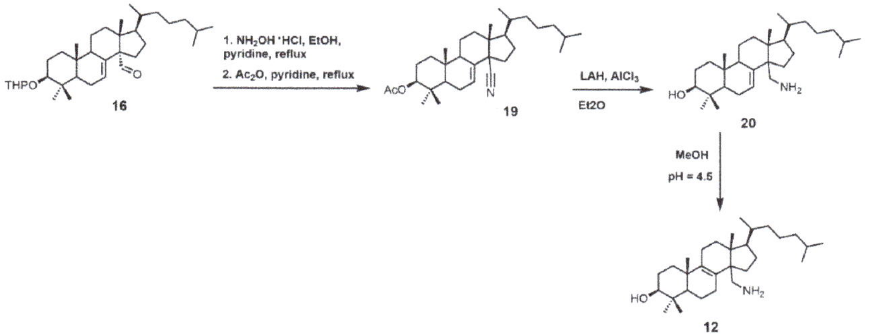

Figure 10. The synthesis of 4,4-dimethyl-14α-aminomethyl-cholest-8-en-3β-ol (**12**).

3. Azole SDM Inhibitors

Azoles are the largest class of SDM inhibitors, and this group of inhibitors is continuously expanding with the creation of new drugs or molecules with drug-like properties. 1,2,4-Triazole fungicides such as difenoconazole (Score® (Syngenta, Basel, Switzerland)), epoxiconazole (Opal® (TRC, North York, ON, Canada)), flusilazole (Punch® (DuPunt, Wilmington, DE, USA)), and so forth are well-known SDM inhibitors used against agricultural relevant fungal diseases, including powdery mildews, rusts, and leaf-spotting fungi from Ascomycetes and Basidiomycetes [17]. Human fungal infections have been treated with antifungal azoles for a long period of time; chlormidazole was the first azole drug, introduced in 1958 for the treatment of topical mycosis [53]. The older antifungal azoles that were predominately discovered in the 1950–1960s have undergone numerous structural modifications to yield the next generation of antifungal azole drugs. In addition, many of these older antifungal azole drugs have reemerged or undergone structural modifications to be used as potential anti-trypanosomiasis drugs. The renaissance of using old antifungal agents for treating or attempting to treat trypanosomiasis was largely driven by large pharmaceutical companies not prepared to invest heavily in neglected diseases that are prevalent in developing countries where there would be no chance of cost recovery [53,54]. Some of the classic azoles used as standards for fungal SDM inhibition include ketoconazole, fluconazole, itraconazole, and posaconazole, and structures of these drugs are shown in Figure 11. Fluconazole (Diflucan®) was released by Pfizer Canada Inc, Kirkland, QC, Canada and was active against *Candida* spp., while itraconazole (Sporanox®) was released from Janssen Pharmaceutica in the early 1990s and was active against both *Candida* spp. and *Aspergillus* spp. [17]. Itraconazole from Janssen Pharmaceutica, Beerse, Belgium was approved for use in 1992, while posaconazole (Noxafil®) from the Schering-Plough Research Institute, Kenilworth, New Jersey, USA was approved by the food and drug administration (FDA) in 2006 [53]. Posaconazole was noted to be more effective than amphotericin B in the treatment of *Aspergillus* spp. infections, and it is structurally similar to itraconazole (Figure 11) [53]. (*R*)-*N*-(1-(2,4-Dichlorophenyl)-2-(1*H*-imidazol-1-yl)ethyl)-4-(5-phenyl-1,3,4-oxadi-azol-2-yl)-benzamide (VNI) (Figure 11), a new-generation imidazole, has been shown to exert a curative effect for both acute and chronic Chagas disease in a murine model with 100% survival and no observable side effects [14,55]. VNI is available at a low cost (<0.10/mg) and has good oral bioavailability, low toxicity, and favorable pharmacokinetics, which makes this compound an attractive candidate for clinical trials to treat patients with Chagas disease [55].

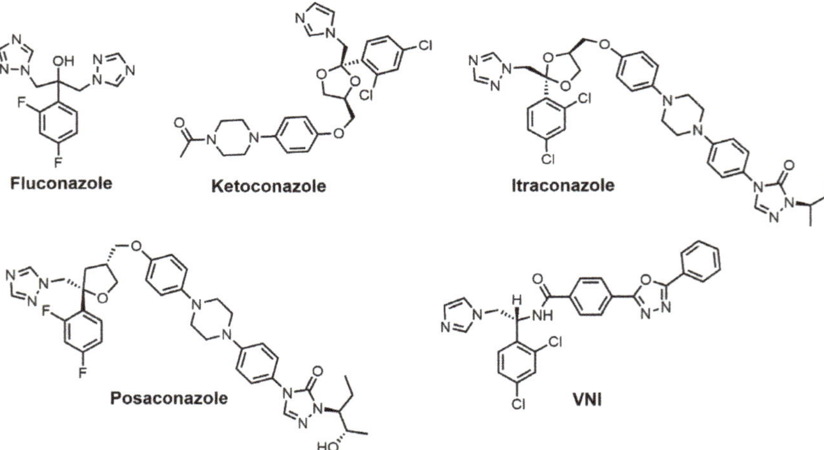

Figure 11. Azole inhibitors of sterol 14α-demethylase (SDM).

Fluconazole is a water-soluble, well-tolerated, and cheap first-generation antifungal drug that penetrates the blood–brain barrier [21]. Fluconazole has reported minimum inhibitory concentrations (MIC_{90}) required to inhibit the growth of various *Candida* spp. isolates by 90% ranging from 2 to 64 mg/mL [56]. Fluconazole was observed to maintain an in vivo effective dose for 50% (ED_{50}) of the murine population with <1.0 mg/kg for 5 days when orally administered in a murine candidiasis model, making it 50 times more potent than ketoconazole [57]. The plasma half-life for fluconazole was 6.1 h, and 75% of the drug was excreted in urine with no changes to its structure [12]. The 2,4-difluorophenyl side was chosen because it was the only phenyl analog that was water-soluble (8 mg/mL), which was needed to enable it to be formulated for intravenous administration. Fluconazole went through safety studies and passed evaluation in humans, where it was shown to very successful in the treatment of *C. albicans* and *C. neoformans* infections but was not as effective against *Aspergillus* infections when compared with itraconazole [12]. Fluconazole has a reported K_d value of 0.23 μM against *T. cruzi* SDM and an EC_{50} value of approximately 40 μM against *T. cruzi*, while fluconazole has a K_d value of 0.34 μM for *T. brucei* SDM [21,58]. Fluconazole has an in vitro IC_{50} value of 8 μM against *T. cruzi*, and a crystal structure of *T. cruzi* SDM cocrystallized with fluconazole has been reported [21,59–61]. A 6 week oral course of fluconazole was shown to be safe and useful for treating leishmaniasis caused by *Leishmania major* [62].

There are several different ways to synthesize fluconazole. One of the popular routes is the oxirane ring opening of 1-[[2-(2,4-difluorophenyl)oxiranyl]methyl]-1*H*-1,2,4-triazole with 1,2,4-triazole and potassium carbonate [17]. One of the more recent syntheses of fluconazole uses a semi-continuous flow method (Figure 12) [39]. 2,4-Difluorobromobenzene (**21**) was reacted with isopropyl magnesium chloride in THF to form a Grignard reagent that was reacted under flow conditions with 1,3-dichloroacetone (**22**) to yield 2-(2,4-difluorophenyl)-1,3-dichloro-2-propanol (**23**) [39]. The synthesis of fluconazole was completed with the nucleophilic attack of intermediate **23** with two 1,2,4-triazoles without using flow chemistry [39].

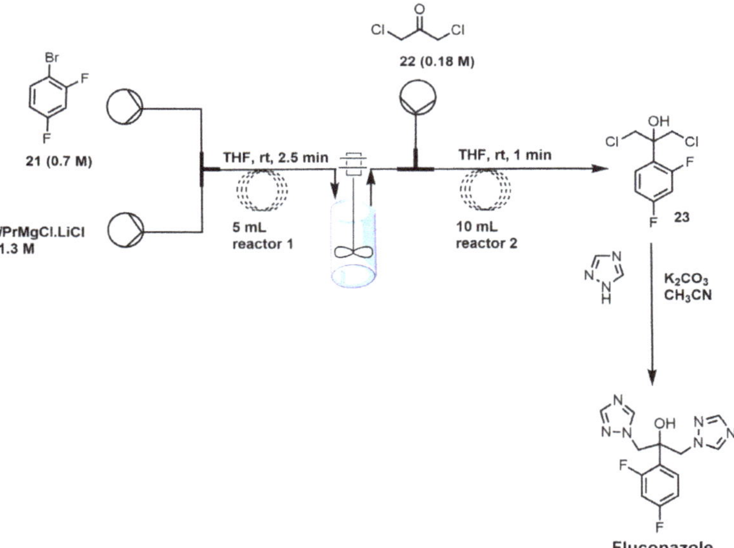

Figure 12. Synthesis of fluconazole using flow-injection methodology (adapted from [39]).

An interesting synthesis of fluconazole is the ^{18}F-labeled synthesis of [4-^{18}F]fluconazole (Figure 13) [38]. Under Friedyl–Crafts acylation conditions, *m*-fluoroacetanilide (**24**) was reacted with chloroacetylchloride to yield *N*-[4-(2-chloroacetyl)-3-fluorophenyl]acetamide, which was reacted with 1,2,4-triazole to produce compound **25** (Figure 13) [38]. Corey–Chaykovsky conditions were used to convert intermediate **25** into an epoxide that was ring-opened by 1,2,4-triazole, and the *N*-acetyl group was removed by acid hydrolysis to form compound **26**. [4-^{18}F]Fluconazole was obtained by a modified Schiemann reaction using 25% tetrafluoroboric acid, sodium nitrite, and potassium [^{18}F]-fluoride [38]. [4-^{18}F]Fluconazole was synthesized to measure the pharmacokinetics of fluconazole in rats by radioactive measurements of excised tissues and in rabbits by positron emission tomography (PET) [38]. A uniform distribution of radioactivity in most organs was observed in both species upon equilibration of [4-^{18}F]fluconazole [38].

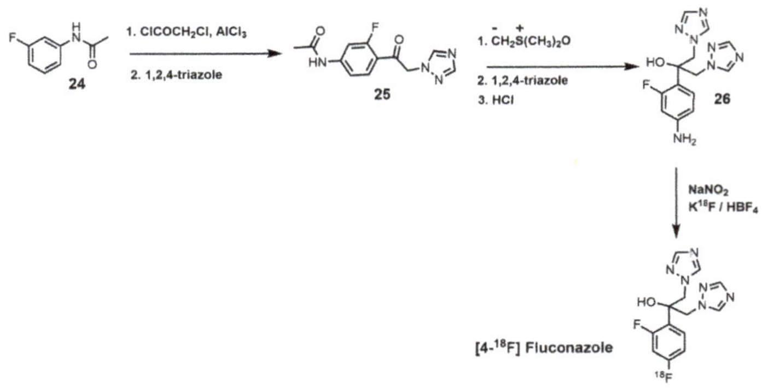

Figure 13. Synthesis of [4-^{18}F]fluconazole.

Ketoconazole developed by Janssen Pharmaceutica was approved by the FDA in 1981 and was the first broad-spectrum antifungal agent with oral activity; its reported MIC_{90} values range from 0.06 to 4 μg/mL against various *Candida spp.* [12,53]. Ketoconazole is highly active against *Trichophyton mentagrophytes* at 0.1 μg/mL, while *Trichophyton rubrum* and *C. neoformans* are inhibited at 1 μg/mL [12]. Ketoconazole has been shown to prolong the survival of mice infected with *T. cruzi*; however, curing effects were not observed [63,64]. Ketoconazole has a reported K_d value of 4.4 μM against cloned *Tb*CYP51 [65]. Ketoconazole at a concentration of 0.1 μM in conjunction with 24(*R*,*S*)-25-epiminolanosterol (EL) at a concentration of 0.3 μM exerts strong antiproliferative effects on *T. cruzi* epimastigotes, leading to >80% inhibition of growth [66].

The synthesis of ketoconazole (Figure 14) started with the ketalization of 3′,5′-dichloroacetophenone (**27**) with glycerine, and the crude product was brominated [31]. Benzylation of the resulting primary alcohol produced a mixture of *cis*/*trans* isomers, and the desired *cis* isomer (**28**) was obtained by crystallization from ethanol [31]. After the bromine atom was displaced by nucleophilic attack by imidazole, the benzoyl ester was cleaved by sodium hydroxide to produce an alcohol that was activated as a mesylate (**29**). Mesylate **29** was reacted with the sodium salt of 1 acetyl-4-(4-hydroxyphenyl)piperazine to produce ketoconazole [31].

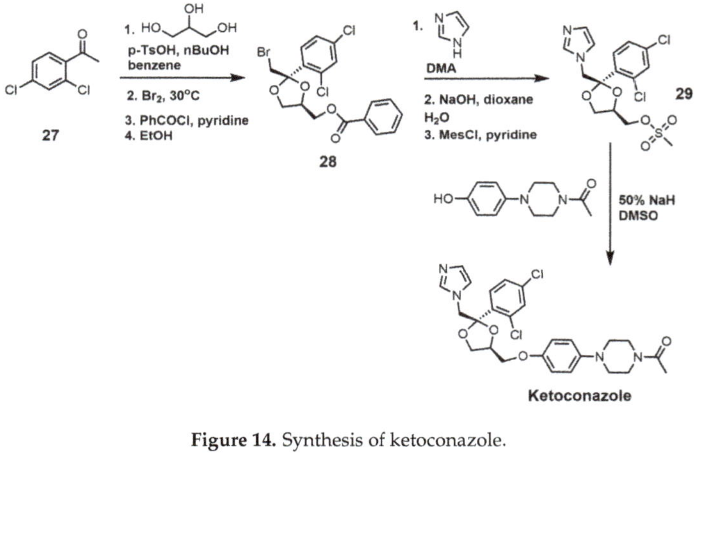

Figure 14. Synthesis of ketoconazole.

Itraconazole has MIC_{90} values ranging from 0.03 to 64 μg/mL against various *Candida* spp. and a MIC_{90} value of 2.7 μg/mL against *Malassezia furfur* [12]. Itraconazole was noted to have potent in vitro activity against *C. neoformans* [67]. Itraconazole was observed to successfully achieve irreversible damage against various fungi, including *C. albicans*, *C. neoformans*, *T. rubrum*, *Paracoccidiodes brasiliensis*, and *Aspergillus fumigatus* [12,68]. Itraconazole absorption is enhanced by food intake, and oral solutions provide higher serum concentrations in comparison to capsules [12,69,70]. Itraconazole is very active in vitro against *Microsporum canis* and other dermatophytes associated with scalp ringworm infections [12]. The selectively index of itraconazole for fungal SDM versus human SDM was reported to be 25, which was much better than the selectively index of ketoconazole [53,71]. Itraconazole was observed to have an ED_{50} value of 1 μM against *T. brucei* parasites, and when combined with 25-azalanosterol (AZAL), parasite death resulted [2]. Itraconazole exhibited in vitro activity against *Leishmania mexicana mexicana* amastigotes in macrophages, and 50% inhibition of ergosterol synthesis was observed at 0.15 μM, which was more potent than that observed for ketoconazole (0.21 μM) [72]. Itraconazole has been reported to successfully treat cutaneous leishmaniasis; however, it was not as effective against *Leishmania braziliensis* [53,73–75]. It is noted that itraconazole has not been able to completely eradicate *T. cruzi* from experimentally infected animals or human patients in studies investigating it as a possible treatment for Chagas disease [76].

The synthesis of itraconazole can be started by reacting compound **30** with sodium hydride and 1 acetyl-4-(4-hydroxyphenyl)piperazine to yield *cis*-1-acetyl-4-[4-[[2-(2,4-dichlorophenyl)-2-(1*H*-1,2,4-triazol-1-ylmethyl)-1,3-dioxolan-4-yl]methoxy] phenyl]-piperazine (Figure 15) [34]. The *N*-acetyl group can be removed by sodium hydroxide, and the unprotected piperazine nitrogen under basic conditions with potassium carbonate can undergo a nucleophilic aromatic substitution with *p*-chloronitrobenzene to yield compound **31**. The aromatic nitro group was reduced to an aniline functional group with hydrogen gas and 5% platinum on carbon, and the aniline derivative was converted into a carbamate with phenyl chloroformate [34]. Reacting the carbamate with hydrazine produced the hydrazine carboxamide **32**. Compound **32** was initially treated with formamidine acetate and then with 2-bromobutane in the presence of potassium hydroxide to yield itraconazole (Figure 15) [34].

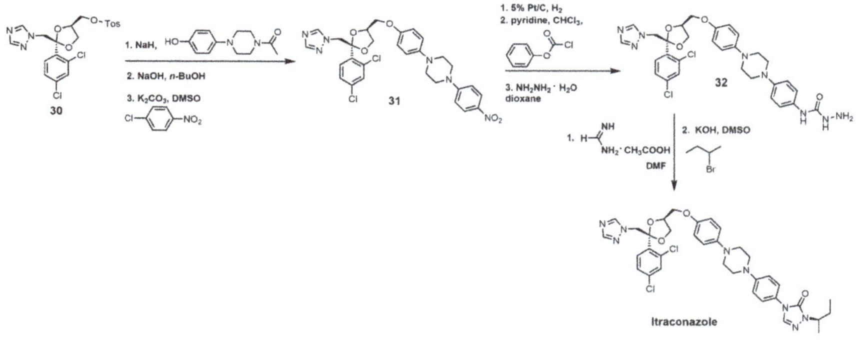

Figure 15. The synthesis of itraconazole.

Posaconazole (Figure 11) has already been registered in the European Union (2005), Australia (2005), and the United States of America (2006) as a prophylactic for invasive fungal infections in addition to azole-resistant candidiasis [77]. Posaconazole was observed to have a MIC value of 20 nM and an IC_{50} value of 14 nM against the epimastigote form of *T. cruzi* and a MIC value of 3 nM and an IC_{50} value of 0.25 nM against the clinically relevant intracellular amastigote form in vitro [78]. Posaconazole demonstrated potent in vivo anti-parasitic activity in a murine model of acute Chagas disease, which was most likely due to its high binding affinity for *Tc*CYP51, good pharmacokinetic properties, and long terminal half-life [77,78]. The crystal structure of posaconazole bound to SDM of *T. brucei* has been solved and provides valuable insights into how posaconazole binds to its target [60]. Posaconazole is active against nitrofuran- and nitroimidazole-resistant *T. cruzi* strains and is able to induce radical parasitological cures in both chronic and acute experimental Chagas disease [76]. An Argentinean woman in Spain who suffered from chronic Chagas disease was successfully treated with posaconazole, although requiring concurrent immunosuppressive therapy [79]. The woman was initially treated with benznidazole, which only resulted in a reduction of *T. cruzi* levels and not an eradication as was achieved with the use of posaconazole [79]. The major drawback of posanconazole is that it is very expensive to produce (>$1000/patient) as it is very low yielding (<1% overall), which will limit its widespread use in developing countries even if it is successful in obtaining approval for treating Chagas disease [55,77,79].

One of the well-established synthetic routes for the synthesis of posaconazole starts with the enzymatic desymmetrization of the homochiral diol (**33**) with hydrolase Novo SP 435 and vinyl acetate, followed by iodocyclization to produce compound **34** (Figure 16) [37,80]. It should be noted that 169 hydrolases were screened in order to make the first step a practical synthetic route [37]. A 1,2,4-triazole displaced the iodine in compound **34**, and the acetate group was replaced with a tosylate to yield compound **35**. Compound **35** was then reacted with compound **36** under basic conditions, and the benzyl group was removed under acidic conditions to yield posaconazole [81,82].

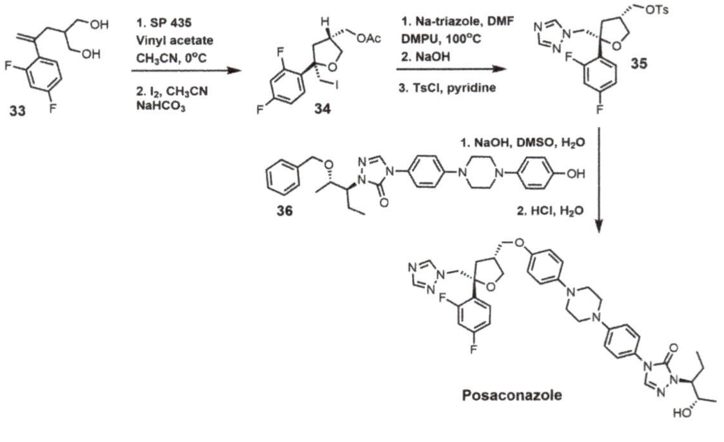

Figure 16. Synthesis of posaconazole.

VNI (Figure 11) is a potent inhibitor of *T. cruzi* SDM and is nontoxic and highly selective [55]. Mice infected with an acute or chronic *T. cruzi* infection were treated orally with VNI at 25 mg/kg for 30 days, which resulted in 100% survival of the infected mice [55]. VNI cocrystallized with both *T. cruzi* and *T. brucei* SDM has been reported [21,83]. The K_d value for VNI against *T. brucei* SDM is 0.37 μM, and it is 0.09 μM against *T. cruzi* SDM [14,83]. VNI exhibits low general cytotoxicity and a potent cellular effect on *T. cruzi* amastigotes, with an EC_{50} value of 1.2 nM, in addition to an excellent selectivity index (human/*T. cruzi* of >50,000) [14]. An advantage of VNI, unlike posaconazole or fluconazole, is that it does not induce *T. cruzi* gene expression, which means it is not as likely to cause drug resistance [14]. As mentioned earlier, VNI is available at a low cost (<$0.10/mg), which makes this compound an attractive candidate for clinical trials [55]. A 2-fluoro-4-(2,2,2-trifluoroethoxy)phenyl derivative of VNI was recently reported to inhibit *C. albicans* and *A. fumigatus* SDM by 89% and 65%, respectively [46]. This 2-fluoro-4-(2,2,2-trifluoroethoxy)phenyl derivative of VNI was observed to have a K_d value of 10 nM against *C. albicans* SDM and of 20 nM against *A. fumigatus*, while posaconazole had a K_d value of 81 nM against *C. albicans* SDM and of 131 nM against *A. fumigatus* [46]. A crystal structure of *A. fumigataus* SDM cocrystallized with this VNI derivative was obtained [46].

VNI can be synthesized in five steps that are scalable to multi-gram quantities with >98% purity as needed for in vivo animal studies (Figure 17) [84]. The racemic epoxide (**37**) was resolved with (*S,S*)-Co(salen) and then opened with imidazole under basic conditions to yield compound **38** [84]. Compound **38** was reacted with diphenylphosphoryl azide to replace the secondary alcohol with an azide group and was subsequently reduced with lithium aluminum hydride to yield amine **39** [84]. VNI was obtained by reacting compounds **39** and **40** under amide bond coupling conditions with 1-[bis(dimethylamino)methylene]-1*H*-1,2,3-triazolo[4,5-b]pyridinium 3-oxide hexafluoro-phosphate (HATU) [84].

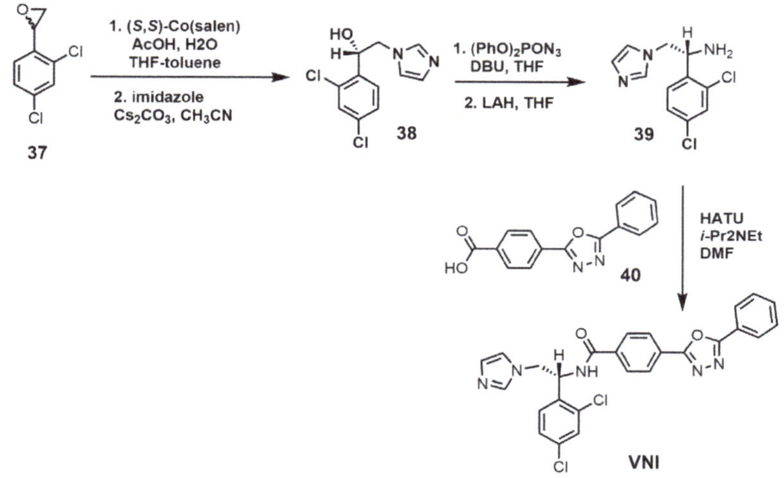

Figure 17. Synthesis of (*R*)-*N*-(1-(2,4-dichlorophenyl)-2-(1*H*-imidazol-1-yl)ethyl)-4-(5-phenyl-1,3,4-oxadiazol-2-yl)-benzamide (VNI).

4. 24-SMT Inhibitors

24-SMT catalyzes a methylation–deprotonation reaction that involves electrophilic alkylations of a double bond at C-24 by a methyl cation originating from *S*-adenosyl methionine (SAM) (Figure 18) [4]. The majority of characterized 24-SMTs are multifunctional and possess a very high substrate specificity, often yielding only a single C-24 methylated olefin product; however, there are a few atypical 24-SMTs that convert substrates to a variety of 24-alkyl(idene) products including *T. brucei* SMT1 and *Glycine max* SMT1 [4]. The various pathways possible with 24-SMT catalysis are outlined in Figure 18.

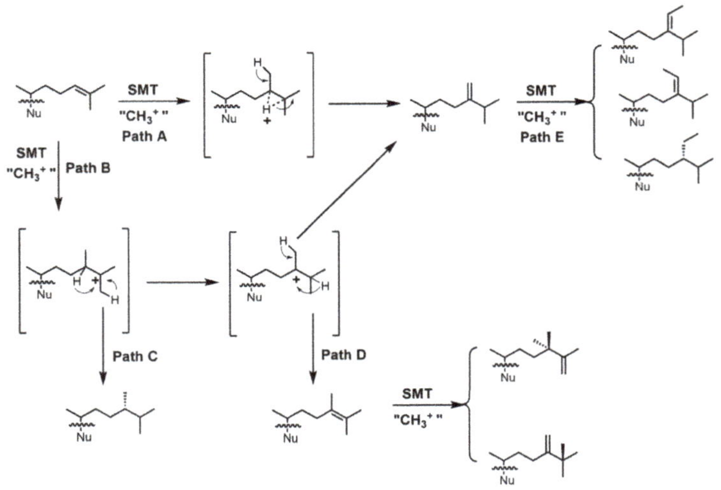

Figure 18. Different C24-alkylation pathways for sterol C24-methyltransferase (24-SMT) substrates (adapted from [4]). Nu: sterol nucleus.

The structures of 24-SMT inhibitors typically have a modified lanosterol side chain (Figure 19).

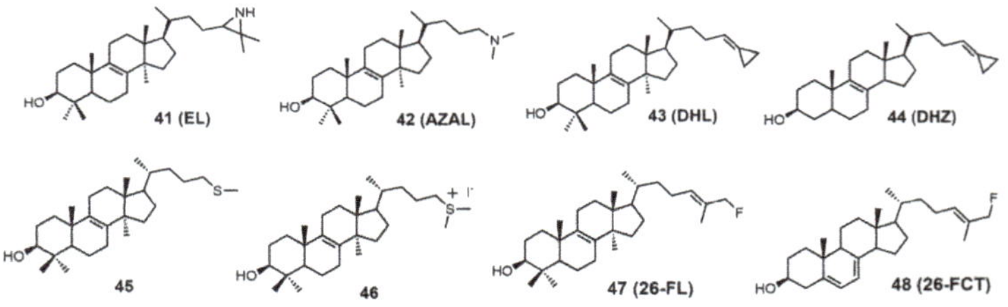

Figure 19. Structures of sterol C24-methyltransferase (24-SMT) sterol-based inhibitors.

Most of the inhibitors outlined in Figure 19 are suicide/irreversible inhibitors that have an electron withdrawing group strategically placed at or near position 25 on the sterol side. The cyclopropylidine derivatives (**43–44**) are spring-loaded electrophiles that are opened irreversibly with nucleophilic attack from 24-SMT (Figure 20).

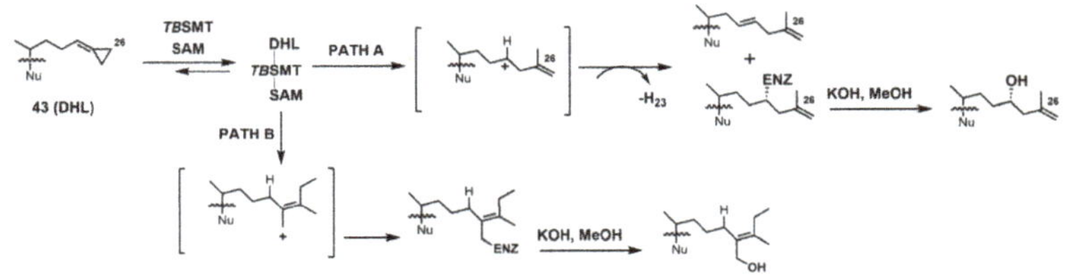

Figure 20. Proposed inhibitory mechanism of *Tb*SMT with 26,27-dehydrolanosterol (DHL) (adapted from Miller et al., 2017). Nu: lanosterol nucleus.

Having a strong electron withdrawing group near the Δ24–25 bond, such as fluorine, affects the intermediate cation formation and timing of the C-24 methylation reaction, promoting partitioning towards irreversible covalent modification over turnover products (Figure 21) [5].

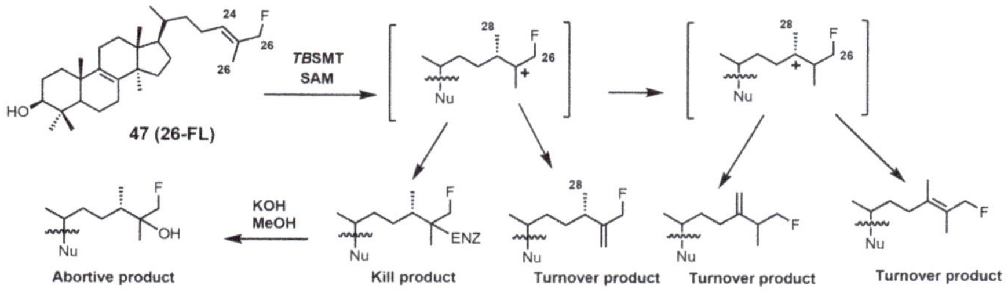

Figure 21. C24-methylation pathway of 26-fluorolanosterol (26-FL) (**47**) with *Tb*SMT. Adapted from [5].

24(*R*,*S*)-25-Epiminolanosterol (EL; **41**) has a reported IC_{50} value of around 0.3 μM against *C. neoformans* and has comparable activity to itraconazole [6]. EL was observed to be a potent non-competitive inhibitor of 24-SMT from sunflower embryos with a K_i value of 3.0 nM, while sitosterol and 24(28)-methylene cycloartenol were observed to be competitive inhibitors, with K_i values of 26 and 14 μM, respectively [85]. EL as 2-tritio-24(*R*,*S*)-25-epiminolanosterol was reported to inhibit 24-SMT of *Gibberella fujikuroi* and was metabolized to 25-aminolanosterol and lanosterol [86]. EL is a potent inhibitor against *C. albicans* 24-SMT with a K_i value of 11 nM and an IC_{50} value of 5 μM [3]. EL added to cultures of rat hepatoma cells (H4) interrupted the conversion of lanosterol to cholesterol [87]. EL caused the accumulation of zymosterol at 45 μM and at 4.5 μM caused the accumulation of desmosterol [87]. H4 rat hepatoma cells seeded into either full growth or lipid-depleted medium containing 22.5 μM EL would not grow unless the media was supplemented with low-density lipoproteins (60 μg/mL) [87].

EL was observed to have a K_i value of 49 nM against *T. brucei* 24-SMT [88]. An amastigote form of *T. cruzi* proliferating in a liver infusion tryptose medium was treated with EL, and growth was completely arrested; lysis occurred at an EL concentration of 6 μM [89]. EL was recently reported as an inhibitor of *Acanthamoeba* spp. trophozoite growth with an IC_{50} value of 7 nM, and in this study, EL yielded 20-fold higher inhibition compared to the reference drug voriconazole [90]. EL exhibited tight biding against both 24- and 28-SMT with K_i values of around 9 nM [90]. EL (**41**) can readily be synthesized directly from unprotected lanosterol with an excess of iodine azide and the addition of LAH (Figure 22) [91].

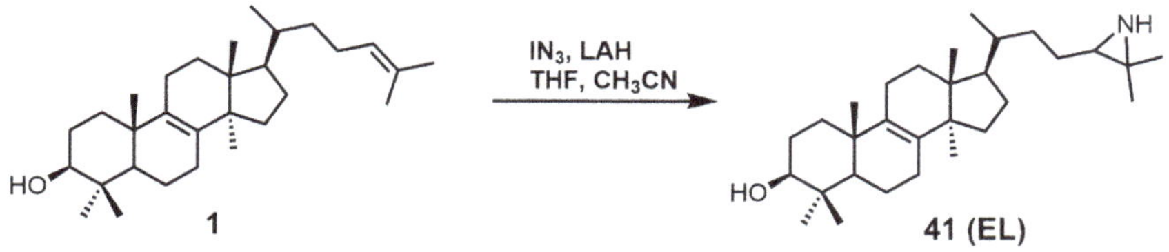

Figure 22. The synthesis of 24(*R*,*S*),25-epiminolanosterol (EL).

There are several different ways EL is thought to inhibit 24-SMT (Figure 23). EL could have its aziridine nitrogen atom protonated (path A) and then form an ammonium salt that can electrostatically interact with a polar amino acid in the active site of 24-SMT (such as a carboxylate group) [92]. Alternatively, EL could be methylated by 24-SMT and donate its *N*-methyl group to the active-site residue of 24-SMT, thereby inactivating it [92].

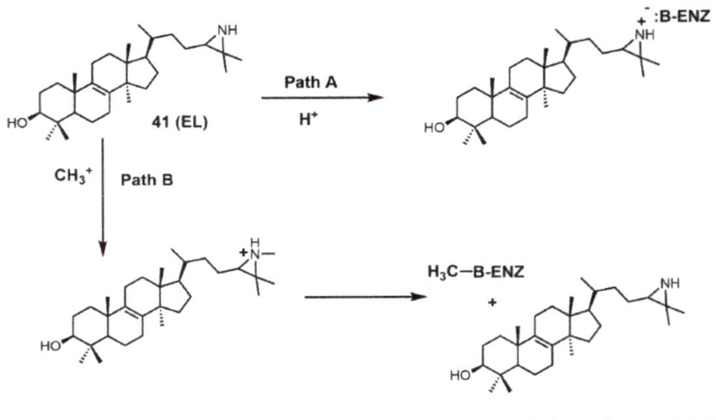

Figure 23. Proposed pathways for inhibition of sterol C24-methyltransferase (24-SMT) with 24(*R*,*S*)-25-epiminolanosterol (EL) (adapted from [92]).

AZAL was reported to have a K_i value of 54 nM and an IC_{50} value of 3 μM against 24-SMT of *C. albicans* [3], while it had a K_i value of 30 nM against 24-SMT from sunflower embryos [3,85]. AZAL was noted to be a potent inhibitor of the ascomycetous fungus *P. brasiliensis* (Pb) with an IC_{50} value of 30 nM and a K_i value of 14 nM against Pb 24-SMT [93]. AZAL was observed to inhibit *P. brasiliensis* growth much more than for the related yeasts *Saccharomyces cerevisiae* and *C. albicans* [93]. AZAL has a K_i value of 39 nM against *T. brucei* 24-SMT and an ED_{50} value of 1 μM against *T. brucei* parasites [2,88]. AZAL failed to inhibit cultured human epithelial cells (HEK) with an ED_{50} value of >50 μM and a therapeutic index of 25 [2]. AZAL has a reported IC_{50} value of approximately 1 μM against both the procylic and bloodstream forms of *T. brucei*, and the bloodstream form of *T. brucei* was not rescued with cholesterol absorption from the host, highlighting the importance of ergosterol in cell proliferation of the parasite [94]. Combing AZAL and itraconazole at ED_{50} concentrations to the bloodstream form of *T. brucei* in lipid-depleted medium resulted in cell death and significant growth inhibition when grown in full growth medium [2].

AZAL can readily be synthesized via three different synthetic routes, which are outlined in Figure 24 [4,95,96]. All three routes involve converting the Δ24–25 double bond of lanosterol into an activated carbonyl group (aldehyde, carboxylic acid, or ester) followed by nucleophilic attack with dimethylamine. Each synthetic route was completed with protecting group removal by LAH.

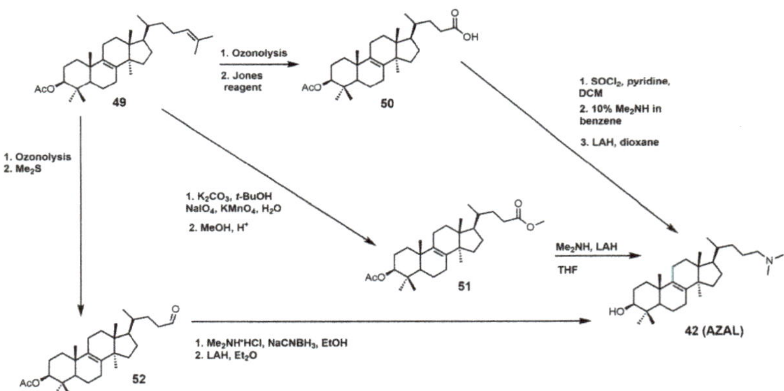

Figure 24. Three synthetic routes used to prepare 25-azalanosterol (AZAL).

26,27-Dehydrolanosterol (DHL; **43**) was reported as an inhibitor of *Acanthamoeba* spp. trophozoite growth, with an IC_{50} value of 6 μM making it a much weaker binder in comparison to EL (IC_{50} of 7 nM) [90]. DHL is metabolized to a favorable substrate, which irreversibly inhibits *Tb*-24SMT, and the ED_{50} values of DHL incubated with procyclic or bloodstream forms of *T. brucei* are 10 and 20 μM, respectively [97]. 26,27-Dehydrozymosterol (DHZ) was reported to inhibit 24-SMT from *C. albicans* with a K_i value of 9 μM and a reported k_{inact} value of 0.03 min^{-1} [3]. [3-^{3}H]26,27-Dehydrozymosterol was noted to inhibit 24-SMT from *S. cerevisiae* with an apparent K_i value of 1.1 μM [98]. DHZ also inhibits SMT2-2 of *Glycine max* (soybean) with a K_i value of 9.3 μM and a K_{inact} value of 0.023 min^{-1} [99], while DHZ irreversibly inhibits *Arabidopsis* SMT2 with a K_i value of 49 μM and a k_{inact} value of 0.009 s^{-1} [100]. *T. brucei* 24-SMT was inhibited by DHZ with a K_i value of 29 μM and a k_{inact} value of 0.26 min^{-1}. Against *T. brucei* 24-SMT, DHZ was noted to be a weaker inhibitor in comparison to EI and AZAL [88]. DHL cannot inhibit *T. brucei* 24-SMT or yeast 24-SMT because "uncharged" 4,4-dimethylsterols cannot bind productively to these types of SMT [88]. 26,27-Dehydrocycloartenol (DHC; **53**) was observed to be a substrate for soybean SMT with K_m and k_{cat} values of 10 mM and 0.018, respectively [101]. There are two main pathways along which DHL can irreversibly inhibit *Tb*SMT (Figure 20) [97]. Path A has a carbocation intermediate on carbon 24, while path A has the intermediate on carbon 27, and both carbocations are reactive with SMT. Either biosynthetic route results in the metabolized DHL to irreversibly bind to *Tb*SMT, and the resulting prosthetic group was hydrolyzed with potassium hydroxide in methanol to yield the C-24 or C-27 alcohol (Figure 23) that was characterized by gas chromatography–mass spectrometry (GCMS).

The 26,27-dehydrosterols can be prepared in a few steps starting with the 3-acetylated sterol (Figure 25). The 24–25 double bond of the corresponding sterol is transformed into an aldehyde by ozonolysis with a reductive workup [102]. The 26,27-cyclopropylidine moiety is installed via a Wittig reaction using the desired sterol aldehyde and the phosphorus ylide formed from cyclopropyltriphenylphosphonium bromide and butyllithium [102]. The acetate group was easily removed with LAH.

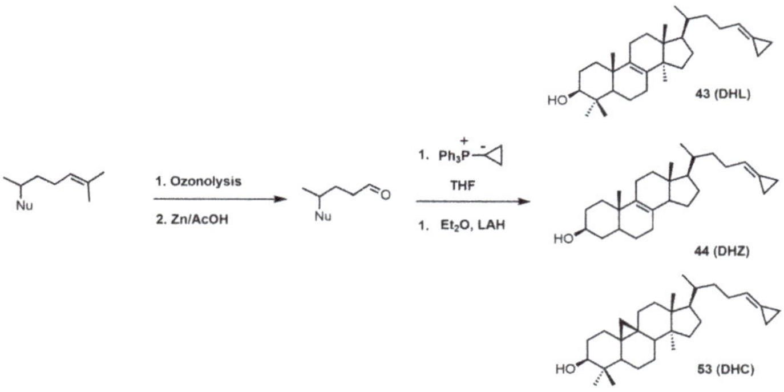

Figure 25. Synthesis of cyclopropylidine sterol derivatives. Nu: sterol nucleus (cycloartenol, lanosterol, or zymosterol).

Another class of 24-SMT inhibitors comprises sulfur-containing sterols such as 25-thialanosterol (**45**) and 25-thialanosterol iodide (**46**) (Figure 19) [103]. 24-SMT from *C. albicans* was irreversibly inhibited with 25-thialanosterol with a K_i value of 5 μM and an apparent k_{inact} value of 0.013 min^{-1}, while the corresponding sulfonium salt was a reversible transition-state inhibitor with a K_i value of 20 nM [103]. 24-Thialanosterol was observed to inhibit *C. albicans* 24-SMT with an IC_{50} value of 20 μM [3]. 25-Thialanosterol iodide has a K_i value of 86 nM against *T. brucei* 24-SMT and an ED_{50} value of 2 μM against both the procyclic and bloodstream forms of the *T. brucei* parasite [2]. 24-SMT can methylate the 25-sulfur atom of 25-thialanosterol to form a sulfonium cation that can act as a methyl donor to irreversibly methylate SMT (Figure 26) [4].

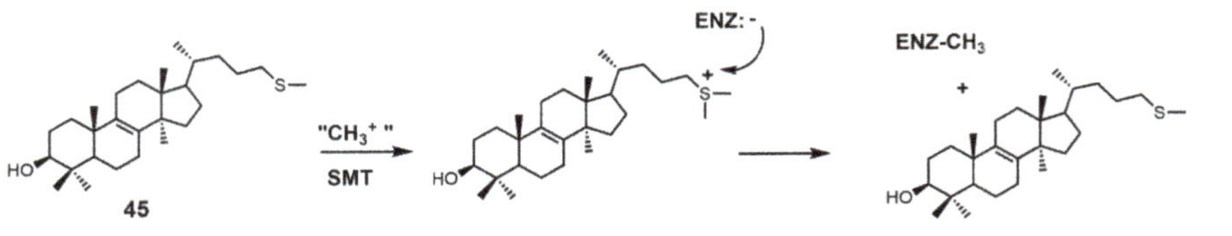

Figure 26. Proposed inhibitory mechanism of SMT with 25-thialanosterol (adapted from [4]).

The synthesis of 25-thialanosterol (**45**) and 25-thialanosterol iodide (**46**) is outlined in Figure 27 [103]. The synthesis started with ozonolysis of acetylated lanosterol (**49**) to yield 3-acetoxylanosta-8,24-dienol-26-al. 3-Acetoxylanosta-8,24-dienol-26-al was then reduced with sodium borohydride to yield an alcohol that was activated as a tosylate. Methanethiolate then displaced the tosylate group to yield 25-thialanosterol (**45**). Treatment of 25-thialansterol with methyl iodide yielded 25-thialanosterol iodide (**46**).

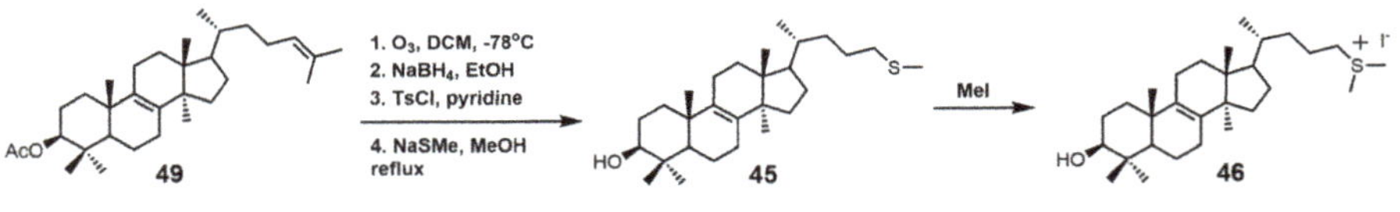

Figure 27. Synthesis of 25-thialanosterol and 25-thialanosterol salt.

26-Fluorolanosterol (26-FL; **47**) is metabolically converted by *T. brucei* into a fluorinated substrate (26-fluorozymosterol) that irreversibly binds to 24-SMT and inhibits ergosterol biosynthesis and growth of both the procyclic and bloodstream forms of *T. brucei* [5]. *T. brucei* cell-based studies were conducted with 26-FL, and IC_{50} values for the procyclic and bloodstream forms were 3 and 16 μM, respectively, while 26-FL had no effect on HEK cell growth at up to 100 μM [5]. In order to further investigate the preferred substrate for *Tb*SMT, 26-fluorocholesta-5,7,24-trienol (26-FCT) was synthesized (Figure 28). 26-FCT was synthesized instead of 26-fluorozymosterol because of the difficulty in obtaining sufficient amounts of zymosterol to enable synthetic manipulation. 26-FCT was observed to be a competitive inhibitor of 24-SMT with respect to zymosterol and exhibited a K_i value of 75 μM [5]. 26-FCT was confirmed to be a metabolite of 26-FL when metabolized by the procyclic form of *T. brucei* by GCMS with an authentic sample of 26-FCT [5]. 26-FCT has an excellent partition ration of 1.08, comparing favorably with eflornithine, which has a partition ratio of 3.3 against L-ornithine decarboxylase [5]. Both 26-FL and 26-FCT exhibit desirable drug characteristics with good specificity and low toxicity, and they might be useful in treating sleeping sickness or other protozoan infections [5]. When 26-FL is methylated by *Tb*SMT, two carbocations can form as short-lived intermediates that can be transformed into various turnover products (Figure 27) [5]. The kill product is where the 24-SMT enzyme has a prosthetic group attached, whereby SMT is irreversibly inhibited.

26-FL was synthesized by the method used to synthesize 26-fluorocycloarentol (Figure 28A) [104]. The 3-hydroxy group of lanosterol was acetylated, and then the 26 methyl group was oxidized to an aldehyde (**54**) with selenium dioxide. The aldehyde was then reduced to the alcohol with sodium borohydride, and then the fluorine atom was installed via DAST. The acetate group was removed by potassium hydroxide in methanol to yield 26-FL.

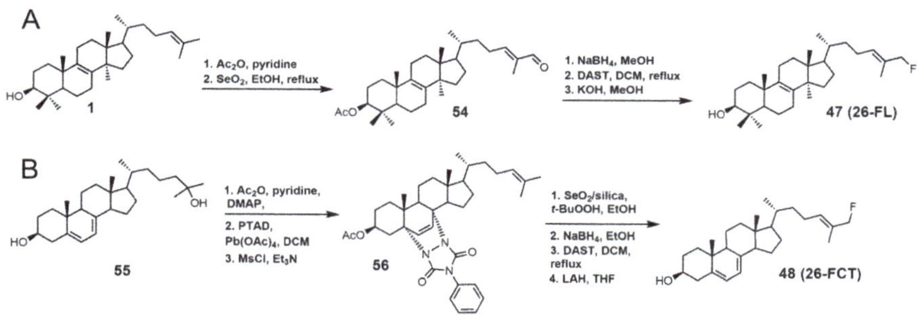

Figure 28. (**A**) Synthesis of 26-fluorolanosterol (26-FL) (**47**); (**B**) synthesis of 26-fluorocholesta-5,7,24-trienol (26-FCT) (**48**).

The synthesis of 26-FCT started with compound **55**, which can be synthesized in eight steps starting with ergosterol (Figure 28B) [105]. Compound **55** was then acetylated, the Δ5 and Δ7 double bonds were protected with 4-phenyl-1,2,4-triazoline-3,5-dione (PTAD), and the Δ24–25 double bond was successfully installed with the use of mesyl chloride and triethylamine to yield compound **56**. The last four steps in the synthesis of 25-FCT were similar to the last four steps in the synthesis of 26-FL, except that global deprotection in the last step was accomplished with LAH.

5. Bifunctional 24-SMT and SDM Inhibitors

Compounds **57** and **58** (Figure 29) were designed to be dual functional SDM and 24-SMT inhibitors [106]. These compounds completely inhibited SDM from rat liver microsomes at 10 μM and showed reasonable in vitro potencies against *C. albicans*, *C. neoformans*, *A. fumigatus*, *T. mentagrophytes*, *Candida. pseudotropicalis*, and *Candida. krusei* with MIC values ranging from <0.1 to 50 mg/mL [106]. Compounds **57** and **58** were then tested in vivo against a murine candidiasis antifungal model, and unfortunately both compounds were ineffective against the induced infection. No explanation was provided as to why these compounds were ineffective in the in vivo model.

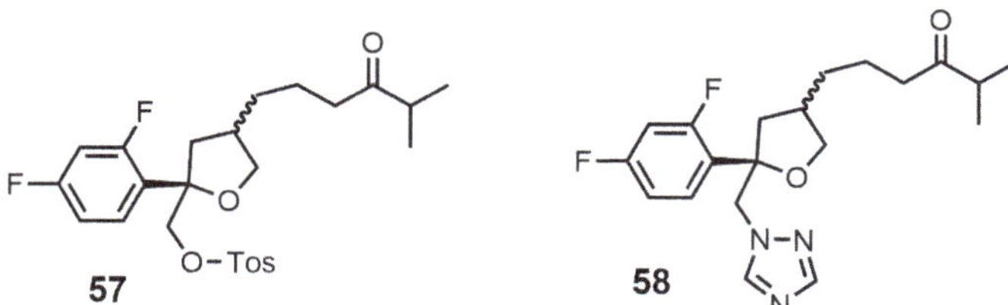

Figure 29. Structures of compounds designed to be bifunctional SDM and sterol C24-methyltransferase (24-SMT) inhibitors.

6. Conclusions

The inhibition of ergosterol biosynthesis in fungi and parasitic protozoa via the inhibition of SDM or 24-SMT with small molecules has been shown to be effective. Azole antifungals that target SDM have already been approved for the treatment of various fungal infections; however, they have not been officially approved to treat protozoan infections, despite various azoles advancing to clinical trials. Posaconazole was investigated in two phase II clinical trials as a possible agent to treat Chagas disease, and on the basis of the results provided thus far, it is unlikely that posaconazole will progress to phase III clinical trials [107–109]. The inhibition of fungal and protozoan 24-SMT with rationally designed molecules that are specific, selective, and non-toxic to humans have the potential to be used as the next generation of drugs to treat fungal infections or neglected tropical diseases that are demonstrating resistance against current therapies. It is anticipated that in the near future, both fungal and protozoan infections will be treated with a combination therapy that utilizes the cocurrent administration of a SDM and a 24-SMT inhibitor.

Funding: This research received no external funding.

Conflicts of Interest: The author declares no conflict of interest.

References

1. Nes, W.D. Biosynthesis of cholesterol and other sterols. *Chem. Rev.* **2011**, *111*, 6423–6451. [CrossRef] [PubMed]
2. Haubrich, B.A.; Singha, U.K.; Miller, M.B.; Nes, C.R.; Anyatonwu, H.; Lecordier, L.; Patkar, P.; Leaver, D.J.; Villalta, F.; Vanhollebeke, B.; et al. Discovery of an ergosterol-signaling factor that regulates *Trypanosoma brucei* growth. *J. Lipid Res.* **2015**, *56*, 331–341. [CrossRef] [PubMed]
3. Ganapathy, K.; Kanagasabai, R.; Nguyen, T.T.M.; Nes, W.D. Purification, characterization and inhibition of sterol C24-methyltransferase from *Candida albicans*. *Arch. Biochem. Biophys.* **2011**, *505*, 194–201. [CrossRef] [PubMed]
4. Liu, J.; Nes, W.D. Steroidal triterpenes: Design of substrate-based inhibitors of ergosterol and sitosterol synthesis. *Molecules* **2009**, *14*, 4690–4706. [CrossRef] [PubMed]
5. Leaver, D.J.; Patkar, P.; Singha, U.K.; Miller, M.B.; Haubrich, B.A.; Chaudhuri, M.; Nes, W.D. Fluorinated sterols are suicide inhibitors of ergosterol biosynthesis and growth in *Trypanosoma brucei*. *Chem. Biol.* **2015**, *22*, 1374–1383. [CrossRef] [PubMed]
6. Nes, W.D.; Zhou, W.; Ganapathy, K.; Liu, J.; Vatsyayan, R.; Chamala, S.; Hernandez, K.; Miranda, M. Sterol 24-C-methyltransferase: An enzymatic target for the disruption of ergosterol biosynthesis and homeostasis in *Cryptococcus neoformans*. *Arch. Biochem. Biophys.* **2009**, *481*, 210–218. [CrossRef] [PubMed]
7. Lepesheva, G.I.; Hargrove, T.Y.; Rachakonda, G.; Wawrzak, Z.; Pomel, S.; Cojean, S.; Nde, P.N.; Nes, W.D.; Locuson, C.W.; Calcutt, M.W.; et al. VFV as a new effective CYP51 structure-derived drug candidate for chagas disease and visceral leishmaniasis. *J. Infect. Dis.* **2015**, *212*, 1439–1448. [CrossRef] [PubMed]
8. Zucca, M.; Scutera, S.; Savoia, D. New chemotherapeutic strategies against malaria, leishmaniasis and trypanosomiases. *Curr. Med. Chem.* **2013**, *20*, 502–526. [CrossRef] [PubMed]
9. Emami, S.; Tavangar, P.; Keighobadi, M. An overview of azoles targeting sterol 14α-demethylase for antileishmanial therapy. *Eur. J. Med. Chem.* **2017**, *135*, 241–259. [CrossRef] [PubMed]
10. Duschank, V.G.; Couto, A.S. Targets and patented drugs for chemotherapy of Chagas disease. In *Frontiers in Anti-Infective Drug Discovery*; Choudhary, M.I., Ed.; Bentham Science Publishers: Oak Park, IL, USA, 2010; Volume 1, pp. 323–408.
11. Simarro, P.P.; Cecchi, G.; Franco, J.R.; Paone, M.; Diarra, A.; Ruiz-Postigo, J.A.; Fevre, E.M.; Mattioli, R.C.; Jannin, J.G. Estimating and mapping the population at risk of sleeping sickness. *PLoS Negl. Trop. Dis.* **2012**, *6*, e1859. [CrossRef] [PubMed]
12. Heeres, J.; Meerpoel, L.; Lewi, P. Conazoles. *Molecules* **2010**, *15*, 4129–4188. [CrossRef] [PubMed]
13. Hoffman, H.L.; Ernst, E.J.; Klepser, M.E. Novel triazole antifungal agents. *Exp. Opin. Investig. Drugs* **2000**, *9*, 593–605. [CrossRef] [PubMed]
14. Hargrove, T.Y.; Kim, K.; Soeiro, M.N.C.; da Silva, C.F.; Batista, D.G.J.; Batista, M.M.; Yazlovitskaya, E.M.; Waterman, M.R.; Sulikowski, G.A.; Lepesheva, G.I. CYP51 structures and structure-based development of novel, pathogen-specific inhibitory scaffolds. *Int. J. Parasitol. Drugs Drug Resist.* **2012**, *2*, 178–186. [CrossRef] [PubMed]
15. Zhou, W.; Song, Z.; Kanagasabai, R.; Liu, J.; Jayasimha, P.; Sinha, A.; Veeramachanemi, P.; Miller, M.B.; Nes, W.D. Mechanism-based enzyme inactivators of phytosterol biosynthesis. *Molecules* **2004**, *9*, 185–203. [CrossRef] [PubMed]
16. Warrilow, A.G.S.; Parker, J.E.; Price, C.L.; Nes, W.D.; Garvey, E.P.; Hoekstra, W.J.; Schotzinger, R.J.; Kelly, D.E.; Kelly, S.L. The investigational drug VT-1129 is a highly potent inhibitor of *Cryptococcus* species CYP51 but only weakly inhibits the human enzyme. *Antimicrob. Agents Chemother.* **2016**, *60*, 4530–4538. [CrossRef] [PubMed]

17. Worthington, P. Sterol biosynthesis inhibiting triazole fungicides. In *Bioactive Heterocyclic Compound Classes: Agrochemicals*; Lamberth, C., Dinges, J., Eds.; Wiley-VCH Verlag GmbH & Co. KGaA: Weinheim, Germany, 2012; pp. 129–145.
18. Parker, J.E.; Warrilow, A.G.S.; Cools, H.J.; Martel, C.M.; Nes, W.D.; Fraaije, B.A.; Lucas, J.A.; Kelly, D.E.; Kelly, S.L. Mechanism of binding of prothioconazole to *Mycosphaerella graminicola* CYP51 differs from that of other azole antifungals. *Appl. Environ. Microbiol.* **2011**, *77*, 1460–1465. [CrossRef] [PubMed]
19. Price, C.L.; Warrilow, A.G.S.; Parker, J.E.; Mullins, J.G.L.; Nes, W.D.; Kelly, D.E.; Kelly, S.L. Novel substrate specificity and temperature-sensitive activity of *Mycosphaerella graminicola* CYP51 supported by the native NADPH cytochrome P450 reductase. *Appl. Environ. Microbiol.* **2015**, *81*, 3379–3386. [CrossRef] [PubMed]
20. Price, C.L.; Parker, J.E.; Warrilow, A.G.S.; Kelly, D.E.; Kelly, S.L. Azole fungicides–understanding resistance mechanisms in agricultural fungal pathogens. *Pest Manag. Sci.* **2015**, *71*, 1054–1058. [CrossRef] [PubMed]
21. Lepesheva, G.I.; Hargrove, T.Y.; Anderson, S.; Kleshchenko, Y.; Furtak, V.; Wawrzak, Z.; Villalta, F.; Waterman, M.R. Structural insights into inhibition of sterol 14α-demethylase in the human pathogen *Trypanosoma cruzi*. *J. Biol. Chem.* **2010**, *285*, 25582–25590. [CrossRef] [PubMed]
22. Tuck, S.F.; Robinson, C.H.; Silverton, J.V. Assessment of the active-site requirements of lanosterol 14α-demethylase: Evaluation of novel substrate analogues as competitive inhibitors. *J. Org. Chem.* **1991**, *56*, 1260–1266. [CrossRef]
23. Frye, L.L.; Cusack, K.P.; Leonard, D.A. 32-Methyl-32-oxylanosterols: Dual-action inhibitors of cholesterol biosynthesis. *J. Med. Chem.* **1993**, *36*, 410–416. [CrossRef] [PubMed]
24. Hargrove, T.Y.; Wawrzak, Z.; Liu, J.; Waterman, M.R.; Nes, W.D.; Lepesheva, G.I. Structural complex of sterol 14α-demethylase (CYP51) with 14α-methylenecyclopropyl-7-24,25-dihydrolanosterol. *J. Lipid Res.* **2012**, *53*, 311–320. [CrossRef] [PubMed]
25. Trzaskos, J.M.; Ko, S.S.; Magolda, R.L.; Favata, M.F.; Fischer, R.T.; Stam, S.H.; Johnson, P.R.; Gaylor, J.L. Substrate-based inhibitors of lanosterol 14α-methyl demethylase: I. Assessment of inhibitor structure-activity relationship and cholesterol biosynthesis inhibition properties. *Biochemistry* **1995**, *34*, 9670–9676. [CrossRef] [PubMed]
26. Trzaskos, J.M.; Fischer, R.T.; Ko, S.S.; Magolda, R.L.; Stam, S.H.; Johnson, P.R.; Gaylor, J.L. Substrate-based inhibitors of lanosterol 14α-methyl demethylase: II. Time-dependent enzyme inactivation by selected oxylanosterol analogs. *Biochemistry* **1995**, *34*, 9677–9681. [CrossRef] [PubMed]
27. Tuck, S.F.; Patel, H.; Safi, E.; Robinson, C.H. Lanosterol 14α-demethylase (P45014DM): Effects of P45014DM inhibitors on sterol biosynthesis downstream of lanosterol. *J. Lipid Res.* **1991**, *32*, 893–902. [PubMed]
28. Trzaskos, J.M.; Magolda, R.L.; Favata, M.F.; Fischer, R.T.; Johnson, P.R.; Chen, H.W.; Ko, S.S.; Leonard, D.A.; Gaylor, J.L. Modulation of 3-hydroxy-3-methylglutaryl-CoA reductase by 14α-flurolanost-7-en-3-ol. *J. Biol. Chem.* **1993**, *268*, 22591–22599. [PubMed]
29. Frye, L.L.; Robinson, C.H. Synthesis of potential mechanism-based inactivators of lanosterol 14α-methyl demethylase. *J. Org. Chem.* **1990**, *55*, 1579–1584. [CrossRef]
30. Lepesheva, G.I.; Zaitseva, N.G.; Nes, W.D.; Zhou, W.; Arase, M.; Liu, J.; Hill, G.C.; Waterman, M.R. CYP51 from *Trypanosoma cruzi*: A phyla-specific residue in the B′ helix defines substrate preferences of sterol 14α-demethylase. *J. Biol. Chem.* **2006**, *281*, 3577–3585. [CrossRef] [PubMed]
31. Heeres, J.; Backx, L.J.J.; Mostmans, J.H.; Van Cutsem, J. Antimycotic imidazoles. Part 4. Synthesis and antifungal activity of ketoconazole, a new potent orally active broad-spectrum antifungal agent. *J. Med. Chem.* **1979**, *22*, 1003–1005. [CrossRef] [PubMed]
32. Heeres, J.; Van Cutsem, J. Antimycotic Imidazoles. 5. Synthesis and antimycotic properties of 1-[[2-aryl-4-(arylalkyl)-1,3-dioxolan-3-yl]methyl]-1H-imidazoles. *J. Med. Chem.* **1981**, *24*, 1360–1364. [CrossRef] [PubMed]
33. Heeres, J.; Hendrickx, R.; Van Cutsem, J. Antimycotic azoles. 6. Synthesis and antifungal properties of terconazole, a novel triazole ketal. *J. Med. Chem.* **1983**, *26*, 611–613. [CrossRef] [PubMed]
34. Heeres, J.; Backx, L.J.J.; Van Cutsem, J. Antimycotic azoles. 7. Synthesis and antifungal properties of a series of novel triazol-3-ones. *J. Med. Chem.* **1984**, *27*, 894–900. [CrossRef] [PubMed]
35. Richardson, K.; Cooper, K.; Marriott, M.S.; Tarbit, M.H.; Troke, P.F.; Whittle, P.J. Discovery of fluconazole, a novel antifungal agent. *Rev. Infect. Dis.* **1990**, *12*, S267–S271. [CrossRef] [PubMed]
36. Saksena, A.K.; Girijavallabhan, V.M.; Wang, H.; Liu, Y.T.; Pike, R.E.; Ganguly, A.K. Concise asymmetric routes to 2,2,4-trisubstituted tetrahydrofurans via chiral titanium imide enolates: Key intermediates towards

synthesis of highly active azole antifungals SCH 51048 and SCH 56592. *Tetrahedron Lett.* **1996**, *37*, 5657–5660. [CrossRef]

37. Saksena, A.K.; Girijavallabhan, V.M.; Lovey, R.G.; Pike, R.E.; Wang, H.; Liu, Y.T.; Pinto, P.; Bennett, F.; Jao, E.; Patel, N.; et al. Advances in the chemistry of novel broad-spectrum orally active azole antifungals: Recent studies leading to the discovery of SCH 56592. *Spec. Publ. R. Soc. Chem.* **1997**, *198*, 180–199. [CrossRef]
38. Livni, E.; Fischman, A.J.; Ray, S.I.; Elmaleh, D.R.; Alpert, N.M.; Weiss, S.; Correia, J.A.; Webb, D.; Dahl, R.; Robeson, W.; et al. Synthesis of ^{18}F-labeled fluconazole and positron emission tomography studies in rabbits. *Nucl. Med. Biol.* **1992**, *19*, 191–199. [CrossRef]
39. Korwar, S.; Amir, S.; Tosso, P.N.; Desai, B.K.; Kong, C.J.; Fadnis, S.; Telgang, N.S.; Ahmad, S.; Roper, T.D.; Gupton, B.F. The application of a continuous Grignard reaction in the preparation of fluconazole. *Eur. J. Org. Chem.* **2017**, 6495–6498. [CrossRef]
40. Aher, N.G.; Pore, V.S.; Mishra, N.N.; Kumar, A.; Shukla, P.K.; Sharma, A.; Bhat, M.K. Synthesis and antifungal activity of 1,2,3-triazole containing fluconazole analogues. *Bioorg. Med. Chem. Lett.* **2009**, *19*, 759–763. [CrossRef] [PubMed]
41. Hoekstra, W.J.; Garvey, E.P.; Moore, W.R.; Rafferty, S.W.; Yates, C.M.; Schotzinger, R.J. Design and optimization of highly-selective fungal CYP51 inhibitors. *Bioorg. Med. Chem. Lett.* **2014**, *24*, 3455–3458. [CrossRef] [PubMed]
42. Yates, C.M.; Garvey, E.P.; Shaver, S.R.; Schotzinger, R.J.; Hoekstra, W.J. Design and optimization of highly-selective, broad spectrum fungal CYP51 inhibitors. *Bioorg. Med. Chem. Lett.* **2017**, *27*, 3243–3248. [CrossRef] [PubMed]
43. Warrilow, A.G.S.; Parker, J.E.; Price, C.L.; Garvey, E.P.; Hoekstra, W.J.; Schotzinger, R.J.; Wiederhold, N.P.; Nes, W.D.; Kelly, D.E.; Kelly, S.L. The tetrazole VT-1161 is a potent inhibitor of *Trichophyton rubrum* through its inhibition of *T. rubrum* CYP51. *Antimicrob. Agents Chemother.* **2017**, *61*, e00333-17/1–e00333-12/11. [CrossRef] [PubMed]
44. Warrilow, A.G.S.; Hull, C.M.; Rolley, N.J.; Parker, J.E.; Nes, W.D.; Smith, S.N.; Kelly, D.E.; Kelly, S.L. Clotrimazole as a potent agent for treating the oomycete fish pathogen *Saprolegnia parasitica* through inhibition of sterol 14α-demethylase (CYP51). *Appl. Environ. Microbiol.* **2014**, *80*, 6154–6166. [CrossRef] [PubMed]
45. Gagnepain, J.; Maity, P.; Lamberth, C.; Cederbaum, F.; Rajan, R.; Jacob, O.; Blum, M.; Bieri, S. Synthesis and fungicidal activity of novel imidazole-based ketene dithioacetals. *Bioorg. Med. Chem.* **2018**, *26*, 2009–2016. [CrossRef]
46. Friggeri, L.; Hargrove, T.Y.; Wawrzak, Z.; Blobaum, A.L.; Rachakonda, G.; Lindsley, C.W.; Villalta, F.; Nes, W.D.; Botta, M.; Guengerich, F.P.; et al. Sterol 14α-demethylase structure-based design of VNI ((*R*)-*N*-(1-(2,4-dichlorophenyl)-2-(1H-imidazol-1-yl)ethyl)-4-(5-phenyl-1,3,4-oxadiazol-2-yl)benzamide)) derivatives to target fungal infections: Synthesis, biological evaluation, and crystallographic analysis. *J. Med. Chem.* **2018**, *61*, 5679–5691. [CrossRef] [PubMed]
47. Choi, J.Y.; Podust, L.M.; Roush, W.R. Drug strategies targeting CYP51 in neglected tropical diseases. *Chem. Rev.* **2014**, *114*, 11242–11271. [CrossRef] [PubMed]
48. Parish, E.J.; Schroepfer, G.J. Sterol synthesis. A simplified method for the synthesis of 32-oxygenated derivatives of 24,25-dihydrolanosterol. *J. Lipid Res.* **1981**, *22*, 859–868. [PubMed]
49. Lepesheva, G.I.; Hargrove, T.Y.; Kleshchenko, Y.; Nes, W.D.; Villalta, F.; Waterman, M.R. CYP51: A major drug target in the cytochrome P450 superfamily. *Lipids* **2008**, *43*, 1117–1125. [CrossRef] [PubMed]
50. Phillips, G.N.; Quiocho, F.A.; Sass, R.L.; Werness, P.; Emery, H.; Knapp, F.F.; Schroepfer, G.J. Sterol biosynthesis: Establishment of the structure of 3-*p*-bromobenzoyloxy-5α-cholest-8(14)-en-15-ol. *Bioorg. Chem.* **1976**, *5*, 1–10. [CrossRef]
51. Gibbons, G.F.; Ramanada, K. Synthesis and configuration at C-15 of the epimeric 5-lanost-8-en-3,15-diols. *J. Chem. Soc. Chem. Commun.* **1975**, *6*, 213–214. [CrossRef]
52. Gaylor, J.L.; Johnson, P.R.; Ko, S.S.; Magolda, R.L.; Stam, S.H.; Trzaskos, J.M. Steroid Derivatives Useful as Hypocholesterolemics. U.S. Patent 5034548, 20 August 1991.
53. Musiol, R.; Kowalczyk, W. Azole antimycotics—A highway to new drugs or a dead end? *Curr. Med. Chem.* **2012**, *19*, 1378–1388. [CrossRef] [PubMed]
54. DiMasi, J.A.; Hansen, R.W.; Grabowski, H.G. The price of innovation: New estimates of drug development costs. *J. Health Econ.* **2003**, *22*, 151–185. [CrossRef]

55. Villalta, F.; Dobish, M.C.; Nde, P.N.; Kleshchenko, Y.Y.; Hargrove, T.Y.; Johnson, C.A.; Waterman, M.R.; Johnston, J.N.; Lepesheva, G.I. VNI cures acute and chronic experimental Chagas disease. *J. Infect. Dis.* **2013**, *208*, 504–511. [CrossRef] [PubMed]
56. Ramos, G.; Cuenca-Estrella, M.; Monzon, A.; Rodriguez-Tudela, J.L. In-vitro comparative activity of UR-9825, itraconazole and fluconazole against clinical isolates of *Candida* spp. *J. Antimicrob. Chemother.* **1999**, *44*, 283–286. [CrossRef] [PubMed]
57. Richardson, K.; Brammer, K.W.; Marriott, M.S.; Troke, P.F. Activity of UK-49,858, a bis triazole derivative, against experimental infections with *Candida albicans* and *Trichophyton mentagrophytes*. *Antimicrob. Agents Chemother.* **1985**, *27*, 832–835. [CrossRef] [PubMed]
58. Lepesheva, G.I.; Ott, R.D.; Hargrove, T.Y.; Kleshchenko, Y.Y.; Schuster, I.; Nes, W.D.; Hill, G.C.; Villalta, F.; Waterman, M.R. Sterol 14α-demethylase as a potential target for antitrypanosomal therapy: Enzyme inhibition and parasite cell growth. *Chem. Biol.* **2007**, *14*, 1283–1293. [CrossRef] [PubMed]
59. Bettiol, E.; Samanovic, M.; Murkin, A.S.; Raper, J.; Buckner, F.; Rodriguez, A. Identification of three classes of heteroaromatic compounds with activity against intracellular *Trypanosoma cruzi* by chemical library screening. *PLoS Negl. Trop. Dis.* **2009**, *3*, e384. [CrossRef] [PubMed]
60. Chen, C.; Leung, S.S.F.; Guilbert, C.; Jacobson, M.P.; McKerrow, J.H.; Podust, L.M. Structural characterization of CYP51 from *Trypanosoma cruzi* and *Trypanosoma brucei* bound to the antifungal drugs posaconazole and fluconazole. *PLoS Negl. Trop. Dis.* **2010**, *4*, e651. [CrossRef] [PubMed]
61. Lepesheva, G.I.; Villalta, F.; Waterman, M.R. Targeting *Trypanosoma cruzi* sterol 14α-demethylase (CYP51). *Adv. Parasitol.* **2011**, *75*, 65–87. [PubMed]
62. Alrajhi, A.A.; Ibrahim, E.A.; De Vol, E.B.; Khairat, M.; Faris, R.M.; Maguire, J.H. Fluconazole for the treatment of cutaneous leishmaniasis caused by *Lesihmania major*. *N. Engl. J. Med.* **2002**, *346*, 891–895. [CrossRef] [PubMed]
63. Urbina, J.A.; Payares, G.; Molina, J.; Sanoja, C.; Liendo, A.; Lazzardi, K.; Piras, M.M.; Piras, R.; Perez, N.; Wincker, P.; et al. Cure of short- and long term experimental Chagas' disease using D0870. *Science* **1996**, *273*, 969–971. [CrossRef] [PubMed]
64. Urbina, J.A.; Payares, G.; Sanoja, C.; Lira, R.; Romanha, A.J. In vitro and in vivo activities of ravuconazole on *Trypanosoma cruzi*, the causative agent of Chagas disease. *Int. J. Antimicrob. Agents* **2003**, *21*, 27–38. [CrossRef]
65. Lepesheva, G.I.; Nes, W.D.; Zhou, W.; Hill, G.C.; Waterman, M.R. CYP51 from *Trypanosoma brucei* is obtusifoliol-specific. *Biochemistry* **2004**, *43*, 10789–10799. [CrossRef] [PubMed]
66. Vivas, J.; Urbina, J.A.; de Souza, W. Ultrastructural alterations in *Tyrpanosoma (Schizotrypanum) cruzi* induced by $\Delta^{24(25)}$ sterol methyl transferase inhibitors and their combinations with ketoconazole. *Int. J. Antimicrob. Agents* **1996**, *7*, 235–240. [CrossRef]
67. Espinel-Ingroff, A.; Shadomy, S.; Gebhart, R.J. In vitro studies with R 51,211 (Itraconazole). *Antimicrob. Agents Chemother.* **1984**, *26*, 5–9. [CrossRef] [PubMed]
68. Borgers, M.; Van de Ven, M.A. Degenerative changes in fungi after itraconazole treatment. *Rev. Infect. Dis.* **1987**, *9*, 33–44. [CrossRef]
69. Lipp, H.-P. Antifungal agents-clinical pharmacokinetics and drug interactions. *Mycoses* **2008**, *51*, 7–18. [CrossRef] [PubMed]
70. Gubbins, P.O. Mould-active azoles: Pharmacokinetics, drug interactions in neutropenic patients. *Curr. Opin. Infect. Dis.* **2007**, *20*, 579–586. [CrossRef] [PubMed]
71. Lamb, D.C.; Kelly, D.E.; Baldwin, B.C.; Kelly, S.L. Differential inhibition of human CYP3A4 and *Candida albicans* CYP51 with azole antifungal agents. *Chem. Biol. Interact.* **2000**, *125*, 165–175. [CrossRef]
72. Hart, D.T.; Lauwers, W.J.; Willemsens, G.; Bossche, H.V.; Opperdoes, F.R. Perturbation of sterol biosynthesis by itraconazole and ketoconazole in *Leishmania mexicana mexicana* infected macrophages. *Mol. Biochem. Parasitol.* **1989**, *33*, 123–134. [CrossRef]
73. Momeni, A.Z.; Jalayer, T.; Emamjomeh, M.; Bashardost, N.; Ghassemi, R.L.; Meghdadi, M.; Javadi, A.; Aminjavaheri, M. Treatment of cutaneous leishmaniasis with itraconazole. Randomized double-blind study. *Arch. Dermatol.* **1996**, *132*, 784–786. [CrossRef] [PubMed]
74. Baroni, A.; Aiello, F.S.; Vozza, A.; Vozza, G.; Faccenda, F.; Brasiello, M.; Ruocco, E. Cutaneous leishmaniasis treated with itraconazole. *Dermatol. Ther.* **2009**, *22* (Suppl. 1), S27–S29. [CrossRef] [PubMed]
75. Minodier, P.; Parola, P. Cutaneous leishmaniasis treatment. *Travel Med. Infect. Dis.* **2007**, *5*, 150–158. [CrossRef] [PubMed]

76. Urbina, J.A. Chemotherapy of Chagas disease. *Curr. Pharm. Des.* **2002**, *8*, 287–295. [CrossRef] [PubMed]
77. Urbina, J.A. Ergosterol biosynthesis for the specific treatment of Chagas disease: From basic science to clinical trials. In *Trypanosomatid Diseases: Molecular Routes to Drug Discovery*, 1st ed.; Jäger, T., Koch, O., Flohé, L., Eds.; Wiley-VCH Verlag GmbH & Co. KGaA: Weinheim, Germany, 2013; pp. 489–514.
78. Benaim, G.; Sanders, J.M.; Garcia-Marchán, Y.; Colina, C.; Lira, R.; Caldera, A.R.; Payares, G.; Sanoja, C.; Burgos, J.M.; Leo-Rossell, A.; et al. Amiodarone has intrinsic anti-*Trypanosoma cruzi* activity and acts synergistically with posaconazole. *J. Med. Chem.* **2006**, *49*, 892–899. [CrossRef] [PubMed]
79. Pinazo, M.; Espinosa, G.; Gállego, M.; López-Chejade, P.L.; Urbina, J.A.; Gascón, J. Case report: Successful treatment with posaconazole of a patient with chronic Chagas disease and systemic lupus erythematosus. *Am. J. Trop. Med. Hyg.* **2010**, *82*, 583–587. [CrossRef] [PubMed]
80. Saksena, A.K.; Girijavallabhan, V.M.; Lovey, R.G.; Pike, R.E.; Wang, H.; Ganguly, A.K.; Morgan, B.; Zaks, A.; Puar, M.S. Highly stereoselective access to novel 2,2,4-trisubstituted tetrahydrofurans by halocyclization: Practical chemoenzymatic synthesis of SCH 51048, a broad-spectrum orally active antifungal agent. *Tetrahedron Lett.* **1995**, *36*, 1787–1790. [CrossRef]
81. Chidambaram, V.S.; Miryala, A.K.; Wadhwa, L. Process for Preparing Posaconazole and Intermediates Thereof. PCT International Application WO2009141837A2, 26 November 2009.
82. Charyulu, P.V.R.; Gowda, D.J.C.; Rajmahendra, S.; Raman, M. Crystalline Forms of Posaconazole Intermediate and Process for the Preparation of Amorphous Posaconazole. U.S. PCT International Application WO2017051342A1, 30 March 2017.
83. Lepesheva, G.I.; Park, H.; Hargrove, T.Y.; Vanhollebeke, B.; Wawrzak, Z.; Harp, J.M.; Sundaramoorthy, M.; Nes, W.D.; Pays, E.; Chaudhuri, M.; et al. Crystal structures of *Trypanosoma brucei* sterol 14α-demethylase and implications for selective treatment of human infections. *J. Biol. Chem.* **2010**, *285*, 1773–1780. [CrossRef] [PubMed]
84. Lepesheva, G.; Christov, P.; Sulikowski, G.A.; Kim, K. A convergent, scalable and stereoselective synthesis of azole CYP51 inhibitors. *Tetrahedron Lett.* **2017**, *58*, 4248–4250. [CrossRef] [PubMed]
85. Nes, W.D.; Janssen, G.G.; Norton, R.A.; Kalinowska, M.; Crumley, F.G.; Tal, B.; Bergenstrahle, A.; Jonsson, L. Regulation of sterol biosynthesis in sunflower by 24(*R*,*S*)25-epiminolanosterol, a novel C-24 methyl transferase inhibitor. *Biochem. Biophys. Res. Commun.* **1991**, *177*, 566–574. [CrossRef]
86. Nes, W.D.; Xu, S.; Parish, E.J. Metabolism of 24(*R*,*S*),25-epiminolanosterol to 25-aminolanosterol and lanosterol by *Gibberella fujikuroi*. *Arch. Biochem. Biophys.* **1989**, *272*, 323–331. [CrossRef]
87. Popjak, G.; Meenan, A.; Parish, E.J.; Nes, W.D. Inhibition of cholesterol synthesis and cell growth by 24(*R*,*S*),25-iminolanosterol and triparanol in cultured rat hepatoma cells. *J. Biol. Chem.* **1989**, *264*, 6230–6238. [PubMed]
88. Zhou, W.; Lepesheva, G.I.; Waterman, M.R.; Nes, W.D. Mechanistic analysis of a multiple product sterol methyltransferase implicated in ergosterol biosynthesis in *Trypanosoma brucei*. *J. Biol. Chem.* **2006**, *281*, 6290–6296. [CrossRef] [PubMed]
89. Urbina, J.A.; Vivas, J.; Lazardi, K.; Molina, J.; Payares, G.; Piras, M.M.; Piras, R. Antiproliferative effects of $\Delta^{24(25)}$ sterol methyl transferase inhibitors on *Trypanosoma (Schizotrypanum) cruzi*: In vitro and in vivo studies. *Chemotherapy* **1996**, *42*, 294–307. [CrossRef] [PubMed]
90. Kidane, M.E.; Vanderloop, B.H.; Zhou, W.; Thomas, C.D.; Ramos, E.; Singha, U.; Chaudhuri, M.; Nes, W.D. Sterol methyltransferase a target for anti-amoebatherapy: Towards transition state analog and suicide substrate drug design. *J. Lipid Res.* **2017**, *58*, 2310–2323. [CrossRef] [PubMed]
91. Parish, E.J.; Nes, W.D. Synthesis of new epiminoisopentenoids. *Synth. Commun.* **1998**, *18*, 221–226. [CrossRef]
92. Nes, W.D. Sterol methyl transferase: Enzymology and inhibition. *Biochim. Biophys. Acta* **2000**, *1529*, 63–88. [CrossRef]
93. Pereira, M.; Song, Z.; Santos-Silva, L.K.; Richards, M.H.; Nguyen, T.T.M.; Liu, J.; Soares, C.M.A.; Cruz, A.H.S.; Ganapathy, K.; Nes, W.D. Cloning, mechanistic and functional analysis of a fungal sterol C24-methyltransferase implicated in brassicasterol biosynthesis. *Biochim. Biophys. Acta* **2010**, *1801*, 1163–1174. [CrossRef] [PubMed]
94. Zhou, W.; Cross, G.A.M.; Nes, W.D. Cholesterol import fails to prevent catalyst-based inhibition of ergosterol synthesis and cell proliferation of *Trypanosoma brucei*. *J. Lipid Res.* **2007**, *48*, 665–673. [CrossRef] [PubMed]
95. Lu, M.C.; Kohen, F.; Counsell, R.E. Hypocholesterolemic agents. 8. Synthesis of 25-azadihydrolanosterol and derivatives. *J. Med. Chem.* **1971**, *14*, 136–138. [CrossRef] [PubMed]

96. Oehlschlager, A.C.; Angus, R.H.; Pierce, A.M.; Pierce, H.D.; Srinivasan, R.J. Azasterol inhibition of 24-sterol methyltransferase in *Saccharomyces cerevisiae*. *Biochemistry* **1984**, *23*, 3582–3589. [CrossRef] [PubMed]
97. Miller, M.B.; Patkar, P.; Singha, U.K.; Chaudhuri, M.; Nes, W.D. 24-Methylenecyclopropane steroidal inhibitors: A Trojan horse in ergosterol biosynthesis that prevents growth of *Trypanosoma brucei*. *Biochim. Biophys. Acta* **2017**, *1862*, 305–313. [CrossRef] [PubMed]
98. Marshall, J.A.; Nes, W.D. Isolation and characterization of an active-site peptide from a sterol methyl transferase with a mechanism-based inhibitor. *Bioorg. Med. Chem. Lett.* **1999**, *9*, 1533–1536. [CrossRef]
99. Neelakandan, A.K.; Song, Z.; Wang, J.; Richards, M.H.; Wu, X.; Valliyodan, B.; Nguyen, H.T.; Nes, W.D. Cloning, functional expression and phylogenetic analysis of plant sterol 24C-methyltransferases involved in sitosterol biosynthesis. *Phytochemistry* **2009**, *70*, 1982–1998. [CrossRef] [PubMed]
100. Zhou, W.; Nes, W.D. Sterol methyltransferase2: Purification, properties, and inhibition. *Arch. Biochem. Biophys.* **2003**, *420*, 18–34. [CrossRef] [PubMed]
101. Song, A.; Zhou, W.; Liu, J.; Nes, W.D. Mechanism-based active site modification of the soybean sterol methyltransferase by 26,27-dehydrocycloartenol. *Bioorg. Med. Chem. Lett.* **2004**, *14*, 33–36. [CrossRef] [PubMed]
102. Jia, Z.; Zhou, W.; Guo, D.; Nes, W.D. Synthesis of rationally designed mechanism-based inactivators of the (*S*)-adenosyl-L-methionine: $\Delta^{24(25)}$-sterol methyl transferase. *Synth. Commun.* **1996**, *26*, 3841–3848. [CrossRef]
103. Kanagasabai, R.; Zhou, W.; Liu, J.; Nguyen, T.T.M.; Veeramachaneni, P.; Nes, W.D. Disruption of ergosterol biosynthesis, growth, and the morphological transition in *Candida albicans* by sterol methyltransferase inhibitors containing sulfur at C-25 in the sterol side chain. *Lipids* **2004**, *39*, 737–746. [CrossRef] [PubMed]
104. Patkar, P.; Haubrich, B.A.; Qi, M.; Nguyen, T.T.M.; Thomas, C.D.; Nes, W.D. C24-Methylation of 26-fluorocycloartenols by recombinant sterol C24-methyltransferase from soybean: Evidence for channel switching and its phylogenetic implications. *Biochem. J.* **2013**, *456*, 253–262. [CrossRef] [PubMed]
105. Fuse, S.; Mifune, Y.; Tanabe, N.; Takahashi, T. Continuous-flow synthesis of activated vitamin D3 and its analogues. *Org. Biomol. Chem.* **2012**, *10*, 5205–5211. [CrossRef] [PubMed]
106. Chung, S.-K.; Lee, K.-W.; Kang, H.I.; Yamashita, C.; Kudo, M.; Yoshida, Y. Design and synthesis of potential inhibitors of the ergosterol biosynthesis as antifungal agents. *Bioorg. Med. Chem.* **2000**, *8*, 2475–2486. [CrossRef]
107. Yang, G.; Lee, N.; Ioset, J.-R.; No, J.H. Evaluation of parameters impacting drug susceptibility in intracellular *Trypanosoma cruzi* assay protocols. *SLAS Discov.* **2017**, *22*, 125–134. [CrossRef] [PubMed]
108. Molina, I.; Prat, J.G.; Salvador, F.; Trevino, B.; Sulleiro, E.; Serre, N.; Pou, D.; Roure, S.; Cabezos, J.; Valerio, L.; et al. Randomized trial of posaconazole and benznidazole for chronic Chagas disease. *N. Engl. J. Med.* **2014**, *370*, 1899–1908. [CrossRef] [PubMed]
109. Morillo, C.A.; Waskin, H.; Sosa-Estani, S.; del Carmen Bangher, M.; Cuneo, C.; Milesi, R.; Mallagray, M.; Apt, W.; Beloscar, J.; Gascon, J.; et al. Benznidazole and posaconazole in eliminating parasites in asymptomatic *T. cruzi* carriers: The stop-Chagas trial. *J. Am. Coll. Cardiol.* **2017**, *69*, 939–947. [CrossRef] [PubMed]

Sample Availability: Samples of the compounds are not available from the authors.

Article

Improved Synthesis of *N*-Methylcadaverine

Kayla N. Anderson, Shiva Moaven, Daniel K. Unruh, Anthony F. Cozzolino * and John C. D'Auria *

Department of Chemistry and Biochemistry, Texas Tech University, Box 41061, Lubbock, TX 79409-1061, USA; kayla.anderson@ttu.edu (K.N.A.); shiva.moaven@ttu.edu (S.M.); daniel.unruh@ttu.edu (D.K.U.)

* Correspondence: anthony.f.cozzolino@ttu.edu (A.F.C.); john.c.dauria@ttu.edu (J.C.D.); Tel.: +1-806-834-1832 (A.F.C.); +1-806-834-7348 (J.C.D.)

Academic Editors: Wenxu Zhou and De-an Guo
Received: 25 April 2018; Accepted: 15 May 2018; Published: 19 May 2018

Abstract: Alkaloids compose a large class of natural products, and mono-methylated polyamines are a common intermediate in their biosynthesis. In order to evaluate the role of selectively methylated natural products, synthetic strategies are needed to prepare them. Here, *N*-methylcadaverine is prepared in 37.3% yield in three steps. The alternative literature two-step strategy resulted in reductive deamination to give *N*-methylpiperidine as determined by the single crystal structure. A straightforward strategy to obtain the mono-alkylated aliphatic diamine, cadaverine, which avoids potential side-reactions, is demonstrated.

Keywords: alkaloid; granatane; *N*-methylcadaverine; *N*-methylpiperidine. reductive deamination

---

1. Introduction

Polyamines (PAs) are abundant in nature. Some of the most common examples include putrescine (1,4-diaminobutane), cadaverine (1,5-diaminopentane), spermidine (*N*-(3-aminopropyl) -1,4-diaminobutane), and spermine (*N,N'*-Bis(3-aminopropyl)-1,4-diaminobutane) [1]. Each of these PAs have been recruited by multiple lineages to perform various biological functions. For instance, all of the abovementioned PAs play a critical role as primary metabolites by mediating fundamental developmental processes [2]. More specifically, PAs in mammals and bacteria participate in the regulation of gene expression and gene transcription [1,3].

Plants utilize PAs for similar functions such as cell proliferation and cell signaling [2]. Additionally, PAs are employed for organ and pollen development [4]. PAs covalently bind to hydroxylcinnamate to form hyrdroxy-cinnamic acid amides (HCAAs), which drive pollen development and the pollen–pistil interaction during fertilization [4]. Contrary to mammals and bacteria, plants also utilize PAs for secondary metabolic purposes, such as stress responses [5].

Due to their sessile nature, plants produce phytoalexins and other specialized metabolites in order to mediate their responses with both the abiotic and biotic forces present in their surrounding environment [5]. Alkaloids comprise a large class of specialized metabolites that play key roles in these interactions. For example, steroidal alkaloids are known to cause inhibition of the fungal species *Phytophthora cactorum*, a known cause of root rot [6]. PAs have been implicated in alkaloid biosynthesis, specifically in piperidine and pyrrolidine alkaloids. Alkaloids are defined as nitrogen containing cyclic compounds. Alkaloids also have significant pharmacological properties. The alkaloids scopolamine and atropine are known for their anticholinergic and antispasmodic properties [7]. Alternatively, the alkaloids pseudopelleterine and *N*-methylpelleterine have historically been used for their anthelminthic (anti-worming) properties [5].

Scopolamine and atropine are compounds that originate from plants of the Solanaceae family. Both compounds belong to the class of alkaloids termed tropane alkaloids. Tropane alkaloids share a common *N*-methyl-8-azabicyclo[3.2.1]-octane core skeleton. Tropane alkaloids can also be categorized as a sub-class of pyrrolidine alkaloids because a pyrrolidine ring is part of the bicyclic structure. The PAs putrescine and spermidine are known intermediates of tropane alkaloid biosynthesis [7,8]. In addition, tropanes are constitutional isomers of granatane alkaloids, containing a one carbon difference in their bicyclic moieties.

Granatane alkaloids are found predominantly in the species *P. granatum* (pomegranate). Granatane alkaloids are a sub-class of piperidine alkaloids, due to the presence of a piperidine ring in their core skeleton. The granatanes include the compounds *N*-methylpelleterine, pelleterine, and the bicyclic compound pseudopelletierine (*N*-methyl-9-azabicyclo[3.3.1]-nonane base structure) (Figure 1). Granatane alkaloids in *P. granatum* originate from the amino acid lysine. The evidence for this biosynthetic origin is based on the incorporation of radio-labeled [2-^{14}C] lysine into the first ring of *N*-methylpelletierine during *in planta* feeding studies [9]. The results of these studies suggest a symmetrical intermediate in the formation of the piperidine ring. The symmetrical polyamine cadaverine is the product of the decarboxylation of lysine (Figure 2). When fed to whole pomegranate plants, radio-labeled [1,5-^{14}C] cadaverine incorporated into the granatanes pelletierine, *N*-methylpelletierine, and pseudopelletierine. Additionally, the mono-methylation of cadaverine is a proposed enzymatic step in granatane alkaloid formation [5] (Figure 2). Currently, feeding studies using *N*-methylcadaverine (**1**) are not possible since commercial sources are not available. Therefore, a synthetic route to producing this compound would aid in the overall understanding of granatane biosynthesis.

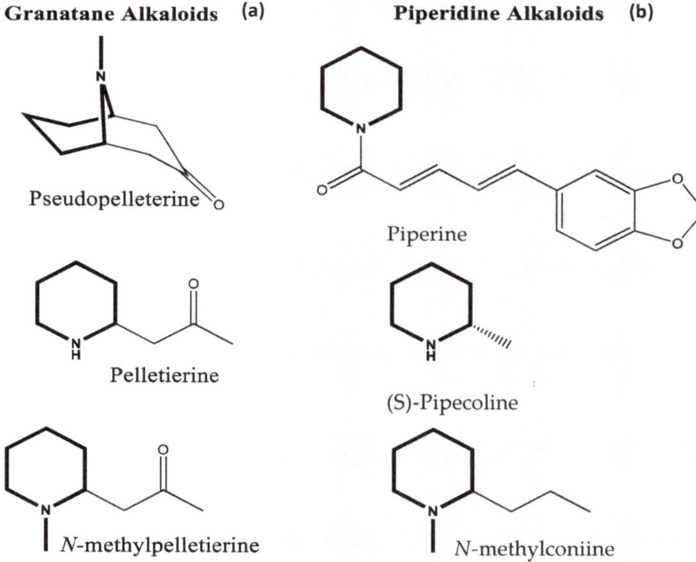

Figure 1. Natural products biosynthesized from *N*-methylcadaverine or *N*-methylpiperidine. The compounds in panel (**a**) are granatane alkaloids. The bolded atoms in panel (**a**) are from the proposed precursor *N*-methylcadaverine. The natural products in panel (**b**) are piperidine alkaloids. The bolded atoms show the piperidine ring from *N*-methylpiperidine.

Piperidine alkaloids compose a broader class of alkaloids also found in plants. Piperidine alkaloids are classified as compounds with a nitrogen containing six-membered core ring structure (Figure 1). However, piperidine alkaloids can be monocyclic or heterocyclic. Piperidine alkaloids are found in black pepper (*Psilocaulon absmile*) and poisonous hemlocks (*Conium maculatum*) (Figure 1) [10]. *N*-methylconiine, as well as other piperidine alkaloids found in *Conium maculatum* have been used for their analgesic abilities [11]. Substituted six-membered *N*-heterocycles are found in numerous natural products and pharmaceutical compounds that are commonly used today, such as the aforementioned *N*-methylconiine [12]. Hameed et al. (1992) utilized *N*-methylpiperidine (**2**) as a starting material for the synthesis of morphine analogs [13]. Alongside the synthesis of **2**, substituted *N*-heterocycles can be further synthesized for pharmaceutical purposes at a lesser cost [13].

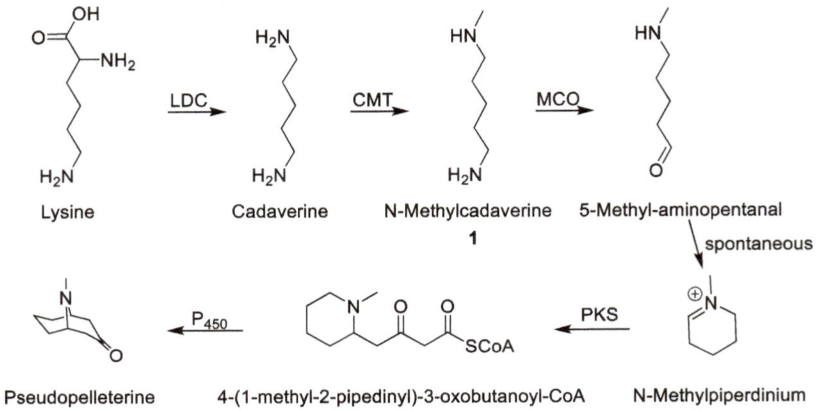

Figure 2. Proposed biosynthetic pathway of granatane alkaloid formation in *Punica grantum* [5]. The abbreviations above the arrows correspond with the presumed enzymes in the granatane pathway. LDC—lysine decarboxylase, CMT—cadaverine methyl-transferase, MCO—methylcadaverine oxidase, PKS—polyketide synthase P450—Cytochrome P450.

A major hindrance to studying alkaloid biosynthesis is the lack of commercially available selectively *N*-methylated polyamines such as: **1**. To perform classical biochemical experiments on piperidine alkaloid producing plants, the synthesis of *N*-methylated polyamines is necessary. Monoalkylation of polyamines can present a challenge in achieving selectivity and also in limiting the extent of methylation [14,15]. Here, the synthesis of *N*-methylcadaverine and *N*-methylpiperidine by reductive amination of a nitrile is reported [16,17].

2. Results and Discussion

2.1. Synthesis of *N*-Methylpiperidinium Chloride (**2**·*HCl*)

In an attempt to prepare *N*-methylcadaverine, the procedure reported by Jourdain, Caron, and Pommelet (Scheme 1) was followed [1]. *N*-Methylbenzylamine and 5-bromovaleronitrile react to yield 5-(benzyl(methyl)amino)pentanenitrile (**3**) (Scheme 1) as reported [1]. Both the reduction of the nitrile to a primary amine and the removal of the benzyl group were reported to proceed by hydrogenation over palladium on charcoal (Pd/C) (Scheme 1). Following the reported procedure, a material was recovered that did not match with **1** spectroscopically [1]. Instead, the ^{1}H and ^{13}C NMR spectra were consistent with a more symmetrical system with only four proton and carbon environments as opposed to the expected six (Figures S3 and S4). The mass spectrum suggested the loss of one amine. The product was concluded to be the cyclized product, *N*-methylpiperidinium chloride (**2**·HCl) (Figure S13). This species can be formed during the hydrogenation reaction, by initial deprotection of the tertiary amine to a secondary amine. This amine can attack the carbon of the Pd-activated nitrile to cyclize while forming the new C-N bond [16,17] (Figure 3). Reduction results first in deamination and finally the formation of **2**·HCl. The crystal structure of the observed product confirmed our conclusion that the synthesized product was **2**·HCl (Figure 4). In the structure, the chloride ion-pairs with the piperidinum through an NH hydrogen bond with a fairly typical Cl-N distance of 3.075 Å and is in close contact with CH hydrogen atoms on adjacent molecules [18,19]. All of the other metrical parameters are as expected.

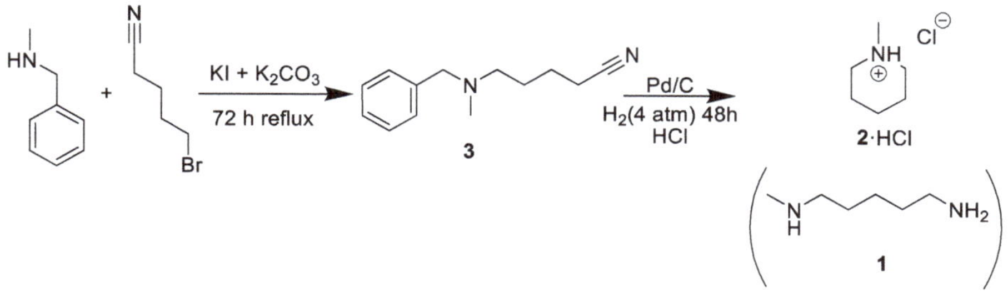

Scheme 1. Synthesis of *N*-methylpiperidine hydrochloride (**2**·HCl) [1]. The proposed product **1**, is shown in parenthesis.

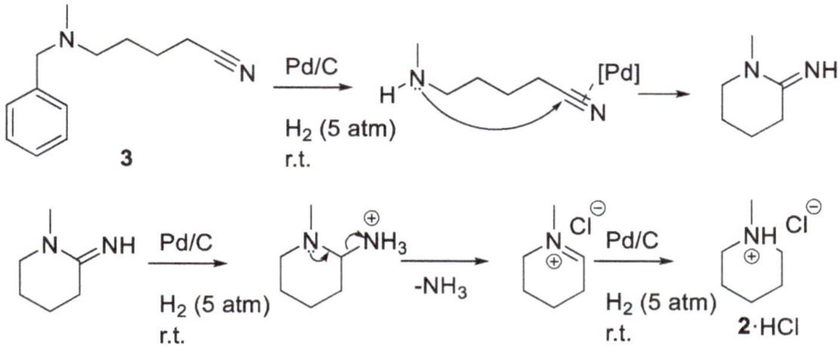

Figure 3. Proposed mechanism of cyclization from compound **3** to the final product **2**·HCl.

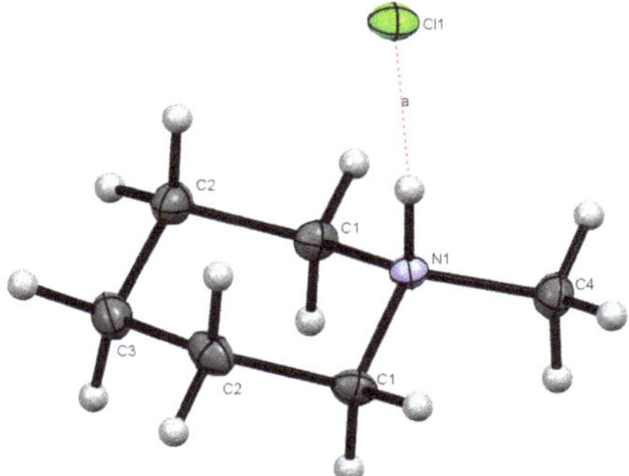

Figure 4. X-ray structure of **2**·HCl. The thermal ellipsoids are represented at 50% probability. Carbon, hydrogen, nitrogen, and chlorine atoms are represented by dark gray, light gray, purple, and green ellipsoids, respectively.

2.2. *Synthesis of N-Methylcadaverine Hydrochloride*, (**1**·2 *HCl*)

To obtain the desired *N*-methylcadaverine, the initial reaction scheme was revised (Scheme 2). The revised reaction still proceeds through **3**. To circumvent cyclization of the product, the nitrile was first reduced to a primary amine with lithium aluminum hydride (LAH). The reduction of the nitrile produced N^1-benzyl-N^1-methylpentane-1,5-diamine (4, Scheme 2). The removal of the benzyl group was subsequently achieved through a 48-h hydrogenation reaction over Pd/C. The desired product is isolated as the hydrochloride salt, **1**·2 HCl.

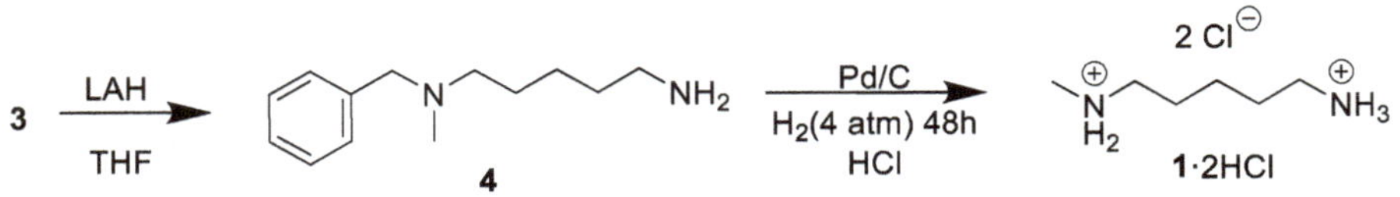

Scheme 2. Synthesis of *N*-methylcadaverine hydrochloride (**1**·2HCl).

3. Materials and Methods

3.1. *General Methods*

The materials *N*-methybenzylamine (>97%, Tokyo Chemical Industry; Portaland, OR, USA), 5-bromovaleronitrile (95%, Santa Cruz Biotechnology; Dallas, TX, USA), Methanol (Fisher ACS grade; Madison, WI, USA), Ethanol (absolute, Pharmco; Toronto, ON, Canada), Potassium Iodide (99%, EMD Chemicals; Burlington, MA, USA), Potassium carbonate anhydrous (99%, EMD Chemicals; Burlington, MA, USA), Magnesium sulfate anhydrous (99%, J.T. Baker; Phillipsburg, NJ, USA), and Palladium/Carbon (10% wet, Oakwood Chemical; Estill, CA, USA) were used as purchased. Anhydrous tetrahydrofuran was obtained by passing HPLC grade THF over a bed of activated molecular sieves in a commercial (LC Technologies Solutions Inc.; Salisbury, MD, USA) solvent purification system (SPS). All NMR spectra were collected using a JEOL ECS 400 MHz NMR spectrometer (JEOL; Tokyo, Japan). All IR spectra were obtained using a Nicolet iS 5 FT-IR spectrometer equipped with a Specac Di Quest ATR accessory (Thermo Scientific; Madsion, WI, USA), high-resolution mass spectra (HRMS) were obtained on a Thermo Exactive MS with an Orbitrap mass analyzer in ESI mode, and CHN analysis were obtained on-site with a Perkin Elmer 2400 Series II CHNS/O Analyzer (Perkin Elmer; Waltham, MA, USA).

3.2. *Synthesis of 5-(Benzyl(methyl)amino)pentanenitrile* (**3**)

N-methylbenzylamine (121.18 $g \cdot mol^{-1}$, 6.05 g, 49.9 mmol) was dissolved in 150 mL of anhydrous ethanol and 10.32 g of potassium carbonate (138.20 $g \cdot mol^{-1}$, 74.67 mmol) and 1.24 g of potassium iodide (166.00 $g \cdot mol^{-1}$, 7.47 mmol) were suspended in the solution. The mixture was brought to reflux and 12.15 g of 5-bromopentanitrile (162.03 $g \cdot mol^{-1}$, 74.98 mmol) dissolved in 50 mL of anhydrous ethanol was added to the suspension dropwise over the course of 3 h. The solution was stirred under reflux for 72 h. Upon cooling, the salts were filtered off and the filtrate was taken to dryness. To the residue was added 100 mL of 2 M HCl solution and unreacted reagents were extracted via ether (3 × 50 mL). The aqueous layer was neutralized with 2 M NaOH solution and the final product was extracted with diethyl ether (3 × 100 mL). The organic solution was dried over $MgSO_4$ and the solvent was removed. A light-yellow liquid (8.45 g) was collected giving **3** in 83.7% yield. **^{1}H-NMR** ($CDCl_3$) = 1.67 (m, 4H); 2.20 (s, 3H); 2.30 (t, 2H) (t, J = 6.6Hz); 2.38 (t, 2H) (t, J = 6.9Hz); 3.48 (s, 2H); 7.24–7.36 (m, 5H). **^{13}C-NMR** ($CDCl_3$) = 139.21 (s, 1C), 129.04 (s, 2C), 128.35 (s, 2C), 127.11 (s, 1C), 127.11 (s, 1C), 119.92 (s, 1C), 62.59 (s, 1C), 55.90 (s, 1C), 42.26 (s, 1C), 26.27 (s, 1C), 23.24 (s, 1C), 16.99 (s, 1C). **FTIR (ATR, cm^{-1})**: 3029 (s, Csp^2–H), 2949 (vs, Csp^3–H) 2245 (vs, C≡N).

3.3. Synthesis of N-Methylpiperidine Chloride (2·HCl)

Compound **3** (203.17 g·mol^{-1}, 3.95 g, 19.4 mmol) was dissolved in 15 mL of methanol and 1 mL of concentrated hydrochloric acid was added to the solution. The solution was transferred to a 250 mL Fisher-Porter bottle and 3.85 g of 10% (w/w) palladium on wet carbon was added to the reactor. The reactor was sealed and charged with 5 atm of H_2. The mixture was stirred at room temperature for 24 h. The mixture was filtered through a bed of Celite to remove the Pd on carbon (Note that the Pd on carbon is pyrophoric at this stage and should not be allowed to dry or be placed in contact with organics). To crystalize the product as the hydrochloride salt, 1 mL of concentrated HCl was added to the solution. The volatile solvents were evaporated using a rotary-evaporator and the residual water was removed under high vacuum to give milky-white crystals (2.62 g, 93.4%). X-ray diffraction quality crystals of **2**·HCl were grown by slow cooling of supersaturated solution of **2**·HCl dissolved in hot acetone. **^{1}H-NMR** (D_2O spiked with acetone-d_5) = 3.31, 3.27 (d, 2H), 2.76 (t, 2H), 2.65 (s, 3H), 1.78, 1.75 (d, 2H), 1.55 (m, 3H), 1.28 (m, 1H). **^{13}C-NMR** (D_2O spiked with Acetone-d_5) = 20.62 (s, 1C), 22.99 (s, 2C), 43.16 (s, 1C), 54.91 (s, 1C). **FTIR (ATR, cm^{-1})**: 3005 (m, N^+–H), 2947 (s, Csp^3–H). **HRMS** (ESI) *m/z*: $[M + H]^+$ Calculated for $C_6H_{14}N$ 100.1821; Found 100.1122.

3.4. Synthesis of N^1-Benzyl-N^1-methylpentane-1,5-diamine (4)

In a 250 mL round bottom flask, 100 mL of degassed anhydrous THF was added under a nitrogen atmosphere and 4.04 g of (**3**) (202.30 g·mol^{-1}, 20.0 mmol) was added with a syringe through a septum. The solution was cooled in an ice bath and 4.54 g of lithium aluminum hydride (37.95 g·mol^{-1}, 119 mmol) was added to the solution under a positive flow of nitrogen. After addition of lithium aluminum hydride, the mixture was allowed to warm to room temperature and subsequently refluxed for 48 h. After 48 h, the mixture was cooled to 0 °C and quenched using Fieser's method [20]. The formed salts were removed by filtration and the remaining solution was taken to dryness to give the desired crude product as a light-yellow liquid. The crude product was further purified by column chromatography (first washed by DCM and next by MeOH–MeCN–Et_3N (4:5:1)) (3.16 g, 76.6%) was collected. **^{1}H-NMR** ($CDCl_3$) = 7.29 (m. 5H), 3.45 (s, 2H), 2.66 (t, 2H), 2.34 (t, 2H), 2.15 (s, 3H), 1.50 (m, 2H), 1.42 (m, 2H), 1.33 (m, 2H). **^{13}C-NMR** ($CDCl_3$) = 139.30 (s, 1C), 129.15 (s, 2C), 128.26 (s, 2C), 126.96 (s, 1C), 62.45 (s, 1C), 57.51 (s, 1C), 42.35 (s, 1C), 42.24 (s, 1C), 33.74 (s, 1C), 27.35 (s, 1C), 24.77 (s, 1C). **FTIR (ATR, cm^{-1})**: 3357 (w, N–H), 3292 (w, N–H), 3025 (s, Csp^2–H), 2930 (s, Csp^3–H).

3.5. Synthesis of N-Methylcadaverine (1·2 HCl)

A methanol (15 mL) solution of (**4**) (206.18 g·mol^{-1}, 1 g, 4.8 mmol) was added to 250 mL Fisher-Porter bottle. Concentrated hydrochloric acid (0.25 mL) and 300 mg of 10% (w/w) palladium on wet carbon were added to the reactor respectively and it was charged with hydrogen gas (60 psi). The mixture was stirred at room temperature for 48 h. The mixture was filtered, and 1 mL of concentrated hydrochloric acid solution was added to this solution and then it was taken to dryness to give *N*-methyl cadaverine as a white solid. (549 mg, 60.2%. **^{1}H-NMR** (D_2O) = 2.79(m, 4H), 2.49 (s, 3H), 1.51 (m, 4H), 1.26 (m. 2H). **^{13}C-NMR** (Acetonitrile-d_6) = 48.93 (s, 1C), 39.36 (s, 1C), 32.86 (s, 1C), 27.21 (s, 1C), 25.31 (s, 1C), 22.84 (s, 1C). **FTIR (ATR, cm^{-1})**: 3011(vw, N^+–H), 2729 (s, N^+–H), 2932 (s, Csp^3–H). **HRMS** (ESI) *m/z*: $[M + H]^+$ Calculated for $C_6H_{17}N_2$ 117.2126; found 117.1386.

4. Conclusions

A synthetic method for mono-methylated polyamines is necessary to continue biochemical analysis of piperidine alkaloid biosynthesis. The three-step method used in this paper allows for the straightforward synthesis of the mono-methylated polyamine, cadaverine, without a possible side reaction to *N*-methylpiperdine.

Supplementary Materials: Images of all ^{1}H-, ^{13}C-NMR and FTIR spectra are available online. The crystal structure information for compound **2**·HCl was deposited with the CCDC as a private communication with deposition number 1542426.

Author Contributions: J.C.D. and A.F.C. conceived and designed the experiments; K.N.A. and S.M. performed the experiments; D.K.U. collected and refined the crystal data; K.N.A., S.M, J.C.D., and A.F.C. analyzed the data and wrote the paper.

Acknowledgments: A.F.C and S.M. are grateful for financial support from the Robert A. Welch Foundation (D-1838, USA). All authors are grateful for financial support from Texas Tech University and the National Science Foundation (NMR instrument grant CHE-1048553). This work was partially funded by the National Science Foundation under grant No. (NSF-1714236) to JCD.

Conflicts of Interest: The authors declare no conflict of interest.

References

1. Jourdain, F.; Caron, M.; Pommelet, J.C. Rapid Synthesis Of Large Enaminolactams, A Novel Class Of Macrocycles. *Synth. Commun.* **1999**, *29*, 1785–1799. [CrossRef]
2. Kusano, T.; Berberich, T.; Tateda, C.; Takahashi, Y. Polyamines: Essential factors for growth and survival. *Planta* **2008**, *228*, 367–381. [CrossRef] [PubMed]
3. Childs, A.C.; Mehta, D.J.; Gerner, E.W. Polyamine-dependent gene expression. *Cell. Mol. Life Sci.* **2003**, *60*, 1394–1406. [CrossRef] [PubMed]
4. Alosi, I.; Cai, G.; Serafini-Fracassini, D.; Del Duca, S. Polyamines in Pollen: From Microsporogenesis to Fertilization. *Front. Plant Sci.* **2016**, *7*, 155. [CrossRef] [PubMed]
5. Kim, N.; Estrada, O.; Chavez, B.; Stewart, C.; D'Auria, J.C. Tropane and Granatane Alkaloid Biosynthesis: A Systematic Analysis. *Molecules* **2016**, *21*, 1510. [CrossRef] [PubMed]
6. Nes, W.D.; Hanners, P.K.; Bean, G.A.; Patterson, G.W. Inhibition of Growth and Sitosterol-Induced Sexual Reproduction in Phytophthora cactorum by Steroidal Alkaloids. *Phytopathology* **1982**, *72*, 447. [CrossRef]
7. Biastoff, S.; Reinhardt, N.; Reva, V.; Brandt, W.; Drager, B. Evolution of putrescine N-methyltransferase from spermidine synthase demanded alterations in substrate binding. *FEBS Lett.* **2009**, *583*, 3367–3374. [CrossRef] [PubMed]
8. Leete, E. Recent Developments in the Biosynthesis of the Tropane Alkaloids. *Planta Med.* **1990**, *56*, 339–352. [CrossRef] [PubMed]
9. O'Donovan, D.G.; Keogh, M.F. Biosynthesis of Piperidine Alkaloids. *Tetrahedron Lett.* **1968**, *9*, 265–267. [CrossRef]
10. Kumar, D.; Singh, V. Study of Heterocyclic Compound Piperidine. *Int. J. Res. Sci. Technol.* **2014**, *3*, 25–28.
11. Madaan, R.; Kumar, S. Screening of Alkaloidal Fraction of Conium maculatum L. Aerial Parts for Analgesic and Antiinflammatory Activity. *Indian J. Pharm. Sci.* **2012**, *74*, 457–460. [CrossRef] [PubMed]
12. Felpin, F.; Lebreton, J. Recent Advances in the Total Synthesis of Piperidine and Pyrrolidine Natural Alkaloids with Ring-Closing Metathesis as a Key Step. *Eur. J. Org. Chem.* **2003**, *2003*, 3693–3712. [CrossRef]
13. Hameed, S.; Saify, Z.S.; Baqar, S.M.; Naqvi, H.; Saeed, M.; Khan, A.; Ahmed, M. Design, Synthesis and Pharmacological Evaluation of N-Methyl Piperidine Derivatives. *Med. J. Islam. World Acad. Sci.* **1992**, *5*, 245–248.
14. Oku, T.; Arita, Y.; Tsuneki, H.; Ikariya, T. Continuous Chemoselective Methylation of Functionalized Amines and Diols with Supercritical Methanol over Solid Acid and Acid−Base Bifunctional Catalysts. *J. Am. Chem. Soc.* **2004**, *126*, 7368–7377. [CrossRef] [PubMed]
15. Enger, R. *N*-Monomethylcadaverine. *Z. Physiol Chem.* **1930**, *189*, 239–242. [CrossRef]
16. Chen, W.; Liu, B.; Yang, C.; Xie, Y. Convenient synthesis of 1,2,3,4-tetrahydroquinolines via direct intramolecular reductive ring closure. *Tetrahedron Lett.* **2006**, *47*, 7191–7193. [CrossRef]
17. Sajiki, H.; Ikawa, T.; Hirota, K. Reductive and catalytic monoalkylation of primary amines using nitriles as an alkylating reagent. *Org. Lett.* **2004**, *6*, 4977–4980. [CrossRef] [PubMed]

18. Steiner, T. Hydrogen-Bond Distances to Halide Ions in Organic and Organometallic Crystal Structures: Up-to-date Database Study. *Acta Crystallogr. B* **1998**, *54*, 456–463. [CrossRef]
19. Desiraju, G.; Steiner, T. *The Weak Hydrogen Bond*; Oxford University Press: Oxford, UK, 1999.
20. Fieser, L.F. *Fieser and Fieser's Reagents for Organic Synthesis*; John Wiley & Sons: New York, NY, USA; Volume 1, ISBN 978-0-471-25875-9.

MDPI
St. Alban-Anlage 66
4052 Basel
Switzerland
Tel. +41 61 683 77 34
Fax +41 61 302 89 18
www.mdpi.com

Molecules Editorial Office
E-mail: molecules@mdpi.com
www.mdpi.com/journal/molecules

www.ingramcontent.com/pod-product-compliance
Lightning Source LLC
Chambersburg PA
CBHW051726210326
41597CB00032B/5628

* 9 7 8 3 0 3 8 9 7 4 1 6 1 *